Human Embryonic Stem Cells

Human Embryonic Stem Cells

Edited by

Arlene Y. Chiu

National Institute of Neurological Disorders and Stroke
National Institutes of Health
Bethesda, MD

and

Mahendra S. Rao

National Institute on Aging
National Institutes of Health
Baltimore, MD

Humana Press
Totowa, New Jersey

Production Editor: Wendy S. Kopf.

Cover design by Patricia F. Cleary.

Cover illustration: Cover photo shows neural cells that have differentiated from human embryonic stem cells. These cells have been immunostained to reveal phenotypic features: red (rhodamine) for neurofilament 200 and green (fluorescein) for GFAP. Nuclei (blue) are visualized by Hoechst staining. Cover photo provided by Dr. Su-Chun Zhang.

For additional copies, pricing for bulk purchases, and/or information about other Humana titles, contact Humana at the above address or at any of the following numbers: Tel.: 973-256-1699; Fax: 973-256-8314; E-mail: humana@humanapr.com

Library of Congress Cataloging in Publication Data
Human embryonic stem cells / edited by Arlene Chiu and Mahendra S. Rao.
    p. cm.
  Includes bibliographical references and index.
  ISBN 1-58829-311-4 (alk. paper); e-ISBN 1-59259-423-9
    1. Human embryo. 2. Embryonic stem cells. I. Chiu, Arlene. II. Rao, Mahendra S.

QP277.H87 2003
612.6'46--dc21

2003041669

Preface

Human embryonic stem cells are derived from the earliest stages of blastocyst development after the union of human gametes. Prior to fertilization, the oocyte first requires timed completion of meiosis. This vital step does not occur throughout a woman's life; rather, oocytes are arrested at the first meiotic division until puberty when small numbers mature competitively during the reproductive years. Maturation is complete at the one-day event of ovulation that occurs in a regular, approximately monthly cycle. In humans, oocytes can be successfully fertilized only during a short period after ovulation; oocytes that are not fertilized are not retained.

Sperm cells mature from spermatic stem cells through a sequential process that has been well characterized. Spermatic stem cells are in turn generated from primordial germ cells set aside during early embryonic development. Generally, tens of millions of sperm are present in an ejaculate of which only one will successfully fertilize the mature oocyte. Sperm that fail to fertilize are discarded.

Fertilization initiates the process of cell differentiation. Because embryonic transcription is not initiated until later, the earliest developmental events are regulated primarily by maternally inherited mRNA. Once the sperm enters the egg, its DNA-associated proteins are replaced by oocyte histones. The two pronuclei become enveloped with oocyte-derived membranes, which fuse and begin the zygote's mitotic cell cycle. Embryonic development starts with a series of cleavages to produce eight undetermined and essentially equivalent blastomeres. The pattern of cleavage is well coordinated by cytoplasmic factors and in mammalian eggs is holoblastic and rotational. Though genomic DNA is inherited from both parents, mitochondrial DNA is inherited from only the mother. Paternal mitochondria transferred at the time of sperm entry are discarded by a little-understood process.

Human eggs, with a diameter of 100 µm, are generally smaller than eggs of other species. They are normally fertilized within the fallopian tubes and undergo cellular division in a defined milieu as they migrate toward the uterus. Over the first few days, cellular division follows a predictable 12–18-h cycle resulting in 2- to 16-cell pre-embryos. The sperm centrosome controls the first mitotic divisions until day 4, when genomic activation occurs within the morula stage. The individual blastomeres are initially totipotent until the morula begins compaction and the cells initiate polarization. During compaction, cell boundaries become tightly opposed and cells are no longer equivalent. Cells in the inner cell mass (ICM) contribute to the embryo proper, whereas cells on the outside contribute to the trophoectoderm. The blastocyst forms approx 24 h after the morula stage by the development of an inner fluid-filled cavity, the blastocell.

Implantation is the process through which the compact zonula is thinned and the blastocyst released to implant. The blastocyst must first hatch from the thinning zona pellucida by alternating expansion and contraction; this process of hatching is critical to further development. Implantation of the hatched blastocyst requires several steps, including apposition, attachment, penetration, and trophoblast invasion, and cannot occur until the first cell specification into trophoectoderm has occurred.

As the trophoblast is developing to form the fetal component of the placenta, the endometrial lining of the maternal uterus is undergoing a decidual reaction to generate the maternal component of the placenta. Simultaneously the inner cell mass undergoes gastrulation, defined as a process of complex, orchestrated cell movements that vary widely among species, but include the same basic movements. These include epiboly, invagination, involution, ingression, and delamination.

Thus, several major developmental events have taken place as the fertilized egg migrates from the site of fertilization (fallopian tubes) to the body of the uterus over a period of 4 d. At all stages the egg is shielded from the external environment, initially by the zonula and subsequently by the trophoectoderm, but is accessible to *ex utero* manipulation (*see* Fig. 1). It is important to emphasize that these critical developmental events have occurred prior to implantation and well before blood vessel growth and heart development. The early stage fertilized egg that has not yet been implanted has been termed a pre-embryo to distinguish it from the implanted embryo. Manipulation of the preimplanted embryo has been feasible for the past three decades, and detailed rules governing the use of blastocysts have been developed.

After implantation, the ICM proliferates and undergoes differentiation. Several results suggest that lineage-specific genes are operating in a totipotent blastocyst cell prior to lineage commitment, and strongly support the concept that stem cells express a multilineage transcriptosome. Most genes (including tissue-specific genes) are

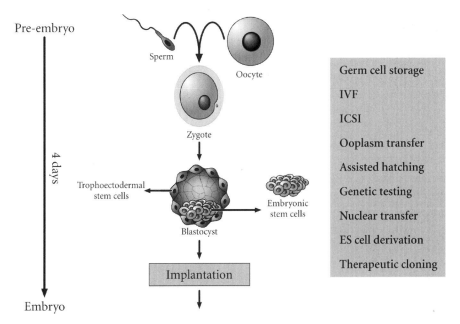

Fig. 1. Many techniques have been devised to manipulate the process of fertilization and maturation prior to implantation (summarized at right). The recent development of techniques to generate embryonic stem cell lines and perform somatic nuclear transfer has increased our ability to understand the process of development and intervene therapeutically. Note that the fertilized egg and early stage zygote are accessible to manipulation prior to implantation.

maintained in an open state with low but detectable levels of transcription with higher levels of specific transcription seen in appropriate cell types. Maintenance of an open transcriptosome in multipotent cells likely requires both the presence of positive factors as well as the absence of negative regulators. Factors that maintain an open transcriptosome include as yet unidentified agents such as demethylases, reprogramming molecules present in blastocyst cytoplasm, and regulators of heterochromatin modeling. Global activators, global repressors, and master regulatory genes play important regulatory roles in switching on or off cassettes of genes, whereas methylation and perhaps small interfering RNA (siRNA) maintain a stable phenotype by specifically regulating the overall transcriptional status of a cell. Allelic inactivation and genome shuffling further sculpt the overall genome profile to generate sex, organ, and cell-type specification.

Few genes have been identified that are required for the maintenance of the epiblast population. *Oct4-/-* embryos die before the egg cylinder stage and embryonic stem (ES) cells cannot be established from *Oct4-/-* cells. Levels of Oct 4 expression are critical to the fate of the cells. Low cell levels lead to differentiation into trophoblast giant cells, whereas high levels cause differentiation into primitive endoderm and mesoderm. FGF4 is required for formation of the egg cylinder and FGF 4-/- embryos fail to develop after implantation and ICM cells do not proliferate in vitro. *Foxd3/Genesis* is another transcription factor that may be required for early embryonic development. TGF-β/SMADs, Wnts, and FGFs are thought to play an important role in the process of gastrulation. BMP4 is essential for the formation of extraembryonic mesoderm and the formation of primordial germ cells. Nodal expression is required for mesoderm expansion, maintenance of the primitive streak, and setting up the anterior–posterior and proximo–distal axis. FGF4 is secreted by epiblast cells and is required for the maintenance of the trophoblast.

As our understanding of early developmental events has increased, our ability to safely manipulate the reproductive process has also increased. In vitro fertilization is now a relatively commonplace procedure that has been performed for more than 20 yr. Today, there are over a thousand established clinics worldwide. Even such technically complex procedures as intracytoplasmic sperm injection (ICSI), ooplasm transfer, assisted hatching, intrauterine genomic analysis, intrauterine surgery, and organ transplants are becoming more commonplace. More then 70 human embryonic cell lines have been established and their ability to differentiate into ectoderm, endoderm, and mesoderm repeatedly demonstrated. Nuclear transfer has become feasible, and the potential of combining ES cell technology with somatic nuclear transfer to clone individuals has caught the attention of people worldwide. At each stage of technological sophistication, profound ethical issues have been raised and publicly debated. Perhaps the most recent technological breakthroughs are the ones that have created the most controversy, primarily because of their potential to be used on a large scale. The ability to generate human ES (hES) cells and the ability to perform somatic nuclear transfer and successfully clone mammalian species raise fears often fueled by limited information.

An additional potential paradigm shift has been the suggestion that pluripotent ES-like cells may exist and indeed may persist into adulthood (*see* Fig. 2). These cells, while differing subtly from ES cells, may be functionally equivalent for therapeutic applications. The possible existence of such adult cells with ES cell properties has fanned the debate and fueled a drive to assess the properties of all classes of pluripotent cells and to understand their underlying differences.

In *Human Embryonic Stem Cells* we invited leaders in the field to present their work in an unbiased way so that readers can assess the potential of stem cells and the current state of the science. The first section covers issues that regulate the use of human pluripotent cells. Chapters 1–3 begin with a summary of the ethical debate surrounding the derivation of human stem cells, and the current policies governing their use in the United States and abroad. The presidential announcement of August 2001 heralded a change in policy enabling federal support of research with hES cells that meet specific criteria. In Chapters 2 and 3, representatives from the National Institutes of Health (NIH) discuss the rules and conditions regulating federal funding, and issues of intellectual property regarding the use of hES cells. Chapter 2 delves into what constitutes "allowable" research and provides a guide to researchers interested in acquiring funding from US federal agencies such as the NIH for studies in this field.

Part II describes the types of human pluripotent cells that are currently being studied, their sources, methods of derivation, and maintenance. Many tissues are constantly renewed by the activities of resident, multipotent precursor or progenitor cells that have the ability to produce several different mature phenotypes. In the well-characterized hematopoietic system, T and B lymphocytes are derived from the lymphoid stem cell, whereas the myeloid stem cell can generate a host of red and white blood cells, including monocytes, eosinophils, platelets, and erythrocytes. However, both the myeloid and lymphoid stem cells are committed precursors, unable to differentiate along other pathways. There are only a few examples of truly pluripotent stem cells with the developmental capacity to generate cells representing all three germ layers (*see* Fig. 2) . Four types of such pluripotent stem cells are discussed in this section. In Chapter 4, Draper, Moore, and Andrews review the tumorigenic origins of embryonal carcinoma (EC) cells and their developmental counterparts, embryonic germ (EG) cells, present in the germinal ridges of young fetuses. Although there are many claims that pluripotent and highly plastic stem cells reside in adult tissues, the best characterized are those present in bone marrow. Cardozo and Verfaille summarize studies demonstrating their pluripotency in Chapter 5.

The high degree of interest in hES cells arise from two properties: their ability to self-renew essentially indefinitely and to be maintained

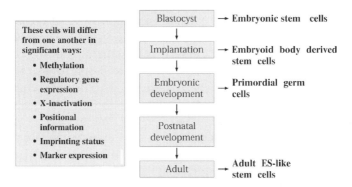

Fig. 2. Many different cell populations have been isolated that appear ES-like in their ability to contribute to chimeras after blastocyst injection, and to differentiate into ectoderm, endoderm, and mesoderm lineages in vitro. These distinct populations, although superficially similar, are likely to differ from each other when examined in more detail.

in an immature state, and their ability to differentiate into a wide range of mature tissues and cells. This enables the same population of cells to be studied under a variety of conditions; their properties, behavior, and fates can be reproduced and predicted. This capacity to standardize, predict, and reproduce results using a particular cell line will in turn greatly enhance the development of treatments and assays. For these reasons, much work is currently focused on methods for the expansion of hES cells and protocols to regulate their differentiation down selective lineages toward defined fates. Procedures for the growth, subcloning, and maintenance of hES cells are presented in Chapters 6 and 7 by three groups that are pioneers in this endeavor.

The following five chapters in Part III focus on specific methods that drive their differentiation into neuroepithelium, pancreatic islet cells, cardiomyocytes, vascular cells, and hematopoietic progenitors. Because much of the groundbreaking work was first conducted on murine ES cells, these initial animal studies are described and compared with the behavior of their human counterparts.

Part IV focuses on the potential uses of human stem cells in a variety of applications. In Chapter 13, Harley and Rao compare the advantages and disadvantages of using hES cells versus stem cells acquired from adult tissues for transplantation therapies. More complex applications to generate cells with the desired genetic composition include genetic manipulation of hES cells (Chapter 14) and somatic cell nuclear transfer (also called therapeutic cloning) (Chapter 15) to produce hES cells with the patient's genetic composition for autologous transplants. In Chapter 16, Kamb and Rao discuss possible uses of human stem cells as tools for drug and gene discovery in vitro, and as therapeutic agents in vivo. The latter include cell, tissue, and organ replacement and regeneration, as well as the use of cells as peptide manufacturers and as delivery systems. They also explore what is needed to generate a donor cell that is universally accepted.

Most cell transplantation treatments will require the oversight of and approval by the Food and Drug Administration (FDA). In Chapter 17, Fink reviews the regulatory role of the FDA in ensuring the safety, purity, potency, and efficacy of new therapies involving human stem cells. Although the main interest in stem cell research resides in their great potential for cell replacement therapy to treat a long list of diseases that are currently incurable, there are at present no stem cell treatments in use, and only a rare few in clinical studies. In the last chapter, Reier and colleagues review preclinical and clinical studies conducted with neuron-like cells derived from one of the best-studied human EC cell lines, the NT-2 cells. This discussion introduces the complexities involved with in vivo studies, and the behavioral and functional analyses following cell transplantation.

Finally, we include a series of appendices that will provide additional information on useful websites, stem cell patents, and examples of Material Transfer Agreements to facilitate the sharing of cells. We hope that the readers will find the contents of *Human Embryonic Stem Cells* useful, and we welcome comments proposing additions or deletions to what we hope will become the standard reference book in the field of hES cell biology.

Arlene Y. Chiu
Mahendra S. Rao

Contents

Contributors

MICHAL AMIT • *Faculty of Medicine, Technion–Israel Institute of Technology, Haifa, Israel*

PETER W. ANDREWS • *Department of Biomedical Science, University of Sheffield, Western Bank, Sheffield, UK*

JESSICA E. ANTOSIEWICZ • *Department of Anatomy, School of Medicine, The Wisconsin Regional Primate Center, University of Wisconsin, Madison, Madison, WI*

NISSIM BENVENISTY • *Department of Genetics, Silberman Institute of Life Sciences, The Hebrew University, Jerusalem, Israel*

MELISSA K. CARPENTER • *Robart's Research Institute, London, Ontario, Canada*

JOSE B. CIBELLI • *Department of Animal Sciences, Michigan State University, East Lansing, MI*

CHRISTINE A. DAIGH • *Department of Anatomy, School of Medicine, The Wisconsin Regional Primate Center, University of Wisconsin, Madison, Madison, WI*

GREGORY J. DOWNING • *Office of Science Policy and Planning, Office of Science Policy, Office of the Director, National Institutes of Health, Bethesda, MD*

JONATHAN S. DRAPER • *Department of Biomedical Science, University of Sheffield, Sheffield, UK*

MICHA DRUKKER • *Department of Genetics, Silberman Institute of Life Sciences, The Hebrew University, Jerusalem, Israel*

DONALD W. FINK, JR. • *Laboratory of Stem Cell Biology/Neurotrophic Factors, Division of Cell and Gene Therapy, Office of Cellular Tissue and Gene Therapies, US-FDA/CBER, Rockville, MD*

SHARON GERECHT-NIR • *Biotechnology Interdisciplinary Unit, Technion–Israel Institute of Technology, Haifa, Israel.*

CALVIN B. HARLEY • *Geron Corporation, Menlo Park, CA*

JOSEPH ITSKOVITZ-ELDOR • *Department of Obstetrics and Gynecology, Rambam Medical Center and the Faculty of Medicine, Technion–Israel Institute of Technology, Haifa, Israel*

ALEXANDER KAMB • *Deltagen Proteomics Inc., Salt Lake City, UT*

xvii

DAN S. KAUFMAN, • *Stem Cell Institute and Department of Medicine, Division of Hematology, Oncology, and Transplantation, University of Minnesota, Minneapolis, MN*

VIRGINIA M.-Y. LEE • *The Center for Neurodegenerative Disease Research, Department of Pathology and Laboratory Medicine, University of Pennsylvania School of Medicine, Philadelphia, PA*

NADYA LUMELSKY • *Diabetes Branch, National Institute of Diabetes and Digestive and Kidney Diseases, Bethesda, MD*

DORIT MANOR • *Department of Obstetrics and Gynecology, Rambam Medical Center, Haifa, Israel*

HARRY MOORE • *Section of Reproductive and Developmental Medicine, The School of Medicine and Biomedical Science, University of Sheffield, Sheffield, UK*

MARTIN F. PERA • *Institute of Reproduction and Development, Monash Medical Centre, Monash University, Clayton, Victoria, Australia*

FELIPE PROSPER • *Servicio de Hematología y Area de Terapia Celular, Clínica Universitaria de Navarra, Pamplona, Spain*

MANI RAMASWAMI • *Department of Molecular and Cell Biology, The University of Arizona, Tucson, AZ*

MAHENDRA S. RAO • *National Institute of Aging, National Institutes of Health, Baltimore, MD*

PAUL J. REIER • *Department of Neuroscience, The McKnight Brain Institute of the University of Florida, Gainesville, FL*

MARK L. ROHRBAUGH • *Office of Technology Transfer, National Institutes of Health, Rockville, MD*

HANNA SEGEV • *Department of Obstetrics and Gynecology, Rambam Medical Center, Haifa, Israel*

JAMES A. THOMSON • *Department of Anatomy, The Wisconsin Regional Primate Center, University of Wisconsin, School of Medicine, Madison, Madison, WI*

JOHN Q. TROJANOWSKI • *The Center for Neurodegenerative Disease Research, and Institute on Aging, Department of Pathology and Laboratory Medicine,University of Pennsylvania School of Medicine, Philadelphia, PA*

STEVE USDIN • *BioCentury Publications, Washington, DC*

MARGARET J. VELARDO • *Department of Neuroscience, The McKnight Brain Institute of the University of Florida, Gainesville, FL*

CATHERINE M. VERFAILLIE • *Stem Cell Institute, University of Minnesota, Minneapolis, MN*

CHUNHUI XU • *Geron Corporation, Menlo Park, CA*

SU-CHUN ZHANG • *The Stem Cell Research Program, Waisman Center, University of Wisconsin, Madison, Madison, WI*

I
POLICY

1
Ethical Issues Associated with Pluripotent Stem Cells

Steve Usdin

1. INTRODUCTION

The announcement in February 1997 by researchers at the Roslin Institute in Edinburgh, Scotland of the successful cloning of a sheep named Dolly from adult mammary cells surprised the scientific community and captured the public's attention, provoking hopeful comments from political leaders, religious figures, and eminent scientists about potential benefits, as well as expressions of fears that it could lead to profound ethical transgressions. The media instantly extrapolated from the creation of a lamb to the potential cloning of humans—one of the first newspaper headlines was "Sheep Clone Raises Alarm Over Humans" *(1)*—and governments scrambled to develop policies to address public concerns. The initial outcry over the potential use of somatic cell nuclear transfer (SCNT) to clone humans was followed by a broader debate about the ethics of the new avenues of medical research that had been opened by new technology, especially the potential to create human embryos that could be used in research and/or therapies.

In November 1998, the nearly simultaneous announcements of the derivation of human embryonic stem (ES) cells and human embryonic germ (EG) cells reignited debate about the ethics of using embryos for research and to create therapies. Papers describing the stem cell advances *(2,2a)* were published as governments, the media and the public around the world already were grappling with the implications of SCNT. Indeed, the stem cell breakthroughs made more tangible some of the theoretical ethical issues raised by the cloning of Dolly.

In the United States, Europe, and elsewhere, institutions that had recently completed studies of the ethical issues raised by SCNT now turned their attention to stem cells. Formal inquiries into the ethics of human pluripotent

From: *Human Embryonic Stem Cells*
Edited by: A. Chiu and M. S. Rao © Humana Press Inc., Totowa, NJ

stem cell research were launched in virtually every country with active biomedical research programs. Although there was a great deal of diversity in the conclusions of these investigations and in the resulting policy decisions, there was widespread agreement regarding the basic ethical issues.

Ethical judgments about the use of embryonic stem cells in research and therapies flow from the status accorded to the embryo. Those who feel that an embryo is a human being, or should be treated as one because it has the potential to become a person, contend that it is unethical to do anything to an embryo that could not be done to a person. At the opposite end of the spectrum, some people have expressed the view that the embryo is nothing more than a ball of cells that can be treated in a manner similar to tissues used in transplantation. An intermediate position has been articulated that ascribes a special status to the embryo that is less than human life but deserving of respect that imposes limits on its ethical use.

Most deliberations about the use of human ES cells and EG cells have started with an examination of concerns about the sources for the two ES and EG cells lines described in the 1998 papers *(2,2a)*: human embryos and aborted fetuses. Most ethical analyses also have considered the context in which the embryos were created or the abortions occurred, focusing especially on the informed consent and motivations of donors. Those supporting the derivation of stem cell lines from human embryos and/or aborted fetuses emphasize the wide range of potential benefits. For example, less than a month after John Gearhart from Johns Hopkins University and James Thomson from the University of Wisconsin first described procedures for creating immortal human EGS and ESG cell lines, respectively, US National Institutes of Health Director Harold Varmus told Congress that the developments were "an unprecedented scientific breakthrough" *(3)*. Although acknowledging that a great deal of work must be done before research on pluripotent stem cells could lead to new therapies, he added that "it is not too unrealistic to say that this research has the potential to revolutionize the practice of medicine and improve the quality and length of life."

The National Institutes of Health (NIH) has stated that research on pluripotent stem cells, including those derived from embryos and cadaveric fetal tissue, "promises new treatments and possible cures for many debilitating diseases and injuries, including Parkinson's disease, diabetes, heart disease, multiple sclerosis, burns and spinal cord injuries" *(4)*. Even ethical analyses that condemn the derivation of stem cells from embryos or aborted fetuses acknowledge that research involving such cells could lead to medical breakthroughs.

2. ARGUMENTS AGAINST THE DERIVATION
OF PLURIPOTENT STEM CELLS FROM HUMAN EMBRYOS

Religious perspectives, especially those concerning the status of the embryo, have been central to many deliberations about stem cell ethics. "Human embryos are not mere biological tissues or clusters of cells; they are the tiniest of human beings. Thus, we have a moral responsibility not to deliberately harm them," according to a statement about the ethics of stem cell research published by the Center for Bioethics and Human Dignity *(5)*, a nonprofit Christian ethics group located in Bannockburn, Illinois that campaigns against embryo research. Although the center's statement, which reflects the views of many US antiabortion groups, acknowledged "that the desire to heal people is certainly a laudable goal," it added that "many have invested their lives in realizing this goal, we also recognize that we are simply not free to pursue good ends via unethical means. Of all human beings, embryos are the most defenseless against abuse. A policy promoting the use and destruction of human embryos would repeat the failures of the past. The intentional destruction of some human beings for the alleged good of other human beings is wrong. Therefore, on ethical grounds alone, research using stem cells obtained by destroying human embryos is ethically proscribed" *(6)*.

The Vatican and institutions representing its views have taken active part in the policy debate. The Catholic perspective is grounded in the belief that embryos are living humans with all of the rights to life that would be accorded to an infant, child, or adult. The conviction that life starts at the instant of conception has been articulated by the Catholic Church in a number of contexts, especially regarding its opposition to abortion. For example, in a 1995 encyclical letters *Evangelium Vitae (7)*, Pope John Paul II stated that "from the time that the ovum is fertilized, a life is begun which is neither that of the father nor the mother; it is rather the life of a new human being with his own growth. It would never be made human if it were not human already. This has always been clear, and … modern genetic science offers clear confirmation. It has demonstrated that from the first instant there is established the program of what this living being will be: a person, this individual person with his characteristic aspects already well determined. Right from fertilization the adventure of a human life begins, and each of its capacities requires time—a rather lengthy time—to find its place and to be in a position to act. Even if the presence of a spiritual soul cannot be ascertained by empirical data, the results themselves of scientific research on the human embryo provide a valuable indication for discerning by the use of reason a personal presence at the moment of the first appearance of a human life: how could a human individual not be a human person?"

The *Evangelium Vitae* encyclical letter adds: "Furthermore, what is at stake is so important that, from the standpoint of moral obligation, the mere probability that a human person is involved would suffice to justify an absolutely clear prohibition of any intervention aimed at killing a human embryo."

The *Evangelium Vitae* encyclical letter preceded both Dolly and the derivation of human EG and ES cells, but Pope John Paul II has subsequently made broad statements condemning human reproductive cloning and embryonic stem cell research. Representatives of the Catholic Church have fleshed out these positions. Richard M. Doerflinger, Associate Director for Policy Development at the United States Conference of Catholic Bishops Secretariat for Pro-Life Activities, outlined the church's views about the derivation and use of ES cells in testimony to a US Senate subcommittee in July 2001 *(8)*. Proceeding from the premise that embryos are human, Doerflinger argued, "Human life deserves full respect and protection at every stage and in every condition. The intrinsic wrong of destroying innocent human life cannot be 'outweighed' by any material advantage—in other words, the end does not justify an immoral means. Acceptance of a purely utilitarian argument for mistreating human life would endanger anyone and everyone who may be very young, very old, very disabled, or otherwise very marginalized in our society."

The Catholic Church rejects contentions that it is acceptable to use surplus embryos from fertility clinics that are scheduled to be destroyed. According to Doerflinger, this reasoning is flawed because it ignores the intrinsic rights of the embryo. "If parents were neglecting or abusing their child at a later stage, this would provide no justification whatever for the government to move in and help destroy the child for research material. We do not kill terminally ill patients for their organs, although they will die soon anyway, or even harvest vital organs from death row prisoners, although they will be put to death soon anyway," he told the Senate Appropriations Subcommittee on Labor, Health, and Education.

In recent years, the Catholic Church has fought strenuously against the concept, articulated by Saint Augustine and Saint Thomas Aquinas that the soul entered the fetus 40 days after conception. This doctrine was repudiated by Pope Pius X in 1869 and replaced by the current doctrine that life begins at conception. In his congressional testimony, Doerflinger attacked a view of ensoulment in which human life is not said to begin until an embryo is placed in a mother's womb. Counter to this view, which he said was held by some members of congress who otherwise shared the Vatican's position on abortion, Doerflinger asserted that an "embryo's development is directed completely from within—the womb simply provides a nurturing environment."

Doerflinger contended that conducting research on embryos or fetal tissue is dehumanizing—both to the lives that he considers have been killed as well as to the society that allows or benefits from their destruction *(9)*. "The human individual, called into existence by God and made in the divine image and likeness, … must always be treated as an end in himself or herself, not merely as a means to other ends. …"

Several religions concur with the Catholic Church's contention that embryo research represents an unacceptable commodification of human life. The commodification argument has led some religious organizations that do not oppose abortion to advocate prohibitions on embryonic stem cell research.

The United Methodist Church, which has affirmed a woman's right to elective abortion, objects to the derivation of stem cells from embryos. It has taken the position that "[d]estroying human embryos for the sole purpose of carrying on scientific research that promises only the possibility of potential treatments, with little concrete evidence of success, again raises profound and disturbing moral and ethical issues" *(10)*. The Methodist Church therefore urged President George W. Bush to impose "an extended moratorium on the destruction of human embryos for the purpose of stem cell or other research." Research practices that require the destruction of an embryo "seem to be destructive of human dignity and speed us further down the path that ignores the sacred dimensions of life and personhood and turns life into a commodity to be manipulated, controlled, patented, and sold," according to Jim Winkler, General Secretary of the United Methodist General Board for Church and Society.

The Church of Scotland has stated that embryo research could be ethical in some circumstances, particularly the pursuit of in vitro fertilization. However, it too opposes ES cell research. In a paper outlining its position *(11)*, the Church of Scotland stated that such research "would reduce the embryo to a mere resource from which convenient parts are taken."

Other religious groups and philosophers also have warned against the prospect that embryonic stem cells may become items of commerce. Through ES and EG cell research, "human life is destroyed for its parts, which can then be bought and sold as a therapeutic agent," according to Daniel McConchie, Director of Media and Policy at the Center for Bioethics and Human Dignity *(12)*.

Opposition to obtaining EG cells from cadaveric fetal tissue is almost entirely based on opposition to abortion. "Catholic teaching rejects all complicity in abortion, and the Church has opposed any collaboration with abortionists (including government collaboration) to obtain tissue for vaccines or other research," according to a US Conference of Catholic Bishops statement about stem cells.

3. ARGUMENTS IN SUPPORT OF THE DERIVATION OF PLURIPOTENT STEM CELLS FROM HUMAN EMBRYOS

3.1. Religious Perspectives

There is a diversity of opinions among religious groups, even among Christian denominations, about the ethics of embryonic stem cell research. For example, the Presbyterian Church's position differs substantially from the Catholic view.

In June 2001, the Presbyterian Church USA adopted a resolution about stem cells that affirmed that embryos are "potential" humans, but that their use in stem cell research is justified by possible medical benefits. "We believe, as do most authorities that have addressed the issue, that human embryos do have the potential of personhood, and as such they deserve respect. That respect must be shown by requiring that the interests or goals to be accomplished by using human embryos be compelling and unreachable by other means. Indications are that human embryonic stem cell research has the potential to lead to lifesaving breakthroughs in major diseases," the Presbyterian resolution *(13)* stated. The resolution added that a prohibition on the derivation of stem cells from embryos "would elevate the showing of respect to human embryos above that of helping persons whose pain and suffering might be alleviated. Embryos resulting from infertility treatment to be used for such research must be limited to those embryos that do not have a chance of growing into personhood because the woman has decided to discontinue further treatments and they are not available for donation to another woman for personal or medical reasons, or because a donor is not available."

Authorities on Jewish law have expressed the view that research involving preimplantation embryos raises few ethical concerns. "Genetic materials outside the uterus have no legal status in Jewish law, for they are not even a part of a human being until implanted in a woman's womb, and even then, during the first 40 days of gestation, their status is 'as if they were simply water,'" Rabbi Elliot N. Dorff, of the University of Judaism, stated in testimony to President Bill Clinton's National Bioethics Advisory Commission (the NBAC) in May 1999 *(14)*. He added that Jewish law allows abortions under specific circumstances, and if a "fetus was aborted for good and sufficient reason within the parameters of Jewish law," it would be permissible to derive stem cells from it.

More recently, leaders of American Reform Jews stated in a July 2001 letter to President George W. Bush that "Jewish authorities have used the concept of *pikuach nefesh*, or the primary responsibility to save human life, which overrides almost all other laws, to approve a broad range of medical experimentation. Cutting off funding for medical research that has such tremendous potential benefits—even where, as here, it raises complex and far-reaching issues—is both immoral and unethical according to our tradition" *(15)*.

The consensus view of both major schools of Islam, Sunni, and Shi'ites is that the fetus does not acquire a soul until some time after conception, usually defined as 120 days. In testimony to the NBAC about the Islamic perspective on embryonic stem cells, Abdulaziz Sachedina, professor of religious studies at the University of Virginia, articulated a framework for defining the status of the embryo that would be acceptable to all schools of thought in Islam:

1. The Koran and the Tradition regard perceivable human life as possible at the later stages of the biological development of the embryo.
2. The fetus is accorded the status of a legal person only at the later stages of its development, when perceptible form and voluntary movement are demonstrated. Hence, in earlier stages, such as when it lodges itself in the uterus and begins its journey to personhood, the embryo cannot be considered as possessing moral status.
3. The silence of the Koran over a criterion for moral status of the fetus (i.e., when the ensoulment occurs) allows the jurists to make a distinction between a biological and a moral person, placing the latter stage after, at least, the first trimester of pregnancy.

Sachedina concluded "in Islam, research on stem cells made possible by biotechnical intervention in the early stages of life is regarded as an act of faith in the ultimate will of God as the Giver of all life, as long as such an intervention is undertaken with the purpose of improving human health."

3.2. Biomedical Research Advocates' Perspectives

Many of the politicians, scientists, and patient advocates who support the derivation of stem cells from embryos contend that embryos are not human beings and that, in practice, the embryos used for stem cell research do not have the potential to become humans, especially the surplus embryos from in vitro fertilization procedures. Others who accord some type of special status to embryos have stated that the probable medical benefits from stem cell research outweigh concerns about destruction of embryos.

Indeed, some scholars have rejected the idea that a binary decision must be made regarding the status the embryo. Gene Outka, a professor of philosophy and Christian ethics at Yale University Divinity School, has outlined a continuum in status and rights from preimplanted fetuses to infants.

In an April 2000 paper commissioned by President George W. Bush's Council on Bioethics *(16)*, he stated that starting from conception, "each entity is a form of primordial human life, a being in its own right." However, before "individuation and implantation, the entity does not yet have the full fledged moral standing that an implanted fetus has. Yet, for its part, the fetus' value is not equally protectible with the pregnant woman's, for she too is an end in herself and potentiality still characterizes the fetus. Equal protectibility holds after the fetus becomes capable of independent existence outside the womb." These claims about status make it possible to accept that the embryo is a form of human life while, at the same time, rejecting the notion that abortion and embryonic stem cell research are morally indistinguishable from murder, according to Outka.

Based on this framework, Outka invoked the "nothing is lost" doctrine to justify the destruction of embryos for research and therapeutic purposes. He defines the doctrine as follows: "One may directly kill when two conditions obtain: (a) the innocent will die in any case; and (b) other innocent life will be saved. ..." Applying the doctrine to embryonic stem cell research, Outka stated that "it is correct to view embryos in reproductive clinics who are bound either to be discarded or frozen in perpetuity as innocent lives who will die in any case, and those third parties with Alzheimer's, Parkinson's, etc., as other innocent life who will be saved by virtue of research on such embryos."

Indeed, some "pro-life" advocates cross over on the issue, asserting that it is possible to condemn and oppose elective abortion while endorsing ES research. In 2001, for example, Republican Senator Orrin Hatch of Utah said that he has "strong pro-life, pro-family values and strongly oppose[s] abortion," yet has concluded "that support of embryonic stem cell research is consistent with and advances pro-life and pro-family values" *(17)*. He said this contention was supported in part by his belief that "a human's life begins in the womb, not in a petri dish or refrigerator."

For Hatch, it is more ethical to use embryos for stem cell research than to discard them. He noted that "in the in vitro fertilization process, it is inevitable that extra embryos are created, embryos that simply will not be implanted in a mother's womb. As these embryos sit frozen in a test tube, outside the womb, under today's technology, there is no chance for them to develop into a person." Hatch added: "To me, the morality of the situation dictates that these embryos, which are routinely discarded, be used to improve and extend life. The tragedy would be in not using these embryos to save lives when the alternative is that they will be destroyed."

Some ethicists contend that the stage of development of the embryo is critical to determining if it can ethically be used in research. The development of the "primitive streak," the point fourteen days after conception when cells differentiate into the embryo and the placenta, is a critical threshold for many ethicists.

Thomas Shannon, professor of religion and social ethics in the Department of Humanities and Arts at Worcester Polytechnic Institute in Massachusetts, asserts that because "differentiation has not yet occurred in the preimplantation embryo, such an entity is not individualized; it therefore lacks a core feature of personhood. This is so for two reasons: First, the cells in this entity have the capacity to become some (pluripotent) or any (totipotent) part of the body—and therefore the preimplantation embryo cannot be understood to be a single individual. For by definition, an individual is an entity that cannot be divided or, if it is, it becomes two halves neither of which can survive on its own. Second, the cells of the preimplantation embryo can be separated without harm to the organism (for example, in preimplantation diagnosis, one or two cells can be removed and examined for genetic disease)." He concludes that, ultimately, research involving preimplantation embryos "is not research on a human person. It is research on our common human nature, and as such is morally justifiable" *(18)*.

Richard O. Hynes, president of the American Society for Cell Biology (ASCB), used a similar argument in congressional testimony in September 2000. He stated that "if the issue is morality, using embryonic cells for potentially life-saving research is greatly preferable to discarding them. Surely we should take advantage of the enormous life-saving potential of the thousands of embryos that are currently frozen and destined for destruction" *(19)*.

Indeed, some scientists and policy-makers contend that it would be unethical to prohibit the derivation of stem cells from embryos. Representing the ASCB, Lawrence S. B. Goldstein, a professor of pharmacology at the Howard Hughes Medical Institute, told congress in April 2000 that there are "serious ethical implications to not proceeding [with ES cell research]. Thus, while I am acutely and personally aware of the well meaning ethical concerns that have been expressed about the sacrifice of embryos to prepare stem cells, I am also cognizant that the embryos in question will be legally and ethically destroyed in any case. We must then ask: Is it ethical to literally throw away the opportunity to allow all people to benefit from the demise of these embryos? How can we justify not pursuing every reasonable means of finding cures for our friends, our parents, and our children, who will suffer and die if we do not find suitable therapies?" *(20)*.

4. NATIONAL ETHICS POLICIES ON PLURIPOTENT STEM CELLS

4.1. United States of America

In the United States, policy deliberations on the ethics of embryo based research were underway well before the cloning of Dolly the sheep. Laws and regulations have been enacted piecemeal, usually in response to public concern about specific experiments or technologies.

In his first presidential campaign, Bill Clinton said he would overturn President George H.W. Bush's de facto prohibition on federal funding of research involving fetal tissue, which had been imposed in 1979 to prevent government funding of in vitro fertilization research. On June 10, 1993, President Clinton signed legislation authorizing NIH to conduct and fund human embryo research. He then issued an order in December 1994 prohibiting NIH from funding the "creation of human embryos for research purposes."

Congress subsequently approved a ban on the use of appropriated funds for the creation of human embryos for research purposes or for research in which human embryos are destroyed. The prohibition was introduced by Republican Representative Jay Dickey of Arkansas as an amendment to the fiscal 1996 Labor Department and Health and Human Services (HHS) Department appropriations bill, and similar amendments have been appended to every subsequent NIH spending bill.

Neither the "Dickey amendment" nor President Clinton's ban applied to privately funded activities.

On February 24, 1997, the day after the first news reports about Dolly, President Clinton asked the NBAC for a study on "the legal and ethical issues associated with the use of this technology," as well as "possible federal actions to prevent its abuse." Clinton noted that although "this technological advance could offer potential benefits in such areas as medical research and agriculture, it also raises serious ethical questions, particularly with respect to the possible use of this technology to clone human embryos." He also announced an immediate ban on federal funding related to attempts to clone human beings.

The NBAC report, released in June 1997, recommended a legislative ban on human reproductive cloning, but it also cautioned that the text of legislation should be carefully drafted to avoid inadvertently prohibiting medical research that was not oriented toward creating a human clone. The report anticipated the potential use of SCNT for therapeutic purposes, stating: "One could imagine the prospect of nuclear transfer from a somatic cell to generate an early embryo and from it an embryonic stem cell line for each individual human, which would be ideally tissue-matched for later transplant purposes. This might be a rather expensive and far-fetched scenario. An alternative scenario would involve the generation of a few, widely used and well characterized human embryonic stem cell lines, genetically altered to prevent graft rejection in all possible recipients."

The NBAC cloning report also anticipated the controversy that would accompany the creation of stem cells from embryos, as well as the possibility of other, less controversial sources of stem cells. "Because of ethical and moral concerns raised by the use of embryos for research purposes it would be far more desirable to explore the direct use of human cells of adult origin to produce specialized cells or tissues for transplantation into patients," the commission noted. "It may not be necessary to reprogram terminally differentiated cells but rather to stimulate proliferation and differentiation of the quiescent stem cells which are known to exist in many adult tissues, including even the nervous system." The advisory panel added that "approaches to cellular repair using adult stem cells will be greatly aided by an understanding of how stem cells are established during embryogenesis."

The NBAC report accelerated congressional and executive branch action on human cloning issues. The US Food and Drug Administration publicly asserted in February 1998 that it had the power to regulate, and therefore to prohibit, the cloning of humans. At the same time, Senate Majority Leader Trent Lott, Republican of Mississippi, attempted to schedule a Senate vote on S.1601, the Human Cloning Prohibition Act.

Republican Senator Bill Frist of Tennessee, who carried influence in the Senate because of his experience as a heart transplant surgeon and cardiac researcher, said S.1601 was needed to put the brakes on science that was running away from appropriate ethical constraints. He told the Senate: "We have to today consider the ethical implications that surround scientific discovery. We must consider the ethical ramifications that might—in certain very narrowly defined and specific arenas—tell us to stop, tell us to slow down before we jump or really leap ahead into the unknown" *(21)*.

Taking the opposite view, the ASCB coordinated the drafting of a public letter signed by 40 Nobel laureates that urged the Senate to defeat S.1601 because it "would impede progress against some of the most debilitating diseases known to man. For example, it may be possible to use nuclear transplantation technology to produce patient-specific embryonic stem cells that could overcome the rejection normally associated with tissue and organ transplantation. Nuclear transplantation technology might also permit the creation of embryonic stem cells with defined genetic constitution, permitting a new and powerful approach to understanding how inherited predispositions lead to a variety of cancers and neurological diseases such as Parkinson's and Alzheimer's diseases."

In a speech on the Senate floor opposing S.1601, California Democratic Senator Dianne Feinstein cited a position paper drafted by the Biotechnology Industry Organization *(22)*. The letter noted that the bill would ban the use of SCNT to generate stem cells, which it characterized as "exciting and potentially revolutionary."

S.1601 was defeated in large measure in February 1998 because a number of antiabortion Republicans voted against it, including Senators Orrin Hatch of Utah, Strom Thurmond of South Carolina, and Connie Mack of Florida. They cited concerns that the legislation could inhibit medical research.

The alliances that emerged to lobby for and against S.1601, as well as many of the arguments invoked by both sides, outlived the debate over the bill and have been central to efforts to enact legislation to regulate stem cell research. Subsequent efforts to legislate federal prohibitions on human reproductive cloning have stalled because sponsors have been unable to bridge the divide between a comprehensive ban, which some members of Congress feel would needlessly preclude certain types of biomedical research, and supporters of a more specific approach, which some members of Congress say would be unenforceable. Similar controversies have prevented Congress from adopting policies to regulate pluripotent stem cell research.

The November 1998 breakthroughs in human stem cells also elicited rapid responses from the NIH and the Clinton White House. Within days of the publication of the initial reports of the isolation of human ES and EG cells, NIH Director Harold Varmus initiated a legal review to determine if NIH and its grantees could side-step the Dickey amendment's ban on embryo destruction by making a of the distinction between the derivation of ES cells and the use of the resulting immortal cell lines *(23)*. In January 1999, Varmus released a legal opinion by the Department of Health and Human Services which concluded that federal funds could be used to support research using human pluripotent stem cells that are derived from human embryos. Varmus told a Senate committee that "the statutory prohibition on human embryo research does not apply to research utilizing human pluripotent stem cells because human pluripotent stem cells are not embryos" *(24)*.

Meanwhile, President Clinton again turned to his bioethics advisors, asking for a study of the ethics of human stem cell research within days of reports that a biotechnology company had created a stem cell "that is part human and part cow" *(25)*. His November 1998 letter to the NBAC also requested "a thorough review of the issues associated with ... human stem cell research, balancing all ethical and medical considerations."

The NBAC's report, released in September 1999, concluded that ES and EG research is ethically and morally acceptable and that the federal government should support both the derivation and use of stem cells from embryos and aborted fetuses. The commission grounded its recommendation in its confidence that research using ES and EG cells would provide great social benefits. "We have found substantial agreement among individuals with diverse perspectives that although the human embryo and fetus deserve respect as forms of human life, the scientific and clinical benefits of stem cell research should not be foregone," the commission stated *(25)*.

The decision by the Department of Health and Human Services to distinguish between the derivation and use of embryonic stem cell lines was based on an interpretation of its legal authority, not an ethical analysis. The NBAC concluded that there was no ethical basis to distinguish between the derivation and use of embryonic stem cells. "Although some may view the derivation and use of ES cells as ethically distinct activities, we do not believe that these differences are significant from the point of view of eligibility for federal funding," the advisory panel said *(26)*.

The NBAC acknowledged that some people believe that "embryos deserve some measure of protection from society because of their moral status as persons," but it concluded that this view should not prevail over the opinions of those who do not view embryos as living beings. A congressional ban on federal funding of research in which an embryo is destroyed "conflicts with several of the ethical goals of medicine, especially healing, prevention, and research—goals that are rightly characterized by the principles of beneficence and nonmaleficence, jointly encouraging the pursuit of each social benefit and avoiding or ameliorating potential harm," according to the NBAC *(25)*.

Meanwhile, the ethics panel found that views about the derivation of stem cells from fetal cadavers were driven almost entirely by opinions about the morality of abortion. "Considerable agreement exists, both in the United States and throughout the world, that the use of fetal tissue in therapy for those with serious disorders, such as Parkinson's disease, is acceptable. Research that uses cadaveric tissue from aborted fetuses is analogous to the use of fetal tissue in transplantation. The rationales for conducting EG research are equally strong, and the arguments against it are not persuasive," the NBAC concluded *(25)*.

As with ES cells derived from embryos, the NBAC emphasized the potential benefits of research using EG cells derived from aborted fetuses and suggested that it would be unethical to discard the tissue without attempting to use it to benefit medical research. ES cell research "is allied with a worthy cause, and any taint that might attach from the source of the cells appears to be outweighed by the potential good that the research may yield," the NBAC stated *(25)*.

"Research using fetal tissue obtained after legal elective abortions will greatly benefit biomedical science and will also provide enormous therapeutic benefits to those suffering from various diseases and other conditions. In our view, there is no overriding reason for society to discourage or prohibit this research and thus forgo an important opportunity to benefit science and those who are suffering from illness and disease—especially in light of the legality of elective abortions that provide access to fetal tissue and of the risks involved in losing these valuable opportunities. Indeed, the consequences of forgoing the benefits of the use of fetal tissue may well be harmful. Moreover, if not used in research, this tissue will be discarded," the NBAC concluded.

Nevertheless, the NBAC added, any decision to allow derivation and use of ES and EG stem cells must be coupled to strong regulations ensuring that donation is contingent on informed consent and is not tainted by inappropriate motivations. Those who opposed the use of tissue from aborted fetuses have warned that a woman's decision to have an abortion could be influenced by a desire to contribute to medical research, possibly to create tissue that could help the donor or a relative, or by financial inducements.

The NBAC considered these conflicts and concluded that a woman's decision to undergo an abortion should be distinct from the decision to donate cadaveric fetal tissue. Therefore, the bioethics panel recommended a prohibition on the donor's ability to direct that the tissue be used for specific purposes. "To assure that inappropriate incentives do not enter into a woman's decision to have an abortion," the advisory committed recommended that "directed donation of cadaveric fetal tissue for EG cell derivation be prohibited." This prohibition is intended to assure "that inappropriate incentives, regardless of how remote they may be, are not introduced into a woman's decision to have an abortion. Any suggestion of personal benefit to the donor or to an individual known to the donor would be untenable and possibly coercive" *(26)*.

The NBAC report lists a number of criteria for ensuring the appropriate informed consent of donors. These include disclosure that ES cell research is not intended to provide medical benefits to embryo donors; making it clear that embryos used in research will not be transferred to any woman's uterus; and making it clear that the research will involve destruction of the embryos. In addition, the NBAC recommended that researchers be prohibited from promising donors that ES cells derived from their embryos would be used to treat specific patients.

The NBAC did not adopt a clear position on the creation of embryos for research purposes, stating that "there is no compelling reason to provide federal funds for the creation of embryos for research at this time." However, when he accepted NBAC's recommendations to allow federal funding of embryonic stem cell research, President Clinton decided to retain the 1994 ban on the use of federal funds for the creation of human embryos for research purposes *(27)*.

Based on the HHS legal analysis and the NBAC report, NIH proceeded with plans to fund human embryonic stem cell research. In December 1999, NIH published draft guidelines for federal support of research involving human pluripotent stem cells, and established a Human Pluripotent Stem Cell Review Group to review documentation of compliance with the guidelines *(28)*. Publication of the final guidelines in August 2000 was accompanied by a notice that NIH would begin accepting applications for grants to conduct research involving human pluripotent stem cells *(29)*.

After the 2000 presidential election, the new administration of George W. Bush said that it would act slowly on stem cell issues. In April 2001, the White House canceled the inaugural meeting of NIH's Human Pluripotent Stem Cell Review Group, effectively placing a hold on any decision to fund human ES cell research *(30)*.

In an August 9, 2001 speech outlining his stem cell policies, President Bush touched on the conflicting perspectives about the status of the embryo that are at the heart of the ethical debate about stem cells. He told the nation that as he thought about the issues, he "kept returning to two fundamental questions: First, are these frozen embryos human life, and therefore, something precious to be protected? And second, if they're going to be destroyed anyway, shouldn't they be used for a greater good, for research that has the potential to save and improve other lives?" Bush said that he received conflicting advice, noting that one researcher told him that a "five-day-old cluster of cells is not an embryo, not yet an individual, but a pre-embryo," whereas an "ethicist dismissed that as a callous attempt at rationalization." The president said that thinking about stem cells "forces us to confront fundamental questions about the beginnings of life and the ends of science. It lies at a difficult moral intersection, juxtaposing the need to protect life in all its phases with the prospect of saving and improving life in all its stages."

President Bush reported that his position was shaped by deeply held beliefs, both about the potential of science to find medical cures and that "human life is a sacred gift from our Creator." To reconcile these beliefs, the president announced a policy under which NIH will be allowed to fund research only on existing stem cell lines, "where the life and death decision has already been made." He also announced the formation of a new council, chaired by Leon Kass, a biomedical ethicist from the University of Chicago, "to monitor stem cell research, to recommend appropriate guidelines and regulations, and to consider all of the medical and ethical ramifications of biomedical innovation."

4.2. United Kingdom

In sharp contrast with the United States, the United Kingdom (UK) has a well-established regulatory framework for embryo research, covering the public and private sectors. The Human Fertilisation and Embryology Act 1990 (the 1990 Act) established parameters for the creation and use of embryos, both in research and treatment, and created the Human Fertilisation and Embryology Authority (the HFEA) as a regulatory body with authority over embryo research.

The 1990 Act permitted the creation and use of embryos up to 14 days old for any of five purposes: promoting advances in infertility treatment; increasing knowledge about the causes of congenital disease; increasing knowledge about the causes of miscarriages; developing more effective techniques of contraception; and developing methods for detecting the presence of gene or chromosome abnormalities in embryos before implantation.

About 48,000 embryos that were originally created for in vitro fertilization were used in research in the UK between August 1991 and March 1998, and 118 embryos were created in the course of research in the same period *(31)*.

The 1990 Act allowed for the addition of new permitted research and therapeutic uses of embryos. Subsequent UK deliberations about pluripotent stem cell policies centered on extending the law to allow research for noncongenital conditions, such as Parkinson's disease. The British government policy debate was strongly influenced by a discussion paper *(32)* published in April 2000 by the Nuffield Council on Bioethics, an independent organization jointly funded by the Nuffield Foundation, The Wellcome Trust, and the Medical Research Council. The Nuffield analysis started from the premise that research involving embryos is already permitted in the United Kingdom for some purposes, such as the development and refinement of in vitro fertilization procedures and diagnostic methods. Thus, according to the council, the relevant question is whether embryonic stem cells pose new issues, not whether embryo research should be permitted. "Research into potential therapies is not qualitatively different from research into diagnostic methods or reproduction. Neither benefits the embryo upon which research is conducted but both may be of benefit to people in the future. Each form of research involves using the embryo as a means to an end but, since we accept the morality of doing so in relation to currently authorised embryo research, there seems to be no good reason to disallow research on the embryo where the aim of the research is to develop therapies for others," the discussion paper stated.

The Nuffield Council concluded that once a decision has been made not to implant an embryo, "it no longer has a future and, in the normal course of events, it will be allowed to perish or be donated for research. We consider that the removal and cultivation of cells from such an embryo does not indicate lack of respect for the embryo. Indeed, such a process could be analogous to tissue donation."

The Nuffield Council did identify an ethical issue regarding the donation of embryos or fetal tissue for immortal cell lines. "Although the establishment of a cell line will involve the destruction of the embryo, the DNA ... in the cell of the embryo has the potential to exist indefinitely in culture. The cells could be ultimately used in a wide range of therapeutic applications and, with DNA testing, such a cell line could theoretically be traced back to the individual embryo donors. Consequently, where specific research regarding the establishment of ES cell line is contemplated, embryo donors should be asked explicitly whether or not they consent to such research and subsequent use of the cell line," the paper recommended.

A similar issue arises in regard to donated fetal cadaveric tissue, according to the Nuffield paper, and, thus, specific consent should be required in order to derive EG cells from aborted fetuses. The report acknowledges the concern that the "potential of EG cells to create valuable cell lines for transplantation raises the possibility that an abortion could be sought with a view to donating cadeveric fetal tissue in return for possible financial or therapeutic benefits." To guard against this possibility, the council recommended that women should not be given payments or any other financial incentives to donate tissue.

At the same time, the Nuffield Council also concluded that directing that donations of fetal tissue be used for EG cell derivation would be ethically acceptable. Its paper suggested that "if specific consent is required for the donation of an embryo where an immortal cell line is to be produced from it, it would be only consistent to require special consent for the production of such a cell line from fetal tissue also. Such consent would be to the possible use of fetal tissue in research on EG cells and could not specify that the primordial fetal cells be used in an individual project such as one that might benefit the woman."

In June 1999 the UK government asked Liam Donaldson, its chief medical officer, to establish an expert advisory group to make policy recommendations. The group's report (the Donaldson report) articulated four basic tenets underpinning the use of embryos in research:

1. The embryo of the human species has a special status but not the same status as a living child or adult.
2. The human embryo is entitled to a measure of respect beyond that accorded to an embryo of other species.

3. Such respect is not absolute and may be weighed against the benefits arising from proposed research.
4. The embryo of the human species should be afforded some protection in law.

Citing previous studies of the ethics of embryo research, the Donaldson report stated that "the issue to be considered is one of balance: whether the research has the potential to lead to significant health benefits for others, and whether the use of embryos at a very early stage of their development in such research is necessary to realise those benefits." It concluded that embryonic stem cell research meets both criteria.

According to the Donaldson report, "the potential benefit of research involving embryonic stem cells in terms of eventual therapeutic possibilities was equally as valuable as those purposes for which embryo research is currently allowed. Indeed, the ultimate benefit to human health of such research could prove to be of far wider application and significance."

The Donaldson report recommended extending the 1990 Act "to allow for research to increase understanding about human disease and its cell-based treatments including that involving the extraction and culture of stem cells from embryos." It also recommended that derivation of stem cells from embryos be contingent on the consent of couples donating embryos. It recommended a review to determine if similar consent should be required for women donating fetal tissue.

In August 2000, the UK government announced that it accepted the Donaldson report's recommendations and would submit legislation necessary to implement them to Parliament.

As in the United States, the UK legislation to allow the derivation of stem cells from embryos and aborted fetuses, and their use in research and medicine, was endorsed by patient groups, the scientific community, and the biopharmaceutical industry. Groups opposing abortion condemned the government's proposed legislation as a step toward the creation of Aldous Huxley's Brave New World. Parliament members from all parties argued for and against the measure in an extensive debate that preceded the vote.

The UK House of Commons voted 366 to 174 on December 19, 2000 to allow researchers to conduct basic or applied research on serious diseases using human embryonic stem cells. Under the law, researchers must obtain permission from the HFEA and adhere to a number of limitations, including a stipulation that embryos used in research can be no more than 14 days old.

On February 28, 2002, the HFEA announced the first approvals of human ES cell research. The two approved applications, from the Centre for Genome Research in Edinburgh and Guy's Hospital, London, both involve research on human embryos to produce stem cell lines that will be placed in a stem cell bank that is being developed by the Medical Research Council.

4.3. Other Countries

A detailed international review of the ethical debates about pluripotent stem cell research is beyond the scope of this chapter. Summaries of the status of embryonic stem cell research regulations in selected countries follow.

4.3.1. Germany

The creation of human embryonic stem cells in Germany remains forbidden by the Embryo Protection Law. In April 2002, the German Bundestag voted to allow the import and use of human embryonic stem cells provided that a new agency permits the import according to specific criteria, such as consent by the genetic parents and the creation of the cells before January 1, 2002. Also, the agency will evaluate whether the research goal only can be achieved through the use of human stem cells.

The German law contains a controversial provision stipulating that German researchers participating in international projects involving human embryonic stem cells can face punishment if the cells have not been generated according to the provisions of German law. Researchers would not be punished under the law if they participated in nonconforming research during a fellowship term abroad. However, some legal experts speculate that researchers could face jail if they provide culture media or send doctoral candidates to such laboratories abroad. Indeed, Wolf-Michael Catenhusen, parliamentary undersecretary in Germany's federal research ministry, told a journalist that the amendment would "make the risk for researchers uncalculable" *(33)*.

4.3.2. Australia

Australia has several groups actively pursuing embryonic stem cell research and the national government has identified pluripotent stem cells as a priority area for investment. In May 2002, Prime Minister John Howard announced the award of US$26 million for the creation of a nonprofit Centre for Stem Cells and Tissue Repair to fund basic research and clinical applications related to stem cells. At the time, Australia did not have consistent national policies or regulations governing the derivation and use of ES or EG cells.

In April 2000, federal, state, and territorial government leaders agreed on terms for stem cell legislation that would apply throughout the country. Researchers would be allowed to create new stem lines from surplus embryos generated in fertility programs, and research would be restricted to embryos created before April 5, 2002, subject to consent from donors. On September 25, 2002, the legislation was passed by the House of Representatives in a 99 to 33 vote and was forwarded to the Senate.

Under Australia's constitution, each state can impose different regulations on issues such as stem cell research. As of November 5, 2002, three states—South Australia, Victoria, and Western Australia—had enacted laws banning the creation of embryonic stem cell lines, but other states had not followed suit.

4.3.3. Singapore

The Singapore government appointed a Bioethics Advisory Committee (BAC) in December 2000 and asked it to develop recommendations about human stem cell research and cloning. In February 2001, the BAC formed a human stem cell research subcommittee, which issued a consultation paper in November 2001 to 39 religious and professional groups. The BAC also commissioned papers from local experts and created a panel of international experts.

In June 2002, the BAC released a report, "Ethical, Legal and Social Issues in Human Stem Cell Research, Reproductive and Therapeutic Cloning" *(34)*, which made the following recommendations:

1. Research involving the derivation and use of stem cells from adult tissues is permissible, subject to the informed consent of the tissue donor.
2. Research involving the derivation and use of stem cells from cadaveric fetal tissues is permissible, subject to the informed consent of the tissue donor. The decision to donate the cadaveric fetal tissue must be made independently from the decision to abort.
3. Research involving the derivation and use of ES cells is permissible only where there is strong scientific merit in, and potential medical benefit from, such research.
4. Where permitted, ES cells should be drawn from sources in the following order: existing ES cell lines, originating from ES cells derived from embryos less than 14 days old; and surplus human embryos created for fertility treatment less than 14 days old.
5. The creation of human embryos specifically for research can only be justified where there is strong scientific merit in, and potential medical benefit from, such research; no acceptable alternative exists, and on a highly selective, case-by-case basis, with specific approval from a proposed statutory body.
6. For the derivation and use of ES cells, there must be informed consent from the donors of surplus human embryos, gametes, or cells.
7. There should be a complete ban on the implantation of a human embryo created by the application of cloning technology into a womb, or any treatment of such a human embryo intended to result in its development into a viable infant.
8. There should be a statutory body to license, control, and monitor all human stem cell research conducted in Singapore, together with a comprehensive legislative framework and guidelines.

9. In obtaining consent from donors of cells, gametes, tissues, fetal materials, and embryos, the information provided to the donors must be comprehensive, and there must not be any inducements, coercion, or undue influence.
10. The legislative and regulatory framework should prohibit the commerce and sale of donated materials, especially surplus embryos. However, researchers should not be prohibited from gaining commercially from the products of research, as well as treatments and therapies developed from the donated materials.
11. The legislative framework should provide that no one shall be under a duty to participate in any manner of research on human stem cells, which would be authorized or permitted by the law, to which he has a conscientious objection.

The government of Singapore has charged its Ministry of Health with the responsibility to create a regulatory framework to license, control, and monitor all human stem cell research conducted in Singapore, based on the BAC's recommendations *(35)*.

REFERENCES

1. Sheep clone raises alarm over humans, *The Sunday Times (London)*, 23 February 1997.
2. Thomson, J. A., Itskovitz-Eldor, J., Shapiro, S. S., et al. (1998) Embryonic stem cell lines derived from human blastocysts, *Science* **282,** 1145–1147.
2a. Shamblott, M. J., Axelman, J., Wang, S., et al. (1998) Derivation of pluripotent stem cells from cultured human primordial germ cells, *PNAS* **95,** 13,726–13,731.
3. Statement of Harold Varmus, Director, National Institutes of Health, before the Senate Appropriations Subcommittee on Labor, Health and Human Services, Education and Related Agencies, December 2, 1998. Available from www.nih.gov/about/director/120298.htm.
4. Press release: NIH publishes final guidelines for stem cell research, August 23, 2000. Available from http://www.nih.gov/news/pr/aug2000/od-23.htm.
5. Center for Bioethics and Human Dignity. Using stem cells from embryos will make human flesh profitable, June 29, 2001. Available from www.cbhd.org/resources/aps/dsm-stemcell.htm.
6. Center for Bioethics and Human Dignity. On human embryos and stem cell research: an appeal for legally and ethically responsible science and public policy, April 26, 2000. Available from www.stemcellresearch.org/statement/statement.htm.
7. Encyclical letter *Evangelium Vitae*, addressed by The Supreme Pontiff John Paul II to the bishops, priests and deacons, men and women, religious lay faithful and all people of good will on the value and inviolability of human life, 1995. Available from www.vatican.va/holy_father/john_paul_ii/encyclicals/documents/hf_jp-ii_enc_25031995_evangelium-vitae_en.html.
8. Testimony of Richard M. Doerflinger on behalf of the Committee for Pro-Life Activities United States Conference of Catholic Bishops before the Subcom-

mittee on Labor, Health and Human Services, and Education Senate Appropriations Committee Hearing on Stem Cell Research, July 18, 2001. Available from www.usccb.org/prolife/issues/bioethic/stemcelltest71801.htm.

9. Doerflinger, R. M. (1999) The ethics of funding embryonic stem cell research: a Catholic viewpoint, *Kennedy Inst. Ethics J.* **9(2),** 137–150.

10. General Board of Church and Society, The United Methodist Church, General Secretary's letter to President Bush to extend moratorium on human embryo stem cell research, July 17, 2001. Available from www.umc-gbcs.org/gbpr118a.

11. Background Paper on the Donaldson Report, Society Religion & Technology Project, and the Board of Social Responsibility of the Church of Scotland, 16 August 2000. Available from dspace.dial.pipex.com/srtscot/clonin41.

12. Using stem cells from embryos will make human flesh profitable, June 29, 2001. Available from www.cbhd.org/resources/aps/dsm-stemcell.

13. Overture 01-50. On adopting a resolution enunciating ethical guidelines for fetal tissue and stem cell research—from the Presbytery of Baltimore, June 16, 2001. Available from www.pcusa.org/ga213/business/OVT0150.

14. National Bioethics Advisory Commission report on ethical issues in human stem cell research, volume III, religious perspectives, p. C3, May 1999. Available from www.georgetown.edu/research/nrcbl/nbac/stemcell3.

15. Letter to President George W. Bush from the leaders of American Reformed Jews, July 2001. Available from uahc.org/reform/rac/news/071601.

16. Outka, G. The ethics of stem cell research, April 2000. Available from www.bioethics.gov/outka.

17. Testimony of The Honorable Orrin G. Hatch before the Senate Appropriations Committee Subcommittee on Labor—HHHS—Stem Cell Research, July 18, 2001. Available from www.senate.gov/~hatch/index.cfm?FuseAction=Topics. Detail&PressRelease id=177442&Month=7&Year=2001.

18. Shannon, T. A. (1998) Remaking ourselves? The ethics of stem-cell research, *Commonweal* **125(21),** 9.

19. Testimony of Richard O. Hynes, Ph.D., President, American Society for Cell Biology to the Labor, Health & Human Services and Education Subcommittee of the Appropriations Committee, United States Senate, September 14, 2000. Available from www.ascb.org/publicpolicy/hynessctest.

20. Testimony of Lawrence S. B. Goldstein, Ph.D., representing the American Society for Cell Biology to the Labor, Health & Human Services and Education Subcommittee of the Appropriations Committee, United States Senate, April 26, 2000. Available from www.ascb.org/publicpolicy/goldsteinsctest.

21. Usdin, S. (1998) Cloning showdown comes quickly, Feb. 9, 1998. *BioCentury* **6(12).**

22. Biotechnology Industry Organization, Statement of the biotechnology industry organization regarding legislation introduced to ban human cloning, Feb. 10, 1998. Available from thomas.loc.gov/cgi-bin/query/C?r105:./temp/~r1057B0Obr.

23. Usdin, S. and Zipkin, I. (1998) NIH wants in, *BioCentury*.

24. Usdin, S. (1999) Lawmakers fire back at Varmus, Feb. 22, 1999, *BioCentury*, **7(12)**.
25. National Bioethics Advisory Commission report on ethical issues in human stem cell research, volume I, p. 88, September 1999. Available from www.georgetown.edu/research/nrcbl/nbac/stemcell.pdf.
26. National Bioethics Advisory Commission report on ethical issues in human stem cell research, executive summary, p. 4, September 1999. Available from www.georgetown.edu/research/nrcbl/nbac/execsumm.
27. White House Statement on Human Stem Cell Research, July 14, 1999.
28. Federal Register Notice: Draft NIH Guidelines for Research Involving Human Pluripotent Stem Cells, December 1999. Available from www.nih.gov/news/stemcell/draftguidelines.htm.
29. NIH Publishes Final Guidelines for Stem Cell Research, August 23, 2000. Available from www.nih.gov/news/pr/aug2000/od-23.htm.
30. Vogel, G. (2001) NIH pulls plug on ethics review, *Science* **292,** 415–416.
31. Chief Medical Officer's Expert Group on Therapeutic Cloning report on "Stem Cells: Medical Progress with Responsibility," p. 7, August 16, 2000. Available from www.doh.gov.uk/cegc/stemcellreport.htm.
32. Nuffield Council on Bioethics (2000) Stem cell therapy: the ethical issues, a discussion paper. Available from www.nuffieldbioethics.org/publications/stemcells/rep0000000299.asp
33. Wess, L. (2002) German stem cell schizophrenia, April 29, 2002. *BioCentury* **10(19)**.
34. Bioethics Advisory Committee (2000) Ethical legal and social issues in human stem cell research, reproductive and therapeutic cloning. Available from www.bioethics-singapore.org/bac/detailed.jsp?artid=33&typeid=1&cid=35&bSubmitBy=false.
35. The Conference on "Beyond Determinism and Reductionism: Genetic Science and the Person," 2002. Available from www.gov.sg/singov/announce/170702tt.htm.

2

A Researcher's Guide to Federally Funded Human Embryonic Stem Cell Research in the United States

Gregory J. Downing

1. INTRODUCTION

The issuance of the first grant awards for human embryonic stem (hES) cell research by the National Institutes of Health (NIH) in 2002 represents a milestone in this emerging field of biomedical investigation. Based on a policy announced on August 9, 2001 by President George W. Bush, scientists can now apply for and receive federal funds for research to use certain hES cell lines in their laboratory investigations. This chapter provides information often sought by researchers regarding federal policies, regulations, procedures, and opportunities for pursuing studies using hES cells. Although research that results in the creation or destruction of human embryos is prohibited from receiving federal support, there are no legal restrictions on performing these research activities with private sources of funding. In 1998, a research group led by Thomson published their privately funded studies on hES cells, which were obtained after removal of the inner cell mass from an early blastocyst *(1)*. Also in 1998, Gearhart and co-workers published their privately funded studies on human embryonic germ (hEG) cells, which were derived from the primordial ridge of early-gestation fetal tissue *(2)*. Even though hES and hEG cells are isolated from distinctly different stages in human development, both cell lines were found to have the capacity for infinite proliferation in an undifferentiated state and development into specialized cells with properties similar to those found in many different types of tissues. Both hES and hEG cells are described as "pluripotent," inferring that the undifferentiated cells are capable of developing into cells typically found in specialized tissues representing all three germ layers (the endoderm, mesoderm, and ectoderm) in the course of normal development of an

From: *Human Embryonic Stem Cells*
Edited by: A. Chiu and M. S. Rao © Humana Press Inc., Totowa, NJ

organism.[a] These discoveries sparked intense interest among the scientific community because of the enormous potential for basic research and development of cell-based therapies, and they led to the establishment of policy for the use of public resources for hES cell research. It is important to note that the presidential decision of August 9, 2001 dealt strictly with hES cell research and did not directly affect the existing guidelines pertaining to the use of federal funds for hEG cell research *(3)*. The main objective of this chapter is to discuss conditions whereby research with hES and hEG cells can be supported by the federal government, and it focuses primarily on the regulations that apply to obtaining funding from the NIH. Other important issues such as those involving patents and intellectual property are addressed in the next chapter.

2. CONDUCTING HUMAN EMBRYONIC STEM CELL RESEARCH IN THE UNITED STATES

Over the past 20 yr, research to develop and use murine *(4,5)* and nonhuman primate embryonic stem cells *(6)* was supported through public and private resources, and this contributed indirectly to fundamental advances in hES cell research. However, none of the innovative work leading to the derivation of hES and hEG stem cell lines to date has been supported with US federal funds; they were developed solely with private resources.[b] Publicly financed stem cell research is widely recognized as important because of the need for extensive basic research on the unique biological properties of hES cells such as studies to understand the regulation of genes involved in the control of proliferation and differentiation. In general, basic research is not the major target of the private sector, which is generally more focused on technology development and cellular engineering. An important point about the funding of hES research in the United States is that whereas there are specific research activities that cannot be carried out with the use of public funds, these same activities *can* be conducted with support from private resources, such as companies, foundations, or universities.[c] A second

---

[a]The term *pluripotent stem cell* has been used to describe cells that are derived from human embryos or from the primordial ridge in fetal tissues. The discussion in this chapter uses the term to refer solely to hES and hEG cells. However, recent scientific evidence suggests that some "adult-type" stem cells may also have pluripotent capabilities, but research activities using them are not subject to the policies, procedures, and guidances presented here.

[b]In this chapter, a distinction is made between federally and publicly sponsored research activities. In some cases, state government funding has been authorized to support research for the development of hES cell lines.

[c]Many states have laws and regulations regarding human embryos and fetal tissue and the use of them in research. Researchers are recommended to confer with institutional administrative officials regarding local and state requirements. A resource for current state embryonic and fetal research laws and pending legislation may be found at the National Conference of State Legislatures website (www.ncsl.org/programs/health/genetics/embfet).

important distinction is the difference in policies regulating the use of public funds for the development or use of hES cell lines versus those regulating hEG cells because the former are derived from human embryos and the latter are from fetal tissue. Investigators who perform research with US federal funding need to understand these key distinctions in order to remain in compliance with regulations and guidances.

2.1. Human Embryo Research

The existing policy governing the use of public resources for hES cell research in the United States is linked to legal interpretations of the federal prohibition of funding for investigations that use human embryos. Since 1993, NIH has been prohibited from supporting research that result in the creation or destruction of human embryos. This prohibition was recently reaffirmed in the congressional appropriations for NIH in fiscal year 2002 as Public Law 107-116 of the Departments of Labor, Health and Human Services, and Education, and Related Agencies Appropriations Act, 2002 Sec. 510. which states:

(a) None of the funds made available in this Act may be used for (1) the creation of a human embryo or embryos for research purposes; or (2) research in which a human embryo or embryos are destroyed, discarded, or knowingly subjected to risk of injury or death greater than that allowed for research on fetuses in utero under 45 CFR 46.208(a)(2) and section 498(b) of the Public Health Service Act (42 U.S.C. 289g(b)).

(b) For purposes of this section, the term "human embryo or embryos" includes any organism, not protected as a human subject under 45 CFR 46 as of the date of the enactment of this Act, that is derived by fertilization, parthenogenesis, cloning, or any other means from one or more human gametes or human diploid cells.

Clearly, federal funds may not be used for the derivation of hES cell lines, a process that requires the destruction of a human embryo. In 1999, a legal interpretation of the prohibition by the Department of Health and Human Services Office of General Council determined that research *using* hES cell lines did not constitute human embryo research as long as federal funds were not used for the actual derivation of the cells. This interpretation opened the door for the development of NIH guidelines, issued in August 2000, pertaining to the use of pluripotent stem cells in federally sponsored research *(3)*. In 2001, a year after these guidelines were issued, President George W. Bush announced a policy that, in effect, eliminated portions of these guidelines covering hES cell research. The next two sections detail current application of these guidelines.

2.2. Use of Human Embryonic Stem Cells in Research

President George W. Bush announced the current policy for federal funding of hES cell research on August 9, 2001. The new policy allowed federal-expenditures for research conducted *only* on cell lines developed prior to the presidential announcement. This means that funding for any new cell lines established by the destruction of an embryo(s) after August 9, 2001 is prohibited. On the other hand, federal funds may now be used for research on existing hES cell lines if (1) the derivation process (which commences with the removal of the inner cell mass from the blastocyst) had already been initiated prior to the announcement on August 9, 2001 and (2) the embryo from which the stem cell line was derived no longer had the possibility of development as a human being.

In addition, President Bush's announcement identified four requirements that must be met regarding the acquisition of the embryo for derivation of the hES cells:

- The embryo was originally created for reproductive purposes.
- The embryo was no longer needed for reproductive purposes.
- Informed consent must have been obtained for the donation of the embryo.
- No financial inducements were provided for donation of the embryo.

In November 2001, an NIH Human Embryonic Stem Cell Registry (escr. nih.gov) was developed to facilitate hES cells research by providing investigators with the list of the hES cells that meet the eligibility criteria and with contact information so that they could acquire cells that are available. Each of the sources of the 78 cell lines currently described on the Registry provided NIH with documents and assurances that they were in compliance with these eligibility criteria. The next chapter provides an in-depth discussion of issues dealing with intellectual property, acquisition costs for cell lines, and material transfer agreements with the sources of cells. Table 1 summarizes a series of steps to guide investigators seeking NIH funding for research using human pluripotent cells.

2.3. Use of Human Embryonic Germ Cells in Research

The presidential decision of August 9, 2001 dealt strictly with hES cells. It did not change the rules for federal funding of hEG cell research; however, their use is still regulated by the NIH Guidelines for Research using Human Pluripotent Stem Cells *(3)*. Despite the fact that no prohibition or policy has ever prevented federal funding of hEG research, it is noteworthy that the first hEG cell lines *(2)* were not developed with the use of federal funds. The NIH guidelines, which apply to the use of hEG cells, are closely aligned with provisions governing the use of human fetal tissue in research,

Table 1
Applying for Federal Funding of Research Using Embryonic Stem Cells and Embryonic Germ Cells[a]

A. Human Embryonic Stem Cell Research

- Include statement in grant proposal indicating which cell lines from NIH Registry (escr.nih.gov) will be used in research activities—cite the NIH identifying code for the cell line(s) to be used in the proposal
- Indicate in the proposal that the research will not include prohibited activities
- Obtain cell lines from eligible sources using material transfer agreement
- Contact institutional officials regarding need for IRB review and approval—refer to OHRP guidance (ohrp.osophs.dhhs.gov/humansubjects/guidance/stemcell)
- Contact appropriate NIH extramural program officer with technical questions about research plan (grants1.nih.gov/grants/stem_cell_contacts)

B. Human Embryonic Germ Cell Research

- Include description of cell lines to be used in research protocol in the grant proposal
- Indicate in the proposal that the research will not include prohibited activities
- Provide documentation of compliance with criteria described in Section II.B.2 of the NIH Guidelines for Research Using Human Pluripotent Stem Cells (www.nih.gov/news/stemcell/stemcellguidelines)
- Submit supplemental application of research assurances to Human Pluripotent Stem Cell Review Group *(7)*
- Contact institutional officials regarding need for IRB review and approval—refer to OHRP guidance (ohrp.osophs.dhhs.gov/humansubjects/guidance/stemcell)
- Contact appropriate NIH extramural program officer with technical questions about research plan (grants1.nih.gov/grants/stem_cell_contacts)

[a]Research activities for basic, in vitro, or animal studies. Future clinical studies with stem cells may have additional submission and approval requirements by NIH and US Food and Drug Administration.

as listed below. The policies regulating hEG and hES cell research differ in five important points:

- Federal funding may be used in the derivation of the hEG cells from fetal tissue.
- Research may be conducted with federal funds for hEG cell lines established *after* August 9, 2001.
- Federal funding may be used only for those hEG cells that are in compliance with the NIH Guidelines for Research Using Pluripotent Stem Cells (Section II.B.2).
- The use of hEG cells in federally sponsored research requires the approval by

an NIH committee known as the Human Pluripotent Stem Cell Review Group (HPSCRG) as described in the NIH Guidelines *(3)*.

• Eligible hEG cells are not listed on the NIH Human Embryonic Stem Cell Registry.

Individuals planning to use hEG cells in research should also consider two important aspects of federal laws and regulations. The first pertains to the legal aspects of human subjects protections that are associated with acquiring human fetal tissue for use in research. Section 42 U.S. Code 289g-2 (www.nih.gov/news/stemcell/42-289g2) describes conditions, prohibitions, and penalties associated with the acquisition of human fetal tissue. Research conducted with human fetal tissue is also subject to state and local regulations, as specified in 45 Code of Federal Regulations 46.210 (www.nih.gov/news/stemcell/45).

If federal funding for conducting hEG cell research is sought, the application must comply with the NIH Guidelines for Research Using Pluripotent Stem Cells *(3)*. An NIH guide notice *(7)* has been issued that describes in detail procedures for submitting materials that must accompany a grant application proposing to use hEG cells. These procedures also apply to funded grantees who would like to include the use of hEG cells in their existing grant awards. The following eight criteria must be fulfilled and documentation submitted to NIH for review by the HPSCRG prior to final approval and award of funding by the NIH:

• An assurance, signed by the responsible institutional official, that the hEG cells were derived from human fetal tissue in accordance with the conditions set forth in Section II.B.2 of the NIH guidelines and that the institution will maintain documentation in support of the assurance.
• A sample informed consent document (with patient identifier information removed) and a description of the informed consent process that meet the criteria for informed consent described in the following paragraph.
• An abstract of the scientific protocol used to derive hEG cells from fetal tissue.
• Documentation of institutional review board (IRB) approval of the derivation protocol.
• An assurance that the hEG cells to be used in the research were or will be obtained through a donation or through a payment that does not exceed the reasonable costs associated with the transportation, processing, preservation, quality control, and storage of the stem cells.
• The title of the research proposal or specific subproject that proposes the use of hEG cells.
• An assurance that the proposed research using hEG cells is not a class of research that is ineligible for NIH funding as set forth in Section III of the NIH guidelines.
• The Principal Investigator's written consent to the disclosure of all material submitted to the NIH as necessary to carry out the public review and other oversight procedures set forth in the NIH guidelines.

There are aspects of the informed consent procedures that must be considered by investigators who have derived hEG cells from fetal tissue, regardless of whether this was carried out with federal funding. Informed consent documents should include the following information:

- A statement that fetal tissue will be used to derive stem cells for research that may include human transplantation research.
- A statement that the donation is made without any restriction or direction regarding the individual(s) who may be the recipient(s) of transplantation of cells derived from the fetal tissue.
- A statement as to whether or not information that could identify the donors of the fetal tissue, directly or through identifiers linked to the donors, will be removed prior to the derivation or the use of human pluripotent stem cells.
- A statement that derived cells and/or cell lines may be kept for many years.
- A disclosure about the possibility that the results of research on the stem cells may have commercial potential, and a statement that the donor will not receive financial or any other benefits from any such future commercial development.
- A statement that the research is not intended to provide direct medical benefit to the donor.

The above materials must be submitted to the HPSCRG by investigators seeking federal funds to use hEG cells in their research before approval for funding can be granted *(7)*. If a research proposal has been determined to merit federal funding, the assurance documents will be reviewed by HPSCRG in a public meeting for compliance with the NIH guidelines. For investigators proposing to use hEG cells that have received prior HPSCRG approval, there is an expedited review and approval process that does not require discussion at a public meeting. Under these circumstances, researchers should provide the HPSCRG with a copy of the approval letter from the source of the cells, a letter specifying their request, and accompanying documents. Researchers planning to use hEG cells in their investigations should contact and consult with the appropriate program officials at NIH before submitting their grant applications and assurance documents.

3. RESEARCH ACTIVITIES THAT CANNOT BE SUPPORTED WITH US FEDERAL FUNDS

The current policy on federally funded research also identifies research activities that are not eligible for support. These prohibited activities, first described in the 2000 NIH guidelines *(6)* (Section III, Areas of Research Involving Human Pluripotent Stem Cells that are Ineligible for NIH Funding), were not affected by the president's August 2001 policy decision and include the following:

Table 2
Types of Stem Cell Research Activities That Are Not Subject
to Special Requirements for Federally Funded Research

- Studies proposing to use animal sources of adult stem cell, embryonic stem cell, or embryonic germ cell lines
- Basic (nonclinical) research studies involving human adult stem cells (such as hematopoietic stem cells)
- Basic research (nonclinical) using nonpluripotent stem cells derived from human fetal tissue (e.g., stem cells found in tissues other than primordial ridge, such as hematopoietic and neuronal stem cells)

- The derivation of pluripotent stem cells from human embryos
- Research in which human pluripotent stem cells are utilized to create or contribute to a human embryo
- Research utilizing pluripotent stem cells that were derived from human embryos created for research purposes, rather than for fertility treatment
- Research in which human pluripotent stem cells are derived using somatic cell nuclear transfer (i.e., the transfer of a human somatic cell nucleus into a human or animal egg)
- Research utilizing human pluripotent stem cells that were derived using somatic cell nuclear transfer (i.e., the transfer of a human somatic cell nucleus into a human or animal egg)
- Research in which human pluripotent stem cells are combined with an animal embryo
- Research in which human pluripotent stem cells are used in combination with somatic cell nuclear transfer for the purposes of reproductive cloning of a human

Investigators who are conducting federally sponsored research and have questions about eligible research activities should consult with their NIH program officer.

4. FEDERAL REQUIREMENTS FOR REVIEW AND APPROVAL BY INSTITUTIONAL REVIEW BOARD OF BASIC RESEARCH INVOLVING USE OF hES AND hEG CELLS

Investigators and institutional officials should be aware of the requirements for IRB assessment and approval of research that proposes to use hES and hEG cells. Recognizing that the use of these cells in human subjects in clinical trials may be years away, researchers are sometimes uncertain whether IRB review is required for their research. For basic research applications that only involve in vitro studies or studies in animals, the human subjects concerns focus on protections for the donor(s) of the embryos or fetal tissue that served as the source of the cell lines.

Recently, the Office of Human Research Protections (OHRP) developed guidelines for researchers and research institutions with information about circumstances that require IRB action for basic hES and hEG cell research applications (Guidance for Investigators and Institutional Review Boards Regarding Research Involving Human Embryonic Stem Cells, Germ Cells and Related Test Articles [ohrp.osophs.dhhs.gov/humansubjects/guidance/ stemcell.pdf]). Under circumstances described later in this section, research supported by the Department of Health and Human Services (HHS) may be subject to HHS human subjects protection regulations described in Title 45 Code of Federal Regulations (CFR) Part 46 "Protection of Human Subjects," including Subpart B, 45 CFR 46.206.

Of concern is the possibility that identifying information (i.e., information that could reveal the identities of the donors of the embryos or fetal tissue), might become available to the researchers who work with cell lines derived from these sources. Some cell lines have been developed "anonymously," without any information that could link the donors to the cells *(1)*. In the case of other cell lines, donor information has been retained, either by the source of the embryos or by the laboratories where the cell lines were originally developed. If federally sponsored research is conducted with cells where identifying information could be made available to investigators, such research is governed by 45 CFR Part 46 because the donors are considered human subjects, and, therefore, IRB review and approval is required. However, 45 CFR Part 46 does not apply if (1) the investigator and the research institution do not have access to the identifiable private information related to the cell line and (2) a written agreement is obtained from the holder of the information linking the donor to the cell line, indicating that such information will not be released to the investigator under any circumstances. If these conditions are met, an institution or an IRB can decide to waive IRB review of the research using the cell line.

Researchers can determine whether a cell line has identifying information linked to it in a number of ways. In many cases, the material transfer agreements between investigators and the sources of the hES and hEG cells will delineate the anonymity of sources or specify the limitations of access by the researcher to identifying information about the donor(s). In addition, researchers should recognize that their institutions and IRBs might require review and approval of basic research use of these cells even under circumstances that are not required by the OHRP guidance. Therefore, investigators are encouraged to contact their institutional research administrative officials to assure compliance with their policies in addition to federal regulations.

As investigators begin to contemplate obtaining federal support for clinical research (i.e., studies conducted on human subjects) using hES and hEG cells, the OHRP guidance provides additional information about other regulations that apply. Although beyond the scope of this discussion, researchers should be aware that in addition to the OHRP guidance for stem cell research, use of these cells in *clinical* research is also subject to the US Food and Drug Administration IRB and informed consent regulations (Title 21 CFR Parts 50 and 56).

5. FEDERAL FUNDING OPPORTUNITIES FOR STEM CELL RESEARCH

Opportunities for NIH funding of research with hES and hEG cells can be found at a central NIH website with information on this subject (www.nih. gov/news/stemcell/index) as well as related topics such as information about the acquisition of cells, contact information for NIH extramural program staff, and frequently asked questions. The NIH also posts requests for applications (RFAs) and program announcements (PAs) soliciting research applications on specific aspects of stem cell research; these are updated weekly in the NIH Guide for Grants and Contracts (grants.nih.gov/grants/ guide/index). Finally, the NIH provides administrative supplements for investigators with existing grants to enable them to acquire the skills and cells for initiating research in this field.

Some investigators have asked under what circumstances can they conduct research in their NIH-funded laboratories on cell lines that do not fulfill all of the federal criteria, such as hES cells that are not listed on the NIH Registry. They may have private resources from industry or foundations to conduct research that is prohibited with federal funds. In laboratories receiving both federal and nonfederal sources of funding, investigators and their staff must segregate allowable and unallowable activities so that the costs incurred by each type of research is charged to the appropriate funding source. For instance, the time and effort of laboratory personnel working on ineligible stem cells may not be charged to any federal grant. Acquisition of equipment, use of cell and tissue culture supplies in the project, and travel to conferences to discuss or present work on prohibited activities likewise may not be supported with federal funds. Finally, it is the institution's responsibility to provide clear instructions to investigators who conduct research that is "unallowable" under federal research funding policy.

6. CONCLUSIONS

Research using hES and hEG cells is now underway in the United States with both public and private support. The NIH is supporting stem cell research by providing extensive information to the research community on the acquisition of stem cells, guidelines and legal requirements, funding opportunities, and technical assistance about the grant application process. A high priority for the NIH has been the development of the science infrastructure to facilitate investigator-initiated basic research studies. This infrastructure includes enhancing the distribution network for cell lines, improving their characterization and quality assurance measures, supporting laboratory training programs, workshops, and conferences, and establishing career development pathways to ensure that a highly skilled scientific workforce is in place. Investigators who apply for federal funding in the United States are encouraged to learn about the framework for conducting hES and hEG cell research. Once fully informed, researchers will find that there are minimal burdens and many opportunities for successful research programs using these unique cells.

REFERENCES

1. Thomson, J. A., Itskovitz-Eldor, J., Shapiro, S. S., et al. (1998) Embryonic stem cell lines derived from human blastocysts, *Science* **282,** 1145–1147.
2. Shamblott, M. J., Axelman, J., Wang, S., et al. (1998) Derivation of pluripotent stem cells from cultured human primordial germ cells, *Proc. Natl. Acad. Sci. USA* **95,** 13,726–13,731.
3. NIH (2000) National Institutes of Health guidelines for research using human pluripotent stem cells. *Fed. Reg.* **65,** 69,951. Available from http://www.nih.gov/news/stemcell/stemcellguidelines.
4. Evans, M. J. and Kaufman, M. H. (1981) Establishment in culture of pluri-potential cells from mouse embryos, *Nature* **292,** 154–156.
5. Martin, G. R. (1981) Isolation of a pluripotent cell line from early mouse embryos cultured in medium conditioned by teratocarcinoma stem cells, *Proc. Natl. Acad. Sci. USA* **78,** 7634–7638.
6. Thomson, J. A., Kalishman, J., Golos, T. G., et al. (1985) Isolation of a primate embryonic stem cell line, *Proc. Natl. Acad. Sci. USA* **92,** 7844–7848.
7. NIH Guide for Grants and Contracts. Procedures for submission of compliance documents to the human pluripotent stem cell review group for the research use of human embryonic germ cells. Available from http://grants1.nih.gov/grants/guide/notice-files/NOT-OD-02-049.

3
Intellectual Property of Human Pluripotent Stem Cells

Mark L. Rohrbaugh

1. INTRODUCTION

Intellectual property arose as a principle implementation issue after President Bush announced his decision on August 9, 2001 to permit federal funding for qualified human embryonic stem cell (hESC) lines. The research community focused its attention on the existing sources of these cells. Questions immediately arose regarding the effect intellectual property rights might have on this new federally funded research enterprise. In particular, concerns were raised as to how the Wisconsin Alumni Research Foundation and its exclusive licensee, Geron Corporation might use their rights in stem cell patents to exercise control over the distribution of hESCs from multiple sources. To what extent and under what terms would the scientific community be able to gain access to these cells to conduct further research and development with government support *(1)*? This issue once again drew public attention to the laws and policies that govern the issuance of patents on human biological materials as well as on fundamental tools used for basic biomedical research. This chapter will review the existing and pending intellectual property rights in hESCs various parties held, the intellectual property rights held by the US government when inventions arise with Government funding, and how the National Institutes of Health (NIH) has worked with the hESC providers to facilitate the availability of these cells to the research community.

2. PATENTS AS A FORM OF INTELLECTUAL PROPERTY

Patents are a time-limited monopoly granted by the government to the owner (the inventor *[2]* or assignee) of an invention as a *quid pro quo* for disclosing and teaching society how to practice the invention *(3)*. In markets such as biomedical products and processes, where a great deal of time and

From: *Human Embryonic Stem Cells*
Edited by: A. Chiu and M. S. Rao © Humana Press Inc., Totowa, NJ

money is needed to bring a product to market, patents also serve as incentives for the owner or licensee to invest in the development and ultimate commercialization of the technology. The term of the monopoly in most countries is 20 years from the earliest filing ("priority") date of the patent application, after which time, the intellectual property reverts to the public. Patents do not confer an affirmative right to use the invention but the right to exclude others from making, using, selling, or importing the invention in the country issuing the patent *(4)*. Thus, the patent owner may not be able to commercialize the invention without controlling local or federal approval (e.g., Food and Drug Administration [FDA] approval to market a drug or vaccine). Without the approval of the patent owner, however, others have no right to use the invention *(5)*.

Patents are a form of *intellectual* property because the rights relate to an intangible application of human ingenuity and cannot be physically possessed like material objects or occupied like real estate *(6)*. A patent does not provide rights in tangible property (i.e., the actual physical embodiment of the technology that might be described and claimed in the patent). That means, one who isolates a particular cell owns that cell even if the patent owner, whose patent describes or "claims" it, can restrict its use. One must thus be careful to distinguish between intellectual property of patent owners from tangible property of those who create or isolate the material claimed in a patent. They may or may not be one and the same party. When they differ, both the owner of the intellectual property and the owner of the tangible property may each have certain rights to control of the tangible property—the first through the enforcement of patent rights, the latter through lawful ownership of the material and control through distribution under contracts or agreements that give others the right to use the actual material. The patent rights are superior to rights in the tangible property, such that one would need permission or a license from both to operate legally with materials created and owned by others. This layering of rights comes into play particularly when researchers obtain cell lines from parties other than the patent owner. The provider of the material can use a contract governing the transfer, such as a Material Transfer Agreement (MTA), to place restrictions on the use of materials, and the patent owner can place restrictions in the terms of a license relative to the intellectual property.

In order to be patentable, an invention must first be the work of human ingenuity rather than a product of nature. It can be "any new and useful process, machine, manufacture, or composition of matter, or any new and useful improvements thereof" *(7)*. Patent laws distinguish unpatentable products of nature from processes and products of human ingenuity. Thus, isolated and purified genes and cells may be patentable, but genes and cells in their natural, unmodified state are not patentable *(8)*. Although some countries' policies on this matter are similar to those of the United States, other countries draw the line between the handiwork of nature and the work of human ingenuity, more toward the direction of requiring the inventor to supply further modification, characterization, and commercial use than under US law (*see* Section 4).

In addition, for an invention to be patentable, the inventor must show that the invention has utility (a real world use) *(9)*, novelty *(10)*, and is not obvious to a theoretical person of general skill in that field or "art" *(11)*. The inventor must describe in the patent application the invention with sufficient particularity to enable someone skilled in the art to make and use the full scope of the invention, set forth the inventor's best mode for using it, and show that the inventor had possession of the invention at the time of filing (the "written description"). The description of invention is called the "specification" *(12)*. The patent concludes with claims that precisely set forth the bounds of the invention. It is the invention encompassed within the four corners of the claims that constitutes the enforceable patent right.

There are three basic types of claims: (1) products (including compositions of matter and manufactures [e.g., cells, genes, or equipment]); (2) methods (e.g., means of making [isolating] cells, genes, or means of using [processing] materials); and (3) products by process (e.g., materials made using a particular claimed process). In addition, patent claims can be independent in form, in that they can stand on their own (e.g., claim 1—a transformed human cell line derived from tumor tissue). Claims can be dependent in form, in that they narrow the scope of an independent claim (e.g., claim 2—the cell line of claim 1 wherein the tumor tissue is a basal cell carcinoma). Patent claims are read in light of the rest of the specification such that, for example, if the inventor defines a term in a particular way, the same meaning of the term will be attached to the term in the claims.

Product claims are considered to be the strongest because materials that are sold or used can more easily be characterized as claimed by a patent than processes that might be hidden from public view and not evident from examining the final product. Moreover, one must only identify one utility or use of a composition of matter in order to obtain claims to it. Others may obtain patents on new, novel, nonobvious uses of that same material. However, the original patent on the composition will dominate the later patents, in that anyone practicing the use patents requires permission of both the use patent owner and the owner of the original composition patent. Product by process claims are more limited from an enforcement standpoint in that they only govern products made by the process claimed in the patent. Proving that a particular material was made in a particular manner may be difficult, especially when there are multiple ways to make the product. Moreover, to prove infringement of any claims, the patent owner or licensee must show that the accused infringing product or process incorporates all of the elements, or limitations, in the claim. Thus, if a patent includes claims to a cell having characteristics A, B, and C (or a process with steps A, B, and C), another party's cells can only infringe if the accuser demonstrates that these cells have all three characteristics (or the method involves all three steps). Otherwise, there is no infringement.

3. PATENTS GOVERNING PLURIPOTENT HUMAN EMBRYONIC STEM CELLS

Dr. James A. Thomson's laboratory at The University of Wisconsin, in collaboration with Dr. Joseph Itskovitz's laboratory at Rambam Medical Center, Technion, in Haifa, Israel captured international attention when they published their findings in *Science* in November 1998 describing the isolation of stable pluripotent human embryonic stem cell lines *(13)*. A month later the US Patent and Trademark Office (PTO) granted US patent number 5,843,780 claiming primate pluripotent ESCs. A few years later, the PTO granted US patent number 6,200,806 specifically claiming pluripotent hESCs *(14)*. Thomson assigned his rights in the patents to the Wisconsin Alumni Research Foundation (WARF), which is a private foundation that manages intellectual property for the University of Wisconsin.

3.1. Primate Pluripotent Embryonic Stem Cells

The Washington Alumni Research Foundation filed Dr. Thomson's first ESC patent application on January 20, 1995, with pending claims to primate ESCs and a method of deriving them *(15)*. A year later, on January 18, 1996, they filed a continuation-in-part *(16)* of this original application and allowed the first application to become abandoned. A continuation-in-part application, by definition, contains new matter not found in the original application *(17)*. The new matter consisted primarily of further characterization of two rhesus ESC lines (R366 and R367) that had been described in the original application and the isolation of a fourth (R394). The original application also described seven marmoset putative ESCs cultured for more than 6 mo and one for over 14 mo. The continuation-in-part included further characterization of rhesus lines R366 and R367, namely normal XY karyotype, proliferation in an undifferentiated state for about 3 mo, and the ability to form teratomas in severe combined imunodeficient (SCID) mice. In this regard, these lines show the characteristics of rhesus line R278.5, which had been more extensively characterized in the original application. Line R394 is described as having normal XX karyotype. The application includes statements that all four of these rhesus lines have similar morphology, do not require leukemia inhibitory factor (LIF) to derive or proliferate, and the particular source of fibroblasts for coculture is not critical *(18)*. The 1998 continuation-in-part application was issued as patent 5,843,780 on December 1, 1998 with essentially the same claims as filed in the original application.

The applicant was successful in overcoming criticism by the PTO that this invention was not obvious in light of the work of Bongso et al. *(19)*, a patent to Hogan *(20)*, and a published international patent application to Dyer et al. *(23)*. In particular, the PTO relied upon the applicant's rebuttal that this prior art does not demonstrate the ability to obtain purified preparations of primate ESCs that are stable in long-term culture (i.e., greater than 1 yr). In addition, a declaration by Thomson on February 12, 1998 provided the examiner with acceptable evidence of the general reproducibility of the technology without "undue experimentation." He noted that, using the same methodology, he had created at least 11 independent rhesus and 8 marmoset ESC lines that appear to have the same characteristics of the original cell lines.

Ultimately, WARF obtained patent 5,843,780 with claims to "a purified preparation of primate embryonic stem cells that (1) is capable of proliferation in an in vitro culture for over 1 yr, (2) maintains a karyotype in which all the chromosomes characteristic of the primate species are present and not noticeably altered through prolonged culture, (3) maintains the potential to differentiate into derivatives of endoderm, mesoderm, and ectoderm tissues throughout the culture, and (4) will not differentiate when cultured on a fibroblast feeder layer." Another claim is directed to "a purified preparation of primate embryonic stem cells wherein cells are negative for the SSEA-1 marker, positive for the SSEA-3 marker, positive for the SSEA-4 marker, express alkaline phosphatase activity, are pluripotent, and have karyotypes that includes the presence of all of the chromosomes characteristic of the primate species and in which none of the chromosomes are noticeably altered." A method claim is directed to "a method of isolating a primate embryonic stem cell line, comprising the steps of (1) isolating a primate blastocyst, (2) isolating cells from the inner cell mass of the blastocyte of (1), (3) plating the inner cell mass cells on embryonic fibroblasts, wherein inner-cell-mass-derived cell masses are formed, (4) dissociating the mass into dissociated cells, (5) replating the dissociated cells on embryonic feeder cells, (6) selecting colonies with compact morphologies and cells with high nucleus-to-cytoplasm ratios and prominent nucleoli, and (7) culturing the cells of the selected colonies." There are also dependent claims that further define these broader independent claims, including a cell line derived by the method of isolating the embryonic stem cells.

It is important to analyze the claim language of any patent carefully because the value and enforceability of a patent derives from the scope of its claims. The inventors based the patent application on experiments demonstrating a technique for the isolation of ESCs from rhesus, a macaque Old World monkey species, and the common marmoset, a New World monkey. However, the patent claims primate ESCs. Under patent law, patent claims are defined in terms of the plain meaning of their terms or the terms as defined by the inventors in the patent specification. The term "primate" is a key term in discerning the scope of this patent. "Primate" by its plain meaning in biology means the order of Primates including Old and New World monkeys, apes, and man. Moreover, the inventors clearly intended the meaning to be this broad because the specification includes humans as an example of primates and intended uses of "primate" stem cells. They make the argument that the invention can be applied in the same way to the scope of primate species, including higher primates such as humans, because the evolutionary difference between Old World and New World species is far greater than the evolutionary distance between humans and rhesus monkeys.

3.2. Human Pluripotent Embryonic Stem Cells

About 1 yr after WARF filed the continuation-in-part application for Thomson's invention, they filed a divisional application to separate specific claims to human cell lines *(24)*. The application is identical in its specification but the claims differ by the replacement of the word "human" for the word "primate" and substituting "maintains a karyotype in which all the chromosomes characteristic of the human species are present and not noticeably altered through prolonged culture" with a phrase that ultimately issued as "maintains a karyotype in which the chromosomes are euploid and not altered through prolonged culture" *(25)*. In addition, the limitation for SSEA-3 positivity in the second independent claim of the primate patent was removed from the human ESC patent. WARF probably used this strategy because not all hESCs test positive, or at least strongly positive, for SSEA-3, and leaving out the requirement for this marker broadens the claims to encompass more cells lines-those that are as well as those that are not positive for SSEA-3. During the prosecution of the patent, the inventors agreed with the examiner's request to add the limitation "pluripotent" to "human embryonic stem cells" because the examiner argued that the term "human embryonic stem cell" denoted totipotency and the ability to form germ cells, and the inventor had neither demonstrated the cells ability to differentiate into germ cells nor could be expected to produce what would be unethical and possibly illegal—the production of adult humans carrying germ cells from these ESCs. This limitation of pluripotency does not devalue the claims because any organism that is totipotent is also pluripotent and the commercial utility of these claims is in producing organs and tissues, namely pluripotential uses.

With this second patent application, WARF specifically sought to strengthen their patent position with respect to human cells by obtaining specific claims for human ESCs even though the claims to primate include human cells. The patent and the PTO file ("prosecution history") make it clear that the PTO found the patent application provided sufficient description of the invention based on monkey ESCs to "teach" others how to obtain human ESCs without undue experimentation, the standard to show "possession" and "enablement" of the claimed invention. In particular, the inventor's submission of evidence in the form of his *Science* paper *(26)* demonstrated to the patent examiner that the methods described in the patent based on monkey ESCs could be used to produce human ESCs. However, the Patent Office argued that these claims could have been included in the original application and thus were not a distinct invention from the primate claims *(27)*. In response, the applicant had to agree to a terminal disclaimer *(28)*, which means that the term of the human ESC patent does not extend beyond the term of the original primate ESC patent, so as not to give the patent owner any greater or longer term of exclusive rights than it would have had in the original patent. WARF filed this patent internationally, and it is pending in Europe. However, WARF did not file the patent in Australia or Japan.

In order to infringe the claims to hESCs, one would have to derive cells either using each step (element) in the method claim and/or use cells with all the characteristics of either or both of the product claims, the one based on biological properties of the cells and/or the other based on markers. All of the hESC lines that qualify for federal funding appear to have been derived using the method claimed by Thomson. Scientists are more likely to develop alternative methods of deriving hESC lines than they are to derive lines with biological properties that differ from the Thomson product claims. For example, ES International, Inc. announced that it had derived new hESCs without using various human nonembryonic feeder cells *(29)*. This new method of derivation appears not to infringe the Thomson method claim because Thomson includes, and thus requires, the element of using embryonic fibroblasts as feeder cells. However, these new cells may infringe one or both of the Thomson product claims. Thus, the two product claims, as opposed to the method claim, will likely provide WARF with stronger patent protection in enforcing their patent against parties who make, use, sell or import new hESC lines, even though they can only be used with private funding.

3.3. Inventions Made with US Government Funding

Thomson's invention and the primate ESC lines reported in the patent were made with funding from the National Center for Research Resources, National Institutes of Health, under a Primate Center Grant *(30)*. The use of US taxpayer funds to conduct research that gives rise to new inventions invokes certain rights and obligations of the institution receiving the funds. Under federal law, namely the Bayh–Dole Act *(31)*, and the implementing regulations *(32)*, the institution has the right to elect title to inventions its scientists make ("conceive or first actually reduce to practice") within the scope of a federal funding mechanism ("subject inventions") *(33)*. In this case, the inventors assigned their rights in the invention to WARF, which elected title to the invention *(34)*. As required by law, WARF granted the federal government a "nonexclusive, nontransferable, irrevocable, paid-up license to practice or have practiced for or on behalf of the United States the subject invention throughout the world" *(35)*. This license allows the federal government to use the invention directly or to authorize its contractors to use it on its behalf. The government has not interpreted its Bayh–Dole license as extending to institutions receiving government grants because grants are an assistance mechanism rather than a mechanism to perform work for or on behalf of the government. Thus, from an intellectual property perspective, the government does not need a license from WARF to derive or use primate ESCs, although it is not permitted to derive hESCs. Like other institutions, however, the government does need the permission of owners of particular ESCs lines to use their proprietary tangible property that might fall within the scope of the Thomson patent claims.

3.4. Licensing of the Thomson Patents

The Bayh–Dole Act gives the institution holding title to the federally funded subject invention a great deal of freedom in licensing the invention to one or more parties who will bring the invention to practical application *(36)*. Under this authority, WARF granted an exclusive license to Geron Corporation, Menlo Park, CA, in 1999 to develop commercial treatments related to six cell types derived from the hESCs. Geron also agreed to sponsor the research in Thomson's laboratory, which ultimately led to the derivation of four hESC lines. In the same year, WARF established WiCell Research Institute, Inc. ("WiCell"), a nonprofit subsidiary to advance stem cell research and distribute ESCs to investigators.

By 2001, however, the parties to the license were at odds over their respective rights. WARF asserted that Geron had not met its diligence requirements for the six cell types already licensed and thus, under the terms of the license, Geron was not entitled to exercise its options to commercialize additional cell types from hESCs. Geron disagreed with WARF assertions, but WARF subsequently took the pre-emptive action of filing suit against Geron on August 13, 2001 *(37)*.

On January 9, 2002, the parties resolved their differences and entered into a new license agreement *(38)*. Under this agreement, Geron retains rights to develop therapeutic and diagnostic products in an exclusive basis from hESC derived neural, cardiomyocyte, and pancreatic islet cells and on a non-exclusive basis from hESC derived hematopoietic, chondrocyte, and osteoblast cells. The new license grants Geron non-exclusive rights to develop research products for hepatocytes, neural cells, hematopoietic cells, osteoblasts, pancreatic islets, and myocytes *(39)*.

The other area of contention was the use of stem cells in "research products," meaning the use of hESCs in commercial research as tools to develop new drugs and therapeutics. This research would include using stem cells to screen chemical compounds for specific useful attributes such as arresting growth (cancer indications) or for toxicity indications, especially as a model of rapidly growing or embryonic human cells.

Finally, WARF and Geron agreed that they would not enforce their stem cell patent rights against academic and government researchers who use the patented technologies for research purposes, such as those who obtain cells from third parties *(40)*. This aspect of the agreement is particularly important for researchers who wish to receive cells from commercial providers of cells from within the United States as well as providers of cells outside the United States, where there are as yet no issued hESC patents with claims to unmodifed hESCs. hESC providers outside the US might otherwise risk patent infringement in shipping cells to US researchers or have cell shipments blocked at the port of entry. WARF and Geron noted that, although they would not seek license fees for non-commercial research, they reserved the right to charge fees to cover the cost of providing their own cell lines.

This announcement was welcomed news in that it opened the door for broader access of hESC lines to the research community. The settlement also resolved the dispute over Geron's license rights, which, in turn, provided certainty to other commercial parties to move forward in negotiating agreements with the appropriate party, Geron and/or WARF, depending on their field of interest. In limiting Geron's breadth of license exclusivity, the settlement opened up to other parties more areas of hESC commercial development.

3.5. Other Patents to Pluripotent Stem Cells

In addition to the Thomson patents, there are other issued patents with claims to pluripotent stem cells. Some involve methods of using hESCs, such as a method of directing the differentiation of hESCs into hematopoietic cells invented by Dr. Kaufman and Dr. Thomson and assigned to WARF *(41)*. The Thomson patents would dominate any patents claiming methods of using them. Any party wishing to use these method claims, at least for commercial purposes, would need a license from WARF/Geron, as appropriate, to practice the dominate claims to the cells themselves. In this case, however, the logistics and administrative burden of obtaining a license is lessened because WARF owns both patents.

Although perhaps the focus of most attention, the hESC patents are not the only issued patents to pluripotent stem cells and their uses. Johns Hopkins University obtained three patents to isolated pluripotent human embryonic germ cells (hEGCs) and methods of using them *(42)*. Dr. Gearhart and Dr. Shamblott made these inventions under the sponsorship of Geron, Inc., which exercised its option to an exclusive license to the intellectual property. The first patent claims hEGCs obtained from primordial germ cells, such as gonadal ridge tissue at 3–13 wk postfertilization, which "exhibit the following culture characteristics during maintenance: (1) dependence on a ligand that binds to a receptor that can heterodimerize with glycoprotein 130 (gp 130) and (2) dependence on a growth factor." The second patent claims methods of producing and maintaining hEGCs in culture, and the third patent claim a method of determining the effect of a compound on hEGC function. A series of three patents assigned by Vanderbilt University claim pluripotent embryonic stem cell technology invented by Dr. Brigid L. M. Hogan. This invention is a composition comprising a pluripotential embryonic stem cell and a fibroblast growth factor, leukemia inhibitory factor, membrane associated steel factor, and appropriate amounts of soluble steel factor. Similar claims replace the ES cells with primordial germ cells or embryonic ectoderm cells in the presence of the other factors, as well as methods of making such pluripotent stem cells *(21)*. The second patent claims a method of screening factors for the ability to promote the formation of embryonic stem cells *(21a)*. A third patent claims an isolated non-murine mammalian pluripotential cell that can be maintained on feeder layers for at least twenty passages, gives rise to embryoid bodies and differentiated cells of multiple phenotypes in monolayer culture "wherein said cell is derived from a primordial germ cell by the process" described in the first patent *(21b)*.

Another US patent, which has garnered controversy in Europe as the "Edinburgh" patent (*see* Section 4), is directed to a method of enriching a population of mammalian cells for stem cells by transforming an animal or a population of cells with an antibiotic-resistance gene linked to a promoter specific for mammalian stem cells. The expression of the antibiotic resistance gene can then be used to preferentially select and maintain stem cells *(43)*. Dr. Smith and Dr. Mountford invented this technology at the University of Edinburgh, Scotland. In addition to the method claims, one claim is directed to transgenic cells with these properties. "Animal" is defined in the patent as including humans and "stem cells" includes cells from a variety of sources such that the patent claims cover human pluripotent stem cells modified by the claimed methods.

4. EUROPEAN PATENT LAWS AND POLICIES

In Europe, the general standards for patentability are similar to the United States. To be patentable, an invention must be "susceptible of industrial application" *(44)* (similar to the US utility requirement), must be novel *(45)*, and involve an inventive step *(46)* (similar to the US nonobviousness requirement). However, methods of treating humans or animals "by therapy and diagnostic methods practiced on the human or animal body" do not have "industrial application," but this provision does not limit the patentability of substances or compositions that could be used to treat humans or animals *(47)*. Like the United States, European law requires that a patent application disclose the invention in sufficient detail and specificity to enable others skilled in the art to practice it *(48)*. Unlike US law, however, European patent law takes into account moral issues in deciding whether to grant a patent *(49)*.

With respect to stem cell patents, the European landscape has diverged from that in the United States. The European Patent Office (EPO) held up patent applications claiming human stem cells until the European Commission (EC) ethics committee, the European Group on Ethics in Science and New Technologies, reviewed this new technology *(50)*. The group issued its recommendation on May 7, 2002, relying primarily on the 1998 European Directive on the Legal Protection of Biotechnological Inventions *(51)*. It concluded that isolated, unmodified stem cells do not qualify as patentable subject matter, especially in regard to their lack of specific "industrial applications," in that the potential uses are innumerable and not well defined. However, stem cells that had been modified by biological manipulation or genetic engineering to fulfill a particular industrial application should constitute patentable subject matter. The Opinion cautioned that the issuance of broad patent claims to stem cells could hinder further research and development leading to new innovations. With respect to process claims, it had no objection to patents involving the use of stem cells as long as the other EPO standards for patentability were met.

A few months later, the EPO affirmed the recommendations of the EC Opinion during an opposition hearing for the "Edinburgh" patent, which included claims to transgenic animal stem cells *(52)*. The Opposition Division amended the patent to remove claims to unmodified animal (including human) stem cells but retained claims to modified stem cells. In particular, they ruled that the patent did not comply with the legal requirement for a written description sufficiently clear for others in the field to replicate the invention nor did it comply with a rule of the European Patent Convention that excludes patents on uses of human embryos for industrial and commercial purposes *(53)*. Thus, it appears that broad claims to hESCs like those issued to WARF are not likely to issue in Europe. Specific claims to modified cells and processes for modifying cells are likely to meet the standards laid out by the EPO. The inability to obtain broad patent claims in Europe will lead inventors to seek more limited claims to particular modified or partially differentiated hESCs. The European marketplace could become more competitive with respect to intellectual property in hESC technologies if multiple parties secure such patents.

5. PATENT PENDING TECHNOLOGIES

The patents issued to WARF governing primate and hESCs appear to be the only issued patents to date in the world governing this technology. Patents relating to unique subpopulations of hESCs, differentiated cell lines derived from hESCs, other pluripotent stem cells, and methods of using hESCs and other pluripotent stem cells are likely to continue to issue. In the United States, the Thomson patent will dominate these new claims to the extent that such claims directly require the use of the Thomson products or methods or a party applies a general method of manipulating cells specifically to Thomson hESCs. Thus, WARF and/or Geron could require commercial institutions to take a license to the Thomson hESCs patents even if such institutions have the right to use subordinate patent claims.

A number of patents are pending around the world as international applications *(54)* that could be filed after preliminary international examination in particular countries. Artecel of Durham, NC has filed one such application based on technology invented by Dr. Wilkison and Dr. Gimble. They claim to have isolated pluripotent stem cells generated from adipose-tissue-derived stromal cells that "have been induced to express at least one phenotypic characteristic of a neuronal, astroglial, hematopoietic progenitor, or hepatic cell" *(55)*.

Other parties have claims to a variety of pluripotent stem cells and their uses. These include a purified preparation of hESCs expressing Oct-4, immunoreactive with markers SSEA-4, GCTM-2 antigen, and TRA 1-60 and a method of preparing them *(56)*, hESCs immunoreactive with SSEA-4, GCTM-2, and TRA 1-60 capable of differentiation into neural progenitor cells *(57)*, other specific hESC lines *(58)*, primate ESCs expressing compatible histocompatibility genes *(59)*, a preparation of hESCs and a method of culturing them *(60)*, a method of cultivating primate ESCs in serum-free medium *(61)*, a method of culturing primate pluripotent stem cells "essentially free" of feeder cells and a conditioned medium for growing them *(62)*, a method of thawing cryopreserved hESCs *(63)*, and a method of differentiating primate ESCs into neural precursor cells *(64)*.

It is important to note that these are pending applications that may or may not issue and the claims of which could be substantially modified in response to rejections during the patent prosecution process (or opposition hearing in the case of the EPO). In addition, some of these pending claims overlap with other pending or issued patents. In most of the world, including Europe, the first applicant to file an application with patentable claims has priority over an applicant that files later with claims to the same invention. However, the United States has a "first-to-invent" system in which the party with competing claims filed on a later date could persevere over an issued or pending application filed on an earlier date if the party filing later can demonstrate that its inventor made the invention first. The administrative process conducted by the PTO for determining the first to invent is called an "interference" *(65)*. During the interference before the Board of Patent Appeals and Interferences, the parties file legal briefs and provide evidence of conception and reduction to practice of their inventions. The Board decides who invented first, but the losing party has the right to appeal the decision to the federal courts. Thus, for several reasons, the ultimate disposition of these various pending patent claims is unpredictable.

6. NIH ROLE IN FACILITATING ACCESS TO STEM CELLS

Shortly after President Bush announced his hESC policy on August 9, 2001, the National Institutes of Health (NIH) began meeting with various parties known to have derived stem cells to explain the policy and to better understand their needs for the further development and distribution of cells.

In addition, NIH staff explained intellectual property laws and policies relating to NIH-funded research. These include the Bayh–Dole Act discussed earlier, guidelines for sponsored research agreements involving the use of NIH funds *(66)*, and the principles and guidelines for sharing biological research resources ("Research Tools Guidelines"), which are a term of award of NIH grants and contracts *(67)*.

6.1. Research Tools Guidelines

The NIH Research Tools Principles and Guidelines were developed to implement recommendations made by a Working Group of the Advisory Committee to then NIH Director, Dr. Harold Varmus. The Committee's recommendations addressed concerns raised by research scientists in the dissemination of unique research resources and the potential competing interests of intellectual property owners and research tool users. The concern is that some patent owners have refused to disseminate research tools to the greater research community or have offered to do so only under terms that potentially harm the research enterprise. Following the issuance of the Guidelines, Congress amended the Bayh–Dole Act to affirm that inventions developed with US public funding are to be used and licensed in a manner to promote free enterprise and commercialization "without unduly encumbering future research and discovery" *(68)*.

The Principles are to (1) ensure academic freedom and publication, (2) ensure appropriate implementation of the Bayh–Dole Act, (3) minimize administrative impediments to academic research, and (4) ensure dissemination of research resources developed with NIH funds. The Guidelines for Implementation give specific guidance to recipients of NIH funding as to how to implement the Principles. The Principles and Guidelines require that researchers share with the greater research community any unique research resources *(69)* generated with NIH funding, including those arising out of the use of proprietary materials imported into NIH-funded research, such as qualified hESC lines.

6.2. Agreements with Stem Cell Providers

Agreements, such as Memoranda of Understanding, Material Transfer Agreements, or Sponsored Research Agreements, governing the transfer of hESCs into NIH-funded research should be consistent with the Research Tools Principles and Guidelines. Such agreements should permit prompt publication of research results. The definition of the "materials" transferred, the hESCs, should not be so broad as to include all modifications or derivatives. Such terms would restrict the recipient's ownership and ability to distribute and license new materials generated using the hESCs transferred. The recipient should not agree to terms that otherwise limit the broad distribution of new unique research tools to the greater research community. Recipients could grant a nonexclusive license to the provider to use new improvements and new uses of the materials. Agreements with nonprofits to use research materials arising from NIH-funded research should not include commercial option rights, royalty reach-through, or product reach-through rights to the provider. The Principles and Guidelines do permit exclusive commercial licensing arrangements for research tools as long as the materials are also provided to the research community, by transfer or sale, without onerous terms such as reach-through provisions.

After President Bush's announcement, it became clear that one of the many hurdles to the distribution of hESCs would be the negotiation of agreements governing the transfer of these cells lines to NIH-funded researchers consistent with the Research Tools Guidelines. It was also apparent that NIH had no authority to negotiate binding agreements on behalf of third parties, namely hESC providers and NIH-funded (extramural) institutions. Thus, NIH decided that its Office of Technology Transfer (OTT) would negotiate agreements with at least some of the hESC providers where it did have authority, namely on behalf of the Public Health Service (PHS) intramural researchers, including researchers working at the NIH, FDA, and Centers for Disease Control and Evaluation (CDC). The agreements would also require that the hESC providers offer no more restrictive terms to NIH-funded nonprofit institutions. Then, extramural institutions could choose to accept the same reasonable terms or attempt to negotiate alternative agreements for themselves to suit their particular needs, policies, or local laws.

The OTT began negotiations first with WARF/WiCell because it owned the dominant patents that would likely govern most, if not all, of the known approved hESCs lines. Furthermore, WiCell expressed an interest in distributing cells broadly to NIH-funded investigators. In relatively quick order, the NIH and WiCell entered into a Memorandum of Understanding (MOU) on September 5, 2001 governing the terms of transfer and use of the Dr. Thomson's five hESC lines. The MOU included as an appendix a "Simple Letter Agreement" to be used with each individual transfer of cells *(70)*. The agreement permits investigators to freely publish the results of their research and provides WiCell and the University of Wisconsin the right to request samples of new materials at no charge made in the course of the research project. WiCell retains ownership of its materials and charges a fee of $5000 for each sample of cells.

WiCell agreed to permit PHS investigators the right to use the Thomson patented technology in nonprofit research *(71)* and granted a royalty-free, noncommercial, research license to suppliers of hESCs to PHS-funded researchers as long as the agreements with such third parties were "no more onerous" than those in the WiCell agreement. WiCell specifically excluded sponsored research where the research sponsor receives commercial rights, other than a grant for noncommercial research purposes, without a license from WiCell or WARF. This provision opened the door for other hESC providers to supply their cell lines without the potential threat that they might infringe the WiCell patents.

The NIH entered into discussions with other hESC providers and entered into MOUs with three additional providers under similar terms to the WiCell agreement. These include an agreement with ES Cell International (ESI) of Melbourne, Australia, on April 5, 2002 governing six cell lines *(72)*, with BresaGen, Inc. of Athens, Georgia, on April 24, 2002 governing four cell lines *(73)*, and with the University of San Francisco on April 26, 2002 governing their two cell lines developed under a sponsored research agreement with Geron, Inc. *(74)*. A total of seventeen cell lines are governed by these agreements. In addition, recipients of infrastructure grant awards must propose an acceptable plan for distribution of their cell lines that comply with the NIH Research Tools Guidelines. Awardees who have not yet entered into an MOU can provide, in the grant application, a model MTA to fulfill this requirement. The greater research community, and ultimately the public health, will benefit from these important research resources that are provided without restrictions on publication of results or with any limitations that might entangle the distribution and ultimate commercialization, where appropriate, of new technologies developed using these hESC lines.

7. THE FUTURE

Intellectual property considerations have been blamed as a major stumbling block in the distribution of many of the hESC lines *(75)*. Patents receive much of the blame, but even without patent protection, owners of proprietary materials decide the terms under which they will distribute their materials. Even so, most of the cell lines that are not yet available for distribution are not available primarily for other reasons. Some investigators working with hESCs point to the need for further characterization and scale-up of cells that were frozen at early stages after separation from the blastocyst. NIH is funding investigators to conduct this characterization and scale-up. Some providers have said that they prefer to share cells with collaborators rather than to devote funds and resources to the broad distribution of their cells. In addition, these cells are difficult to grow and investigators need training and possible collaboration with experts before they are likely to be successful in propagating hESC lines. Other countries and individual institutions are still developing policies governing the distribution of these unique resources. Although it may not be appropriate for NIH to enter into model MOUs and MTAs with all the other providers, it continues to work with all providers and researchers interested in receiving funding for hESCs to develop mechanisms for sharing of cell lines and support mechanisms to ensure that as many of the cell lines are made available to researchers as possible. Significant progress has been made since 2001 in making hESCs available to researchers *(76)*.

In time, research with pluripotent embryonic stem cells will likely lead to fundamental new discoveries and breakthrough therapies for debilitating diseases. As these new technologies developed, the number and complexity of patents will likely increase with patents governing specific methods to stimulate differentiation of stem cells, factors used to modify differentiation, and new fully or partially differentiated cells with therapeutic utility. Individual companies are likely to acquire patents and each build an intellectual property portfolio that permit them to develop specific technologies to treat specific diseases. Basic research on hESCs is likely to inform the science of fetal, cord blood, and adult stem cells so that, in time, these other cells may be manipulated in new ways to treat disease as well. In these respects, this emerging field will parallel other breakthrough, platform technologies. Most importantly, the general public will obtain the greatest medical benefits from the fruits of basic and clinical research, and commercial enterprise will thrive if all parties work together to permit researchers to perform noncommercial research with stem cells without imposing harmful encumbrances.

REFERENCES

1. Stolberg, S. G. (2001) Patent on Stem Cell Puts US Officials in Bind, NY Times, Aug. 17, p. 1.
2. "Inventor" will be used throughout in a generic sense to mean one or more inventor. When more than one inventor makes an invention, each inventor has an undivided interest in the invention and resulting patent and can assign his or her rights in the patent to one or more parties, the "assignee(s)."
3. Haight, J. C. (2002) Technology Transfer, in *Principles and Practice of Clinical Research*, (Gallin, J. I., ed.), Academic Press, San Diego, CA; Goldstein, J. A. and Folod, E. (2002) Human gene patents. *Acad. Med.* **77(12),** 1315–1328.
4. There is a limited exception in the infringement statute that permits the use of a patented drug for purposes of submission of a regulatory filing with the Food and Drug Administration. 35 U.S.C. § 271.

5. There is a limited right recognized by common law to use a patented technology for research purposes, "the research exemption," but it does not apply to any research directed to a particular practical purpose. The US Court of Appeals for the Federal Court recently ruled on the scope of the research exemption in Madey vs Duke, 307 F. 3d 1351 (Fed. Cir. 2002). *See* Mueller, J. M. (2001) No dilettante affair: rethinking the experimental use exception to patent infringement for biomedical research tools, *Wash. Law Rev.,* **76,** 1–66. In most European countries, national laws acknowledge a traditional research exemption allowing noncommercial researchers to practice a patented technology without obtaining a license. *See* ref. *50.*

6. An inventor may have intellectual property in an invention, and material embodiments of that invention, that may not meet the legal standard of patentability in a particular country. In addition to patents, there are other forms of intellectual property as well, such as copyrights and trademarks.

7. 35 U.S.C. § 101.

8. *Diamond v. Chakrabarty,* 100 S.Ct. 2204 (1980); *see* U.S. Department of Commerce, Manual of Patent Examining Procedure § 2105, Patentable Subject Matter—Living Subject Matter (8th ed) US GPO, Washington DC; Woessner, W. D. (2001) The evolution of patents on life—transgenic animals, clones and stem cells, *J. Pat. and Trademark Off. Soc'y,* **83,** 830–844.

9. 35 U.S.C. § 101; U.S. Patent and Trademark Office Utility Examination Guidelines, January 5, 2001, *Fed. Reg.* **66,** 1092–1099.

10. 35 U.S.C. § 102.

11. 35 U.S.C. § 103.

12. 35 U.S.C. § 112.

13. Thomson, J. A., Itskovitz-Eldor, J., Shapiro, S. S., et al. (1998) Embryonic stem cell lines derived from human blastocyts, *Science* **282,** 1145–1147; Weiss, R. (1998) A crucial cell is isolated, multiplied, *Washington Post,* A1, November 16.

14. U.S patent 6,200,806, issued on March 13, 2001.

15. U.S. patent application 08/376,327.

16. U.S. patent application 08/591,246 filed under 35 U.S.C. § 120, 37 C.F.R. § 1.53(b)(2).

17. 37 C.F.R. § 1.53(b).

18. U.S. patent 5,843,780, column 16, lines 3–31.

19. Bongso et al. (1994) Isolation and culture of inner cell mass cells from human blastocysts, *Human Reprod.* **9(1),** 2110–2117. Describes the isolation of human ESCs using methods similar to those used to establish mouse ESCs but with a lifetime of no more than 24 days.

20. U.S. patent 5,453,357, "Pluripotential embryonic stem cells and methods of making same," filed on October 8, 1992 and issued on September 26, 1995, describing the isolation of SCs from germ cells of non-mouse embryos and their growth on feeder layers dependent on LIF or steel factor to maintain the cells in an undifferentiated state. This invention was made under a grant from the National Institutes of Health.

21. U.S. patent 5,670,372, issued on September 23, 1997 as a divisional patent of 5,453,357.

22. U.S. patent 5,690,926, issued November 25, 1997 as a continuation-in-part of the patent application that was issued as 5,453,357. Vanderbilt licensed these three patents to Plurion, which has entered into negotiations with BresaGen to give them exclusive rights to this intellecutal property. BresaGen Press Release, Nov. 18, 2002. Plurion Deal—Stem Cell Commercialization, available at www.bresagen.com.au/news/PRO72.html. Accessed May 3, 2003.
23. International patent application WO 94/03585, published February 17, 1994, describing the culture of ESCs from various animals maintained in an undifferentiated stated on a chicken embryonic fibroblast feeder layer.
24. 35 U.S.C. § 121. U.S. patent application 09/106,390, filed on January 18, 1996. Sets of claims in one application can be split off into separate patent applications when "two or more distinct inventions are claimed in one application."
25. In the application, the term was originally "maintains a normal karyotype through prolonged culture."
26. *See* ref. *14.*
27. *See* ref. *23.*
28. 37 C.F.R. § 1.321. A terminal disclaimer is required when the PTO determines that two patents owned by the same party contain substantially the same invention, termed "double patenting."
29. ES International Press Release, "ES Cell International announces significant advance in human stem cell therapy," August 5, 2002; www.escellinternational.com
30. NIH grant number RR P51-00167; US Patent 5, 843,780
31. Codified at 35 U.S.C. § 200 *et seq.* Note that although the Act addresses inventions made by nonprofit organizations and small businesses, Executive Order 12591 (April 10, 1987) extended its provisions to large businesses receiving funds under a federal grant or contract.
32. 37 C.F.R. Part 401.
33. 35 U.S.C. § 201(e); 35 U.S.C. § 202(a).
34. 35 U.S.C. § 202(c)(7). Nonprofit organizations cannot assign subject inventions without approval of the federal agency, except when the assignment is to an organization, like WARF, that has as one of its primary functions the management of inventions. The assignee must abide by all the relevant obligations that the funding recipient would otherwise have.
35. 37 C.F.R. § 401.14.
36. 35 U.S.C. § 202(c)(7). The statute imposes only a few restrictions on an institution's ability to license its inventions. For example, institutions licensing inventions made with government funding must preferentially grant licenses to small US businesses. 35 U.S.C. § 204. Any party obtaining a license to a government funded invention must agree to substantially manufacture in the United States any product sold in the United States using the licensed invention.
37. WARF asked the court to issue a declaratory judgment in its favor with respect to the status and terms of the license.Geron continues aggressive development of embryonic stem cell technology. (2001) *Business Wire,* November 1.
38. Pollack, A. (2002) University resolves dispute on stem cell patent license, *New York Times,* C11, January 10.

39. Press Release. Geron Corporation and Wisconsin Alumni Research Foundation Resolve Lawsuit and Sign New License Agreement, January 9, 2002, available at http://www.geron.com/pr_20020109. Accessed May 3, 2003.

40. Testimony of Carl Gulbrandsen before Senate Appropriations Subcommittee on Labor, Health and Human Services, Education, August 1, 2001, available at http://olpa.od.nih.gov/hearings/107/session1/reports/stem_cell.asp. Accessed May 3, 2003.

41. US patent 6,280,718, Kaufman, D. S. and Thomson, J. A., Hematopoietic differentiation of human pluripotent stem cells.

42. US patents 6,090,622, Human embryonic pluripotent stem cells, issued July 18, 2000; and a pair of patents filed as one original application entitled "Human embryonic germ line cells and methods of use" (patent 6,245,566, issued June 12, 2001, and patent 6,331,406, issued December 18, 2001).

43. U.S. patent 6,146,888 issued on November 14, 2000, "Method of enriching for mammalian stem cells."

44. European Patent Convention, Article 57 (www.epo.co.at/legal/epc/e/ma1.). The Convention, signed in Munich in 1973, created the European Patent Organization and currently includes 20 European countries, including all 15 members of the European Union.

45. European Patent Convention, Article 54. The requirement is similar to the US novelty requirement except that anything published prior to the application date obviates novelty in Europe, but in the US there is a 1-yr grace period prior to the original application date for public disclosures and uses. 35 U.S.C. § 102.

46. European Patent Convention, Article 56.

47. European Patent Convention, Article 52(4)

48. European Patent Convention, Article 83(d)

49. Inventions contrary to "ordre public" or morality are not patentable. European Patent Convention, Article 53(a). In the U.S., courts have taken into account public policy concerns in considering the enforcement of claims in an issued patent. *City of Milwaukee v. Activated Sludge*, 69 F.2d 577, 593 (7th Cir. 1934), where the court refused to enforce a injunction against the City of Milwaukee for infringing a valid patent because that would have shut down the sewage treatment plant at significant immediate impact to public health.

50. Scheinfeld, R. C. and Bagley, P. H. (2001) The current state of embryonic stem cell patents, *NY Law J.,* Sept. 26, pp. 38.

51. European Group on Ethics in Science and New Technologies, Opinion number 16, "Ethical Aspects of Patenting Inventions Involving Human Stem Cells." *See* europa.eu.int/comm/european_group_ethics/avis3_en.

52. European patent EP 0695351, "Isolation, selection and propagation of animal transgenic stem cells"; opposition oral proceedings were held in Munich July 22–24, 2002; an opposition is a public proceeding of the EPO where third parties can present arguments as to why a patent found suitable for issue should not issue. The patent claims can be modified by the EPO as a result of the opposition hearing. European Patent Convention, Articles 99–105.

53. Edinburgh patent limited after European Patent Office opposition hearing, EPO Press Releases, July 24, 2002.
54. Under the Patent Cooperation Treaty (PCT), one can file a patent internationally within 1 yr of filing a patent in a member country (most countries in the world) and reserve the right to file the patent in any of the member countries following an initial examination of prior art and patentability; *see* 35 U.S.C. § 351–376, 37 C.F.R. § 1.431 *et seq.*
55. US App 20010033834, published October 25, 2001, "Pleuripotent stem cells generated from adipose tissue-derived stromal cells and uses thereof" based on U.S. provisional patent application 60/185,338 filed on February 26, 2000. International application WO 0162901. *Also see* their report of collaboration with Dr. Henry Rice at Duke University in which they differentiate adipose-derived stem cells into neuronal cells. Erickson, G. R., Gimble, J. M., Franklin, D. M., et al. (2002) Chrondrogenic potential of adipose tissue-derived stromal cells in vitro and in vivo. *Biochem. Biophys. Res. Commun.* **290(2),** 763–769.
56. International patent application WO 200027995 based on 2 Australian priority applications filed as early as November 9, 1998 and assigned to Hadasit Medical Research Services and Development Co. Ltd. (Israel), Monash University (Australia), and National University of Singapore, Bongso, A., Fong, C., Pera, M.F., et al. Novel undifferentiated human embryonic stem cells which are useful as a source of novel gene products."
57. International patent application WO 200168815 and US patent application 20020068045 filed March 14, 2001 claiming priority to three Australian patent applications filed as early as March 14, 2000 and assigned to Hadasit Medical Research Services and Development Co. Ltd. (Israel), Monash University (Australia), and National University of Singapore, Reubinoff, Benjamin E., Pera, Martin F., and Ben-Hur, T., Embryonic stem cells and neural progenitor cells derived therefrom."
58. Thomson, J. US patent applications 09/420,482 and 09/522,030, filed October 15 and October 19, 1999, respectively.
59. Thomson, J. US patent application 09/786,174, "Primate embryonic stem cells with compatible histocompatibility genes," filed February 28, 2001.
60. Pera, M. US patent application 20020022267 based on 2 Australian patent applications filed as early as June 20, 2000, "Methods of culturing embryonic stem cells and controlled differentiation"
61. Thomson, J. US patent application 09/522,030, "Serum-free cultivation of primate embryonic stem cells, filed March 9, 2000.
62. Carpenter, M. K., Funk, W. D., and Thies, R. S. U.S. patent application 20020019046, filed June 21, 2001 and assigned to Geron Corp., "Direct differentiation of human pluripotent stem cells and characterization of differentiated cells," also filed internationally WO 200188104 with priority to 4 U.S. applications filed in the U.S. beginning in May 17, 2000.
63. Lim, J., Park, S., and Kim, E. US patent application 20020045259, filed August 30, 2001 and assigned to Maria Biotech Co., Ltd, Seoul, Korea, "Human embryonic stem cells derived from frozen-thawed embryo" also filed internationally WO 200218549 with priority to 3 Korean applications beginning August 30, 2000.

64. Zang, S., Thomson, J., and Duncan, I. US patent application 09/970,382, "Method of in vitro differentiation of transplantable neural precursor cells from primate embryonic stem cells," filed October, 3, 2001.
65. 35 U.S.C. § 135.
66. National Institutes of Health (1994) Considerations for recipients of NIH research grants and contracts, *Fed. Reg.*, **59**, 55674–55679.
67. National Institutes of Health (1999) Sharing biomedical research resources: principles and guidelines for recipients of NIH research grants and contracts, *Fed. Reg.*, **64**, 72,090–72,096. http://ott.od.nih.gov/NewPages/Rtguide_final.
68. Technology Transfer Commercialization Act of 2000, U.S. Public Law 106–404 enacted November 1, 2000, 35 U.S.C. § 200; *also see* Statement on Stem Cell Research by Maria C. Freire, Ph.D., Director, Office of Technology Transfer, NIH, Before the Committee on Appropriations, Senate Subcommittee on Labor, Health and Human Services, Education and Related Agencies, January 12, 1999. http://www.hhs.gov/asl/testify/t010801.
69. The term is meant to be interpreted broadly "to embrace the full range of tools that scientists use in the laboratory, including cell lines, monoclonal antibodies, reagents, animal models, growth factors, combinatorial chemistry and DNA libraries, clones and cloning tools (such as polymerase chain reaction [PCR]), methods, laboratory equipment and machines." The terms "research tools" and "materials" are used interchangeably with "unique research resources."
70. www.escr.nih.gov/WARF/WicellMOU.pdf
71. The US Government also has a government use license under Bayh–Dole to use patent 5,843,780 for and on behalf of the US government. *See* ref. *34*.
72. www.escr.nih.gov/WARF/WicellMOU. Accessed May 3, 2003.
73. www.nih.gov/news/stemcell/BresaGenMOU. Accessed May 3, 2003.
74. www.nih.gov/news/stemcell/UCSF. Accessed May 3, 2003.
75. Rosenberg, D. (2002) Stem cells slow progress, *Newsweek* p. 8, August 12.
76. Zerhouni, E. (2003) Stem cell programs, *Science*, **300**.

REFERENCE ADDENDUM

The United State Code (U.S.C.), the compendium of US federal laws, can be found at uscode.house.gov/usc. The Code of Federal Regulations (C.F.R.) can be found at www.access.gpo.gov/nara/cfr/cfr-table-search.The Federal Register (Fed. Reg.), which contains announcements of federal policy and new regulations, can be accessed at http://www.access.gpo.gov/su_docs/aces/aces140. The U.S. Patent Office we page www.uspto.gov has a patent section (click on patents on the home page) from which you can search issued and published patent applications (published since March 15, 2001 based on filings 18 mo prior to publication date). The European Patent Office website (european-patent-office.org) has search capabilities under "search engines and index," or one can access an international search engine directly ep.espacenet.com/. At the bottom of the page, one can access international (PCT) applications. The European Patent Convention can be found at www.european-patent-office.org/legal/epc/e/ma1.

II

Types of Pluripotent Cells

4

Embryonal Carcinoma Cells
The Malignant Counterparts of ES and EG Cells

Jonathan S. Draper, Harry Moore, and Peter W. Andrews

1. INTRODUCTION

Teratomas are histologically complex tumors containing a wide variety of distinct tissues, often in disorganized patterns, but sometimes with more or less recognizable tissue architecture and, occasionally, with recognizable organs. Teeth and hair are especially well known in dermoid cysts, the benign teratomas of the ovary. These tumors have excited interest and imagination throughout history, with their occasional obvious resemblance to abnormal embryos—hence the name, teratoma, from the Greek term for "monster." In earlier times, different societies took them as mystical signs, sometimes as bad omens, sometimes to predict good fortune (e.g., see ref. *1*). However, by the late 19th and early 20th centuries, pathologists began to develop more rational concepts of the causes and origins of teratomas, better rooted in the developing understanding of embryogenesis. Fascinating accounts of the natural history of these tumors and the development of ideas about their origins are provided by Damjanov and Solter *(2)* and Wheeler *(3)*. The study of teratomas in the laboratory mouse and the identification of embryonal carcinoma (EC) cells as the pluripotent stem cells of teratocarcinomas provided a starting point that eventually led to the development of embryonic stem (ES) cell lines as tools for developmental biology and the recognition of their potential use in regenerative medicine.

2. TERATOMAS AND TERATOCARCINOMAS: SUBSETS OF GERM CELL TUMORS

Teratomas occur in a number of sites throughout the body and in various manifestations. Some occur in newborn infants, notably at the base of the

From: *Human Embryonic Stem Cells*
Edited by: A. Chiu and M. S. Rao © Humana Press Inc., Totowa, NJ

spine and in the gonads, but these are rare. Considerably more common are the benign ovarian cysts of young women. Although these jumbled arrays of tissues can grow to a sufficiently large size as to be life threatening, they do not possess the features of invasion and metastasis that typify malignant tumors. However, rather more rarely, similar tumors occur in the testes of young men and these are invariably highly malignant *(4,5)*. Such malignant forms of teratomas are known as teratocarcinomas. The prognosis for patients with these cancers was generally poor until the advent of chemotherapy based on *cis*-platinum. Now, testicular teratocarcinomas are among the most successfully treated solid tumors, with cure rates as high as 95% even in moderately advanced cases *(6,7)*.

The common occurrence of teratomas and teratocarcinomas in the gonads led to the early concept that they originate from germ cells and so they are commonly grouped together with several other gonadal tumors as "germ cell tumors" (GCTs). Teratocarcinomas are only one manifestation of GCT, but it clear they are closely related to the other histological types of GCT, which often occur in complex mixtures. In the testis, almost all tumors appear to be GCTs, with tumors only rarely arising from Sertoli or Ledig cells *(5)*. Whether the extragonadal teratomas and teratocarcinomas are also of germ cell origin remains a matter of frequent debate, although it is widely held that those occurring in adults arise from germ cells that have migrated abnormally during embryogenesis *(8,9)*.

Testicular GCT are usually classified as seminomas and nonseminomas *(10,11)*. Seminomas are relatively homogenous tumors containing cells that resemble primordial germ cells (PGCs). By contrast, nonseminomatous germ cell tumors (NSGCTs) are histologically heterogeneous. These may contain teratoma elements with a variety of somatic tissues, as well as undifferentiated primitive cells, known as embryonal carcinoma cells. Early on, it was proposed that EC cells are the malignant stem cells that give rise, by differentiation, to the somatic elements of the tumors. Such combined NSGCTs, containing both teratoma and EC elements, comprise the teratocarcinomas. In some cases, EC elements alone may occur without the presence of somatic tissues, presumably because such EC cells have lost their ability to differentiate. As well as teratoma components and EC cells, testicular NSGCTs may also contain cells that resemble those of the extraembryonic membranes, the trophoblast and yolk sac that envelop the developing fetus. Such choriocarcinoma and yolk sac carcinoma cells respectively occasionally occur alone but are more commonly found together with teratocarcinoma components. Confusingly, NSGCTs may also contain elements of seminoma. It is thus evident that all of these cell types, which can be recognized as counterparts of the cells and tissues occurring during normal embryonic development, are closely related to one another in a way strongly suggestive that they have a common origin and that histogenesis in GCT presents a caricature of normal embryonic development.

In humans, GCTs are typically aneuploid, often with a roughly hypotriploid chromosome number but containing many chromosomal rearrangements. Confirmation of a common origin for the disparate elements of GCTs has been provided by the finding that different elements of a single tumor may contain the same marker chromosomes *(12)*. Karyology has also shown that many benign ovarian teratomas most likely arise from meiotic germ cells. The cells of these tumors commonly show centromeric homozygosity, suggesting an origin from meiotic cells that have progressed through meiosis I, but in which meiosis II has been suppressed *(13,14)*. However, this is not always the case, although evidence of crossing over still indicates an origin from meiotic cells *(15)*. By contrast, testicular teratocarcinomas probably arise from mitotic germ cells prior to meiosis. If these male tumors arose from meiotic cells, they would either have an XX or YY karyotype; because YY cells lacking an X chromosome would not be viable, only XX tumors would be seen. However, testicular GCTs generally contain at least one Y chromosome, arguing strongly for a mitotic origin *(12,16,17)*.

Among the multiple chromosomal rearrangements seen in testicular GCTs, one or more isochromosomes of the short arm of chromosome 12, i(12p), are very commonly present *(18,19)*. In tumors where such isochromosomes are not observed, there is usually evidence of amplification of elements from chromosome 12p *(9,20)*. Thus, it seems likely that a gene, or genes, located on chromosome 12p plays a role in GCT development and/or progression. However, the identity of these loci remains obscure. Other loci elsewhere in the genome have also been suggested from karyology and from genetic analysis of familial cases of GCTs. These candidate loci include a possible tumor suppressor gene on chromosome 12q *(21,22)*, and an X-chromosomal locus that appears to contribute strongly to familial cases involving affected sib-pairs *(23)*. However, the X-chromosomal locus cannot contribute to that group of familial cases that involve affected fathers and sons, and no convincing evidence of the identity of any of these loci has been forthcoming.

3. TERATOMAS OF THE LABORATORY MOUSE

Teratomas occur sporadically in many species, but they are generally too rare to provide useful experimental insights into the biology of these tumors. An exception is the laboratory mouse. In 1954, Stevens and Little reported that about 1% of males of the 129 strain develop spontaneous testicular teratomas *(24)*. Although they do not occur spontaneously with any useful frequency in other mouse strains, they can be induced in the testes of several strains by transplantation of genital ridges from embryos between 11 and 13.5 d of development to ectopic sites, typically the testis capsule of adult males *(25–27)*. They can also be induced in almost any strain by ectopic transplantation of earlier embryos, particularly at the egg cylinder stage, about 7 d of gestation *(28–31)*. In each situation, either benign teratomas or retransplantable teratocarcinomas containing an EC component may occur. However, in the case of embryo transplants, the genotype and immune status of the host seems to be a primary determinant of whether teratomas or teratocarcinomas are formed *(32)*. For example, C3H embryos transplanted to C3H hosts mostly form teratocarcinomas, but they form teratomas if the host is rendered immunodeficient by neonatal thymectomy and sublethal irradiation. Similarly, teratomas form after transplantation of BALB/c embryos to BALB/c *nu/nu* mice, which lack a thymus, whereas they yield teratocarcinomas when transplanted to BALB/c hosts.

The testicular teratomas of 129 mice can be found in the seminiferous tubules of the developing gonad as early as 13 d of development *(33)*, shortly after the PGCs have arrived in the genital ridge, about 11 d of development, after migration from the extra embryonic yolk sac, via the hindgut *(34)*. To test the PGC origin of these tumors, Stevens bred the Steel mutation *(Sl)* into the 129 strain. PGCs do not survive migration in homozygous *Sl/Sl*, and, indeed, genital ridges from *Sl/Sl* homozygotes did not yield teratomas *(35)*. The short period when transplantation of genital ridges can result in the formation of teratomas and the fact that these are largely confined to males suggest that it is some aspect of the interaction of newly arrived PGCs with the male genital ridge that contributes to their susceptibility to transformation. The "default" state for PGCs seems to be to enter meiosis, which is inhibited by the environment of the male genital ridge; PGCs arriving in a male genital ridge remain in mitosis and eventually enter mitotic arrest, whereas in female genital ridges, or if arrested in extragonadal sites, they enter meiosis *(36)*. It could be that a failure to regulate the mitosis to meiosis switch in an appropriate manner contributes to the transformation of these cells.

By way of comparison, the switch from mitosis to meiosis in the nematode worm, *Caenorhabditis elegans*, is controlled by a factor encoded by the *Lag2* gene, produced by a somatic cell of the gonad. In the absence of *Lag2*, the PGCs in *C. elegans* enter premature meiosis *(37)*. On the other hand, a gain-of-function mutation of the gene *glp-1*, which encodes the receptor for *Lag2*, causes the PGCs to remain permanently in mitosis and to form a tumor, which has been suggested to be analogous to GCTs of mice and humans *(38)*. Perhaps the male genital ridge factor that inhibits entry into meiosis in mice, and presumably humans, is a homolog of *Lag2*, a member of the DSL family of cell-surface-associated ligands for the receptors encoded by the *Notch* family, of which *glp*-1 is a member. The *Notch* genes act as proto-oncogenes in a number of other cancers (e.g., *Notch1* in T-cell leukemias *[39]*), and we have speculated that dysfunction of a putative DSL-Notch signaling between migratory PGCs and the male genital ridge might underlie GCT development in mammals *(40)*.

3.1. Cell Lines from Murine Teratocarcinomas: Definition of the Embryonal Carcinoma Cell

Cell lines from teratocarcinomas were first established in vitro in the laboratory of Ephrussi *(41,42)*. Subsequently, EC cell lines were derived and characterized by several groups *(43–48)*. Many retained the ability to differentiate and form teratocarcinomas when transplanted back into a mouse host even after extended periods of culture. Some of these pluripotent lines were also able to differentiate in vitro. Several, but not all, EC lines had to be cultured on feeder layers of mitotically inactivated transformed mouse fibroblasts to maintain an undifferentiated phenotype *(46,47)*. Differentiation occurred when the cells were removed from the feeders, especially if they were grown in suspension when they aggregated to form structures known as embryoid bodies with an inner core of EC cells surrounded by a layer of visceral endoderm. These embryoid bodies then continued to differentiate and a complex array of differentiated cells would grow out if they were allowed to reattach to a substrate.

Many EC cells can be induced to differentiate by specific chemicals and, often, the direction of differentiation depends on the chemicals used. For example, the otherwise nullipotent EC line, F9, differentiates into parietal endoderm when cultured with retinoic acid and cAMP *(49,50)*. On the other hand, when F9 cells were cultured in suspension in the presence of retinoic acid, they formed embryoid bodies in which the outer layer of cells resembles visceral endoderm while the inner cells retained an EC phenotype *(51)*. Other agents (e.g., dimethyl sulfoxide [DMSO] and hexamethylene bisacetamide [HMBA]) also induce EC cell differentiation *(52,53)*. In the case of the P19 EC cell line, retinoic acid induces neural differentiation, whereas cardiac muscle is induced by DMSO *(54)*.

The existence of established EC cell lines enabled the properties of these cells to be characterized and defined in detail. Previously, the injection of single cells isolated from teratocarcinoma tumors into new hosts had been used to confirm that the many differentiated cells found in these tumors can arise from a single stem cell *(55)*. Likewise, it was shown that clonal lines in vitro could differentiate into multiple cell types, also confirming the stem cell nature of EC cells *(46,47)*. Detailed characterization of these defined EC cell lines, particularly with respect to surface antigens, revealed their similarity to cells of the inner cell mass (ICM) of the blastocyst stage of the early embryo *(56,57)*. That this similarity was indeed functional was shown by the formation of chimeric mice with normal tissues derived from EC cells after their transfer to blastocysts, which were then reimplanted into the uterus of pseudopregnant female mice *(58–60)*.

Thus, EC cells appear to be the malignant counterparts of embryonic ICM cells. Their sometimes restricted potential for differentiation when compared to ICM cells may be attributed to their adaptation to tumor growth. It seems likely that mutations that limit differentiation will provide a selective advantage for malignant stem cells because their differentiated derivatives typically possess a limited growth potential and are not malignant themselves. However, after culturing ICM cells under appropriate conditions in vitro, lines of cells, termed embryonic stem (ES) cells, have been derived with a similar phenotype to tumour-derived EC cells, but typically with much greater potential for differentiation and an ability to form all tissues, including germ cells, of chimeric mice formed after their transfer to blastocysts *(61,62)*. Nevertheless, even ES cells, if cultured under less than optimal conditions, frequently acquire subtle changes that limit their ability to chimerize embryos and particularly to chimerize the germ line.

Murine EC and ES cells were originally studied with the notion that they would provide tools for investigating the mechanisms of cell differentiation during embryogenesis. However, over the past 20 yr, most of the use of mouse ES cell technology has been directed toward production of transgenic mice, not for answering questions of fundamental cell biology pertinent to ES cells *per se*. Nevertheless, some studies of the mechanisms that regulate mouse ES cell differentiation have been conducted. For example, the crucial role of specific discrete levels of Oct4 expression for the maintenance of an undifferentiated phenotype of mouse ES cells has been identified *(63)*. Thus, if the level of Oct4 expression in mouse ES cells is reduced, differentiation into trophectoderm ensues, whereas if levels are raised, then differentiation to extraembryonic endoderm is promoted. Further, the maintenance of an undifferentiated phenotype depends on activation of the STAT3 signaling pathway and on the inhibition of the MAPK/ERK pathway, following interaction of leukemia inhibitory factor (LIF) with its receptor *(64–66)*. However, because LIF does not appear to influence the behavior of human ES cells *(67,68)*, it remains uncertain to what extent these lessons from the laboratory mouse will apply to human ES cells.

4. HUMAN EC CELL LINES

Human GCTs differ in certain respects from those of the laboratory mouse, and the question has always been raised as to whether the murine teratocarcinomas provide an appropriate model for the human tumors. For example, seminomas do not occur in the laboratory mouse, whereas trophoblastic differentiation, sometimes manifested as choriocarcinomas, is common in human GCTs but is not typically seen in murine teratocarcinomas. Indeed, the late ICM and primitive ectoderm to which murine EC cells are usually equated appear to have lost the capacity for trophectodermal differentiation in that species. A further difference is the gross aneuploidy of human GCT, whereas the murine tumors are typically euploid or only have limited karyotypic changes.

Human GCTs were maintained as xenograft tumors in hamster cheek pouches during the 1950s *(69)*, but those lines are no longer extant. Several groups derived further lines in vitro during the 1970s *(70–73)*. Although these exhibited some similarities to murine EC cells, it quickly became apparent that there were substantial differences also.

A comparative study of several cell lines derived from testicular GCTs *(74)* and a more detailed analysis of one of these *(75)* highlighted several of these similarities and differences. For example, their morphology and growth patterns are similar, as both tend to grow in clusters of tightly packed cells with relatively little cytoplasm and prominent nucleoli, and both express high levels of alkaline phosphatase *(43,76)*. On the other hand, there were differences. For example, trophoblastic differentiation by human EC cells occurs in culture *(77)*, but it is not generally seen in cultures of murine EC cells.

Particularly marked differences were evident in the patterns of surface antigens of human and mouse EC cells. For example, human EC cells commonly express the class 1 major histocompatibility complex (MHC) antigens, human leukocyte antigens (HLAs), whereas murine EC cells do not *(75,78–80)*. Another example is provided by the stage-specific embryonic antigens (SSEA1, SSEA3, and SSEA4). The SSEA epitopes are all carried by cell surface glycolipids: SSEA3 and SSEA4 are associated with globoseries oligosaccharides, whereas SSEA1 is associated with a type 2 polylactosamine carbohydrate. The globoseries and lactoseries core structures of these glycolipids are differentially synthesized from the same precursor, lactosylceramide, depending on the relative activities of specific glycosyltransferases *(81)*. SSEA1, which is a characteristic surface marker of murine EC cells *(82–84)*, is not expressed by human EC cells, although it is expressed by some cells after differentiation *(74,75,80,85)*. On the other hand, the two embryonic antigens SSEA3 and SSEA4, expressed on cleavage stage mouse embryos but not by mouse EC or ICM cells *(86,87)*, were found to be expressed by human EC cells *(75,85,88)*.

These differences are reflected in the expression of these antigens by ES cells and early embryos of these species. Thus, human ES cells and ICM cells also express SSEA3 and SSEA4, but not SSEA1, in contrast to their murine counterparts *(89,90)*, whereas human ES cells also express the class 1 MHC antigens *(89)*.

The developmental significance of these differences in antigen expression between mouse and human EC cells is unclear. On the one hand, these antigens appear to be tightly regulated, and it has been reported that the SSEA1 epitope, also known as Le^X, plays a role in compaction at the morula stage *(91,92)*. On the other hand, evidence has been advanced that, at least in fish embryos, glycolipid expression can be all but eliminated during early embryogenesis, with no significant influence on subsequent development *(93)*. Certainly, these antigens may play a role in pathology associated with early development. Thus, the SSEA3 and SSEA4 epitopes are members of the P blood group system and both epitopes are expressed on red blood cells *(94)*. A very small number of individuals lack the ability to synthesize these antigens. Women who then lack the P blood group antigens have a high rate of spontaneous abortions, which may be because of an immune reaction to antigens such as SSEA3 and SSEA4 expressed on the very early embryo *(95)*.

Another set of antigens that have been identified in human EC cells are keratan sulfate-associated epitopes, notably TRA-1-60 and TRA-1-81 *(96,97)* as well as GCTM2 *(98)* and K21 and K4 *(99)*. It appears that these epitopes are widely expressed by human EC cells and, indeed, some, notably TRA-1-60, have been shown to be useful serum markers in germ cell tumor patients as they are shed from the EC cell *(100,101)*. As with the other antigenic markers, these, too, are expressed by human ES and ICM cells *(89,90)*.

4.1. Differentiation of Human EC Cells

Few human EC cell lines are able to differentiate into well-recognizable cell types. In part, this might reflect their development in tumors, because an ability to differentiate could provide a selective disadvantage for stem cells, whereas the acquisition of an inability to differentiate would provide a strong selective advantage. Human EC cells are highly aneuploid and it is easy to envisage that genetic changes might occur to inhibit their differentiation. Indeed, cell hybrid studies with both mouse and human EC cells support this idea and suggest that a loss of pluripotency results from the loss of key gene functions *(102,103)*. This facet of their biology makes the definition of EC cells difficult because the ability to differentiate is a key diagnostic feature, whereas, of course, their special interest for developmental biologists lies in their ability to differentiate. Nevertheless, several human EC cell lines that can differentiate have been isolated. One line, GCT27, requires maintenance on feeder layers and differentiates into a wide range of cell types when removed from the feeders *(104–107)*. Others include NCCIT *(108,109)*, NCG.R3 *(110,111)*, and TERA2 *(70,80)*. In general, these pluripotent human EC cells exhibit the same patterns of marker expressions as those EC cells that either do not differentiate or exhibit only very limited capacity for differentiation.

4.2. TERA2: A Pluripotent Human EC Cell Line

TERA2 is perhaps the most widely studied pluripotent human EC cell line. It was derived from a lung metastasis in a patient suffering from a testicular teratocarcinoma *(70)*. The cultures are frequently heterogeneous, but they contain cells that closely resemble human EC cells in cultures derived from other patients *(80)*. However, TERA2 EC cells form well-differentiated teratocarcinomas containing multiple cell types, most notably glandular structures and neural elements, when grown as xenografts in immunosuppressed mice. Moreover, in culture, they differentiate extensively when exposed to retinoic acid *(112)*, a feature that is not common among human EC cell lines *(113)*. A particularly robust subline of TERA2, NTERA2, was derived by Andrews et al. *(80)* from a xenograft tumor of TERA2 produced in a *nu/nu* athymic (nude) mouse. Proof of the stem cell nature of these cells was provided by the isolation of single-cell clones of NTERA2, which express characteristics common to other human EC cells such as 2102Ep and produced similar teratocarcinoma xenografts to the original TERA2 and NTERA2 lines.

Not only do NTERA2 EC cells form teratomas when grown as xenografts in nude mice, but when cultured in the presence of 10^{-5} or 10^{-6} *M* all-*trans*-retinoic acid, the EC cell phenotype of NTERA2 cells is rapidly lost while they acquire markedly different growth patterns and cellular morphology *(112)*. Expression of EC markers such as SSEA3, SSEA4, and TRA-1-60 is largely extinguished within 2 wk of initial exposure to retinoic acid, as several other antigens, notably ganglioseries glycolipids, appear on the surface of the cells *(114)*. Generally, even after only 2–3 d in medium containing retinoic acid, almost all the cells have committed to differentiate, and within 2–3 wk, EC cells are not detectable in the cultures. As differentiation progresses, neurons appear during the second week of retinoic acid induction, marked by the expression of neurofilament proteins *(112,115)*. These neurons comprise only 2–5% of all the differentiated cells, but they are most obvious and prominent differentiated cells. They can be purified from the cultures *(116)* and express exhibit a wide variety of neuronal characteristics *(117)*, including tetrodotoxin-sensitive sodium channels and regenerative membrane *(118)*, as well as glutamate receptors and voltage-gated calcium channels *(119)*. Several groups have recently studied the potential of these teratocarcinoma-derived neurons for repairing damage to the central nervous system (CNS). It has been reported that NTERA2-derived neurons will integrate functionally to correct neural defects such as those resulting from stroke *(120–123)*. Experiments to implant these EC-cell-derived neurons into human stroke patients have also been reported *(124)*.

The differentiation of NTERA2 EC cells is marked not only by changes in surface antigen expression but also by the appearance of susceptibility to infection and replication of human cytomegalovirus (HCMV) *(125)* and human immunodeficiency virus *(126)*, neither of which will grow in the undifferentiated EC cells. The failure of HCMV to replicate in undifferentiated EC cells is because of their repression of the major immediate early promoter (MIEP) *(127,128)*. This is of interest not only for its significance for virology biology but also in the light of general difficulties found in using the HCMV MIEP to drive stable gene expression in transfected human EC and ES cells.

Differentiation is further marked by changes in the expression of a large number of genes. The *Hox* family includes one group of genes that are induced following retinoic acid exposure *(129,130)*. In the mammalian genome, the *Hox* genes are encoded by four separate clusters related in organization and expression to those that occur in invertebrates such as *Drosophila*. Generally, the pattern of expression of *Hox* genes along the anterior posterior axis of the developing embryo is related to their position in the cluster, with the 3' *Hox* genes having a more anterior pattern of expression to the 5' genes. During the differentiation of NTERA2 cells, several *Hox* genes are induced by retinoic acid in a concentration-dependent manner, with genes at 3' ends of the clusters being inducible to maximum levels by low concentrations of retinoic acid ($10^{-7} M$), whereas the 5' genes tended to require higher concentrations of 10^{-5} or $10^{-6} M$ retinoic acid for maximal induction. These results may be related to the postulated role of retinoic acid in patterning anterior-posterior-axis of the developing embryo.

Many other genes also show marked regulation during NTERA2 differentiation. Thus, *Oct4* is characteristically expressed by EC and ES cells and its expression is markedly reduced following retinoic acid induction of NTERA2 cells. At the same time, a number of other genes are induced. Among these is a member of the *Wnt* family, *Wnt13(Wnt2b)*, that is not expressed by NTERA2 or other human EC cells, but it is induced strongly upon retinoic acid induction *(131)*. Although a function of the *Wnt* genes in regulating human EC cell differentiation has not yet been demonstrated, several *Wnt* genes have been shown to influence the differentiation of mouse EC cells *(132,133)*. Members of the *Frizzled* family of genes that encode putative receptors for the Wnt proteins are also expressed in various patterns during NTERA2 differentiation and we have speculated that this may indicate a possible role for Wnt signaling in regulating the type of cells that are generated during differentiation *(134)*.

The molecular events that occur during the differentiation of NTERA2 cells into neurons appear to follow those that occur during embryogenesis *(135)*. For example, the expression of nestin, a marker of committed, proliferating neuroblasts, peaks after about 3 d of exposure to retinoic acid. Expression of NeuroD1, a transcription factor that, within the developing CNS, marks post-mitotic neuroblasts, peaks a few days later and is eventually followed by the appearance of transcripts typical to terminally differentiated neurons such as neuron-specific enolase and synaptophysin appear.

The differentiation of NTERA2 EC cells can also be induced by several other agents in addition to retinoic acid, particularly hexamethylene bisacetamide (HMBA) *(136)* and the bone morphogenetic proteins (BMPs) *(137)*. The cells induced by these agents are mostly quite different from those induced with retinoic acid, and neural differentiation is much less marked in HMBA- and BMP-induced cultures. On the other hand, there are overlaps in the type of the cells induced. For example, smooth muscle actin is expressed after induction with BMP7, but it is also expressed at lower levels after retinoic acid induction. Also, although many of the surface antigen markers typical of retinoic acid induction are not induced in the early stages of HMBA induction, several such antigens are eventually expressed by a few HMBA-induced cells and a few neurons are occasionally observed.

The nature of the non-neural cells seen in differentiating cultures of NTERA2 whether induced by retinoic acid, HMBA, or BMP7 have never been fully characterized. We have found that *Brachyury* is expressed by human EC cells, including NTERA2, even though we have never noted the induction of cells corresponding to skeletal muscle, which would indicate mesodermal differentiation *(138)*. The appearance of smooth muscle actin might suggest the differentiation of smooth muscle cells, and we have also noted the appearance of cartilage in xenografts of NTERA2 cells and the expression of collagen II in culture after retinoic acid induction *(103)*. Rather than mesodermal differentiation, however, we favor the notion that such smooth muscle and cartilage might arise via neural crest differentiation, but this possibility remains to be explored.

5. PRIMORDIAL GERM CELLS
AND EMBRYONIC GERM CELL LINES

The evident origin of teratocarcinomas from PGCs has focused interest on the behavior of PGC cultured in vitro. Early investigations demonstrated that mouse PGCs survive and proliferate when cultured on mouse fibroblast feeders *(139)*. Later, this technique was refined with the supplementation of growth factors to enable the cells to proliferate indefinitely *(140,141)*. However, under these conditions, the cells acquired a phenotype similar to EC and ES cells and were called embryonic germ (EG) cells. For mouse EG cell proliferation stem cell factor (SCF) is crucial, together with basic fibroblast growth factor (bFGF) and LIF. The latter activates the gp130 pathway to inhibit differentiation while inducing colony formation *(142)*. Murine EG cells have been derived from PGCs before and after their migration to the gonad and from a variety of mouse strains. When injected into blastocysts, EG cells colonize every cell lineage of the developing embryo. Human EG cells have also been produced *(143)* from gonadal ridge and mesentery tissue of 5- to 9-wk-old fetuses obtained after medical termination of pregnancy. The derivation of the human EG cell lines was essentially similar to the technique used for the mouse, although human cells are apparently more resistant to disaggregation with trypsin/EDTA than mouse cells. Both XX and XY cell lines have been generated.

Both murine and human EG cells are characteristically positive for tissue-nonspecific alkaline phosphatase and *Oct4*, and they express cell surface markers that are characteristic for pluripotent ES cells of their respective species, although there are some potentially important differences. Thus, human EG cells display the surface antigens SSEA3, SSEA4, TRA-1-60, and TRA-1-81. Curiously, however, both human and mouse EG cell express SSEA1, an antigen that is generally not expressed by human EC or ES cells *(85,89)* or on ICM cells of the human blastocyst, where it is expressed by the trophectoderm *(90)*. On the other hand, SSEA1 is expressed by human and murine gonadal PGCs *(88)*.

Perhaps a more important difference is the epigenetic condition of EG cells compared to that of ES cells *(144)*. Chimeras generated from EG cells can show phenotypic abnormalities of fetal overgrowth, and skeletal malformations not normally found with chimeras produced from ES cells. Moreover, if somatic cells fuse with EG cells, they undergo erasure of imprints, whereas these imprints are retained in somatic cells that are fused to ES cells *(145,146)*. This can be explained in part by changes in genetic imprinting and nuclear reprogramming that occurs in the embryo and early fetus, which is reflected in the ICM cells and PGCs having different epigenetic states. When PGCs begin migration from the yolk sac, they contain genomic imprints and one of the two X chromosomes is inactivated in the female. However, considerable epigenetic modifications occur as PGCs enter the genital ridge with erasure of imprints and with genome wide demethylation in both male and females and reactivation of the inactive X chromosome in female PGC. Subsequently, on initiation of gametogenesis, new imprinting occurs mainly during oogenesis. In the zygote, further marked differences in epigenetic modification of the parental genomes occur.

Some mouse EC cells have been derived by transplantation of early embryos, whereas others have originated in genital ridges, either spontaneously or following ectopic transplantation. It might be that the embryo-derived lines would resemble ES cells in their imprinting status, whereas those of genital ridge origin would more closely resemble EG cells. To our knowledge, differences between murine EC cell lines of different origins have not been examined systematically. In one study, the introduction of imprinted human H19 genes into embryo-derived mouse EC cells (P19) led to epigenetic reprogramming but without erasure of the primary imprint that marked the parental origin suggestive of similarities to an ES state *(147)*. On the other hand, biallelic expression of H19 and IGF2, two genes subject to imprinting, has been reported in human testicular NSGCT *(148)*. Such a pattern would be consistent with a similarity to EG cells and an origin from PGCs in which erasure of the inherited imprints had already occurred.

The mechanisms by which PGCs are prevented from acquiring a somatic cell fate and maintain totipotency are just beginning to be unraveled. A recent study by single-cell analysis of murine PGCs indicate that an interferon-inducible transmembrane protein, encoded by *fragilis*, promotes germ cell competence and a homotypic association that demarcates putative germ cells from somatic cells. This is followed by the expression of *stella*, coding for a highly basic protein with a putative DNA-binding region. These cells are initially destined for a mesodermal, fate but concomitant with the repression of homeobox genes, the PGCs escape a somatic cell fate. *Stella* continues to be expressed in the germ cell lineage and might be critical for the general pluripotent state *(149)*.

6. PLURIPOTENT CELLS OF THE EMBRYO AND THE ADULT: SIBLINGS OR COUSINS?

Clear similarities exist between human EC cells and ES and EG cells. It is currently unclear whether a relationship exists between pluripotent or plastic adult stem cells recently described and pluripotent cells found in the developing embryo.

Biological dogma has, until recently, held that there are only a few adult stem cell varieties, which appear to contribute to fixed lineages. For example, bone marrow cells were known to produce blood and bone derivatives, but not other cell types, and keratinocyte stem cells contribute solely to the intergumentary system. It was assumed that only a limited number of cells were truly pluripotent during the life cycle of an organism: ES, EG and, on rare occasions, their tumor-derived analogs, EC cells. Against this notion, however, a number of recent studies have suggested that it is possible to remove cells from several locations in which they were previously thought to be quiescent or, at the very least, highly restricted in the cell lineages they could produce and "transdifferentiate" them into lineages that are disparate from the tissue or even the germ layer origin from which they originate *(150)*.

Extragonadal GCTs, which retain many of the histological and karyotypic characteristics of gonadal germ cell tumors, occur in both neonates and adults *(151)*. Many adult extragonadal GCTs share elements in common with gonadal GCTs, including characteristic chromosomal changes, histological phenotype, the inclusion of seminoma, and, most likely, a common origin from primordial germ cells. However, there also exists a second group of extragonadal teratomas that are typically nearly euploid and do not contain seminoma. Tumors of this group, which can be found in variety of sites, are proposed to have a distinct origin from embryonic material that has escaped regulation by organizers of early development, lying quiescent after birth in the majority of adults and only rarely reactivating to form extragonadal teratomas. It could be that these cells are also the rare cells characterized as pluripotent "adult stem cells" that can be reactivated if they are removed and cultured in vitro. Certainly, very few cells extracted from adult tissue show the ability to transdifferentiate into other lineages *(152)*, and the pluripotency observed in adult stem cells appears to be gained by reacquisition. That a reacquisition of pluripotency is required is consistent with the idea that the adult environment holds these cells in a quiescent state and they can be reactivated by the changes that are connected with transfer to in vitro culture conditions and subsequent maintenance in an environment that is devoid of the native signals that hold the cells in stasis. Alternatively, reacquisition of pluripotency may arise because of in vitro culture conditions that actively promote reactivation. This idea is supported by work from Verfaillie's group *(152)*, who report that multipotent adult progenitor cells (MAPCs) derived from the bone marrow of adult mice require an extended period of in vitro culture at relatively low-cell-density before displaying pluripotency.

7. CONCLUSION

It seems highly likely that ES and EG cells, and possibly even EC cells, will play a major role in providing specific cell types for transplantation therapies used in regenerative medicine. It is also evident that these cells provide unique tools for understanding the genetic and cellular interactions that guide early human development. Although model systems (e.g., the laboratory mouse) can indicate key paradigms for early mammalian development, it is clear that human embryogenesis differs in a variety of respects, so that a full understanding can only be achieved by the use of human model systems. The recently identified adult stem cells may also provide valuable information pertinent to development and oncology as well as providing potential tools for regenerative medicine. However, compared to ES cells, they are currently ill-defined and unproven in many key characteristics.

ACKNOWLEDGMENTS

This work was supported in part by grants from the Wellcome Trust and Yorkshire Cancer Research.

REFERENCES

1. Ballantyne, J. W. (1894) The teratological records of Chalden, *Teratologia* **1,** 127–142.
2. Damjanov, I. and Solter, D. (1974) Experimental teratoma, *Curr. Topics Pathol.* **59,** 69–130.
3. Wheeler, J. E. (1983) History of teratomas, in *The Human Teratomas: Experimental and Clinical Biology* (Damjanov, I., Knowles, B. B., and Solter, D., eds.), Humana, Clifton, NJ, pp. 1–22.
4. Dixon, F. S. and Moore, R. A. (1952) Tumors of the male sex organs, *Atlas of Tumor Pathology*, vol. 8, Armed Forces Institute of Pathology, Washington, DC, Fascicles 31b and 32.
5. Mostofi, F. K. and Price, E. B. (1973) Tumours of the male genital system, in *Atlas of Tumor Pathology*, *Second Series*, Armed Forces Institute of Pathology, Washington, DC, Fascicle 8.
6. Einhorn, L. H. (1987) Treatment strategies of testicular cancer in the United States, *Int. J. Androl.* **10,** 399–405.
7. Stoter, G. (1987) Treatment strategies of testicular cancer in Europe, *Int. J. Androl.* **10,** 407–415.
8. Oosterhuis, J. W., Castedo, S. M., and de Jong, B. (1990) Cytogenetics, ploidy and differentiation of human testicular, ovarian and extragonadal germ cell tumours, *Cancer Surv.* **9,** 320–332.
9. Geurts van Kessel, A., Suijkerbuijk, R. F., Sinke, R. J., et al. (1993) Molecular cytogenetics of human germ cell tumours: i(12p) and related chromosomal anomalies, *Eur Urol.* **23,** 23–28.

10. Damjanov, I. (1990) Teratocarcinoma stem cells, *Cancer Surv.* **9,** 303–319.
11. Damjanov, I. (1993) Pathogenesis of testicular germ cell tumors, *Eur. Urol.* **23,** 2–7.
12. Martineau, M. (1969) Chromosomes in human testicular tumors, *J. Pathol.* **99,** 271–281.
13. Linder, D. (1969) Gene loss in human teratomas, *Proc. Natl. Acad. Sci. USA* **63,** 699–704.
14. Linder, D. (1983) The origin of teratomas, in *The Human Teratomas: Experimental and Clinical Biology* (Damjanov, I., Knowles, B. B., and Solter, D., eds.), Humana, Clifton, NJ, pp. 67–80.
15. Carritt, B., Parrington, J. M., Welch, H. M., et al. (1982) Diverse origins of multiple ovarian teratomas in a single individual, *Proc. Natl. Acad. Sci. USA* **79,** 7400–7404.
16. Atkin, N. B. (1973) Y bodies and similar fluorescent chromocentres in human tumors including teratomata, *Br. J. Cancer* **27,** 183–189.
17. Wang, N., Perkins, K. L., Bronson, D. L., and Fraley, E. E. (1981) Cytogenetic evidence for premeiotic transformation of human testicular cancers, *Cancer Res.* **41,** 2135–2140.
18. Atkin, N. B. and Baker, M. C. (1982) Specific chromosome change i(12p) in testicular tumors, *Lancet* **ii,** 1349.
19. Atkin, N. B. and Baker, M. C. (1983) I(12p): specific chromosomal marker in seminoma and malignant teratoma of the testis?, *Cancer Genet. Cytogenet.* **10,** 199–204.
20. Suijkerbuijk, R. F., Sinke, R. J., Meloni, A. M., et al. (1993) Overrepresentation of chromosome 12p sequences and karyotypic evolution in i(12p)-negative testicular germ-cell tumors revealed by fluorescence in situ hybridization, *Cancer Genet. Cytogenet.* **70,** 85–93.
21. Murty, V. V., Houldsworth, J., Baldwin, S., et al. (1992) Allelic deletions in the long arm of chromosome 12 identify sites of candidate tumor suppressor genes in male germ cell tumors, *Proc. Natl. Acad. Sci. USA* **89,** 11,006–11,010.
22. Murty, V. V., Renault, B., Falk, C. T., et al. (1996) Physical mapping of a commonly deleted region, the site of a candidate tumor suppressor gene, at 12q22 in human male germ cell tumors, *Genomics* **35,** 562–570.
23. Rapley, E. A., Crockford, G. P., Teare, D., et al. (2000) Localization to Xq27 of a susceptibility gene for testicular germ-cell tumours, *Nature Genet.* **24,** 197–200.
24. Stevens, L. C. and Little, C. C. (1954) Spontaneous testicular teratomas in an inbred strain of mice, *Proc. Natl. Acad. Sci. USA* **40,** 1080–1087.
25. Stevens, L. C. and Hummel, K. P. (1957) A description of spontaneous congenital testicular teratomas in strain 129 mice, *J. Natl. Cancer. Inst.* **18,** 719–747.
26. Stevens, L. C. (1967) The biology of teratomas, *Adv. Morphol.* **6,** 1–31.
27. Stevens, L. C. (1970) Experimental production of testicular teratomas in mice of strains 129, A/He and their F$_1$ hybrids, *J. Natl. Cancer Inst.* **44,** 929–932.
28. Stevens, L. C. (1970) The development of transplantable teratocarcinomas from intratesticular grafts of pre and post implantation mouse embryos, *Dev. Biol.* **21,** 364–382.

29. Solter, D., Skreb, N., and Damjanov, I. (1970) Extrauterine growth of mouse egg-cylinders results in malignant teratoma, *Nature* **227,** 503–504.
30. Solter, D., Dominis, M., and Damjanov, I. (1979) Embryo-derived teratocarcinomas: I. The role of strain and gender in the control of teratocarcinogenesis, *Int. J. Cancer* **24,** 770–772.
31. Solter, D., Dominis, M., and Damjanov, I. (1981) Embryo-derived teratocarcinoma. III Development of tumors from teratocarcinoma-permissive and non-permissive strain embryos transplanted to F1, hybrids, *Int. J. Cancer* **28,** 479–483.
32. Solter, D. and Damjanov, I. (1979). Teratocarcinomas rarely develop from embryos transplanted into athymic mice, *Nature* **278,** 554–555.
33. Stevens, L. C. (1964) Experimental production of testicular teratomas in mice, *Proc. Natl. Acad. Sci. USA* **52,** 654–661.
34. Bendel-Stenzel, M., Anderson, R., Heasman, J., and Wylie, C. (1998) The origin and migration of primordial germ cells in the mouse, *Semin. Cell Dev. Biol.* **9,** 393–400.
35. Stevens, L. C. (1967) Origin of testicular teratomas from primordial germ cells in mice, *J. Natl. Cancer Inst.* **38,** 549–552.
36. McLaren, A. and Southee, D. (1997) Entry of mouse embryonic germ cells into meiosis, *Dev. Biol.* **187,** 107–113.
37. Crittenden, S. L., Troemel, E. R., Evans, T. C., and Kimble, J. (1994) GLP-1 is localized to the mitotic region of the *C. elegans* germ line, *Development* **120,** 2901–2911.
38. Berry, L. W., Westlund, B., and Schedl, T. (1997) Germ-line tumor formation caused by activation of *glp-1,* a *Caenorhabditis elegans* member of the Notch family of receptors, *Development* **124,** 925-936.
39. Ellisen, L. W., Bird, J., West, D. C., et al. (1991) *Tan-1,* the human homologue of the *Drosophila Notch* gene is broken by chromosomal translocation in T-lymphoblastic neoplasms, *Cell* **66,** 649–661.
40. Gokhale, P. J., Eastwood, D., Walsh, J., and Andrews, P. W. (1998) The possible role of Notch gene expression in germ cell tumour development and progression, in *Germ Cell Tumours IV* (Jones, W. G., Appleyard, I., Harnden, P., et al., eds.), John Libbey, London, pp. 69–71.
41. Finch, B. W. and Ephrussi, B. (1967) Retention of multiple developmental potentialities by cells of a mouse testicular teratocarcinomas during prolonged culture in vitro and their extinction upon hybridization with cells of permanent lines, *Proc. Natl. Acad. Sci. USA* **57,** 615–621.
42. Kahn, B. W. and Ephrussi, B. (1970) Developmental potentialities of clonal in vitro cultures of mouse testicular teratoma, *J. Natl. Cancer Inst.* **44,** 1015–1029.
43. Bernstine, E. G., Hooper, M. L., Grandchamp, S., and Ephrussi, B. (1973) Alkaline phosphatase activity in mouse teratoma, *Proc. Natl. Acad. Sci. USA* **70,** 3899–3903.
44. Evans, M. J. (1972) The isolation and properties of a clonal tissue culture strain of pluripotent mouse teratoma cells, *J. Embryol. Exp. Morphol.* **28,** 163–176.

45. Jakob, H., Boon, T., Gaillard, J., Nicolas, J.-F., and Jacob, F. (1973) Tératocarcinoma de la souris: isolement, culture, et proprieties de cellules à potentialities multiples, *Ann. Microbiol. Inst. Pasteur* **124B,** 269–282.

46. Martin, G. R. and Evans, M. J. (1974) The morphology and growth of a pluripotent teratocarcinomas cell line and its derivatives in tissue culture, *Cell* **2,** 163–172.

47. Martin, G. R. and Evans, M. J. (1975) Differentiation of clonal lines of teratocarcinomas cells: formation of embryoid bodies in vitro, *Proc. Natl. Acad. Sci. USA* **72,** 1441–1445.

48. Nicolas, J-F., Dubois, P., Jakob, H., Gaillard, J., and Jacob, F. (1975) Tératocarcinome de la souris: différenciation en culture d'une lignée de cellules primitives à potentialities multiples, *Ann. Microbiol. Inst. Pasteur* **126A,** 3–22.

49. Strickland, S. and Mahdavi, V. (1978) The induction of differentiation in teratocarcinomas stem cells by retinoic acid, *Cell* **15,** 393–403.

50. Strickland, S., Smith, K. K., and Marotti, K. R (1980) Hormonal induction of differentiation in teratocarcinoma stem cells: generation of parietal endoderm by retinoic acid and dibutyryl cAMP, *Cell* **21,** 347–355.

51. Hogan, B. L. M., Barlow, D. P., and Tilly, R. (1983) F9 teratocarcinoma cells as a model for the differentiation of parietal and visceral endoderm in the mouse embryo, *Cancer Surv.* **2,** 115–140.

52. Jakob, H., Dubois, P., Eisen, H., and Jacob, F. (1978) Effets de l'hexaméthylènebisacétamide sur la différenciation de cellules de carcinome embryonnaire, *C. R. Acad. Sci. (Paris)* **286,** 109–111.

53. McBurney, M. W., Jones-Villeneuve, E. M., Edwards, M. K., and Anderson, P. J. (1982) Control of muscle and neuronal differentiation in a cultured embryonal carcinoma cell line, *Nature* **299,** 165–167.

54. Jones-Villeneuve, E. M., McBurney, M. W., Rogers, K. A., and Kalnins, V. I. (1982) Retinoic acid induces embryonal carcinoma cells to differentiate into neurons and glial cells, *J. Cell Biol.* **94,** 253–262.

55. Kleinsmith, L. J. and Pierce, G. B. (1964) Multipotentiality of single embryonal carcinoma cells, *Cancer Res.* **24,** 1544–1552.

56. Artzt, K., Dubois, P., Bennett, D., Condamine, H., Babinet, C., and Jacob, F. (1973) Surface antigens common to mouse cleavage embryos and primitive teratocarcinoma cells in culture, *Proc. Natl. Acad. Sci. USA* **70,** 2988–2992.

57. Jacob, F. (1978) Mouse teratocarcinoma and mouse embryo, *Proc. Ry. Soc. Lond. B* **201,** 249–270.

58. Brinster, R. L. (1974) The effect of cells transferred into the mouse blastocyst on subsequent development, *J. Exp. Med.* **140,** 1049–1056.

59. Papaioannou, V. E., McBurney, M. W., Gardner, R. L., and Evans, M. J. (1975) Fate of teratocarcinoma cells injected into early mouse embryos, *Nature* **258,** 70–73.

60. Mintz, B. and Illmensee, K. (1975) Normal genetically mosaic mice produced from malignant teratocarcinoma cells, *Proc. Natl. Acad. Sci. USA* **72,** 3585–3589.

61. Evans, M. J. and Kaufman, M. H. (1981) Establishment in culture of pluripotential cells from mouse embryos, *Nature* **292,** 154–156.
62. Martin, G. R. (1981) Isolation of a pluripotent cell line from early mouse embryos cultured in medium conditioned by teratocarcinoma stem cells, *Proc. Natl. Acad. Sci. USA* **78,** 7634–7636.
63. Niwa, H., Miyazaki, J., and Smith, A. G. (2000) Quantitative expression of Oct-3/4 defines differentiation, dedifferentiation or self-renewal of ES cells, *Nature Genet.* **24,** 372–376.
64. Niwa, H., Burdon, T., Chambers, I., and Smith, A. G. (1998) Self-renewal of pluripotent embryonic stem cells is mediated via activation of STAT3, *Genes Dev.* **12,** 2048–2060.
65. Burdon, T., Stracey, C., Chambers, I., Nichols, J., and Smith, A. (1999) Suppression of SHP-2 and ERK signaling promotes self-renewal of mouse embryonic stem cells, *Dev. Biol.* **210,** 30–43.
66. Boeuf, H., Hauss, C., De Graeve, F., Baran, N., and Kedinger, C. (1997) Leukemia inhibitory factor-dependent transcriptional activation in embryonic stem cells, *J. Cell Biol.* **138,** 1207–1217.
67. Thomson, J. A., Itskovitz-Eldor, J., Shapiro, S. S., et al. (1998) Embryonic stem cell lines derived from human blastocysts, *Science* **282,** 1145–1147.
68. Reubinoff, B. E., Pera, M. F., Fong, C. Y., Trounson, A., and Bongso, A. (2000) Embryonic stem cell lines from human blastocysts: somatic differentiation in vitro, *Nature Biotechnol.* **18,** 399–404.
69. Pierce, G. B., Verney, E. L., and Dixon, F. J. (1957) The biology of testicular cancer I. Behavior after transplantation, *Cancer Res.* **17,** 134–138.
70. Fogh, J. and Trempe, G. (1975) New human tumor cell lines, in *Human Tumor Cells In Vitro* (Fogh, J., ed.), Plenum, New York, pp. 115–159.
71. Hogan, B., Fellous, M., Avner, P., and Jacob, F. (1977) Isolation of a human teratoma cell line which expresses F9 antigen, *Nature* **270,** 515–518.
72. Cotte, C. A., Easty, G. C., and Neville, A. M. (1981). Establishment and properties of human germ cell tumors in tissue culture, *Cancer Res.* **41,** 1422–1427.
73. Wang, N., Trend, B., Bronson, D. L., and Fraley, E. E. (1980) Nonrandom abnormalities in chromosome 1 in human testicular cancers, *Cancer Res.* **40,** 796–802.
74. Andrews, P. W., Bronson, D. L., Benham, F., Strickland, S., and Knowles, B. B. (1980). A comparative study of eight cell lines derived from human testicular teratocarcinoma, *Int. J. Cancer* **26,** 269–280.
75. Andrews, P. W., Goodfellow, P. N., Shevinsky, L., Bronson, D. L., and Knowles, B. B. (1982) Cell surface antigens of a clonal human embryonal carcinoma cell line: morphological and antigenic differentiation in culture, *Int. J. Cancer* **29,** 523–531.
76. Benham, F. J., Andrews, P. W., Bronson, D. L., Knowles, B. B., and Harris, H. (1981) Alkaline phosphatase isozymes as possible markers of differentiation in human teratocarcinoma cell lines, *Dev. Biol.* **88,** 279–287.
77. Damjanov, I. and Andrews, P. W. (1983) Ultrastructural differentiation of a clonal human embryonal carcinoma cell line in vitro, *Cancer Res.* **43,** 2190–2198.

78. Bronson, D. L., Andrews, P. W., Solter, D., Cervenka, J., Lange, P. H., and Fraley, E. E. (1980) A cell line derived from a metastasis of a human testicular germ-cell tumor, *Cancer Res.* **40**, 2500–2506.

79. Andrews, P. W., Bronson, D. L., Wiles, M. V., and Goodfellow, P. N. (1981) The expression of major histocompatibility antigens by human teratocarcinoma derived cells lines, *Tissue Antigens* **17**, 493–500.

80. Andrews, P. W., Damjanov, I., Simon, D., et al. (1984) Pluripotent embryonal carcinoma clones derived from the human teratocarcinoma cell line Tera-2: differentiation in vivo and in vitro, *Lab. Invest.* **50**, 147–162.

81. Chen, C., Fenderson, B. A., Andrews, P. W., and Hakomori, S.-I. (1989) Glycolipid-glycosyltransferases in human embryonal carcinoma cells during retinoic acid-induced differentiation, *Biochemistry* **28**, 2229–2238.

82. Solter, D. and Knowles, B. B. (1978) Monoclonal antibody defining a stage-specific mouse embryonic antigen (SSEA-1), *Proc. Natl. Acad. Sci. USA* **75**, 5565–5569.

83. Gooi, H. C., Feizi, T., Kapadia, A., Knowles, B. B., Solter, D., and Evans, M. J. (1981) Stage specific embryonic antigen involves $\alpha 1\rightarrow 3$ fucosylated type 2 blood group chains, *Nature* **292**, 156–158.

84. Kannagi, R., Nudelman, E., Levery, S. B., and Hakomori, S. (1982) A series of human erythrocyte glycosphingolipids reacting to the monoclonal antibody directed to a developmentally regulated antigen, SSEA-1, *J. Biol. Chem.* **257**, 14,865–14,874.

85. Andrews, P. W., Casper, J., Damjanov, I., et al. (1996) Comparative analysis of cell surface antigens expressed by cell lines derived from human germ cell tumors, *Int. J. Cancer* **66**, 806–816.

86. Shevinsky, L., Knowles, B. B., Damjanov, I., and Solter, D. (1982) Monoclonal antibody to murine embryos defines a stage-specific embryonic antigen expressed on mouse embryos and human teratocarcinoma cells, *Cell* **30**, 697–705.

87. Kannagi, R., Cochran, N. A., Ishigami, F., et al. (1983) Stage-specific embryonic antigens (SSEA-3 and -4) are epitopes of a unique globo-series ganglioside isolated from human teratocarcinoma cells, *EMBO J.* **2**, 2355–2361.

88. Damjanov, I., Fox, N., Knowles, B. B., Solter, D., Lange, P. H., and Fraley, E. E. (1982) Immunohistochemical localization of murine stage-specific embryonic antigens in human testicular germ cell tumors, *Am. J. Pathol.* **108**, 225–230.

89. Draper, J. S., Pigott, C., Thomson, J. A., and Andrews, P. W. (2002) Surface antigens of human embryonic stem cells: changes upon differentiation in culture, *J. Anat.* **200**, 249–258

90. Henderson, J. K., Draper, J. S., Baillie, H. S., et al. (2002) Preimplantation human embryos and embryonic stem cells show comparable expression of stage-specific embryonic antigens, *Stem Cells* **20**, 329–337.

91. Bird, J. M. and Kimber, S. J. (1984) Oligosaccharides containing fucose linked $\alpha(1\rightarrow 3)$ and $\alpha(1\rightarrow 4)$ to *N*-acetylglucosamine cause decompaction of mouse morale, *Dev. Biol.* **104**, 449–460.

92. Fenderson, B. A., Zehavi, U., AND Hakomori, S. (1984) A multivalent lacto-*n*-fucopentaose III–lysyllysine conjugate decompacts pre-implantation-stage mouse embryos, while the free-oligosaccharide is ineffective, *J. Exp. Med.* **160,** 1591–1596.

93. Fenderson, B. A., Ostrander, G., Hausken, Z., Radin, N. S., and Hakomori, S. (1992) A ceramide analogue (PDMP) inhibits glycolipids synthesis in fish embryos, *Exp. Cell Res.* **198,** 362–366.

94. Tippett, P., Andrews, P. W., Knowles, B. B., Solter, D., and Goodfellow, P. N. (1986) Red cell antigens P (globoside) and Luke: identification by monoclonal antibodies defining the murine stage-specific embryonic antigens -3 and -4 (SSEA-3 and -4), *Vox Sang.* **51,** 53–56.

95. Race, R. R. and Sanger, R. (1975) *Blood Groups in Man*, 6th ed., Blackwell Scientific, Oxford, pp. 169–171.

96. Andrews, P. W., Banting, G. S., Damjanov, I., Arnaud, D., and Avner, P. (1984) Three monoclonal antibodies defining distinct differentiation antigens associated with different high molecular weight polypeptides on the surface of human embryonal carcinoma cells, *Hybridoma* **3,** 347–361.

97. Badcock, G., Pigott, C., Goepel, J., and Andrews P. W. (1999) The human embryonal carcinoma marker antigen TRA-1-60 is a sialylated keratan sulphate proteoglycan, *Cancer Res.* **59,** 4715–4719.

98. Pera, M. F., Blasco-Lafita, M. J., Cooper, S., Mason, M., Mills, J., and Monaghan, P. (1988) Analysis of cell-differentiation lineage in human teratomas using new monoclonal antibodies to cytostructural antigens of embryonal carcinoma cells, *Differentiation* **39,** 139–149.

99. Rettig, W. J., Cordon-Cardo, C., Ng, J. S., Oettgen, H. F., Old, L. J., and Lloyd, K. O. (1985) High-molecular-weight glycoproteins of human teratocarcinoma defined by monoclonal antibodies to carbohydrate determinants, *Cancer Res.* **45,** 815–821.

100. Marrink, J., Andrews, P. W., van Brummen, P. J., et al. (1991) TRA-1-60: a new serum marker in patients with germ cell tumors, *Int. J. Cancer* **49,** 368–372.

101. Gels, M. E., Marrink J, Visser, P., et al. (1997) Importance of a new tumor marker TRA-1-60 in the follow-up of patients with clinical state I non-seminomatous testicular germ cell tumors, *Ann. Surg. Oncol.* **4,** 321–327.

102. Andrews, P. W. and Goodfellow, P. N. (1980) Antigen expression by somatic cell hybrids of a murine embryonal carcinoma cell with thymocytes and L cells, *Somat. Cell Genet.* **6,** 271–284.

103. Duran, C., Talley, P. J., Walsh, J., Pigott, C., Morton, I., and Andrews, P. W. (2001) Hybrids of pluripotent and nullipotent human embryonal carcinoma cells: partial retention of a pluripotent phenotype, *Int. J. Cancer* **93,** 324–332.

104. Pera, M. F., Cooper, S., Mills, J., and Parrington, J. M. (1989) Isolation and characterization of a multipotent clone of human embryonal carcinoma cells, *Differentiation* **42,** 10–23.

105. Roach, S., Cooper, S., Bennett, W., and Pera, M. F (1993) Cultured cell lines from human teratomas: windows into tumor growth and differentiation and early human development, *Eur. Urol.* **23,** 82–88.

106. Roach, S., Schmid, W., and Pera, M. F. (1994) Hepatocytic transcription factor expression in human; embryonal carcinoma and yolk sac carcinoma cell lines: expression of HNF-3alpha in models of early endodermal cell differentiation, *Exp. Cell Res.* **215**, 189–198.

107. Pera, M. F. and Herzfeld, D. (1998) Differentiation of human pluripotent teratocarcinomas stem cells induced by bone morphogenetic protein-2, *Reprod. Fertil. Dev.* **10**, 551–555.

108. Teshima, S., Shimosato, Y., Hirohashi, S., et al. (1988) Four new human germ cell tumor cell lines, *Lab. Invest.* **59**, 328–336.

109. Damjanov, I., Horvat, B., and Gibas, Z. (1993) Retinoic acid-induced differentiation of the developmentally pluripotent human germ cell tumor-derived cell line, NCCIT, *Lab. Invest.* **68**, 202–232.

110. Hata, J., Fujita, H., Ikeda, E., Matsubayashi, Y., Kokai, Y., and Fujimoto, J. (1989) Differentiation of human germ cell tumor cells, *Hum. Cell* **2**, 382–387.

111. Umezawa, A., Maruyama, T., Inazawa, J., Imai, S., Takano, T., and Hata, J. (1996) Induction of mcl1/EAT, Bcl-2 related gene, by retinoic acid or heat shock in the human embryonal carcinoma cells, NCR-G3, *Cell Struct. Funct.* **21**, 143–150.

112. Andrews, P. W. (1984) Retinoic acid induces neuronal differentiation of a cloned human embryonal carcinoma cell line in vitro, *Dev. Biol.* **103**, 285–293.

113. Matthaei, K., Andrews, P. W., and Bronson, D. L. (1983) Retinoic acid fails to induce differentiation in human teratocarcinoma cell lines that express high levels of cellular receptor protein, *Exp. Cell Res.* **143**, 471–474.

114. Fenderson, B. A., Andrews, P. W., Nudelman, E., Clausen, H., and Hakomori, S.-I. (1987) Glycolipid core structure switching from globo- to lacto- and ganglio-series during retinoic acid-induced differentiation of TERA-2-derived human embryonal carcinoma cells, *Dev. Biol.* **122**, 21–34.

115. Lee, V. M.-Y. and Andrews, P. W. (1986) Differentiation of NTERA-2 clonal human embryonal carcinoma cells into neurons involves the induction of all three neurofilament proteins, *J. Neurosci.* **6**, 514–521.

116. Pleasure, S. J., Page, C., and Lee, V. M.-Y. (1992) Pure, post-mitotic, polarized human neurons derived from Ntera2 cells provide a system for expressing exogenous proteins in terminally differentiated neurons, *J Neurosci.* **12**, 1802–1815.

117. Pleasure, S. J. and Lee, V. M. Y. (1993) NTERA-2 cells a human cell line which displays characteristics expected of a human committed neuronal progenitor cell, *J. Neurosci. Res.* **35**, 585–602.

118. Rendt, J., Erulkar, S., and Andrews, P. W. (1989) Presumptive neurons derived by differentiation of a human embryonal carcinoma cell line exhibit tetrodotoxin-sensitive sodium currents and the capacity for regenerative responses, *Exp. Cell Res.* **180**, 580–584.

119. Squires, P. E., Wakeman, J. A., Chapman, H., et al. (1996) Regulation of intracellular Ca2+ in response to muscarinic and glutamate receptor antagonists during the differentiation of NTERA2 human embryonal carcinoma cells into neurons, *Eur. J. Neurosci.* **8**, 783–793.

120. Borlongan, C. V., Tajima, Y., Trojanowski, J. Q., Lee, V. M., and Sanberg, P. R. (1998) Transplantation of cryopreserved human embryonal carcinoma-derived neurons (NT2N cells) promotes functional recovery in ischemic rats, *Exp. Neurol.* **149,** 310–321.

121. Hurlebert, M. S., Gianani, R. I., Hutt, C., Freed, C. R., and Kaddis, F. G. (1999) Neural transplantation of hNT neurons for Huntington's disease, *Cell Transplant.* **8,** 143–151.

122. Kleppner, S. R., Robinson, K. A., Trojanowski, J. Q., and Lee, V. M. (1995) Transplanted human neurons derived from a teratocarcinoma cell line (NTERA2) mature, integrate and survive for over 1 year in the nude mouse brain, *J. Comp. Neurol.* **357,** 618–632.

123. Philips, M. F., Muir, J. K., Saatman, K. E., et al. (1999) Survival and integration of transplanted postmitotic human neurons following experimental brain injury in immunocompetent rats, *J. Neurosurg.* **90,** 116–124.

124. Kondziolka, D., Wechsler, L., Goldstein, S., et al. (2000) Transplantation of cultured human neuronal cells for patients with stroke, *Neurology* **55,** 565–569.

125. Gönczöl, E., Andrews, P. W., and Plotkin, S. A. (1984) Cytomegalovirus replicates in differentiated but not undifferentiated human embryonal carcinoma cells, *Science* **224,** 159–161.

126. Hirka, G., Prakesh, K., Kawashima, H., Plotkin, S. A., Andrews, P. W., and Gönczöl, E. (1991) Differentiation of human embryonal carcinoma cells induces human immunodeficiency virus permissiveness which is stimulated by human cytomegalovirus coinfection, *J. Virol.* **65,** 2732–2735.

127. LaFemina, R. and Hayward, G. S. (1986) Constitutive and retinoic acid-inducible expression of cytomegalovirus immediate-early genes in human teratocarcinoma cells, *J. Virol.* **58,** 434–440.

128. Nelson, J. A., Reynolds-Kohler, C., and Smith, B. A. (1987) Negative and positive regulation by a short segment in the 5'-flanking region of the human cytomegalovirus major immediate-early gene, *Mol. Cell. Biol.* **7,** 4125–4129.

129. Simeone, A., Acampora, D., Arcioni, L., Andrews, P. W., Boncinelli, E., and Mavilio, F. (1990) Sequential activation of human HOX2 homeobox genes by retinoic acid in human embryonal carcinoma cells, *Nature* **346,** 763–766.

130. Bottero, L., Simeone, A., Arcioni, L., et al. (1991) Differential activation of homeobox genes by retinoic acid in human embryonal carcinoma cells, in *Recent Results in Cancer Research, Vol 123, Pathobiology of Human Germ Cell Neoplasia* (Oosterhuis, J. W., Walt, H., and Damjanov, I., eds.), Springer-Verlag, Berlin, pp. 133–143.

131. Wakeman, J. A., Walsh, J., and Andrews, P. W. (1998) Human Wnt-13 is developmentally regulated during the differentiation of NTERA-2 pluripotent human embryonal carcinoma cells, *Oncogene* **17,** 179–186.

132. Lako, M., Lindsay, S., Lincoln, J., Cairns, P. M., Armstrong, L., and Hole, N. (2001) Characterisation of Wnt gene expression during the differentiation of murine embryonic stem cells in vitro: role of Wnt3 in enhancing haemato-poietic differentiation, *Mech. Dev.* **103,** 49–59.

133. Smolich, B. D. and Papkoff, J. (1994) Regulated expression of *Wnt* family members during neuroectodermal differentiation of P19 embryonal carci-

noma cells: overexpression of *Wnt-1* perturbs normal differentiation-specific properties, *Dev. Biol.* **166,** 300–310.

134. Walsh, J. and Andrews, P. W. (2002) Expression of Wnt and Notch pathway genes in a pluripotent human embryonal carcinoma cell line and embryonic stem cells, *Acta Pathol. Micrbiol. Immunol. Scand.* **111,** 663–677.

135. Przyborski, S. A., Morton, I. E., Wood, A., and Andrews, P. W. (2000) Developmental regulation of neurogenesis in the pluripotent human embryonal carcinoma cell line NTERA-2, *Eur. J. Neurosci.* **12,** 3521–3528.

136. Andrews, P. W., Nudelman, E., Hakomori, S.-I., and Fenderson, B. A. (1990) Different patterns of glycolipid antigens are expressed following differentiation of TERA-2 human embryonal carcinoma cells induced by retinoic acid, hexamethylene bisacetamide (HMBA) or bromodeoxyuridine (BUdR), *Differentiation* **43,** 131–138.

137. Andrews, P. W., Damjanov, I., Berends, J., et al. (1994) Inhibition of proliferation and induction of differentiation of pluripotent human embryonal carcinoma cells by osteogenic protein-1 (or bone morphogenetic protein-7), *Lab. Invest.* **71,** 243–251.

138. Gokhale, P. J., Giesberts, A. N., and Andrews, P. W. (2000) Brachyury is expressed by human teratocarcinoma cells in the absence of mesodermal differentiation, *Cell Growth Differ.* **11,** 157–162.

139. Donovan, P. J., Stott, D., Cairns, L. A., Heasman, J., and Wylie, C. C. (1986) Migratory and postmigratory mouse primordial germ cells behave differently in culture, *Cell* **44,** 831–838.

140. Dolci, S., Williams, D. E., Ernst, M. K., et al. (1991) Requirement for mast cell growth factor for primordial germ cell survival in culture, *Nature* **352,** 809–811.

141. Matsui, Y., Zsebo, K., and Hogan, B. L. M. (1992) Derivation of pluripotential embryonic stem cells from murine primordial germ cells in culture, *Cell* **70,** 841–847.

142. Koshimizu, U., Taga, T., Watanabe, M., et al. (1996) Functional requirement of gp130 mediated signaling for growth and survival of mouse primordial germ cells in vitro and derivation of embryonic germ (EG) cells, *Development* **122,** 1235–1242.

143. Shamblott, M. J., Axelman, J., Wang, S., et al. (1998) Derivation of pluripotent stem cells from cultured human primordial germ cells, *Proc. Natl. Acad. Sci. USA* **95,** 13,726–13,731.

144. Surani, A. (2001) Reprogramming of genome function through epigenetic inheritance, *Nature* **414,** 122–128.

145. Tada, M., Tada, T., Lefebvre, L., Barton, S. C., and Surani, M. A. (1997) Embryonic germ cells induce epigenetic reprogramming of somatic nucleus in hybrid cells, *EMBO J.* **16,** 6510–6520.

146. Tada, T. and Tada, M. (2001) Toti-/pluripotent stem cells and epigenetic modification, *Cell Struct. Funct.* **26,** 149–160.

147. Mitsuya, K., Meguro, M., Sui, H., et al. (1998) Epigenetic reprogramming of the human H19 gene in mouse embryonic cells does not erase the primary parental imprint, *Genes Cells* **3,** 245–255.

148. van Gurp, R. J., Oosterhuis, J. W., Kalscheuer, V., Mariman, E. C., and Looijenga, L. H. (1994) Biallelic expression of the H19 and IGF2 genes in human testicular germ cell tumors, *J. Natl. Cancer Inst.* **86,** 1070–1074.
149. Saltou, M., Barton, S. C., and Surani, M. A. (2002) A molecular programme for specification of germ cell fate in mice, *Nature* **418,** 293–300.
150. Clarke, D. and Frisen, J. (2001) Differentiation potential of adult stem cells, *Curr. Opin. Genet. Dev.* **11,** 575–580.
151. Oosterhuis, J. W., Looijenga, L. H., van Echten, J., and de Jong, B. (1997) Chromosomal constitution and developmental potential of human germ cell tumors and teratomas, *Cancer Genet. Cytogenet.* **95,** 96–102.
152. Jiang, Y., Jahagirdar, B. N., Reinhardt, R. L., et al. (2002) Pluripotency of mesenchymal stem cells derived from adult marrow, *Nature* **418,** 41–49.

5

Human Pluripotent Stem Cells from Bone Marrow

Felipe Prosper and Catherine M. Verfaillie

1. INTRODUCTION

Stem cells have become a major area of discussion both in scientific journals and in the lay press *(1)*. The ethical implications of using embryonic stem (ES) cells along with new scientific evidence of the greater potential of adult stem cells have stimulated the debate regarding the need for human ES cells for therapies *(2,3)*. In the next chapter, we will discuss some of the existing evidence that pluripotent adult stem cells can be derived from bone marrow (BM) or other postnatal tissues. We will not limit our discussion to human cells, as most of the studies so far suggesting that pluripotent adult stem cells exist are based on animal models.

Stem cells have been defined as clonogenic cells that can undergo self-renewal as well as differentiation to committed progenitors and, eventually, to functional differentiated tissues. Stem cells can be subdivided according to their potential in totipotent stem cells capable of giving rise to both embryonic and extraembryonic tissues, pluripotent stem cells, that may differentiate into tissues derived from any of the three germ layers, or multipotent stem cells, with a more limited differentiating capacity *(4–7)*.

It is generally accepted that ES cells are pluripotent. In contrast, it is generally accepted that adult stem cells are multipotent but not pluripotent. However, recent studies describing adult stem cell plasticity have led to intense discussions of whether some or all adult stem cells may have the same pluripotent capacity as ES cells. Although there is yet no official definition of stem cell plasticity, it could be defined as the capacity of a given cell to acquire morphological and functional characteristics of a tissue different than the one from which the cell is originally derived *(7–9)*. True stem cell plasticity should include the following criterion: a single tissue-specific adult stem cell (e.g., a hematopoietic stem cell [HSC]), thought to

From: *Human Embryonic Stem Cells*
Edited by: A. Chiu and M. S. Rao © Humana Press Inc., Totowa, NJ

be committed to a given cell lineage can, under certain microenvironmental conditions, acquire the ability to differentiate to cells of a different tissue. The differentiated cell types should be functional in vitro and in vivo, and the engraftment should be robust and persistently in the presence (and absence) of tissue damage.

Adult stem cells have been found in most tissues, including hematopoietic *(10)*, neural *(11)*, epidermal *(12)*, gastrointestinal *(13)*, skeletal muscle *(14)*, cardiac muscle *(15)*, liver *(16)*, pancreas *(17)*, or lung tissue *(18)* (for review of organ specific stem cells *see* special issue of *Journal of Pathology*, volume 197 [2002]). Until recently, it was accepted that tissue-specific cells could only differentiate into cells present in the tissue of origin. However, this concept has been challenged by recent studies suggesting that cells originating from one germ layer (e.g., mesoderm) can generate tissues derived from a second germ layer (e.g., ectoderm or endoderm) *(8,19)* (*see* Fig. 1). In addition, several recent studies have suggested that adult pluripotent stem cells obtained from the BM *(20,21)* and from the brain *(22,23)*, may be capable of giving rise to tissues derived from all three germ layers. However, this unexpected plasticity of adult stem cells has been questioned by the observation of in vitro cell fusion between BM or neural stem cells (NSCs) and ES cells *(24,25)*, and by the fact that some experiments have not been reproduced despite the intense efforts of different groups of investigators *(26,27)*.

2. ADULT STEM CELLS

Multipotent adult stem cells obtained from human BM have been used therapeutically for at least 50 yr: Allogeneic BM and peripheral blood (PB) transplants have definitively demonstrated that HSCs in these sources contribute to long-term and complete hematopoiesis after myeloablation *(28,29)*. More recent reports have indicated that BM or PB cells may contribute to regeneration of other organs such as the heart, liver, muscle or brain, even in humans *(30,31)*. Different types of stem cell can be isolated from the BM, including HSCs, mesenchymal stem cells (MSCs), side population (Sp) cells, and, very recently, multipotent adult progenitor cells (MAPCs). It is unclear to what extent all these populations are different or overlapping.

2.1. Hematopoietic Stem Cells

Hematopoietic stem and progenitor cells have been defined by different in vitro and in vivo assays. Human long-term repopulating progenitors can be measured using the nonobese diabetic severe combined immunodeficient

(NOD-SCID) mouse model (SCID-repopulating cells or SRCs) as well as culture assays such as the long-term culture-initiating cells (LTC-ICs), the extended long-term culture-initiating cells (E-LTC-ICs), or the myeloid–lymphoid-initiating cells (ML-ICs) *(32–35)*. It is also possible to identify different populations of BM/PB progenitor cells by flow cytometry, best characterized in mouse models in which a hierarchy of stem and progenitor cells has been established based on phenotypic markers *(5,36)*. HSCs measured in this way are multipotent adult stem cells; they are capable of giving rise to any hematopoietic cell at the clonogenic level and can self-renew extensively.

A number of studies have suggested that HSCs might, under certain circumstances, be more potent than expected. However, aside from the studies by Krause et al. *(20)* and Grant et al. *(37)*, none of the published studies have shown that single HSCs can indeed acquire a phenotype other than hematopoietic cells. In addition, except for the studies by Orlic et al. *(38)*, Lagasse et al. *(39)* and Grant et al. *(37)*, no study has shown that engraftment of cells from BM in tissues other than blood leads to functional repopulation of the second tissue. Finally, most studies, except the study by Lagasse et al. and Orlic et al., showed only low levels of engraftment.

Bone marrow and mobilized PB HSCs and progenitor cells may contribute to angiogenesis and vasculogenesis. Both in vitro and in vivo studies have demonstrated that CD34+ cells contain not only HSCs but also endothelial progenitor cells, which supports the hypothesis of a common hemangioblast progenitor cell *(40)*. The hemangioblast was identified as a single ES-derived Flk1+ cell that differentiates, depending on the culture conditions, into CD34$^+$ endothelial cells or CD34$^+$ hematopoietic cells. Until recently, such a cell has not been identified in postnatal life. However, a recent study has shown that single purified HSCs might have functional hemangioblast activity during retinal neovascularization *(37)*. Angioblasts or endothelial progenitor cells, on the other hand, are endothelial progenitors that, like HSCs, can be mobilized into the PB by growth factors and identified based on expression of surface antigens such as AC133 and VEGFR-2 also found on HSCs *(41–44)*. In animal models, VEGFR2 positive circulating endothelial progenitors (CEPs) are recruited into tumors and contribute to vasculogenesis, suggesting new clinical strategies to block tumor growth *(45,46)*. In mouse models of limb ischemia, endothelial progenitors are mobilized from the BM and contribute to neo-angiogenesis in the ischemic limb *(41,47,48)*. Likewise, in a rat model of cardiac ischemia, mobilized CD34$^+$ CD117^{bright} AC133$^+$ VEGFR2$^+$ cells contribute to repair of the rat heart after myocardial infarction by the remarkable capacity to

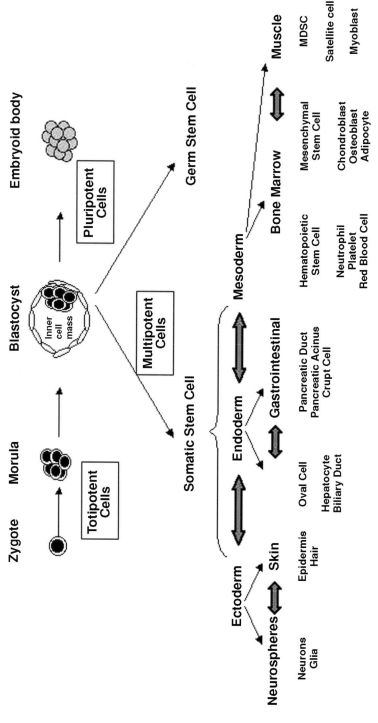

Fig. 1. Model of hierarchy of stem cells according to their potential. Embryonic stem cells are pluripotent and give rise both to somatic and germinal stem cells. Traditionally, adult stem cells derived from one germ layer differentiate into tissues derived from the same germ layer. This concept has been recently challenged (*see* Section 1). Hatched arrows indicate examples of studies that have suggested that stem cells originating from one tissue/germ layer can differentiate into cells of tissue germ layer of different origin.

generate new capillaries within the infarct zone *(42)*. Although neoangiogenesis by mobilized PB stem cells could be demonstrated, no generation of new cardiac muscle cells (myocytes) was observed in this study.

The potential of BM or PB cells to differentiate into cardiac muscle has been suggested by several studies. Some investigators have used BM mononuclear cells either directly implanted into the heart or by percutaneous infusion *(49)*, and even a small number of patients with myocardial infarction treated with coronary angioplasty have received autologous BM into their coronary arteries with some functional benefit *(50)*. Because there is no information regarding the nature of the cells that may contribute to cardiac muscle regeneration, the relevance of these studies is, at best, very limited. More compelling evidence for the capacity of BM cells to regenerate cardiac muscle stems from the group of Orlic et al. *(38)*. These authors demonstrated in a mouse model of myocardial infarction that direct injection into the heart of Lin⁻kit⁺ BM cells results in more than half of the infarcted area being colonized by donor cells within 9 d. Donor cells acquired phenotypic characteristics of cardiomyocytes and contributed to improved cardiac function and improved survival of the animals. In a subsequent study, this same group of investigators demonstrated that cardiac infarct size was significantly reduced and cardiac function significantly improved in animals in which stem cell factor (SCF) and granuolocyte-colony stimulating factor (G-CSF) were used to mobilize HSCs and progenitors from the BM. This was associated with increased proliferation of cardiac myoblasts, smooth muscle cells, and endothelial cells in the infarcted tissue, leading the authors to conclude that progenitor cells for these three tissues were mobilized from the BM. Although these studies suggest that cardiac defects as a result of ischemia may improve by local infusion of or mobilization of progenitors from the BM, the etiology of the cell(s) responsible for this effect is not clear. Despite the fact that these cells can be found in the HSC-rich Lin⁻kit⁺ fraction of BM, proof that HSC transdifferentiate into myoblasts is still lacking.

The contribution of adult stem cells to cardiac regeneration in humans has been recently suggested in a cardiac transplant model *(51)*. Male recipients were transplanted with a heart from a female donor. After transplant, recipient-derived cardiomyocytes, arterioles, and capillaries were detected in the transplanted heart as determined by the presence of the Y chromosome in up to 7–10% of the cardiomyocytes *(51)*. The origin of the engrafted recipient-derived cells was not determined. This study has not been reproduced, and, in fact, other investigators have reported that the level of recipient engraftment in the same model is only 0.016% *(52)*.

Bone marrow cells may also differentiate into hepatocytes and contribute, in vivo, to the regeneration of the liver *(39,53–55)*. Based on the fact that oval cells express antigens traditionally associated with HSCs (c-kit, flt-3, Thy-1, and CD34), it was suggested that perhaps BM HSCs could give rise to oval cells and hepatocytes. The study by Petersen et al. *(56)* was the first to indicate that oval cells and mature hepatocytes can be derived from circulating BM cells in a rat model. Identification of hepatic cells of donor origin in female rats transplanted with BM cells from male rats was based on the presence of Y chromosome and coexpression of hepatic markers by the same cell. An independent approach using a whole-liver transplant model in which Lewis rats expressing major histocompatibility complex (MHC) class II antigen L21-6 received a whole liver from Brown Norway rats that were negative for L21-6 showed similar results, suggesting that biliary duct engraftment was derived from circulating BM cells *(56)*. Following a similar approach based on gender-mismatched, fluorescence-activated (FACS)-sorted CD34⁺Lin⁻ BM cells or unselected BM cells from male mice were transplanted in female recipients. Even in a liver without obvious damaged, a 1–2% engraftment (hepatocytes from donor origin) was detected. In fact, 200 CD34⁺Lin⁻ cells provided the same degree of engraftment as 20,000 unfractionated BM cells. Because these studies were done with unfractionated BM mononuclear cells, the authors could not conclude that oval cells and hepatocytes were derived from HSCs. Indeed, BM may contain cells with characteristics of oval cells, the progenitor for hepatic and biliary epithelial cells *(57)*. Thus, contribution of BM-derived cells to liver regeneration could be the result of the infusion of hepatic oval cells present in the BM sample.

The remarkable capacity of BM cells to regenerate the liver has recently been reported in a model of severe liver damage *(39)*. Mice bearing the mutation for the fumaryl acetoacetic hydrolase (FAH$^{-/-}$ mice) develop liver failure unless they are maintained on 2-(2-nitro-4-fluoromethylbenzoyl)-1,3-cyclohexanedione (NTBC), which blocks the pathway upstream by inhibiting 4-hydroxyphenylpyruvate dioxygenase. When highly purified KTLS (c-kit^{high}Thy^{low}Lin-Sca-1$^+$) HSCs from wild-type LacZ$^+$ male mice were transplanted into lethally irradiated FAH$^{-/-}$ animals, foci of donor hepatocytes expressing LacZ and FAH could be detected in the FAH$^{-/-}$ recipients upon removal of NTBC, and donor BM cells restored liver function, keeping the recipient animals alive *(39)*. However, despite the fact that as few as 50 cells could reconstitute the liver, the study did not prove that a single clonogenic HSC can, in this circumstance, give rise to both blood and liver.

The potential of human BM cells to differentiate into hepatocytes in humans has been suggested by two recent studies *(53,54)*. In both studies, a mismatched gender approach was employed: Female patients who had previously received a BM graft from a male donor were examined for cells of donor origin using a DNA probe specific for the Y chromosome by *in situ* hybridization. In a second group of patients, Y-chromosome-positive cells were sought in female livers grafted into male patients but which were later removed for recurrent disease. In both sets of patients, Y-chromosome-positive hepatocytes were readily identified. The number of Y-chromosome-positive hepatocytes into human liver was highly variable and appeared to correlate with the severity of liver damage, with up to 40% of hepatocytes and cholangiocytes of BM origin in patients who received a BM graft, and of host origin in patients with severe cirrhosis who received a liver graft *(54)*. A second recent study in patients undergoing allogeneic G-CSF mobilized PB transplantation using male donor cells into female recipients has shown that between 4% and 7% of the hepatocytes in the recipient are donor origin derived *(31)*.

The potential of BM HSCs to acquire characteristics of adult neurons and glial cells in vitro and also in vivo in mouse models has recently been described *(19,58–61)*. Mezey et al. *(19)* studied the in vivo potential of BM-derived cells to differentiate into neurons in PU-1⁻ homozygous mice, which lack the PU.1 ETS transcription factor required for normal hematopoiesis. At birth, PU-1 knockout mice lacked macrophages, neutrophils mastocytes, T-cells, B-cells, and osteoclasts and required a BM transplant to survive. Female PU-1$^{-/-}$ mice were transplanted with male wild-type mice. In addition to normal hematopoietic reconstitution, the authors observed that up to 4.3% of cells in the central nervous system (CNS) expressed the Y chromosome and 2.3% coexpressed the neuronal marker NeuN *(19)*. When BM cells were grown in culture for several weeks, the authors also noted expression of the neural progenitor antigen, nestin, suggesting that BM cells can differentiate into neural progenitors. Lack of expression of neural markers in BM cells before transplant and without in vitro culture indicates that BM cells do not contain neural progenitors that could be responsible for CNS engraftment. Similarly, Eglitis and Mezey detected significant numbers of microglia (F4/80 positive) and astrocytes (GFAP positive) of BM origin in the brains of recipient female mice 6 wk after transplantation of male BM *(58)*. As we have previously mentioned, the fact that unselected populations of BM cells were used along with the small fraction of "transdifferentiated" cells, identified solely based on the Y chromosome but not on a functional basis does not provide solid ground for neural differentiation of BM HSCs.

That BM-derived stem cells can contribute to other tissues such as lung *(62)*, kidney *(30)*, gastrointestinal epithelium *(63)*, or skin *(20)* has been suggested by several studies, but the evidence is far from being compelling.

In most instances, it is based on the analysis of the presence of the Y chromosome in the context of gender-mismatched transplant, and the level of engraftment is in many cases below 1%.

2.2. Mesenchymal Stem Cells

Bone marrow also contains stem cells that have been termed mesenchymal stem cells (MSCs) (also marrow stromal cells) *(64)*. In 1976, Friedenstein et al. described that fibroblastic colonies could be obtained from BM cells selected by their adherence to plastic in the presence of fetal calf serum *(65,66)*. Further work indicated that these cells could differentiate into bone, cartilage, and adipocytes in vitro and were transplantable *(67–69)*. Plastic adherent cells are a very heterogeneous population and about 30% of these cells can be considered MSCs *(64)*. More recent studies by Caplan and colleagues have identified a set of surface markers that are expressed by MSCs, including SH2, SH3, CD29, CD44, CD71, CD90, and CD106. A number of hematopoietic markers are not present on MSCs (CD34, CD45, or CD14) *(70,71)*. In vitro, human and mouse MSCs can differentiate into functional mesodermal-derived tissues. MSCs differentiate into osteoblasts, chondroblasts, adipocytes, and skeletal myoblasts *(70,72)*. Ring cloning *(70)* as well as single-cell sorting strategies *(73)* have been used to demonstrate that single MSCs give rise to all of these different lineages. A recent study also suggested that MSCs differentiate in cardiac myoblasts: Injection of MSC into an infarcted pig heart resulted in differentiation to cells staining positive for certain cardiac muscle markers and improvement of cardiac function *(49,74)*.

When human MSCs were transplanted by intraperitoneal injection into preimmune and immune fetal sheep, contribution to different tissues was detected by polymerase chain reaction (PCR) for human-specific β_2-microglobulin *(75)* and by immunohistochemical staining for chondrocytes, adipocytes, myocytes, cardiomyocytes, and stromal marrow cells, and thymic stroma combined with *in situ* hybridization for human specific DNA. The potential of MSCs to differentiate into mesenchymal lineages has been exploited clinically in patients with osteogenesis imperfecta undergoing allogeneic BM transplant. Similar to what was shown for murine BM cells *(76)*, transplantation of whole BM was associated with osteoblast engraftment between 1% and 2% and improvement in symptoms associated with their underlying disease *(77)*.

Several studies have suggested that MSCs might also differentiate into neuroectoderm-derived tissues *(78–80)*. In vitro, both rat and human MSCs can be induced to express neuronal markers such as enolase, NeuN, neurofilament, and tau *(79)*. However, these studies did not demonstrate that cells with neuroectodermal staining profile also acquired functional characteristics of neurons or glial cells. When MSCs were injected into the ventricle of neonatal mice, cells migrate throughout the forebrain and cerebellum without disrupting the brain architecture and might acquire markers of astrocytes and possibly neurons *(78,79)*, although again no functional data are available. In one study, MSCs were transplanted in a mouse model of Parkinson's disease. Survival of cells and acquisition of tyrosine hydroxylase expression was found. Furthermore, animals transplanted with MSCs intrastriatally had improved functional recovery compared with saline injected animals *(80)*.

Despite their mesodermal multipotentiality and possible neuroectodermal differentiation ability, MSCs do not differentiate into endoderm-derived tissues and, thus, cannot be considered pluripotent. MSCs may prove to be a very useful resource for clinical applications in a number of diseases, both for regenerative therapy and as a gene therapy vehicle *(71)*.

2.3. Side Population Cells

A small population of cells can be isolated from BM (and other organs of several species) using FACS on the basis of exclusion of the fluorescent dye Hoechst 33342 *(81,82)*. BM small population (Sp) cells are known to contain HSCs in man, mouse, and other species. In addition, several studies have suggested that Sp cells may also contain cells that differentiate into other cell types and integrate in multiple tissues in vivo. For instance, Gussoni et al. showed that BM Sp cells contribute to dystrophin-positive skeletal muscle in mdx mice *(83)*, even though the levels of Sp-derived engraftment were only 1–4%. Jackson et al. showed that BM Sp cells can also differentiate into cells with cardiac muscle characteristics and endothelium in a mouse model of myocardial infarction *(81)*.

2.4. Neural Stem Cells

One of the most impressive initial observations of "transdifferentiation" was the study from Bjorson et al. *(26)*, in which they demonstrated that adult neurospheres from ROSA26 mice maintained in culture for 1–40 passages could give rise to hematopoietic colonies in vitro and when transplanted into irradiated Balb/c, they can contribute to hematopoiesis. These studies have not been reproduced by other groups despite extensive attempts *(27)*. Using fetal mouse neural stem cells, Morshead et al. have failed to demonstrate a single event of hematopoietic reconstitution. However, they have observed that neurospheres grown in culture for prolonged periods of time exhibit altered growth characteristics that may perhaps underlie the abnormal differentiation seen by Bjornson et al., but also raises questions about the biological integrity of the long-term cultured NSCs.

Rietze et al. suggested that most NSC capacity resides in the PNA^{lo} HSA^{lo} NSCs *(84)*. About 80% of these cells would be NSCs and, in fact, 68% of the NSC capacity from unsorted cells in the periventricular region would proceed from this population. PNA^{lo} HSA^{lo} NSCs can acquire characteristics consistent with muscle cells when cultured with the C2C12 cell line. The possibility of fusion between the C2C12 and the PNA^{lo} HSA^{lo} NSCs is obvious.

2.5. Muscle Stem Cells

Studies from Jackson et al. *(85)* published 3 yr ago suggested that muscle-derived stem cells had hematopoietic potential. When these authors, as well as other groups, re-examined their results, it was clear that the hematopoietic potential within the muscle was the result of hematopoietic stem cells and not because of presence of derived pluripotent stem cells or transdifferentiation of muscle stem cells *(86,87)*. Recent studies from other groups have also suggested that muscle stem cells may be more potent that expected *(88–90)*. However, the same limitation as in other plasticity studies can be applied; namely plasticity was not assessed at the clonal level, as the investigators used heterogenous cell populations.

In a very interesting report, Qu-Petersen et al. isolated different populations of muscle-derived stem cells from mice based on the adhesion and proliferation characteristics of the cells *(89)*. One of these populations named muscle-derived stem cells (MDSCs) was characterized by expression of muscle markers (MyoD) and hematopoietic markers (Sca-1 and CD34) as well as lack of expression of MHC class I. MDSCs could be maintained for more than 60 population doublings without evidence of genomic abnormalities, and when transfected with *LacZ* and transplanted into the muscle of mdx mice, they contributed to generation of muscle progenitors from MDSCs, indicating their self-renewal capacity. MDSCs could differentiate in vitro but also in vivo to endothelium, muscle, and neural lineages *(89)*. Because of their potential and phenotypic characteristics, MDSCs may be related to multipotent adult progenitor cells (MAPCs). In fact, studies from our group indicate that MAPCs may be isolated not only from BM but also from brain and muscle *(23)*. Once more, the lack of clonality analysis and the need for in vitro culture for isolation of MDSCs establish limitations to these results.

2.6. Skin Progenitor Cells

Stem cells with the ability to differentiate into cells of two germ layers have been isolated from the dermis of mouse and man *(91)*. These cells could be maintained for more that 12 mo in culture without differentiation and can be instructed to undergo differentiation into neuroectodermal (neurons and glial cells) or mesodermal lineages (adipocytes and smooth muscle) in vitro. Their potential to differentiate into ectoderm- and mesoderm-derived tissues was demonstrated at the clonal level. However, there is no evidence of in vivo multipotentiality of skin progenitor (SKP) cells or that the differentiated tissues are functional. Furthermore, the percentage of cells that acquire morphological and phenotypic characteristics of neural or mesodermal tissues is less that 10%.

3. PLURIPOTENT ADULT STEM CELLS

A number of recent reports have provided some evidence that pluripotent stem cells can be obtained from adult tissues—that is, cells derived from one germ cell layer capable of giving rise to tissues derived from all three germ layers *(20–22)*. Because of the biological and clinical relevance of this concept, we will discuss some of the strengths and caveats of these studies.

3.1. Pluripotent NSCs

Clarke et al. *(22)* injected neurospheres from ROSA26 mice into chick embryos and demonstrated both by expression of LacZ and by the specific mouse epitope H2-K[b] that almost 25% of the surviving embryos were chimeric not only in neural tissues but also in mesoderm and endoderm tissues. When cells obtained from disassociated neurospheres or whole neurospheres from ROSA26 mice were injected into mouse blastocysts, contribution of injected cells to embryos was detected in between 1% and 12% of the animals. The contribution of ROSA26 derived neural stem cells was seen in several embryonic organs, including CNS, heart, liver, intestine, and other tissues. However, no animals were evaluated after birth, making evaluation of the function of donor-derived tissue impossible.

3.2. Pluripotent Hematopoietic Cells

Krause et al. *(20)* transplanted female mice with male BM HSC labeled with PKH26. PKH26-labeled transplanted cells were harvested from the BM 48 h after infusion and single PKH26-positive cells were again transplanted in a second recipient. Thus, pluripotent stem cells, that were CD34+Sca1– were enriched based on their ability to home in vivo. The self-renewal capacity of these cells is supported by long-term repopulation in secondary female recipients with male hematopoietic cells. A single male hematopoietic stem cell transplanted in a female mouse also gave rise to donor multilineage engraftment, including epithelial cells of the liver, lung, gastrointestinal tract, and skin. Engraftment was defined based on the coexpression of the Y chromosome and the presence of epithelial markers (anticytokeratin monoclonal antibodies). Levels of engraftment varied from a maximum of 20% in the lung to lower than 0.2% in the large bowel. However, no functional evidence of engraftment was shown.

3.3. Multipotent Adult Progenitor Cells

We have recently described the isolation and expansion of a rare population of pluripotent cells from the BM named MAPCs that copurifies within MSC *(21,92–94)* *(see* Fig. 2).

Multipotent adult progenitor cells (MAPCs) can proliferate for over 100 population doublings without senescence. Between population doubling 30 and 120, MAPCs become a very uniform population of cells characterized by expression of telomerase activity, which results in the maintenance of telomere length despite continuous proliferation. Mouse *(21)* and human *(92)* MAPC phenotypes are similar but different from that of murine hematopoietic stem cells with transdifferentiation potential *(20,38,39,81)* or from MSCs (71). MAPCs do not express CD34, CD44, MHC I or II, CD45, or c-kit. Mouse MAPCs do express low levels of Flk-1, Sca-1, and Thy-1 and higher levels of CD13 and SSEA-1 (mouse) and SSEZ-4 (human) *(95)*. MAPCs express transcription factors characteristic of embryonic stem cells such as oct-4 and rex-1. However, the level of expression is about 1000-fold less that in embryonic stem (ES) cells *(21)*.

Multipotent adult progenitor cells have been isolated in humans, mice, rats, and dogs from adult bone marrow and we also have evidence that other organs such as muscle and brain may contain MAPCs *(23)*. A population with similar phenotypic characteristics and potential has been identified in mouse muscle (muscle derived stem cells [MDSCs]) *(89)*.

Multipotent adult progenitor cells differentiate functionally into tissues derived from all three germ layers, both in vivo and in vitro. In vitro, mMAPCs and hMAPCs can be induced to differentiate into limb-bud and visceral mesoderm-derived tissues (bone, cartilage, adipocytes, skeletal muscle, hematopoietic stroma, and endothelial cells) *(92)*. It has not been

---

Fig. 2. *(opposite page)* Single MAPCs differentiate into tissues derived from all three germ layers, both in vivo and vitro. *From left to right:* MAPC cultured in vitro differentiate into phenotypic and functionally defined cells of the mesoderm (bone, skeletal muscle, cartilage, or fat), ectoderm (glia or neurons), or endoderm (hepatocytelike cells). MAPC injected into nonobese diabetic severe combined immunodeficient (NOD-SCID) mice contribute to tissues in which cell turnover is commonly seen; specifically, progeny can be detected in the hematopoietic system, liver, intestine, and lung epithelium. A single mouse MAPC injected into a blastocyst can give rise to up to 45% chimerism.

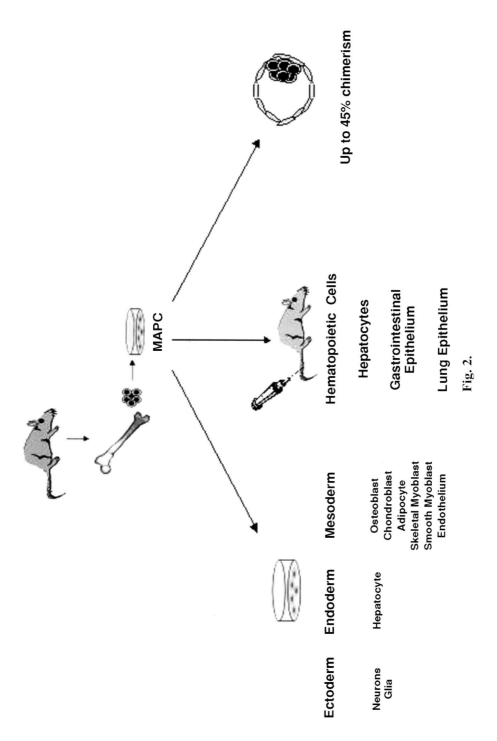

Ectoderm   Endoderm   Mesoderm   Hematopoietic Cells

Neurons   Hepatocyte Osteoblast Hepatocytes
Glia   Chondroblast
                       Adipocyte   Gastrointestinal
                       Skeletal Myoblast Epithelium
                       Smooth Myoblast
                       Endothelium Lung Epithelium

MAPC

Up to 45% chimerism

Fig. 2.

101

possible to induce differentiation of MAPCs into mature hematopoietic cells in vitro or cardiac myoblasts. MAPCs can also differentiate into hepatocytelike cells *(94)* and function as mature hepatocytes, as suggested by the fact that MAPC-derived hepatocytelike cells produce urea and albumin, have phenobarbital-inducible cytochrome p450, can uptake low-density lipoprotein, and store glycogen *(94)*. Differentiation of MAPCs into ectoderm-derived tissues, namely neurons (dopaminergic, gabaergic, serotoninergic), astrocytes, and oligodendrocytes, in vitro has also been demonstrated *(21)*. MAPCs cultured in fibronectin-coated wells without PDGF-BB and EGF but with bFGF acquired morphologic and phenotypic characteristics of astrocytes (glial acidic fibrillary protein [GFAP]), oligodendrocytes (galactocerebroside [GalC]), and neurons (NF-200). We have never found these markers within the same cell. When MAPCs were cultured sequentially with bFGF, FGF-8, and brain-derived neurotrophic factor (BDNF), a more mature phenotype is seen, with 20–30% of cells expressing markers of dopamine-containing neurons (dopa-decarboxylase [DDC]) and tyrosine hydroxylase (TH positive), 20% of serotonin-containing (serotonin positive) neurons, and 50% of γ-aminobutyric acid (GABA)-containing (GABA positive) neurons. Polarization of MAPC-derived neurons can also be demonstrated *(21)*, indicating that MAPC-derived neurons may develop normal function. Using retroviral marking, we demonstrated that a single MAPC can give rise in vitro to tissues derived from all three germ layers *(21,92)*.

We have injected ROSA26-derived MAPCs into NOD-SCID and examined the level of engraftment by either immunochemistry or quantitative PCR for β-gal. Engraftment was hematopoietic tissues (blood, BM, and spleen) and epithelium of the lung, liver, and intestine of all recipient animals and was similar in animals analyzed 4–24 wk after transplantation, and levels ranged between 1% and 9%. No contribution was seen to skeletal or cardiac muscle, tissues in which, in contrast to epithelium, little or no cell turnover is seen in the absence of tissue injury. No significant engraftment of mMAPCs was seen in the brain.

The most conclusive proof of the pluripotency of MAPCs has been obtained by the injection of 1–12 ROSA26-derived mMAPCs into a blastocyst of a C57BL/6 mouse. Chimerism was detected in 33% of the mice derived from blastocysts in which 1 mMAPC had been injected and in 80% of mice in which 10–12 mMAPCs had been injected. The level of chimerism ranged between 0.1% and 45%. Contribution of the ROSA26-derived MAPCs in chimeric animals to many, it not all, tissues, included brain, retina, lung, myocardium, skeletal muscle, liver, intestine, kidney, spleen, bone marrow, blood, and skin. β-Gal cells expressed markers typical for the tissue in which they had incorporated *(21)*.

Finally, MAPCs contribute to in vivo neoangiogenesis *(93)*. When human MAPC-derived endothelial cells were infused into NOD-SCID mice 3–5 d after subcutaneous implantation of Lewis lung carcinoma, between 12% and 45% of the tumor vessel and endothelial cells within the tumor stained positive for human-derived cells, indicating that MAPCs contribute in vivo to neoangiogenesis and differentiate into fully functional endothelium *(93)*.

Despite all our efforts, we cannot yet prospectively isolate MAPC, and selection depends on long-term culture in very specific conditions. Fusion events, although very unlikely, have not been definitively ruled out in our in vivo models.

4. PLASTICITY, CELL FUSION, AND ADULT STEM CELLS

Despite the large amount of evidence suggesting the existence of adult pluripotent stem cells or even the potential of tissue-specific stem cells to "transdifferentiate" (stem cell plasticity), serious concerns have been raised by two very recent reports that suggest that most, if not all, studies in which plasticity of adult stem cells had been suggested may be explained by fusion between adult stem cells and resident tissue specific cells *(24,25)*. In the first study, the authors isolated NSCs from mice and marked the cells with GFP and a puromycin-resistant gene under the control of the Oct-4 promoter (a gene expressed by ES cells) and cocultured them with ES cells that had been transduced with a gene that confers resistance to the antibiotic hygromycin. After 2–4 wk in culture, they isolated several colonies that expressed GFP and were resistant to puromycin (these colonies had characteristics of pluripotent stem cells), suggesting transdifferentiation. However, the cells were also resistant to hygromycin. The only explanation was that ES cells and NSCs had fused. The fusion event occurred at a very low frequency (1:100,000) and resulted in tetraploid cells that expressed markers of ES cells and could generate unbalanced chimerism when injected into blastocysts *(24)*. The second study took a similar approach and cocultured ES cells with BM cells obtained from transgenic mice (GFP and puromycin-resistant gene). In this case, GFP colonies of embryonic-stem-like cells could be isolated from cultures in which puromycin had been added to eliminate ES cells. These cells formed teratomas when injected into NOD-SCID mice and were shown to be tetraploid and the consequence of fusion between ES cells and BM cells. Again, fusion was detected at a frequency of 2–11 clones per 10^6 BM cells *(25)*.

Because proof of "transdifferentiation" or pluripotency has relied solely on the presence of the Y chromosome in a female transplanted with male cells or the presence of markers such as GFP or LacZ , the description of cell fusion has lead to doubts regarding the plasticity of adult stem cells.

Although none of the studies, including ours, has addressed directly the possibility of in vivo or in vitro fusion as an explanation for transdifferentiation events, several observations strongly suggested that this may not always be the case: (1) Fusion has been observed in vitro, but there is no evidence that it happens in vivo to any significant degree. If fusion events were common and would yield cells with no proliferative disadvantage, a number of fusion events should be detectable in females that were pregnant, which is not the case, except for the liver, as a result of transfer of some HSCs present in fetal blood that have access to the circulation of the mother. Further, the reports of chimerism based on a gender-mismatched transplant should show a number of cells with more than one X and Y chromosome and, again, that has not been the case *(31,51)*. (2) The low frequency of fusion events is not consistent with the degree and homogeneity with which chimerism is observed after transfer of MAPCs into blastocysts. (3) The speed and robustness with which engraftment is seen in animals without selectable pressure *(21,93)* or even after organ damage *(39,96)* also argues against the idea that adult stem cell engraftment and differentiation in postnatal animals is caused by fusion. (4) In vitro studies in which MAPCs cultured in the absence of tissue-specific cells or ES cells differentiate into cells of three germ layers show that in vitro behavior cannot be attributed to cell fusion.

The mechanism underlying pluripotency or transdifferentiation is not known (*see* Fig. 3). It is possible that pluripotent stem cells persist after birth in the BM or even in multiple organs. These embryonic remnants may, under certain in vitro culture conditions, acquire proliferative potential, and when stimulated either in vitro or in response to local cues in vivo, they differentiate into the tissues in which they engraft. It is not unreasonable to think that only under very special circumstances in vivo they would undergo proliferation-differentiation.

A second possibility is that a tissue-specific stem cell may undergo genetic reprogramming in culture similar to that which occurs in the "cloning process" induced by the singular culture conditions. These phenomena would account for the concept of "transdifferentiation" or, in other words, organ-specific stem cells could overcome their intrinsic restrictions when exposed to a novel microenvironment *(97,98)*. Odelberg et al. have recently demonstrated that terminally differentiated muscle myotubes can be induced to "dedifferentiate" and "redifferentiate" into cells that express markers for chondrogenic, osteogenic, or adipogenic tissues by overexpression of the msx-1 transcription factor, supporting the idea that reprogramming may be induced in differentiated cells and may explain some of the transdifferentiation observations *(99)*.

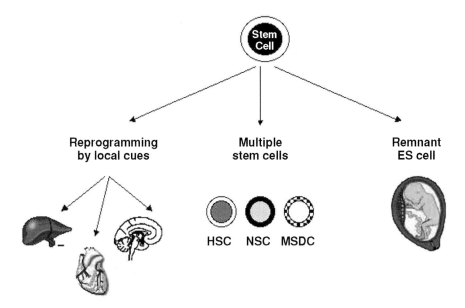

Fig. 3. Possible explanations for presumed adult stem cell plasticity. *From left to right:* MAPC may be stem cells that under specific circumstance (culture, in vivo cues) may be reprogrammed (dedifferentiate and redifferentiate). Stem cells harvested from different tissues are heterogeneous (i.e., stem cells derived from different germ layers coexist in different organs/tissues). Cells with pluripotent characteristics might persist even after the initial steps of embryological development.

Finally, it is possible that some of the plasticity can be attributed to the presence of multiple stem cells in BM, or other organs. Then, cells responsible for giving rise to different tissues are, in fact, multiple stem cells with different potentials (muscle stem cells, neural stem cells, liver stem cells). Although still of potential clinical usefulness, this mechanism would no longer qualify as stem cell plasticity.

5. SUMMARY

A large number of studies has been published suggesting that pluripotent stem cells may be obtained from adult tissues and/or that certain cells may have the capacity to transdifferentiate. The mechanism by which these cells exist and are able to differentiate into specific tissues are unclear, but very likely, they are related to local cues present in the different organs. The field of stem cell plasticity is a new field with remarkable biological, clinical, and ethical implications. However, the evidence that pluripotency and trans-differentiation exists is not always fully proven and, therefore, not yet definitive and conclusive. Some of the initial reports when examined more carefully have been reinterpreted *(85,86)*, whereas some studies have not been reproduced *(26,27)*, leading to skepticism in the scientific community. Standards will be required to define stem cell plasticity and likely should include demonstration at the clonal level of multilineage and functional differentiation in vitro and robust, sustained and functional multilineage engraftment in vivo, along with exclusion of fusion events. These standards are accepted by most researchers and scientists working in this field *(6–8,100)*. Notwithstanding all of these facts, this is an exciting time to be involved in stem cell research with great challenges and also great expectations for treatment of patients with incurable diseases.

REFERENCES

1. Weissman, I. L. (2002) Stem cells—scientific, medical, and political issues, *N. Engl. J. Med.* **346(20),** 1576–1579.
2. Vogel, G. (2001) Stem cell policy. An embryonic alternative, *Science* **292(5523),** 1822.
3. Vogel, G. (2001) Stem cell policy. Can adult stem cells suffice?, *Science* **292(5523),** 1820–1822.
4. Blau, H. M., Brazelton, T. R., and Weimann, J. M. (2001) The evolving concept of a stem cell: entity or function?, *Cell* **105(7),** 829–841.
5. Weissman, I. L. (2000) Stem cells: units of development, units of regeneration, and units in evolution, *Cell* **100(1),** 157–168.
6. Weissman, I. L., Anderson, D. J., and Gage, F. (2001) Stem and progenitor cells: origins, phenotypes, lineage commitments, and transdifferentiations, *Annu. Rev. Cell Dev. Biol.* **17,** 387–403.
7. Anderson, D. J., Gage, F. H., and Weissman, I. L. (2001) Can stem cells cross lineage boundaries?, *Nat. Med.* **7(4),** 393–395.
8. Poulsom, R., Alison, M. R., Forbes, S. J., and Wright, N. A. (2002) Adult stem cell plasticity, *J. Pathol.* **197(4),** 441–456.
9. Vogel, G. (2000) Can old cells learn new tricks?, *Science* **287(5457),** 1418–1419.
10. Bonnet, D. (2002) Haematopoietic stem cells, *J. Pathol.* **197(4),** 430–440.
11. Gage, F. H. (2000) Mammalian neural stem cells, *Science* **287(5457),** 1433–1438.
12. Watt, F. M. (1988) Epidermal stem cells in culture, *J. Cell Sci.* **10(Suppl.),** 85–94.
13. Marshman, E., Booth, C., and Potten, C. S. (2002) The intestinal epithelial stem cell, *Bioessays* **24(1),** 91–98.
14. Seale, P. and Rudnicki, M. A. (2000) A new look at the origin, function, and "stem-cell" status of muscle satellite cells, *Dev. Biol.* **218(2),** 115–124.
15. Hughes, S. (2002) Cardiac stem cells, *J. Pathol.* **197(4),** 468–478.
16. Forbes, S., Vig, P., Poulsom, R., Thomas, H., and Alison, M. (2002) Hepatic stem cells, *J. Pathol.* **197(4),** 510–518.

17. Bonner-Weir, S. and Sharma, A. (2002) Pancreatic stem cells, *J. Pathol.* **197(4)**, 519–526.
18. Otto, W. R. (2002) Lung epithelial stem cells, *J. Pathol.* **197(4)**, 527–535.
19. Mezey, E., Chandross, K. J., Harta, G., Maki, R. A., and McKercher, S. R. (2000) Turning blood into brain: cells bearing neuronal antigens generated in vivo from bone marrow, *Science* **290(5497)**, 1779–1782.
20. Krause, D. S., Theise, N. D., Collector, M. I., et al. (2001) Multi-organ, multi-lineage engraftment by a single bone marrow-derived stem cell, *Cell* **105(3)**, 369–377.
21. Jiang, Y., Jahagirdar, B. N., Reinhardt, R. L., et al. (2002) Pluripotency of mesenchymal stem cells derived from adult marrow, *Nature* **418**, 41–49.
22. Clarke, D. L., Johansson, C. B., Wilbertz, J., et al. (2000) Generalized potential of adult neural stem cells, *Science* **288(5471)**, 1660–1663.
23. Jiang, Y., Vaessen, B., Lenvik, T., et al. (2002) Multipotent progenitor cells can be isolated from postnatal murine bone marrow, muscle, and brain, *Exp. Hematol.* **30**, 896–904.
24. Ying, Q. L., Nichols, J., Evans, E. P., and Smith, A. G. (2002) Changing potency by spontaneous fusion, *Nature* **416(6880)**, 545–548.
25. Terada, N., Hamazaki, T., Oka, M., et al. (2002) Bone marrow cells adopt the phenotype of other cells by spontaneous cell fusion, *Nature* **416(6880)**, 542–545.
26. Bjornson, C. R., Rietze, R. L., Reynolds, B. A., et al. (1999) Turning brain into blood: a hematopoietic fate adopted by adult neural stem cells in vivo, *Science* **283(5401)**, 534–537.
27. Morshead, C. M., Benveniste, P., Iscove, N. N., and van der Kooy, D. (2002) Hematopoietic competence is a rare property of neural stem cells that may depend on genetic and epigenetic alterations, *Nat. Med.* **8(3)**, 268–273.
28. Armitage, J. O. (1994) Medical progress: bone marrow transplantation, *N. Engl. J. Med.* **330(12)**, 827–838.
29. Tabbara, I. A., Zimmerman, K., Morgan, C., and Nahleh, Z. (2002) Allogeneic hematopoietic stem cell transplantation: complications and results, *Arch. Intern. Med.* **162(14)**, 1558–1566.
30. Poulsom, R., Forbes, S. J., Hodivala-Dilke, K., et al. (2001) Bone marrow contributes to renal parenchymal turnover and regeneration, *J. Pathol.* **195(2)**, 229–235.
31. Korbling, M., Katz, R. L., Khanna, A., et al. (2002) Hepatocytes and epithelial cells of donor origin in recipients of peripheral-blood stem cells, *N. Engl. J. Med.* **346(10)**, 738–746.
32. Lapidot, T., Fajerman, Y., and Kollet, O. (1997) Immune-deficient SCID and NOD/SCID mice models as functional assays for studying normal and malignant human hematopoiesis, *J. Mol. Med.* **75(9)**, 664–673.
33. Kobari, L., Giarratana, M. C., Pflumio, F., Izac, B., Coulombel, L., and Douay, L. (2001) CD133+ cell selection is an alternative to CD34+ cell selection for ex vivo expansion of hematopoietic stem cells, *J. Hematother. Stem Cell Res.* **10(2)**, 273–281.
34. Punzel, M., Wissink, S. D., Miller, J. S., Moore, K. A., Lemischka, I. R., and Verfaillie, C. M. (1999) The myeloid–lymphoid initiating cell (ML-IC) assay

assesses the fate of multipotent human progenitors in vitro, *Blood* **93(11),** 3750–3756.

35. Prosper, F., Stroncek, D., and Verfaillie, C. M. (1996) Phenotypic and functional characterization of long-term culture initiating cells (LTC-IC) present in peripheral blood progenitor collections of normal donors treated with G-CSF, *Blood* **88,** 2033–2042.

36. Morrison, S. J., Uchida, N., and Weissman, I. L. (1995) The biology of hematopoietic stem cells, *Annu. Rev. Cell Dev. Biol.* **11,** 35–71.

37. Grant, M. B., May, W. S., Caballero, S., et al. (2002) Adult hematopoietic stem cells provide functional hemangioblast activity during retinal neovascularization, *Nat. Med.* **8(6),** 607–612.

38. Orlic, D., Kajstura, J., Chimenti, S., et al. (2001) Bone marrow cells regenerate infarcted myocardium, *Nature* **410(6829),** 701–705.

39. Lagasse, E., Connors, H., Al-Dhalimy, M., et al. (2000) Purified hematopoietic stem cells can differentiate into hepatocytes in vivo, *Nat. Med.* **6(11),** 1229–1234.

40. Flamme, I. and Risau, W. (1992) Induction of vasculogenesis and hematopoiesis in vitro, *Development* **116(2),** 435–439.

41. Asahara, T., Masuda, H., Takahashi, T., et al. (1999) Bone marrow origin of endothelial progenitor cells responsible for postnatal vasculogenesis in physiological and pathological neovascularization, *Circ. Res.* **85(3),** 221–228.

42. Kocher, A. A., Schuster, M. D., Szabolcs, M. J., et al. (2001) Neovascularization of ischemic myocardium by human bone-marrow-derived angioblasts prevents cardiomyocyte apoptosis, reduces remodeling and improves cardiac function, *Nat. Med.* **7(4),** 430–436.

43. Shi, Q., Rafii, S., Wu, M. H., et al. (1998) Evidence for circulating bone marrow-derived endothelial cells, *Blood* **92(2),** 362–367.

44. Peichev, M., Naiyer, A. J., Pereira, D., et al. (2000) Expression of VEGFR-2 and AC133 by circulating human CD34(+) cells identifies a population of functional endothelial precursors, *Blood* **95(3),** 952–958.

45. Lyden, D., Hattori, K., Dias, S., et al. (2001) Impaired recruitment of bone-marrow-derived endothelial and hematopoietic precursor cells blocks tumor angiogenesis and growth, *Nat. Med.* **7(11),** 1194–1201.

46. Davidoff, A. M., Ng, C. Y., Brown, P., et al. (2001) Bone marrow-derived cells contribute to tumor neovasculature and, when modified to express an angiogenesis inhibitor, can restrict tumor growth in mice, *Clin. Cancer Res.* **7(9),** 2870–2879.

47. Asahara, T., Murohara, T., Sullivan, A., et al. (1997) Isolation of putative progenitor endothelial cells for angiogenesis, *Science* **275(5302),** 964–967.

48. Takahashi, T., Kalka, C., Masuda, H., et al. (1999) Ischemia- and cytokine-induced mobilization of bone marrow-derived endothelial progenitor cells for neovascularization, *Nat. Med.* **5(4),** 434–438.

49. Tomita, S., Li, R. K., Weisel, R. D., et al. (1999) Autologous transplantation of bone marrow cells improves damaged heart function, *Circulation* **100(19 Suppl.),** II247–II256.

50. Strauer, B. E., Brehm, M., Zeus, T., et al. (2001) Intracoronary, human autologous stem cell transplantation for myocardial regeneration following myocardial infarction, *Dtsch. Med. Wochenschr.* **126(34–35),** 932–938 (in German).

51. Quaini, F., Urbanek, K., Beltrami, A. P., et al. (2002) Chimerism of the transplanted heart, *N. Engl. J. Med.* **346(1),** 5–15.

52. Laflamme, M. A., Myerson, D., Saffitz, J. E., and Murry, C. E. (2002) Evidence for cardiomyocyte repopulation by extracardiac progenitors in transplanted human hearts, *Circ. Res.* **90(6),** 634–640.

53. Alison, M. R., Poulsom, R., Jeffery, R., et al. (2000) Hepatocytes from non-hepatic adult stem cells, *Nature* **406(6793),** 257.

54. Theise, N. D., Nimmakayalu, M., Gardner, R., et al. (2000) Liver from bone marrow in humans, *Hepatology* **32(1),** 11–16.

55. Theise, N. D., Badve, S., Saxena, R., et al. (2000) Derivation of hepatocytes from bone marrow cells in mice after radiation-induced myeloablation, *Hepatology* **31(1),** 235–240.

56. Petersen, B. E., Bowen, W. C., Patrene, K. D., et al. (1999) Bone marrow as a potential source of hepatic oval cells, *Science* **284(5417),** 1168–1170.

57. Mitaka, T. (2001) Hepatic stem cells: from bone marrow cells to hepatocytes, *Biochem. Biophys. Res. Commun.* **281(1),** 1–5.

58. Eglitis, M. A. and Mezey, E. (1997) Hematopoietic cells differentiate into both microglia and macroglia in the brains of adult mice, *Proc. Natl. Acad. Sci. USA* **94(8),** 4080–4085.

59. Wu, Y. P., McMahon, E., Kraine, M. R., et al. (2000) Distribution and characterization of GFP(+) donor hematogenous cells in Twitcher mice after bone marrow transplantation, *Am. J. Pathol.* **156(6),** 1849–1854.

60. Brazelton, T. R., Rossi, F. M., Keshet, G. I., and Blau, H. M. (2000) From marrow to brain: expression of neuronal phenotypes in adult mice, *Science* **290(5497),** 1775–1779.

61. Priller, J., Persons, D. A., Klett, F. F., et al. (2001) Neogenesis of cerebellar Purkinje neurons from gene-marked bone marrow cells in vivo, *J. Cell Biol.* **155(5),** 733–738.

62. Kotton, D. N., Ma, B. Y., Cardoso, W. V., et al. (2001) Bone marrow-derived cells as progenitors of lung alveolar epithelium, *Development* **128(24),** 5181–5188.

63. Brittan, M., Hunt, T., Jeffery, R., et al. (2002) Bone marrow derivation of pericryptal myofibroblasts in the mouse and human small intestine and colon, *Gut* **50(6),** 752–757.

64. Prockop, D. J. (1997) Marrow stromal cells as stem cells for nonhematopoietic tissues, *Science* **276(5309),** 71–74.

65. Friedenstein, A. J., Gorskaja, J. F., and Kulagina, N. N. (1976) Fibroblast precursors in normal and irradiated mouse hematopoietic organs, *Exp. Hematol.* **4(5),** 267–274.

66. Friedenstein, A. J., Deriglasova, U. F., Kulagina, N. N., et al. (1974) Precursors for fibroblasts in different populations of hematopoietic cells as detected by the in vitro colony assay method, *Exp. Hematol.* **2(2),** 83–92.

67. Owen, M. and Friedenstein, A. J. (1988) Stromal stem cells: marrow-derived osteogenic precursors, *Ciba Found. Symp.* **136,** 42–60.
68. Ashton, B. A., Allen, T. D., Howlett, C. R., Eaglesom, C. C., Hattori, A., and Owen, M. (1980) Formation of bone and cartilage by marrow stromal cells in diffusion chambers in vivo, *Clin. Orthoped.* **151,** 294–307.
69. Keating, A., Singer, J. W., Killen, P. D., et al. (1982) Donor origin of the in vitro haematopoietic microenvironment after marrow transplantation in man, *Nature* **298(5871),** 280–283.
70. Pittenger, M. F., Mackay, A. M., Beck, S. C., et al. (1999) Multilineage potential of adult human mesenchymal stem cells, *Science* **284(5411),** 143–147.
71. Deans, R. J. and Moseley, A. B. (2000) Mesenchymal stem cells: biology and potential clinical uses, *Exp. Hematol.* **28(8),** 875–884.
72. Makino, S., Fukuda, K., Miyoshi, S., et al. (1999) Cardiomyocytes can be generated from marrow stromal cells in vitro, *J. Clin. Invest.* **103(5),** 697–705.
73. Gronthos, S., Graves, S. E., Ohta, S., and Simmons, P. J. (1994) The STRO-1+ fraction of adult human bone marrow contains the osteogenic precursors, *Blood* **84(12),** 4164–4173.
74. Toma, C., Pittenger, M. F., Cahill, K. S., Byrne, B. J., and Kessler, P. D. (2002) Human mesenchymal stem cells differentiate to a cardiomyocyte phenotype in the adult murine heart, *Circulation* **105(1),** 93–98.
75. Liechty, K. W., MacKenzie, T. C., Shaaban, A. F., et al. (2000) Human mesenchymal stem cells engraft and demonstrate site-specific differentiation after in utero transplantation in sheep, *Nat. Med.* **6(11),** 1282–1286.
76. Nilsson, S. K., Dooner, M. S., Weier, H. U., et al. (1999) Cells capable of bone production engraft from whole bone marrow transplants in nonablated mice, *J. Exp. Med.* **189(4),** 729–734.
77. Horwitz, E. M., Prockop, D. J., Fitzpatrick, L. A., et al. (1999) Transplantability and therapeutic effects of bone marrow-derived mesenchymal cells in children with osteogenesis imperfecta, *Nat. Med.* **5(3),** 309–313.
78. Kopen, G. C., Prockop, D. J., and Phinney, D. G. (1999) Marrow stromal cells migrate throughout forebrain and cerebellum, and they differentiate into astrocytes after injection into neonatal mouse brains, *Proc. Natl. Acad. Sci. USA* **96(19),** 10,711–10,716.
79. Woodbury, D., Schwarz, E. J., Prockop, D. J., and Black, I. B. (2000) Adult rat and human bone marrow stromal cells differentiate into neurons, *J. Neurosci. Res.* **61(4),** 364–370.
80. Li, Y., Chen, J., Wang, L., Zhang, L., Lu, M., and Chopp, M. (2001) Intracerebral transplantation of bone marrow stromal cells in a 1-methyl-4-phenyl-1,2,3,6-tetrahydropyridine mouse model of Parkinson's disease, *Neurosci. Lett.* **316(2),** 67–70.
81. Jackson, K. A., Majka, S. M., Wang, H., et al. (2001) Regeneration of ischemic cardiac muscle and vascular endothelium by adult stem cells, *J. Clin. Invest.* **107(11),** 1395–1402.
82. Goodell, M. A., Rosenzweig, M., Kim, H., et al. (1997) Dye efflux studies suggest that hematopoietic stem cells expressing low or undetectable levels of CD34 antigen exist in multiple species, *Nat. Med.* **3(12),** 1337–1345.

83. Gussoni, E., Soneoka, Y., Strickland, C. D., et al. (1999) Dystrophin expression in the mdx mouse restored by stem cell transplantation, *Nature* **401(6751),** 390–394.
84. Rietze, R. L., Valcanis, H., Brooker, G. F., Thomas, T., Voss, A. K., and Bartlett, P. F. (2001) Purification of a pluripotent neural stem cell from the adult mouse brain, *Nature* **412(6848),** 736–739.
85. Jackson, K. A., Mi, T., and Goodell, M. A. (1999) Hematopoietic potential of stem cells isolated from murine skeletal muscle, *Proc. Natl. Acad. Sci. USA* **96(25),** 14,482–14,486.
86. McKinney-Freeman, S. L., Jackson, K. A., Camargo, F. D., Ferrari, G., Mavilio, F., and Goodell, M. A. (2002) Muscle-derived hematopoietic stem cells are hematopoietic in origin, *Proc. Natl. Acad. Sci. USA* **99(3),** 1341–1346.
87. Kawada, H. and Ogawa, M. (2001) Bone marrow origin of hematopoietic progenitors and stem cells in murine muscle, *Blood* **98(7),** 2008–2013.
88. Seale, P., Asakura, A., and Rudnicki, M. A. (2001) The potential of muscle stem cells, *Dev. Cell* **1(3),** 333–342.
89. Qu-Petersen, Z., Deasy, B., Jankowski, R., et al. (2002) Identification of a novel population of muscle stem cells in mice: potential for muscle regeneration, *J. Cell Biol.* **157(5),** 851–864.
90. Asakura, A., Komaki, M., and Rudnicki, M. (2001) Muscle satellite cells are multipotential stem cells that exhibit myogenic, osteogenic, and adipogenic differentiation, *Differentiation* **68(4–5),** 245–253.
91. Toma, J. G., Akhavan, M., Fernandes, K. J., et al. (2001) Isolation of multipotent adult stem cells from the dermis of mammalian skin, *Nat. Cell Biol.* **3(9),** 778–784.
92. Reyes, M., Lund, T., Lenvik, T., Aguiar, D., Koodie, L., and Verfaillie, C. M. (2001) Purification and ex vivo expansion of postnatal human marrow meso-dermal progenitor cells, *Blood* **98(9),** 2615–2625.
93. Reyes, M., Dudek, A., Jahagirdar, B., Koodie, L., Marker, P. H., and Verfaillie, C. M. (2002) Origin of endothelial progenitors in human postnatal bone marrow, *J. Clin. Invest.* **109(3),** 337–346.
94. Schwartz, R. E., Reyes, M., Koodie, L., et al. (2002) Multipotent adult pro-genitor cells from bone marrow differentiate into functional hepatocyte-like cells, *J. Clin. Invest.* **109(10),** 1291–1302.
95. Thomson, J. A., Itskovitz-Eldor, J., Shapiro, S. S., et al. (1998) Embryonic stem cell lines derived from human blastocysts, *Science* **282(5391),** 1145–1147.
96. Kocher, M., Muller, R. P., Staar, S., and Degroot, D. (1995) Long-term survival after brain metastases in breast cancer, *Strahlenther. Onkol.* **171,** 290–295.
97. Rideout, W. M., 3rd, Eggan, K., and Jaenisch, R. (2001) Nuclear cloning and epigenetic reprogramming of the genome, *Science* **293(5532),** 1093–1098.
98. Reik, W., Dean, W., and Walter, J. (2001) Epigenetic reprogramming in mam-malian development, *Science* **293(5532),** 1089–1093.
99. Odelberg, S. J., Kollhoff, A., and Keating, M. T. (2000) Dedifferentiation of mammalian myotubes induced by msx1, *Cell* **103(7),** 1099–1109.
100. Wells, W. A. (2002) Is transdifferentiation in trouble?, *J. Cell Biol.* **157(1),** 15–18.

6

Protocols for the Isolation and Maintenance of Human Embryonic Stem Cells

Melissa K. Carpenter, Chunhui Xu, Christine A. Daigh, Jessica E. Antosiewicz, and James A. Thomson

1. INTRODUCTION

Human embryonic stem (ES) cell lines have been derived from the inner cell mass of preimplantation embryos by culturing the cells on mouse embryonic feeder cells (1,2). In these conditions, human ES cells show remarkable proliferative capacity and stability in long-term culture (3) and have the capacity to differentiate into cell types from all three germ layers both in vitro and in vivo (1,2). Therefore, human ES cells may be a source of cells for cell therapies, drug screening, and functional genomics applications.

These cells can be maintained in long-term culture on feeders as well as using feeder-free conditions. In mouse ES cultures, the feeder layer can be replaced by the addition of leukemia inhibitory factor (LIF) to the growth medium, which maintains the cells in the undifferentiated state (4,5). The maintenance of undifferentiated human ES cells is not facilitated by the presence of LIF (1,2). The exact role of the feeder cells in the human ES cultures is unclear. The specific factors that promote ES cell self-renewal have not yet been identified and could include secreted, membrane-bound, or matrix-associated factors. In this chapter, we describe the isolation of human ES cells and procedures for maintaining the cells on feeder layers or in feeder-free conditions. Cells cultured under either of these conditions maintain normal karyotype, have similar proliferation rates, and express similar markers, including SSEA-4, TRA-1-61, TRA-1-80, alkaline phosphatase, hTERT, and OCT-4. In addition, cells maintained in either condition have the capacity to differentiate into cell types from the three germ layers both in vitro and in vivo (1,6). Here, we present protocols for the derivation and maintenance of human ES cells.

From: *Human Embryonic Stem Cells*
Edited by: A. Chiu and M. S. Rao © Humana Press Inc., Totowa, NJ

2. MATERIALS

2.1. Solutions and Media

2.1.1. Stocks

1. Collagenase IV solution (200 units/mL). Dissolve 20,000 units of collagenase IV (Gibco-BRL/Invitrogen cat. no. 17104-019) in 100 mL knockout Dulbecco's modified Eagle's medium (DMEM). Add all components to a 250-mL filter unit (0.22 μM, Corning, cellulose acetate, low protein binding) and filter. Aliquot and store at –20°C until use. *Note*: In our hands, 200 units/mL is usually 1 mg/mL.
2. DMEM, high-glucose, without glutamine (Gibco-BRL/Invitrogen cat. no. 11965-092).
3. Fetal bovine serum (FBS) (Hyclone cat. no. 30071-03).
4. Gelatin (0.5%). Add 100 mL of 2% gelatin (Sigma cat. no. G1393) and 300 mL water for embryo transfer (Sigma cat. no. W1503) into a 500-mL filter unit (0.22 μM, Corning, cellulose acetate, low protein binding) and filter. Store at 4°C.
5. L-Glutamine solution (200 mM) (Gibco-BRL/Invitrogen cat. no. 25030-081). Make aliquots of 10 mL and store at –20°C.
6. Guinea pig serum as a source of complement. This is available commercially (Gibco-BRL/Invitrogen cat. no. 19195-015) or it can be prepared from fresh guinea pig blood. There is significant lot variation and new batches should be tested, aliquoted, and stored at –70°C.
7. Human basic fibroblast growth factor, recombinant (hbFGF) (10 μg/mL). Dissolve 10 μg hbFGF (Gibco-BRL/Invitrogen cat. no. 13256-029) in 1 mL phosphate-buffered saline (PBS) with 0.2% bovine serum albumin (BSA). Filter the solution using a 0.22 μM, Corning, cellulose acetate, low protein-binding filter. When handling hbFGF, prewet all pipet tips, tubes, and the filter with PBS + 0.2% BSA. hbFGF is very sticky and this will prevent some loss of the bFGF. Store stock at –20°C or –80°C (for long-term storage, keep stocks at –80°C). Store thawed aliquots at 4°C for up to 1 mo.
8. Knockout DMEM (Gibco-BRL/Invitrogen cat. no. 10829-018).
9. Knockout serum replacement (Gibco-BRL/Invitrogen cat. no. 10828-028).
10. Matrigel. Either growth-factor-reduced Matrigel (Becton Dickinson cat. no. 356231) or regular Matrigel (Becton Dickinson cat. no. 354234) can be used for coating plates. To prepare Matrigel aliquots, slowly thaw Matrigel at 4°C overnight to avoid the formation of a gel. Add 10 mL of cold knockout DMEM to the bottle containing 10 mL Matrigel. Keeping the mixture on ice, mix well with a pipet. Aliquot 1–2 mL into each prechilled tube; store at –20°C.
11. β-Mercaptoethanol (1.43 M) 14.3 M β-mercaptoethanol (Sigma cat. no. M 7522) is diluted 1:10 in PBS and stored at –20°C or –80°C in 40-μL aliquots. Aliquots are thawed and used immediately; do not reuse.
12. Nonessential amino acids (Gibco-BRL/Invitrogen cat. no. 11140-050).
13. D-PBS without Ca^{2+}Mg^{2+} (Gibco-BRL/Invitrogen cat. no. 14190-144).

14. Pronase (0.5%) (from *Streptomyces griseus*, Sigma P5147) in DMEM. Store filter-sterilized aliquots at –20°C.
15. Rabbit anti-BeWo cell serum. Prepare serum by bleeding a rabbit at 10 d after the last of three i.v. injections, administered at 2-wk intervals, of 2×10^7 BeWo cells (ATTC cat. no. CCL-98). Allow the blood to clot; collect serum and heat inactivate at 56°C for 30 min. Store in aliquots at –70°C.
16. Trypsin/EDTA (0.05% trypsin, 0.53 m*M* EDTA) (Gibco-BRL/Invitrogen cat. no. 25300-054).

2.1.2. Media

1. Mouse embryonic fibroblast (MEF) medium. Add all of the following medium components to 500 mL filter unit (0.22 µ*M*, Corning, cellulose acetate, low protein binding) and filter. Store at 4°C and use within 1 mo.
   a. 450 mL DMEM.
   b. 50 mL FBS.
   c. 5 mL of 200 m*M* L-glutamine (final concentration 2 m*M*).
2. Human ES medium. Add all of the following medium components to 500-mL filter unit (0.22 µ*M*, Corning, cellulose acetate, low protein binding) and filter. Store at 4°C for no longer than 2 wk.
   a. 400 mL knockout DMEM.
   b. 100 mL knockout serum replacement.
   c. 5 mL nonessential amino acids.
   d. 2.5 mL of 200 m*M* L-glutamine (final concentration of 1 m*M*).
   e. 35 µL of 0.14 *M* β-mercaptoethanol (final concentration of 0.1 m*M*).
3. Differentiation medium. The differentiation medium is made by replacing knockout serum replacement with 20% FBS (not heat inactivated) in the human ES serum-free medium described above.
4. Cryopreservation media for MEFs and ES cells on feeders. Add medium components a and b below to a 100-mL filter unit (0.22 µ*M*, Corning, cellulose acetate, low protein binding), filter and mix with component c (without filtration).
   a. 60 mL knockout DMEM (for ES cells) or 60 mL DMEM (for MEFs).
   b. 20 mL FBS.
   c. 20 mL dimethyl sulfoxide (DMSO) (Sigma cat. no. D2650).
5. Cryopreservation medium for ES cells on Matrigel. Add medium components a and b below to a 100-mL filter unit (0.22 µ*M*, Corning, cellulose acetate, low protein binding), filter and mix with component c (without filtration).
   a. 40 mL knockout DMEM.
   b. 40 mL knockout serum replacement.
   c. 20 mL DMSO.

2.2. Tissue Culture Plates and Flasks

We use six-well plates (Falcon cat. no. 3046) for ES culture; T75 flasks (Corning cat. no. 430641), T150 flasks (Corning cat. no. 430825), and T225 flasks (Corning cat. no. 431082) for MEF culture, six-well low-attachment plates (Corning, cat. no. 29443-030) for embryoid bodies, and four-well plates for immunosurgery and thawing human ES cells (Nunc cat. no. 176740).

2.3. Incubators

All cells are maintained under sterile conditions in a humidified incubator in a 5% CO_2–95% air atmosphere at 37°C.

3. METHODS

3.1. Mouse Embryonic Fibroblasts Preparation

3.1.1. Harvesting MEFs

This protocol is modified from methods described by Robertson *(7)*.

1. Inject approx 0.5 mL avertin i.p. into a 13- or 14-d pregnant mouse (we use CF-1 mice). When mouse is anesthetized, perform a cervical dislocation.
2. Saturate abdomen with 70% ethanol and pull back the skin to expose the peritoneum. With sterile tools, cut open the peritoneal wall to expose the uterine horns. Remove the uterine horns and place them in a 10-cm dish. Wash three times with 10 mL PBS without $Ca^{2+}Mg^{2+}$.
3. Cut open each embryonic sac with scissors and release the embryos into the dish.
4. Using two pairs of watchmaker's forceps, remove the placenta and membranes from the embryo. Once they have been removed, dissect out the viscera. Place the embryos in a clean Petri dish and wash three times with 10 mL PBS.
5. Transfer two embryos to a 35-mm culture dish. With curved iris scissors or a sterile razor blade, *finely* mince the tissue.
6. Add 2 mL trypsin/EDTA to minced tissue and incubate tissue for 5–10 min at 37°C.
7. Inactivate trypsin/EDTA by adding 5 mL MEF medium and transfer contents to a 15-mL tube.
8. Dissociate cells by trituration. Allow larger pieces to settle to bottom of tube and remove and plate the supernatant in 15 mL MEF medium. We plate approximately two embryos in each T150 flask. Rock flask back and forth to ensure even distribution of the cells and place in the incubator.
9. Split at 1:2 when cells are 80–85% confluent. This is usually the next day in our lab.
10. When cells reach 80–85% confluence at this passage (passage 1), they can be used for feeders or can be frozen (*see* Section 3.1.2.).
11. As with any new cell in the lab, a representative sample should be tested for mycoplasma and sterility.

3.1.2. Cryopreservation of MEFs

1. Prelabel all cryovials with the following information: cell line, passage number, number or surface area of cells frozen, date, and initials.
2. Aspirate MEF medium from flask.

3. Wash cells once with PBS without $Ca^{2+}Mg^{2+}$ (2–3 mL/T75 and 5–10 mL/T150).
4. Add trypsin/EDTA to cells (1.0 mL per T75 and 1.5 mL per T150) and rock flask back and forth to evenly distribute the solution. Incubate for approx 5 min at 37°C.
5. Detach cells from the plate by pipetting off or tapping the flask against the heel of your hand.
6. Neutralize trypsin/EDTA with MEF medium (5 mL per T75, 10 mL per T150).
7. Pipet to break up clumps of cells. If clumps remain, add suspension to a 50-mL tube and allow the chunks to settle out.
8. Perform cell count of the cell suspension to determine the number of vials you will be freezing down. We usually freeze 10–20 million cells per 1 mL in a vial.
9. Pellet the cells by centrifugation for 5 min at 100g.
10. Resuspend pellet in 0.5 mL of DMEM with 20% FBS per vial. (This is one-half the final volume required for freezing.)
11. Dropwise, add an equivalent volume (0.5 mL per vial) of cryopreservative medium and mix. The DMSO concentration is now 10%.
12. Place 1 mL of cell mixture into each freezing vial.
13. Rapidly transfer the cells to a Nalgene freezing container (Fisher cat. no. 15-350-50) and place at –70°C overnight. Transfer cells to liquid nitrogen the next day for long-term storage. Alternatively, cells can be frozen using a controlled-rate freezer and transferred to liquid nitrogen at the completion of the freeze cycle.

3.1.3. Thawing and Maintaining MEFs

1. Remove vial from liquid nitrogen.
2. Do a quick thaw in a 37°C water bath. Carefully swirl vial in 37°C water bath, being careful not to immerse the vial above the level of the cap.
3. When just a small crystal of ice remains, sterilize the outside of the vial with 95% EtOH. Allow EtOH to evaporate before opening the vial.
4. Gently pipet the cell suspension up and down once and place it into a 15 mL conical tube.
5. Add 10 mL warm MEF medium to the tube dropwise to reduce osmotic shock.
6. Centrifuge the cell suspension at 250g for 5 min.
7. Remove the supernatant.
8. Resuspend the cell pellet in 10 mL (T75) or 20 mL (T150) culture medium and add to the flask. Plating density will need to be determined for each lot. We use approx 5×10^4 cells/cm^2.
9. Place in 37°C incubator.
10. Replace the medium the next day and split the MEFs 2–3 d after thawing, when they become 80–85% confluent.
11. The MEFs are split 1:2 every other day with trypsin, keeping cells sub-confluent. MEFs are only used through passage 5.

3.1.4. Irradiating and Plating MEFs

1. Coat flasks or plates with gelatin by adding 0.5% gelatin at 15 mL per T225, 10 mL per T150, 5 mL per T75 and 1 mL per well of six-well plates and

incubate at 37°C for at least 1 h. Use flasks or plates 4 h to 1 d after coating. Remove gelatin solution immediately before use.

2. Aspirate medium from MEF culture.
3. Wash cells once with PBS without Ca^{2+}/Mg^{2+} (2–3 mL per T75, 5–10 mL per T150).
4. Add trypsin/EDTA (1.0 mL per T75 and 1.5 mL per T150), and incubate at 37°C until cells are rounded up (usually 3–10 min). To loosen cells, either tap flask against the heel of your hand or pipet them off.
5. Add 9 mL MEF medium; collect cells in tube.
6. Add the cell suspension to a 15-mL conical tube and pipet several times to dissociate the cells. Perform a cell count by mixing cell suspension thoroughly and removing 10 μL. Add this to 10 μL trypan blue and mix well. Add ≤10 μL of the cell suspension and trypan blue mix to the hemacytometer. Count the phase bright cells in two of the 4 × 4 squares. Do not include dead cells in the count; these pick up the trypan blue. Calculate the cell number as follows:

$$\text{Total cell number} = (\text{Total cells counted in two } 4 \times 4 \text{ squares}) \div \qquad (1)$$
$$(\text{two } 4 \times 4 \text{ squares}) \times (\text{two trypan blue dilution factor}) \times$$
$$(1 \times 10^4 \text{ cell dilution factor}) \times (\text{total mL of cell suspension})$$

Example: You have just finished counting your sample of 23-mL cell suspension and saw 432 live cells in the two 4 × 4 squares. Your calculation will look like:

$$432 \text{ cells} \div 2 \times 2 \times (1 \times 10^4) = 4.32 \times 10^6 \text{ cells/mL} \times (23 \text{ mL}) = \qquad (2)$$
$$99.36 \times 10^6 \text{ total cells.}$$

7. Irradiate cells at 40–80 Gy. This number is variable between fibroblast sources—the goal is to irradiate them enough to stop them from proliferating, but not enough to kill them.
8. Spin down cells at 250*g* for 5 min and discard supernatant.
9. Resuspend cells in 10–30 mL MEF medium. Plate at 18,750 cells/cm^2 for feeder cultures and 56,000 cells/cm^2 for cultures to be used for CM. The final volumes should be 3 mL/well for a six-well plate, 10 mL per T75, 20 mL per T150, and 50 mL per T225.
10. When placing the plate in the incubator, gently shake the plate left to right and back to front to obtain an even distribution of cells.
11. Irradiated MEFs can be used 5 h to 7 d after plating.

3.2. Initial Derivation of Human ES Cell Lines

3.2.1. Immunosurgery

The procedure for the isolation of the inner cell mass (ICM) by immunosurgery is adapted from Solter and Knowles *(8)*.

1. Transfer blastocyst to a 0.5-mL Pronase solution in a four-well plate and incubate until the zona pellucida is removed (1–3 min).
2. Wash the blastocyst twice by transfer to new wells containing 0.5 mL DMEM with 10% FBS for 1 min each.

3. Transfer the blastocyst to rabbit anti-BeWo antiserum diluted with DMEM. The dilution will vary with the serum batch, but it should be in the range of 1:10 to 1:100. Place in incubator for 30 min.
4. Wash the blastocyst twice by transfer to new wells containing 0.5 mL DMEM with 10% FBS for 1 min each.
5. Transfer blastocyst to a new well containing guinea pig complement diluted 1:10 in DMEM. Place in incubator for 30 min.
6. Wash the blastocyst twice by transfer to new wells containing 0.5 mL DMEM with 10% FBS for 1 min each.
7. Remove lysed trophectoderm by gentle pipetting with a narrow-bore micropipet. The micropipet is hand-drawn from a Pasteur pipet over a small flame to be just larger than the diameter of the inner cell mass (ICM).

3.2.2. Initial Plating and Expansion

1. Transfer the isolated ICM to a layer of mitotically inactivated mouse embryonic fibroblasts in appropriate ES cell medium, and place in incubator. During the initial derivation of human ES cell lines (*see* refs. *1* and *2*), DMEM with 20% serum was used. However, the improved culture of human ES cell lines in medium supplemented with knockout serum replacement suggests that this medium may improve the initial derivation.
2. At 9–15 d, the ICM-derived mass is mechanically split into small clumps with a Pasteur pipet, hand-drawn over a flame to a fine point, and the clumps are individually transferred to a new well with fresh mitotically inactivated embryonic fibroblasts. Mechanical splitting is continued until the individual colonies are expanded to cover most of a well; then, enzymatic splitting is used for routine passage (*see* Section 3.3.).

3.3. Culture of Human ES Cells on Fibroblast Feeder Layers

1. Plate mitotically inactivated MEFs at a density of 18,750 cells/cm^2, as described in section 3.1.4.
2. Warm collagenase solution to 37°C in a water bath.
3. Aspirate medium from human ES cell culture plate.
4. Add collagenase to plates (0.5 mL/well of four-well plate or 1.0 mL/well of six-well plate).
5. Incubate at 37°C with 5% CO_2 for 7–10 min. The cells are ready when the edges of the colonies are rounded up and curled away from the MEFs on the plate.
6. Using a 5-mL pipet, gently scrape and wash the colonies off the plate.
7. Transfer cell suspension to a 15-mL conical tube.
8. Break up the colonies by pipetting up and down against the bottom of the tube until there is a suspension of clumps of approx 50–100 cells.
9. Spin cells at 150g for 5 min.
10. Aspirate supernatant and wash cells with 3 mL human ES medium.
11. Spin at 150g for 5 min.
12. While the cells are spinning, prepare the feeder layers by aspirating the MEF medium, adding human ES medium (including 4 ng/mL hbFGF) to the feeder layers (2 mL/well of a six-well plate).
13. Aspirate the wash medium.

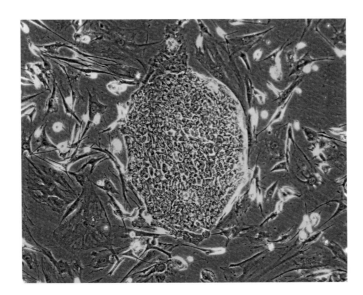

Fig. 1. Human ES cell colony maintained on MEFs.

14. Resuspend cells in an appropriate volume (*see* Section 3.3.1., Note 1).
15. Plate cells by adding 0.4 mL per well of a six-well plate until the last 0.5–0.6 mL remain. Add the last remaining volume dropwise to each well to ensure even distribution.
16. Make sure the cells are evenly distributed across the entire well (*see* Section 3.3.1.).
17. Place gently in incubator. Again, make sure cells are not disturbed.

3.3.1. Notes on Culture of Human ES Cells on Fibroblast Feeder Layers

1. Human ES cells generally have to be split approximately once per week, or when the plates are 70–90% confluent and the colonies are fairly large (*see* Fig. 1). Such colonies will individually fill a field of view when observed with a 10× objective. Splitting ratios vary, but for a 70–90% confluent plate, a 1:3 to 1:4 split is appropriate.
2. Always check MEFs for viability and contamination before splitting.
3. Only use MEFs within 7 d of plating.
4. Make sure that the human ES cells are evenly distributed throughout the plate. If the majority of the cells attach in the middle, they may differentiate or may need to be split sooner.
5. We recommend learning human ES cell culture on MEFs before moving to Matrigel.
6. DMEM/F12 (GibcoBRL/Invitrogen/Invitrogen cat. no. 11330-057) can be substituted for the knockout DMEM for growing human ES cells on fibroblast feeders.

3.4. Culture of Human ES Cells
with Matrigel and Conditioned Medium

3.4.1. Preparation of Conditioned Medium

A schematic diagram of this culture system is shown in Fig. 2.

1. Plate irradiated MEFs at 56,000 cells /cm^2 in MEF medium as described in Section 3.1.4. The final volumes should be 3 mL/well for a six-well plate, 10 mL per T75, 20 mL per T150, and 50 mL per T225.
2. To generate conditioned medium (CM), replace MEF medium with human ES medium supplemented with 4 ng/mL hbFGF (0.4 mL/cm^2) 1 d before use.
3. Collect CM from feeder flasks or plates after overnight incubation and add an additional 8 ng/mL hbFGF.
4. Add fresh human ES serum replacement medium containing 4 ng/mL hbFGF (0.4 mL/cm^2) to the feeders.
5. The MEFs can be used for 1 wk, with CM collection once every day.

3.4.2. Preparation of Matrigel-Coated Plates

1. Slowly thaw Matrigel aliquots at 4°C for at least 2 h to avoid the formation of a gel.
2. Dilute the Matrigel aliquots 1:15 in cold knockout DMEM.
3. Add 1 mL of Matrigel solution to coat each well of a six-well plate.
4. Incubate the plates 1–2 h at room temperature, or at least overnight at 4°C. The plates with Matrigel solution can be stored at 4°C for 1 wk.
5. Before use, remove Matrigel.

3.4.3. Passage of Human ES Cells on Matrigel

1. Aspirate medium from human ES cells, and add 1 mL of 200 units/mL collagenase IV per well of a six-well plate.
2. Incubate 5–10 min at 37°C in incubator. Incubation times will vary among different batches of collagenase; therefore, determine the appropriate incubation time by examining the colonies. Stop incubation when the edges of the colonies start to pull away from the plate.
3. Aspirate the collagenase and wash once with 2 mL PBS.
4. Add 2 mL of CM into each well.
5. Gently scrape cells with a cell scraper or a 10-mL pipet to collect most of the cells from the well, and transfer cells into a 15-mL tube.
6. Gently dissociate cells into small clusters (50–500 cells) by gently pipetting. Do not triturate cells to a single-cell suspension.
7. Remove Matrigel-containing solution from the plates and wash once with knockout DMEM.
8. Seed the cells into each well of Matrigel-coated plates. The final volume of medium should be 4 mL per well. In this system, the human ES cells are maintained at high density. At confluence (usually 1 wk in culture), the cells will be

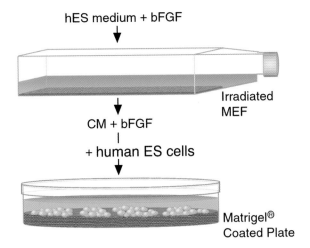

hES medium + bFGF

Irradiated
MEF

CM + bFGF

+ human ES cells

Matrigel®
Coated Plate

Fig. 2.

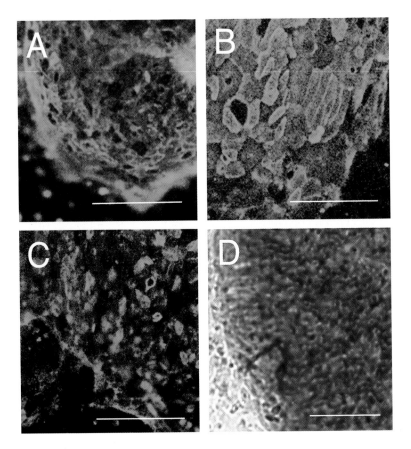

Fig. 3.

at 300,000–500,000 cells/cm^2. We find that optimal split ratio is 1:3 to 1:6. Using these ratios, the seeding density is approx 50,000–150,000 cells/cm^2.

9. Return the plate to the incubator. Be sure to gently shake the plate left to right and back to front to obtain even distribution of cells.
10. The day after seeding, undifferentiated cells are visible as small colonies. Single cells in between the colonies will begin to differentiate. As the cells proliferate, the colonies will become large and compact, representing the majority of surface area of the culture dish.

3.4.4. Daily Maintenance of Feeder-Free Culture

1. Collect CM from feeders, filter using a 0.2-μm filter, and add hbFGF to a final concentration of 8 ng/mL.
2. Feed human ES cells with 4 mL CM supplemented with hbFGF for each well of six-well plates every day.
3. Passage when cells are 100% confluent. At this time, the undifferentiated cells should represent at least 80% of the surface area. The cells in between the colonies of undifferentiated cells appear to be stromalike cells (*see* Fig. 4). The colonies (but not the stromalike cells) show positive immunoreactivity for SSEA-4, TRA-1-60, TRA-1-81, and alkaline phosphatase (*see* Fig. 3).

3.4.5. Notes on Feeder-Free Culture

1. When feeding cells, only prepare the amount of medium needed each time. For each aliquot of medium, add the appropriate amount of hbFGF (4–8 ng/mL). Place this aliquot in a H$_2$O bath until warm. Use immediately. Do not heat up the entire bottle of medium, as the knockout DMEM and knockout serum replacement do not tolerate repeated warming and cooling.
2. The CM can be filtered using a 500-mL filter unit (0.22 μM, Corning, cellulose acetate, low protein binding) and stored at –20°C before use.
3. Other matrix proteins also support human ES growth in MEF CM (*6*). Cells can be maintained on laminin-, fibronectin-, or collagen IV-coated plates. Cultures on laminin contained a high proportion of undifferentiated colonies, which continued to display ES-cell-like morphology after long-term passage. However, cells maintained on fibronectin or collagen IV had fewer colonies that displayed appropriate undifferentiated ES morphology.

Fig. 2. (*previous page*) Feeder-free human ES culture. MEF cells are irradiated to 40 Gy and seeded into a flask for preparation of conditioned medium (CM). After at least 4 h, the medium is exchanged with human ES medium. CM is collected daily and supplemented with an additional 8 ng/mL of hbFGF before addition to the human ES cells on Matrigel.

Fig. 3. (*previous page*) Expression of surface markers on human ES cells in feeder-free condition. SSEA-4 (**A**), TRA-1-60 (**B**), and TRA-1-81 (**C**) were expressed in H1 ES cells grown on Matrigel in MEF CM. Alkaline phosphatase enzyme activity was strongly positive in colonies of undifferentiated human ES cells (**D**). Scale bars represent 100 μm.

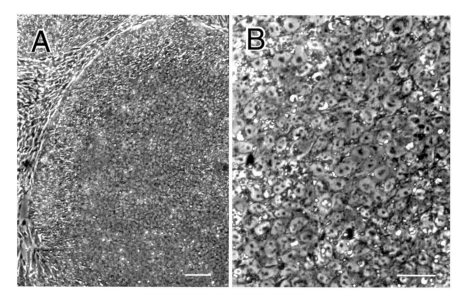

Fig. 4. (A) Morphology of colonies of human ES cells maintained on Matrigel in MEF CM. **(B)** Higher magnification of undifferentiated human ES cells within the colony. Scale bars represent 200 μm **(A)** and 50 μm **(B)**.

4. We found that cell lines such as STO (immortal mouse embryonic fibroblast cell line) and NHG190 (triple drug resistant, GFP-positive mouse embryonic cell line transfected with hTERT; unpublished data) can produce effective CM *(6)*. Routine culture medium for STO cells is MEF medium supplemented with 0.1 mM nonessential amino acids. NHG190 medium is MEF medium supplemented with an additional 10% FBS. Cells can be harvested and irradiated at 40 Gy, counted, and seeded at about 38,000/cm^2 for NHG190 cells or 95,000/cm^2 for STO cells. After at least 4 h, exchange the medium with human ES medium (0.5 mL/cm^2) for production of CM.

3.5. Freezing Human ES Cells

Cells maintained on feeders or in feeder-free condition can be successfully cryopreserved. The procedure is essentially the same for both types of cultures.

1. Treat cells with collagenase for approx 7 min at 37°C (until edges of colonies are curling up).
2. With a 5-mL pipet, *gently* pipet and scrape colonies from plate. Add cell suspension to a 15-mL centrifuge tube and gently break up colonies. It is important to be gentle in this step, as "chunkier" colonies will thaw better than single cells. Ideally, colonies meant for freezing are left slightly larger than they would be for splitting.

3. Centrifuge 5 min at 150*g*.
4. Resuspend pellet (gently) in 3 mL ES medium to wash away collagenase.
5. Centrifuge 5 min at 150*g*.
6. Resuspend pellet (gently) in 0.25 mL DMEM with 20% FBS per vial (This is one-half the final volume required for freezing). For feeder-free cultures, knockout serum replacement can be substituted for FBS.
7. Dropwise, add an equivalent volume (0.25 mL per vial) of cryopreservative medium and mix. The resulting DMSO concentration is 10%.
8. Place 0.5 mL of cell mixture in each freezing vial.
9. Rapidly transfer the cells to a Nalgene freezing container (Fisher cat. no. 15-350-50) and place immediately at –70°C overnight (do not leave cells in DMSO at room temperature for long periods of time). Transfer cells to liquid nitrogen the next day for long-term storage. Alternatively, cells can be frozen using a controlled-rate freezer and transferred to liquid nitrogen at the end of the freeze cycle.

3.6. Thawing Human ES Cells

This procedure can be used successfully for cells grown on feeders or maintained in feeder-free conditions.

1. Remove human ES cells from the liquid-nitrogen storage tank.
2. Thaw cryovial by gently swirling in a 37°C water bath until only a small ice pellet remains, being careful not to completely submerge the cryovial under water.
3. Completely submerge the cryovial in 95% ethanol. Allow the vial to air-dry before opening vial.
4. Very gently pipet cells from the vial into a 15-mL conical centrifuge tube.
5. Slowly, add 9.5 mL of medium dropwise to reduce osmotic shock. While adding medium, gently mix the cells in the tube by gently tapping the tube with a finger.
6. Centrifugate at 150*g* for 5 min.
7. Wash cells by resuspending in 3 mL medium.
8. Centrifuge at 150*g* for 5 min.
9. Resuspend in 2 mL and add 0.5 mL per well of a four-well plate that is already plated with MEFs or that has been coated with Matrigel.
10. Change medium daily; however, it may take 2 wk before cells are ready to be expanded.

3.7. Formation of Embryoid Bodies

In vitro differentiation can be induced by culturing the human ES cells in suspension to form embryoid bodies. We have induced differentiation of human ES into neurons, cardiomyocytes, and endoderm cells by using the following procedure.

1. Aspirate medium from human ES cells and add 1 mL/well of collagenase IV (200 units/mL) into six-well plates.

2. Incubate for 5 min at 37°C in an incubator.
3. Aspirate the collagenase IV and wash once with 2 mL PBS.
4. Add 2 mL of differentiation medium into each well.
5. Scrape cells with a cell scraper or pipet and transfer cells to one well of a low-attachment plate (1:1 split). Cells should be collected in clumps. Add 2 mL of differentiation medium to each well to give a total volume of 4 mL per well. Depending on the density of the hES cells, the split ratio for this procedure can vary.
6. After overnight culture in suspension, ES cells form floating aggregates known as embryoid bodies (EBs).
7. To change the medium, transfer EBs into a 15-mL tube and let aggregates settle for 5 min. Aspirate the supernatant, replace with fresh differentiation medium (4 mL/well), and transfer to low-attachment six-well plates for further culture.
8. Change medium every 2–3 d. During the first days, the EBs are small with irregular outlines; they increase in size by d 4. The EBs can be maintained in suspension for more than 10 d. Alternatively, EBs at different stages can be transferred to tissue culture plates for further induction of differentiation.

REFERENCES

1. Thomson, J. A., Itskovitz-Eldor, J., Shapiro, S. S., et al. (1998) Embryonic stem cell lines derived from human blastocysts, *Science* **282**, 1145–1147.
2. Reubinoff, B. E., Pera, M. F., Fong, C. Y., Trounson, A., and Bongso, A. (2000) Embryonic stem cell lines from human blastocysts: somatic differentiation in vitro, *Nature Biotechnol.* **18**, 399–404.
3. Amit, M., Carpenter, M. K., Inokuma, M. S., et al. (2000) Clonally derived human embryonic stem cell lines maintain pluripotency and proliferative potential for prolonged periods of culture, *Dev. Biol.* **227**, 271–278.
4. Smith, A. G., Heath, J. K., Donaldson, D. D., et al. (1988) Inhibition of pluripotential embryonic stem cell differentiation by purified polypeptides, *Nature* **336**, 688–690.
5. Williams, R. L., Hilton, D. J., Pease, S., et al. (1988) Myeloid leukaemia inhibitory factor maintains the developmental potential of embryonic stem cells, *Nature* **336**, 684–687.
6. Xu, C., Inokuma, M. S., Denham, J., Golds, K., Kundu, P., Gold, J. D., and Carpenter, M. K. (2001) Feeder-free growth o undifferentiated human embryonic stem cells on defined matrices with conditioned medium, *Nature Biotechnol.* **19**, 971–974.
7. Robertson, E. J. (1987) *Teratocarcinomas and Embryonic Stem Cells: A Practical Approach* (Approach, T. a. E. S. c. A. P., ed.), IRL, Washington, DC.
8. Solter, D. and Knowles, B. B. (1975) Immunosurgery of mouse blastocyst, *Proc. Natl. Acad. Sci USA* **72**, 5099–5102.

7

Subcloning and Alternative Methods for the Derivation and Culture of Human Embryonic Stem Cells

Michal Amit, Hanna Segev, Dorit Manor, and Joseph Itskovitz-Eldor

1. INTRODUCTION

Human embryonic stem (ES) cell lines may have broad applications, including the study of development and the differentiation process, lineage commitment, self-maintenance, and precursor cell maturation. They may also serve as models in research done on the functions of genes and proteins, drug testing, and drug toxicity. The first human ES cells were derived by Thomson and colleagues *(1)* from the inner cell mass (ICM) of surplus blastocysts donated by couples undergoing in vitro fertilization treatments. These lines met most of the criteria for ES cell lines listed in Table 1, but their clonality was not tested in that study. Also, the ability of human ES cells to contribute to embryonic development in chimeric embryos cannot be examined for obvious ethical reasons. Since the first report on human ES cell derivation, several other groups have reported the derivation of additional lines *(2–4)*. At present, there are more than 70 human ES cell lines in several laboratories around the world, according to a list published by the National Institutes of Health (NIH; www.nih.gov/news/stemcell/index). Although the NIH list does not offer full information on all the lines fulfilling all the ES cell criteria listed in Table 1, it suggests that the derivation of human ES cells is a reproducible procedure with reasonable success rates.

This chapter focuses on alternative methods for the derivation and maintenance of human ES cell lines, the derivation of genetically compromised human ES cell lines, and the subcloning of human ES cells parental lines.

From: *Human Embryonic Stem Cells*
Edited by: A. Chiu and M. S. Rao © Humana Press Inc., Totowa, NJ

Table 1
Main Features of ES Cells

1. Derived from the preimplantation embryo
2. Pluripotent, capable of differentiating into representative cells from all embryonic germ layers
3. Immortal, with prolonged proliferation at the undifferentiated stage (self-maintenance), and expression of high telomerase activities
4. Maintaining normal karyotype after prolonged culture
5. Ability to contribute to all three embryonic germ layers, including the germ line, following injection into blastocysts
6. Expressing unique markers like transcription factor *Oct-4* or cell surface markers like SSEA-3, SSEA-4, TRA-160, and TRA-181
7. Clonogenic (i.e., each individual cell possessing the above characteristics)

2. DERIVATION OF THE I SERIES

Embryonic stem cell lines are usually derived by immunosurgery, during which the trophoblast layer of the blastocyst is selectively removed and the intact ICM is further cultured on mitotically inactivated mouse embryonic fibroblasts (MEFs). This process is illustrated in Fig. 1. In our laboratory two human ES cell lines, I-3 and I-6, were derived using rabbit anti-human whole antiserum, and a further pluripotent line, I-4, was derived by the mechanical dissection and removal of the trophoblast layer with 27G needles. In our experience, human ES cell lines may also be derived without the complete removal of the trophoblast. In some cases, when the trophoblast is only partially removed, the ICM continues to grow with the remaining surrounding trophoblast as a monolayer. When the ICM reaches sufficient size, it is selectively removed and propagated. The ES cell lines developed in our laboratory have been grown continuously for over 1 yr (more than 60 passages); I-6 is still continuously growing and has reached 130 passages. Lines I-6 and I-3 were found to express high levels of telomerase activity after being continuously cultured for more than 6 mo. After more than 20 passages of continuous culture, karyotype analysis revealed that I-4 and I-3 are normal XX lines and I-6 has a normal XY karyotype. As previously reported on other human ES cell lines *(1,2)*, these cells strongly expressed surface markers that are typical of primate ES cells: stage-specific embryonic antigen-4 (SSEA-4), tumor rejection antibody (TRA)-1-60, and TRA-1-81, with weakly positive staining for SSEA-3 and negative staining for SSEA-1. The overall success rate of derivation was 60%, which is consistent with other reports on human ES cell line derivation *(1,2)*. Table 2 summarizes the characteristics of the new human ES cell lines.

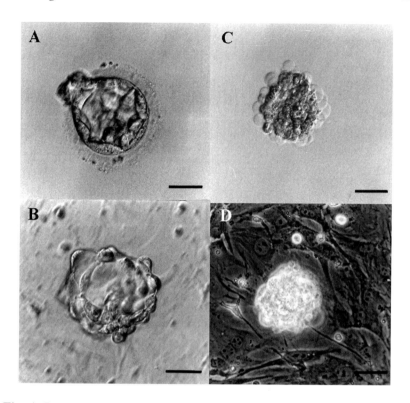

Fig. 1. Immunosurgery of a human blastocyst. (**A**) Donated human embryo produced by in vitro fertilization, post-PGD, which continued to develop to the blastocyst stage. (**B**) Human blastocyst after zona pellucida removal by Tyrode's solution, during exposure to rabbit anti-human whole antiserum. (**C**) The same embryo after exposure to guinea pig complement. (**D**) The intact ICM immediately after immunosurgery on mitotically inactivated MEF feeder layer. Bar = 50 μm. (Reprinted with permission from ref. *4*.)

Following injection into severe combined immunodeficient (SCID)-beige mice, the I-3 and I-6 lines created teratomas that contained representative tissues from all three embryonic germ layers. The I-4 line has not yet been examined. All three lines formed embryoid bodies (EBs) after their removal from the feeder layer or when grown in crowded cultures for more than 3 wk. Thus, these cells answer the criteria for human ES cells.

3. ALTERNATIVE METHOD FOR THE DERIVATION OF PLURIPOTENT HUMAN ES CELL LINES: THE J SERIES

As mentioned earlier, ES cell lines are traditionally derived by either immunosurgery or the mechanical separation of the ICM from the embryo

Table 2
Characterization of Human ES Cell Lines I and J

	I-3	I-4	I-6	J-3	J-6
Year of derivation	7/2000	7/2000	7/2000	1/2001	7/2002
Embryos[a]	Frozen, blastocyst	Frozen, blastocyst	Frozen, blastocyst	Fresh, morula	Fresh, blastocyst
Karyotype	Normal XX	Normal XX	Normal XY	Normal XY	Normal XX
Formation of teratomas	+	To be done	+	+	To be done
EB formation	+	+	+	+	+
TRA-1-60	+++	To be done	+++	To be done	To be done
RRA-1-81	+++	To be done	+++	To be done	To be done
SSEA-1	–	To be done	–	To be done	To be done
SSEA-3	+	To be done	+	To be done	To be done
SSEA-4	+++	To be done	+++	To be done	To be done
Continuous culture	Over 112 passages	Over 60 passages	130 passages and still going	Over 116 passages	Over 54 passages

[a]All embryos were donated by couples undergoing an in vitro fertilization treatment at Rambam Medical Center. The couples voluntarily sign a consent form, which complies with the NIH guidelines.

at the blastocyst stage. Soon after implantation, the ICM develops into a layer of primitive endoderm, which gives rise to the extraembryonic endoderm and a layer of primitive ectoderm, which gives rise to the embryo proper and to some extraembryonic derivatives *(5)*. After implantation and gastrulation, cells become progressively restricted to a specific lineage. It should be noted, however, that these pluripotent ES cells proliferate and replicate in the intact embryo only for a limited period of time before they commit to specific lineages. The pluripotency of human postimplantation embryonic cells, between the time of implantation and the gastrulation process, has never before been examined.

In order to explore the pluripotency of postimplantation embryonic cells, we used an alternative method to derive pluripotent cell lines. In this method, the trophoectoderm layer is not removed. The trophoblast-surrounded ICM is plated on MEFs. In some cases, the embryo continues to develop as a complete embryo and creates a small cyst. Several days later, the cyst and surrounding trophoblast cells are removed, and the pluripotent-resembling cells are further cultured and propagated. The resulting cell line, J-3, exhibits the main features of human ES cell lines: It grows at the undifferentiated stage for over a year, expresses high levels of telomerase activity, creates EBs when grown in suspension, sustains normal XY karyotype, expresses surface markers typical of human ES cells, and differentiates into the three germ layers following injection into SCID-beige mice. Thus, although the J-3 cell line was derived from an embryo in the "postimplantation" stage, it remained pluripotent and immortal. An additional line, J-6, was derived using the same method. Although this line requires further characterization, the method of deriving these unique postimplantationlike cell lines seems to be reproducible. The features of lines J-3 and J-6 are listed in Table 2.

One of the possibilities is that the J-3 line represents a primitive ectoderm cell population. Rathjen and colleagues demonstrated a method for homogenous differentiation of mouse ES cells into early primitive ectodermlike (EPL) cells using conditioned medium of Hep G2 cells *(6)*. J-3 cells organize as a columnar epithelium when grown on MEFs as undifferentiated cells, whereas the human ES cells organized as pails of cells at the same culture conditions. J-3 cell nuclei are smaller; therefore, their nucleus to cytoplasm ratio is reduced compared with that of a human ES cell.

Another characteristic of the EPL cells is their tendency to differentiate into nascent mesodermal tissue sooner and at higher levels compared to mouse ES cell under the same differentiation culture conditions *(7)*. In teratomas formed from J-3 cells, we could see a variety of mesodermal tissues, such as cartilage tissues, bone tissues differentiated directly from mesenchymal tissues, and smooth or stratified muscle tissues, all in a higher rate than in ES cell-line-derived teratomas.

4. DERIVATION OF HUMAN ES CELL LINES HARBORING SPECIFIC GENETIC DEFECTS

Another possible source of donated embryos for human ES cell derivation is embryos from the preimplantation genetic diagnosis (PGD) program. PGD allows couples who are carriers of genetic diseases to examine the embryos before implantation and to retrieve the healthy embryos only. When the in vitro fertilized embryo reaches the six to eight cell stage, one cell is removed and analyzed for the existence of genetic defects either by polymerase chain reaction (PCR) or by fluorescence *in situ* hybridization (FISH). PGD-donated embryos will allow us to create human ES cell lines that harbor different genetic defects and follow the expression of these diseases during differentiation. Such a process may contribute to the development of drugs or gene therapy designed to treat these genetic diseases.

In our experience, embryos donated from couples undergoing PGD continue to develop in vitro to the blastocyst stage. The zona pellucida (ZP) of these embryos remains fractured after the cell to be analyzed is removed by biopsy. In some cases, during culture to the blastocyst stage, parts of the trophoblast cells seep through the fracture in the ZP. These cells can interrupt the removal of the ZP using pronase or Tyrode's solution, but this could be overcome by the mechanical removal of the ZP. An example of a post-PGD embryo that continued to develop to the blastocyst stage is illustrated in Fig. 1A.

Line J-3, described earlier, was obtained from a surplus embryo that underwent PGD for cystic fibrosis. This cell line was found to be heterozygous for the W1282X mutation—the most common type of cystic fibrosis-causing mutation among Ashkenazy Jews *(8)*. Like other human ES cell lines, the J-3 line meets all of the ES cell criteria listed in Table 1. Two additional ES cell lines were derived from embryos, which underwent PGD. The first line was heterozygous for the Gorlin mutation and the second line was heterozygous for the metachromatic leukodystrophy disease. These lines are still at an early passage and need to be fully characterized for ES cell features.

Gene therapy is often based on targeted correction, using small fragments of a corrected region for the gene. The availability of a line heterozygous of the W1282X cystic fibrosis mutation enables us to develop a targeted correction model for this common mutation *(9)*. The availability of human pluripotent cell lines carrying a mutation of cystic fibrosis may offer a suitable system for investigating the nature of this disease and its progress. Such an understanding may help in the development of both drug and gene therapy models for cystic fibrosis and other genetic diseases.

5. DERIVATION OF HUMAN ES CELL SUBCLONES

As discussed earlier in this chapter, ES cell lines are derived from the clump of cells in the ICM, which may not represent a homogenous cell population. The criteria for pluripotency usually include the derivation of the stem cell line from a single cloned cell. This eliminates the possibility that several distinct committed multipotential cell types are present in the culture and that, together, they account for the variety of differentiated derivatives produced.

In addition to proving the pluripotency of single human ES cells, single-cell clones may have further applications. They form homogeneous cell populations, which may be instrumental in the development of research models based on gene knockout or targeted recombination. The transfected or knockout cells could be cloned and analyzed individually; clones that express the desired genotype could be further cultured and used for research.

5.1. Karyotype Stability

Any future use of human ES cells for scientific or therapeutic purposes will depend on their ability to proliferate for long periods as undifferentiated cells, without losing their developmental potential or karyotypic stability. All of the reports on human ES cell line derivation specifically state that their karyotypes remain normal after prolonged proliferation *(1,2)*. Two of the human ES cell lines, I-3 and I-6, were tested after 105 and 89 passages of prolonged culture, respectively; Each of the examined cells demonstrated normal karyotype. There are two reports on the karyotype instability of human ES cell parental lines *(10,11)*. Amit et al. *(10)* examined the karyotype of the H-9 parental line after 7, 8, 10, and 13 mo of continuous culture. In only one case, after 7 mo of continuous culture, 4 out of the 20 cells examined demonstrated abnormal karyotypes. Eiges and colleagues *(11)* reported on two cells with trisomy in a stably transfected clone. It is possible that a subpopulation with abnormal karyotype will acquire a selective growth advantage and take over the culture. Therefore, the periodical cloning of cultured human ES cells may be needed. According to the data presented in this section on karyotype stability, the need for cloning in order to maintain a euploid population of human ES cells will be infrequent and will have no influence on the long-term expansion of human ES cells in culture.

In our experience, another advantage of single-cell clones is that they are easier to grow and manipulate in comparison to the parental lines.

5.2. Methods

Several culture media were tested in order to clone the parental human ES cell lines; medium supplemented with either FBS or serum replacement and either with or without human recombinant basic fibroblast growth factor (bFGF) *(10)*. The serum-free growth conditions supplemented with bFGF have been found to be the suitable ones for the clonal derivation of human ES single-cell lines. The resulting colonies of ES cells in the different cloning conditions are illustrated in Fig. 2. Addition of leukemia inhibitory factor (LIF) or forskolin has no beneficial effect on the cloning rates.

In order to derive the first single-cell clones of human ES cells, H-9 cells were trypsinized to single cells. Each individual cell was plated in a separate well in 96-well plates and grown in serum-free growth conditions. After approx 2 wk of growth, the resulting colonies were passaged and propagated. Consequently, two clonally derived human ES cell lines, H-9.1 and H-9.2, were derived *(10)*. These two single-cell clones proliferated continuously for a period of at least 8 mo after the clonal derivation, maintained a stable and normal karyotype, differentiate in vivo in teratomas into advanced derivatives of all three embryonic germ layers, and expressed high levels of telomerase activity.

To date, eight single-cell clones from five different parental ES cell lines (H-1, H-9, H-13 *[1]*, I-3 and I-6 *[4]*) and one clone from the J-3 pluripotent line have been derived in our laboratory. A summary of all existing single-cell clones is presented in Table 3. Most parental lines have the same cloning efficiency (0.5%). One line, H-1, has a lower cloning efficiency of 0.16% and we were not able to clone line I-4. In fact, all attempts to grow the I-4 line in the serum-free culture conditions failed. During the first passage in the serum-free culture conditions, almost 80% of the cells died. In any passaging technique, the remaining cells did not survive a second passage. However, in a serum-containing culture condition, the I-4 grows with normal survival rates for over 60 sequential passages. The variability between the cell lines with respect to the mechanism underlining the biological versatility of the parental lines and their subclones in their ability for self-renewal and differentiation remains to be determined.

All nine single-cell clones fulfilled the main criteria described for human ES cell lines (Table 1). The clones' karyotype remained normal after more than 6 mo of prolonged culture. All clones formed EBs in suspension, and following injection into SCID mice, they created tertomas containing representative tissues of the three embryonic germ layers (examples are illustrated in Fig. 3).

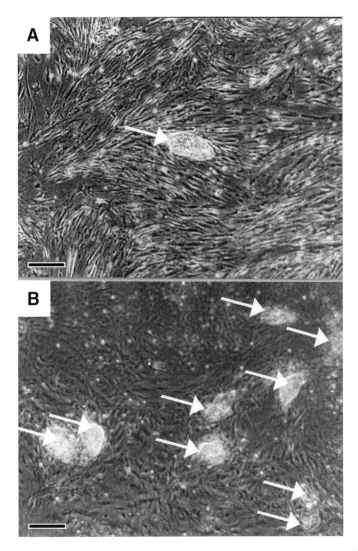

Fig. 2. Resulting colonies of H-9 ES cells in different cloning conditions. **(A)** Resulting colony cloned using medium supplemented with FBS; **(B)** Resulting colonies cloned using medium supplemented with serum replacement and bFGF. Bar: 10 µm.

Preliminary results indicate that there may be a difference between various single-cell clones and the tendency of parental lines to differentiate in a specific direction. In the culture conditions described by Kehat et al., 8% of the EBs formed from H-9.2 single-cell clone beats *(12)*. In the same culture condition, only less than 1% of the EBs formed by H-9 parental line beats (unpublished data). A recent publication by Xu and colleagues *(13)* reported

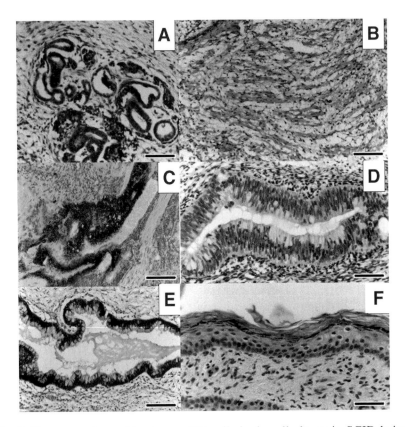

Fig. 3. Teratomas formed by human ES cell single-cell clones in SCID-beige mice. (**A**) Tubules interspersed with structures resembling fetal glomeruli, clone H-9.2; (**B**) embryonal myotubes, clone H-13.1; (**C**) cells producing melanin, clone H-1.1; (**D**) respiratory epithelium, clone H-1.1; (**E**) Mucus-secreting surface epithelium resembling the stratified epithelium found in the stomach, clone H-13.2; (**F**) skinlike epithelium facing a lumen, clone H-9.2 H&E stain. Bar: 50 μm.

on additional culture conditions, including supplement 5-azacytidine, which is suitable for the differentiation of human ES cells into cardiomyocytes. In this differentiation system, H-9.2 clone creates the highest percentage of beating EBs (70%) compared to the parental line H-9 and additional lines (H-1 and H-7).

6. CULTURE OF HUMAN ES CELL LINES IN ANIMAL-FREE CONDITIONS

Human ES cells may also be directly applied in cell-based therapies. In order to clinically use human ES cells, they should correspond with the Food and Drug Administration (FDA) guidelines. One of the major concerns

Table 3
Main Characteristics of the Existing Single-Cell Clones

	Karyotype	EB formation	Formation of teratomas	Continuous culture (mo)	Cloning efficiency of the parental line
H-1.1	Normal XY	+	+	6	1/600
H-9.1	Normal XX	+	+	8	1/400
H-9.2	Normal XX	+	+	8	1/400
H-13.1	Normal XY	+	+	6	1/400
H-13.2	Normal XY	+	+	6	1/400
I-3.2	Normal XX	+	+	6	1/400
I-3.3	Normal XX	+	+	6	1/400
I-6.2	Normal XY	+	+	6	1/400
J-3.2	Normal XY	+	+	10	1/400

related to the application of human ES cells in cell replacement therapy will be the exposure of these cells to retroviruses or other pathogens. These are potentially present in the mouse feeder layer or fetal bovine serum (FBS) with which these cells are derived and grown. Overcoming this problem requires that the human ES cells be derived and cultured in an entirely animal-free environment.

Richards and colleagues offered an animal-free system for the growth and formation of human ES cell lines *(14)*. In their culture system, human ES cells can be grown using a coculture with aborted human fetal-derived feeder layers or human adult fallopian tube epithelial feeder layers. Supplement of 20% human serum replaced the FBS commonly used in the culture of human ES cells. These culture conditions were found to be suitable for human ES cell line derivation. The line derived on a human fallopian tube feeder shares a similar morphology with other human ES cells and expresses *Oct 4* and other ES-cell-specific markers, thus fulfilling the criteria of ES cells.

Exploring the area of animal-free culture systems for human ES cells, we have demonstrated that human foreskin feeder layers support the growth of human ES cells in serum-free conditions. After more than 60 passages (more than 200 doublings), the three human ES cell lines grown on the foreskin feeders (I-3, I-6, and H-9) exhibit all human ES cell characteristics, including teratomas formation, EB creations, expression of typical surface markers, and normal kryotypes. The morphology of a human ES cell colony growing on foreskin is illustrated in Fig. 4A. Nine different foreskin lines that were derived and tested succeeded in supporting a prolonged and undifferentiated proliferation of human ES cells. The cells were grown under serum-free conditions using serum replacement and basic fibroblast growth factor, known to support human ES cell growth *(10a)*.

Human foreskin has several advantages. Unlike aborted fetal fibroblasts, which can grow to reach a certain limited passage, human foreskin lines can grow to reach passage 42. In our experience, high-passage human foreskin feeders can support the growth of human ES cells and can still be frozen and thawed at high efficiency. These feeders may therefore have an advantage when large-scale growth of human ES cells is concerned. Furthermore, the donation of foreskin from circumcised babies has no ethical implications such as those accompanying the donation of aborted fetus. The ability of foreskin feeders to support the *derivation* of human ES cell lines has yet to be determined.

The ideal solution to an animal-free method for growing human ES cells will be the ability to grow these cells on a matrix using serum-free medium.

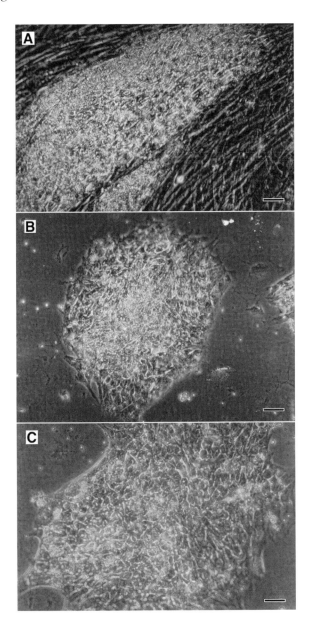

Fig. 4. Human ES cells colonies growing in different culture conditions. **(A)** Human ES cell colony from I-3 line growing 21 passages on mitotically inactivated human foreskin feeder layer (bar = 50 μm); **(B)** human ES cell colony forms I-6 line after several passages on MEF matrix (bar = 75 μm); **(C)** human ES cell colony forms I-6 line after several passages on Matrigel (bar = 75 μm).

Xu et al. demonstrated a culture system in which human ES cells were grown on Matrigel, laminin, or fibronectin using 100% MEF-conditioned medium *(15)*. Richards et al. tested the ability of human feeders or MEF-conditioned media to sustain a continuous undifferentiated proliferation of human ES cells grown on collagen I, human extracellular matrix, Matrigel, or laminin *(14)*. They found that these culture conditions were less suitable for growing human ES cells than with a human feeder layer. In both culture systems, the human ES cells may still be exposed to animal pathogens through the conditioned medium. Another disadvantage of these culture systems is the requirement for the parallel growth of the feeder lines. In our experience, culturing human ES cells on Matrigel or MEF matrix using MEF-conditioned medium is doable but not trivial. We were able to grow ES cells in these conditions for over 15 passages while maintaining all ES cell features. An example of a human ES cell colony growing on a MEF matrix is illustrated in Fig. 5B and on Matrigel in Fig. 5C. The feeder-free culture system for human ES cells still have to be improved in order to be used for the derivation of human ES cell lines. A further discussion on the feeder free culture of human ES cells appears in Chapter 16.

ACKNOWLEDGMENTS

The authors thank Kohava Shariki and Victoria Marguletz for their technical assistance. We thank Professor Raymond Coleman for his invaluable advice and assistance and Hadas O'Neill for editing the manuscript. The research was partly supported by the Fund for Medical Research and Development of Infrastructure and Health Services, Rambam Medical Center and the Technion Research Fund (TDRF).

REFERENCES

1. Thomson, J. A., Itskovitz-Eldor, J., Shapiro, S. S., et al. (1998) Embryonic stem cell lines derived from human blastocysts, *Science* **282,** 1145–1147 [erratum in *Science* (1998) **282,** 1827].
2. Reubinoff, B. E., Pera, M. F., Fong, C., Trounson, A., and Bongso, A. (2000) Embryonic stem cell lines from human blastocysts: somatic differentiation in vitro, *Nat. Biotechnol.* **18,** 399–404.
3. Lanzendorf, S. E., Boyd, C. A., Wright, D. L., Muasher, S., Oehninger, S., and Hodgen, G. D. (2001) Use of human gametes obtained from anonymous donors for the production of human embryonic stem cell lines, *Fertil. Steril.* **76,** 132–137.
4. Amit, M. and Itskovitz-Eldor, J. (2002) Derivation and spontaneous differentiation of human embryonic stem cells, *J. Anat.* **200,** 225–232.

5. Gardner, R. L. (1982) Investigation of cell lineage and differentiation in the extraembryonic endoderm of the mouse embryo, *J. Embryol. Exp. Morphol.* **68,** 175–198.

6. Rathjen, J., Lake, J. A., Bettess, M. D., Washington, J. M., Chapman, G., and Rathjen, P. D. (1999) Formation of primitive like cell population, EPL cells, from ES cells in response to biologically derived factors, *J. Cell Sci.* **112,** 601–612.

7. Lake, J. A., Rathjen, J., Remiszewski, J., and Rathjen, P. D. (2000) Reversible programing of pluripotent cell differentiation, *J. Cell Sci.* **113,** 555–566.

8. Shoshani, T., Augarten, A., Gazit, E., et al. (1992) Association of a nonsense mutation (W1282X), the most common mutation in the Ashkenazi Jewish cystic fibrosis patients in Israel, with presentation of severe disease, *Am. J. Hum. Genet.* **50,** 222–228.

9. Colosimo, A., Goncz, K. K., Novelli, G., Dallapiccola, B., and Gruenert, D. C. (2001) Target correction of a defective selectable marker gene in human epithelial cells by small DNA fragments, *Mol. Ther.* **3,** 178–185.

10. Amit, M., Carpenter, M. K., Inokuma, M. S., et al. (2000) Clonally derived human embryonic stem cell lines maintain pluripotency and proliferative potential for prolonged periods of culture, *Dev. Biol.* **227,** 271–278.

10a. Amit, M., Margulets, V., Segev, H., et al. (2003) Human feeder layers for human embryonic stem cells, *Biol. Reprod.* **68,** 2150–2156.

11. Eiges, R., Schuldiner, M., Drukker, M., Yanuka, O., Itskovitz-Eldor, J., and Benvenisty, N. (2001) Establishment of human embryonic stem cell-transfected clones carrying a marker for undifferentiated cells, *Curr. Biol.* **11,** 514–518.

12. Kehat, I., Kenyagin-Karsenti, D., Snir, M., et al. (2001) Human embryonic stem cells can differentiate into myocytes with structural and functional properties of cardiomyocytes, *J. Clin. Invest.* **108,** 407–414.

13. Xu, C., Police, S., Rao, N., and Carpenter, M. K. (2002) Characterization and enrichment of cardiomyocytes derived from human embryonic stem cells, *Cir. Res.* **91,** 501–508.

14. Richards, M., Fong, C. Y., Chan, W. K., Wong, P. C., and Bongso, A. (2002) Human feeders support prolonged undifferentiated growth of human inner cell masses and embryonic stem cells, *Nature Biotechnol.* **20,** 933–936.

15. Xu, C., Inokuma, M. S., Denham, J., et al. (2001) Feeder-free growth of undifferentiated human embryonic stem cells, *Nature Biotechnol.* **19,** 971–974.

III
DIFFERENTIATION

8

Differentiation of Neuroepithelia from Human Embryonic Stem Cells

Su-Chun Zhang

1. INTRODUCTION

Neural tissue is derived from the embryonic ectoderm. The initial step in the generation of the vertebrate nervous system is the specification of neuroepithelia from ectodermal cells—a process known as neural induction. The specified neuroepithelia in the midline dorsal ectoderm rapidly grow into a pseudostratified layer of neural plate, which later folds and closes to form the neural tube, the rudiment of the central nervous system (CNS). The lateral lip of the neural plate detaches when the neural plate closes and gives rise to neural crest derivatives. In mouse, neural plate forms at embryonic day 7, whereas in humans, it develops around embryonic day 18 *(1)*.

How neuroepithelia are induced from ectoderm is primarily investigated using stage-specific *Xenopus* and chick embryos. In mammals, these studies are compromised by experimental inaccessibility to early embryos. Embryonic stem (ES) cells are the in vitro counterparts of the inner cell mass of a preimplantation embryo at the blastocyst stage. They are, at least theoretically, capable of giving rise to all cell types that constitute an animal. Hence, ES cells provide a simple in vitro alternative for molecular and cellular analyses of neural induction in mammals. The establishment of continuous human ES (hES) cell lines *(2,3)* has sparked remarkable enthusiasm, with the expectation that these cells will be used not only to model aspects of early human development but also to generate selective lineage cells for regenerative medicine.

Realization of the enormous potential of hES cells depends on our ability to direct them into specific pathways and then to support the survival and differentiation of individual somatic stem/progenitor cells. To date, the vast majority of differentiation protocols, including those of neural differentiation from murine or human ES cells, employ embryoid body formation and

From: *Human Embryonic Stem Cells*
Edited by: A. Chiu and M. S. Rao © Humana Press Inc., Totowa, NJ

treatment with a common morphogen, retinoic acid. Consequently, the differentiated products are a mixture of cells from all three embryonic germ layers and/or cells at various developmental stages such as cells ranging from neural progenitors to mature neurons and glia. From a therapeutic perspective, cells of interest may be isolated from this mixture through selective culture conditions. From the standpoint of developmental biology, these empirical approaches are not sophisticated enough to reflect the cell lineage development in vivo. Insights from developmental studies examining how hES cells take on one cell fate at a particular stage will be essential to apply hES cells to regenerative medicine.

2. NEURAL DIFFERENTIATION FROM MOUSE ES CELLS

2.1. Neural Differentiation: From a Therapeutic Perspective

The selection of culture systems for generating a specific cell lineage varies considerably depending on objectives. These culture systems vary from clonal to mixed cultures and can use chemically-defined medium or medium containing unknown components such as cell-line-conditioned media or sera. Both genetic manipulation and epigenetic induction have been used. From a therapeutic perspective, the goal is to easily obtain the cells of choice at high purity and in a large quantity. The most commonly used approach for neural differentiation from mouse ES cells is the formation of embryoid bodies and treatment of these ES cell aggregates with retinoic acid (RA). This method was optimized by Gottlieb and colleagues *(4)* based on successful induction of neuronal cells from teratocarcinoma cells *(5)*. The procedure involves culturing ES cell aggregates in suspension in the regular ES cell growth medium without the growth factor leukemia inhibitory factor (LIF) for 4 d, followed by the addition of RA (1 μM) for another 4 d. Hence, this procedure is often regarded as a 4–/4+ protocol *(4)*. ES cells treated in this manner yield a good proportion (38%) of neuronal cells upon differentiation *(4)*. Other RA-induced neural differentiation protocols are variables of the 4–/4+ protocol *(6–10)*. Although RA is important during development, particularly in rostral/caudal patterning *(11)*, little evidence suggests that RA in these protocols acts to induce neural specification from ES cells. RA is a strong morphogen that appears to push ES cells toward postmitotic neurons, although RA-treated cultures contain neural cells at various developmental stages including neural progenitors *(12)*. Additionally, RA is known to induce a caudal fate *(13)*. Whether neurons generated from RA-treated ES cells are committed to a caudal fate is not clearly defined.

Fibroblast growth factor 2 (FGF2) is a survival and proliferation factor for early neuroepithelial cells. Based on this fact, McKay and colleagues have designed a chemically defined culture system to selectively support the survival and proliferation of neuroepithelial cells that are differentiated from ES cells *(14)*. Recent studies have shown that FGF2 plays an important role in neural induction *(15,16)*. It is possible that in the FGF2-treated ES cell differentiation model, FGF2 may also play a role in inducing neuroepithelial differentiation from ES cells. Hence, this model, to some degree, reflects early neural development. In contrast to the RA-induced neural differentiation process, treatment of ES cells with FGF2 produces a high proportion of neuroepithelial cells (80%) at a synchronous developmental stage *(14)*. These neural precursor cells do not appear to commit to a regional fate and can differentiate into various types of neurons and glia in response to environmental cues *(14,17–19)*.

In addition to the commonly used approaches outlined, conditioned media from mesoderm-derived cell lines have been used to effectively promote the generation of neural cells from ES cells. The rationale behind it is that signals from mesodermal cells are required to induce neural specification from ectoderm during early development. Although the identity of the neural-promoting molecules in the conditioned media remains unknown, these approaches allow for the production of large numbers of specialized neural cells. In particular, Sasai and colleagues are able to induce mouse ES cells to differentiate into a large proportion of midbrain dopamine neurons using a bone marrow stroma cell line (PA6 cells) *(20)*. Remarkably, the conditioned medium from the same cell line induces differentiation of dopamine neurons from nonhuman primate ES cells *(21)*. It would be interesting to see if human ES cells can be similarly induced to generate dopamine neurons. In addition, Rathjen et al. *(22)* use a conditioned medium from a hepatic cell line to induce mouse ES cells to differentiate into neuroepithelial cells through an intermediate stage called primitive ectoderm like cells.

2.2. Neural Specification: From a Developmental Standpoint

Embryonic stem cells sit on top of the cell lineage tree. Hence, ES cells provide an ideal system to analyze early embryonic induction, especially cell lineage specification in mammals. Understanding how an individual cell lineage is specified will be instrumental for effective generation of a particular cell type for therapy. Studies of mouse ES cell neural differentiation appear to have gone through a detour; only recently have the ES cells been applied to understand early neural specification. Using a clonal culture system, Tropepe et al. *(23)* have shown that mouse ES cells transform into primitive neural stem cells within hours in the absence of any exogenous inductive factors. This suggests that mouse ES cells have an innate tendency to become neural cells without the presence of inhibitory signals from neighboring cells. This is further supported by the observation that neural differentiation is enhanced in the ES cells that lack Smad4, a key component in the bone morphogenetic protein (BMP) signaling pathway. This study demonstrates that ES cells can be modeled to study early mammalian development, especially cell lineage specification. It is also advantageous from a therapeutic perspective that these clonally derived cells are a pure population of cells specified to a neural fate. Nevertheless, the efficiency of the neural differentiation in this system is relatively low, with only 0.2% of the ES cells being transformed to primitive neural stem cells *(23)*. Studies using mixed cultures of mouse ES cells also suggest the involvement of anti-BMP signaling in early neural specification in mice *(20,24)*. Most recently, Jessell and colleagues *(25)* have developed a system to guide mouse ES cells to differentiate into spinal motor neurons step by step based on the current understanding of motoneuron development. This procedure entails formation of embryoid bodies and induction of neuroectodermal cells by RA, caudualization of the ES-derived neuroectodermal cells by RA, and, finally, induction of motoneurons by the ventralizing molecule sonic hedgehog *(25)*. This study reinforces the importance of developmental insights in advancing the potential of ES cells.

2.3. Isolation of ES-Derived Neural Cells

The neural differentiation protocols summarized in Sections 2.1. and 2.2., with the exception of the clonal differentiation culture, yield a mixed population of cells including neural cells. This means there is a need to isolate the cells of interest from the mixture. A simple and efficient way of separating cells of choice is immunoseparation based on known cell surface epitopes. For example, Rao and colleagues sort neural progenitor cells from mouse ES-cell-differentiated progenies based on the fact that neural progenitors express an embryonic form of neural cell adhesion molecule (ENCAM) *(26)*. However, cell surface markers are not always available for neural cells at various stages. An alternative is to isolate these cells based on their physiochemical properties. McKay and colleagues use a medium that preferentially promotes the survival and proliferation of neural precursor cells *(14)*. A more complicated approach is genetic selection. By coupling identifiable markers such as green fluorescent protein gene under the control of a cell-type specific promoter, cells of choice can be isolated through immunosorting *(19,25,27)*.

3. NEURAL DIFFERENTIATION FROM HUMAN ES CELLS

Differentiation of neural cells from hES cells is in its initial stage. Little information is available to generalize the process. To date, protocols used to differentiate hES cells into neural cells are mainly a modification of those worked out for deriving neural cells from mouse ES cells. However, hES cells possess different properties than their mouse counterparts *(2,28)*. It is not surprising that some modification will be necessary to generate neural lineage from hES cells. As in mouse ES cell differentiation, the initial process is the aggregation of ES cells, the so-called embryoid body formation. The difficulty in making healthy embryoid bodies (EBs) in their initial studies has led Reubinoff et al. *(29)* to grow human ES cells in a high-density for a prolonged period so that the ES cells spontaneously differentiate into various cell types, including neural cells. Neural precursor cells differentiated in this manner form clusters characteristic of neurospheres. These neurosphere-like clusters could be mechanically dissected with a micropipet and expanded as neurospheres in suspension cultures. Carpenter et al. *(30)*, on the other hand, used the traditional approach to derive neural precursor cells. ES cells are first aggregated to form EBs, which are subsequently treated with RA to induce neural differentiation. As discussed earlier with mouse ES cells, RA treatment leads to the differentiation of neural cells from progenitor to mature stages. In order to select neural precursor cells, immunoseparation (magnetic sorting) was performed based on the fact that a certain population of neuronal progenitor cells expresses the embryonic form of NCAM *(30)*. What differs from mouse ES cell neural differentiation is that (10 times) more RA appears necessary to generate neural cells *(30,31)*.

From the standpoint of developmental biology, the above approaches have little relevance to embryonic development. From a therapeutic perspective, these approaches are far from efficient. Whether these hES-cell-derived neural progenitors can be further directed to specialized neurons or glia of therapeutically interest remains to be seen. The key is to direct the pluripotent ES cells to a cell lineage of interest at a particular developmental stage. Because many of the principles of cell fate specification are highly conserved, insights from developmental biology of other vertebrate animals and neural differentiation of mouse ES cells will be useful.

Neurogenesis in humans remains unexplored. It is generally accepted that embryonic induction including neural specification is a conserved process throughout evolution. Thus, it is not unreasonable to expect that neural specification in humans is similar to that in other vertebrates. The prevailing view on neural induction is that neuroectoderm is differentiated from ectoderm as long as the activity of transforming growth factor family such as BMPs is inhibited (reviewed in ref. *32*). In other words, neuroepithelia are the default fate of ectoderm during gastrulation. This theory receives substantial support from work performed in *Xenopus* and chick embryos in which cell fate can be followed continuously. It may also be true in mammals because mouse ES cells transform to primitive neural stem cells in a clonal culture that removes inhibitory signals from neighboring cells. The emerging idea is that neural induction may take place before gastrulation and members of the FGF family play a key role *(15,16)*. What is different is that, in humans, neural differentiation process takes place in a substantially longer period than in mice. Practically, cloning efficiency of human ES cells is substantially lower than that of mouse ES cells. Considering that cellular development is not only dependent on extrinsic signals but also on intrinsic programs *(33)*, we have designed a chemically defined culture system to induce human ES cells toward a neural fate, which mimics in vivo neural development in terms of timing, spatial organization, and, potentially, the mechanism of neural specification.

Neural differentiation is initiated through the aggregation of ES cells in suspension for 4 d. These aggregates are traditionally termed "embryoid bodies" because, in a long-term culture, they develop into a cystic cavity surrounded by three germ layers, resembling an embryo at gastrulation. It is speculated that cell–cell interaction in the aggregate may initiate a differentiation process, which positions the cells into an inside–outside pattern. This can be seen in EBs in which the ectoderm resides interiorly, the endoderm sits exteriorly, and the mesoderm is in the middle *(34,35)*. Even the aggregates that do not form cystic cavities appear to possess positional information, as neuroectodermal cells always develop in the center of an ES cell aggregate that has been cultured in suspension for 4 d *(see* below). In mouse, a simple aggregation of ES cells appears to induce a transition of ES cells to epiblasts or primitive ectoderm cells *(22,36)*. The primitive ectodermal stage is critical for embryonic induction and patterning. Because embryonic induction including neural specification takes place between blastula and gastrula stages, it would be counterproductive to use cyst-forming EBs to "guide" specific lineage differentiation. Thus, we apply neural inducing agents into ES cell aggregates that have been cultured for 3–4 d. These are solid cell aggregates without any cystic cavity. Hence, the name EB is somewhat misleading. In practice, mouse ES cells are usually trypsinized and placed in a bacteriological-grade Petri dish or in hanging drops so that individual ES cells aggregate together. As human ES cells survive better in aggregates, we have designed an approach to simply "lift" the ES cell colonies from the feeder layer or Matrigel by treatment with a low concentration of dispase or collagenase instead of trypsin. The detached ES cell colonies form round aggregates when placed in a suspension culture *(37)*.

Most of the lineage induction protocols employ the addition of morphogens or growth factors to the ES cell aggregates in suspension cultures.

This is technically straightforward, as it is a simple extension of the suspension culture. It, however, has some drawbacks. As mentioned earlier, an extended culture of ES cell aggregates in suspension often leads to cyst formation, resulting in a differentiation culture that is not controlled. An unusually high concentration of morphogens or growth factors is required in order for the factors to reach cells in the aggregates *(4,25,30,31)*. Depending on the size of the aggregates, cells on the surface and those inside the aggregates will have a varied degree of exposure to morphogens, thus creating a wide range of cell lineages or cells at various developmental stages. Because of the cluster nature, it is impossible to visualize the continual change of cell morphology in response to treatments. To overcome these drawbacks and to preserve the cellular interactions within the aggregate (or colony) at the same time, we plated the ES cell aggregates onto a culture dish in a chemically defined medium. In this way, cells in the aggregate formed a colony of monolayer cells in a low-density culture within 48 h. These cells were morphologically indistinguishable from ES cells grown on feeder layer or Matrigel. In the presence of FGF2, however, cells in the colony center transformed to small elongated cells, whereas those in the periphery gradually became flattened (*see* Fig. 1A,B). This small, columnar cell population expanded in the presence of FGF2 and organized into rosette formations by 7 d after plating the aggregates (Fig. 1C). These rosette formations are reminiscent of the early neural tube viewed from coronal sections (Fig. 1D). Hence, the small, columnar cells in the rosettes are likely neuroepithelial cells. This is confirmed by their expression of the early neuroepithelial markers, nestin and Musashi-1 *(37)*.

Considering human ES cells are equivalent to inner cell mass of a 5- to 6-d-old embryo, rosette formation in vitro translates into d 16–18 in a human embryo, the time when neuroectoderm begins to develop *(1)*. Thus, the in vitro neural specification recapitulates in vivo neural ectoderm formation with respect to temporal development, suggesting that the intrinsic program of neural specification is preserved in this culture system. This notion is substantiated by our observation that the temporal course of neural rosette formation from rhesus monkey and mouse ES cells is consistent with that of neuroectoderm specification in monkey and mouse embryos respectively *(38;* Zhang, unpublished data). The in vitro-generated neuroepithelial cells, identified by columnar morphology and expression of neuroepithelial markers nestin and Musashi-1, invariably organize into neural-tube-like rosettes in the center of the colony. These neural rosettes segregate themselves from the surrounding non-neural cells in extended culture *(37)*. Hence, the spatial arrangement of neuroepithelial cells in the adherent colony culture mirrors

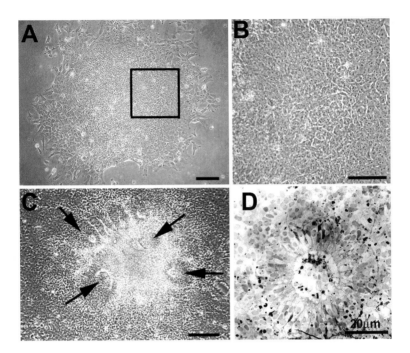

Fig. 1. Neural specification of hES cells. (**A**) Phase-contrast photograph of a colony of monolayer hES cells (H9, p22) formed after an aggregate was plated onto a culture flask in the neural induction medium *(37)* in the presence of FGF2 for 3 d. Cells in the colony center began to transform into a small columnar shape. (**B**) Enlarged view of the inset area in (**A**), showing that cells in the colony center (left side) are small and columnar in morphology, whereas those in the periphery (right side) resemble ES cells but with more cytoplasm. (**C**) Seven days after the ES cell aggregate was plated, the columnar cells in the colony center developed into multiple rosette formations (arrows). (**D**) Cross-section (1 μm, stained with toluidine blue) of these rosette formations indicates that the columnar cells organize into a neural-tube-like structure. ([**C**] and [**D**] are reproduced from Fig. 1B of ZhanBg et al. *[37]* with permission). Bar = 100 μm.

positional organization in an embryo. Therefore, the ES cell neural differentiation culture system offers an ideal model to dissect the effect of signaling molecules on neural specification in humans.

Because the neuroepithelial cells aggregate together as rosette formations and the aggregate delineates from surrounding non-neural cells in extended culture, the neural rosettes are separated from the surrounding cells by treatment with a low concentration of dispase. Neural rosette aggregates detach faster than surrounding cells and, hence, are differentially separated. Contaminated cells can be removed by a differential adhesion step as the flat cells attach to regular plastic culture surface faster. Neuroepithelial cells isolated in this way contain at least 95% of the cells that are positive for nestin. The separated neural population comprises over 70% of the total ES cell derivatives *(37)*, suggesting the high efficiency of the differentiation paradigm.

4. PROPERTIES OF hES-CELL-GENERATED NEURAL CELLS

Neural progenies differentiated from hES cells differ from each other depending on the protocols used. Similar to mouse ES cells, RA treatment of hES cells results in the generation of neural cells at various developmental stages *(30,31)*, as does the spontaneous differentiation from hES cells in high-density cultures *(29)*. In both cases, however, neural precursor cells can be isolated based on their expression of neural cell adhesion molecules on the cell surface *(30)* or on their characteristic morphological features *(29)*. In contrast, FGF treatment induces a synchronized differentiation and/or proliferation of hES to neural precursor cells without mature neurons and glia *(37)*. Although the nature of those neural precursors remains to be clarified, they appear to be at different developmental stages. Neural precursor cells generated from RA treatment and spontaneous differentiation express polysialylated neural cell adhesion molecule (PSA-NCAM) and proliferate in response to both FGF2 and epiderrmal growth factor (EGF) *(29,30)*, whereas those induced by FGF2 do not express PSA-NCAM and do not require EGF for proliferation *(37)*. However, when expanded in suspension or grown as dissociated cells in an adherent culture in the presence of FGF2, the NCAM-negative cells will express NCAM *(37)*, suggesting that the NCAM-negative cells are at an earlier developmental stage than the NCAM-positive cells. Given the fact that the FGF2-induced neural precursors appear first, at the time equivalent to the birth of neuroectoderm in vivo, and that these cells invariably organize into a neural-tube-like rosette formation, it is likely that the NCAM-negative precursors are neuroectoderm cells. Whether NCAM– and NCAM+ populations of neural precursor cells have different capacities of population expansion and downstream lineage differentiation is presently unknown.

The hES-derived neural precursors can be expanded for at least several weeks while they still generate neurons and glia in culture *(29,37)*. Generally, these precursors differentiate predominantly into neuronal cells within 2–3 wk of differentiation culture. Glial differentiation is a delayed process (*see* Fig. 2). This temporal pattern of neuronal and glial differentiation from neural precursors is similar to what has been seen in mouse neuroectodermal cells *(39)*, suggesting that the intrinsic neural differentiation program is essentially the same as in rodents.

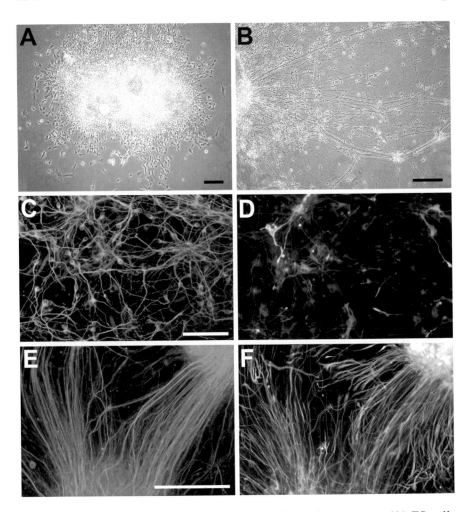

Fig. 2. In vitro differentiation of hES-derived neural precursors. (**A**) ES cell-derived neural rosettes (H1.1, p68), when plated onto laminin substrate in the differentiation medium in the absence of FGF2 *(37)*, generated cells of epithelial morphology, some of which began to extend out neurites at 3 d. (**B**) After 2 wk in differentiation culture, neurons with multiple long neurites developed and the neurites connected to each other forming networks. Immunostaining after 3 wk of differentiation indicates that the majority of cells are β_{III}-tubulin+ neurons (**C**) and that only a few cells are GFAP+ astrocytes (**D**). After 45 d of differentiation, many more GFAP+ astrocytes (**F**) appeared along with NF200+ neurites (**E**). Bar = 100 μm.

Therapeutic application of hES derivatives relies on the functional property of the in vitro-generated cells. Mouse ES-cell-derived neural precursor cells have been shown to further generate specialized neurons and glial cells.

The ES-generated neurons possess similar electrophysiological characteristics as neurons in primary cultures *(4)*. After transplantation into animal models of neurological disorders, the mouse ES-cell-generated neural precursors contribute to functional recovery of the grafted animals *(19,40)*. Electrophysiological analyses by Carpenter et al. *(30)* indicate that hES-cell-derived neurons express voltage-gated potassium and sodium currents when depolarized. These neurons can also fire action potentials in response to depolarizing stimuli. Therefore, hES-cell-derived neurons are electrophysiologically active.

In vivo analyses of the hES-derived neural cells are at the proof-of-concept stage. Transplantation of hES-derived neural precursors into the ventricles of neonatal mouse brains yields a widespread migration of the grafted cells. These cells incorporate into both neurogenic regions, such as the hippocampus and the rostral migratory pathway, and non-neurogenic areas, such as the cerebral cortex. More importantly, they differentiate into mature neurons and glia, which are indistinguishable from endogenous cells unless otherwise marked *(29,37)*. These results suggest that the in vitro-generated neural precursors, similar to their mouse counterparts, are able to mature in response to normal developmental cues. It is worth noting that the neural precursors are generated from different hES cell lines using very different approaches in different laboratories. Hence, the engraftability and responsiveness to local cues may be inherent to neural cells at a particular developmental stage.

Whether the hES-cell-derived neural cells incorporate into adult brain and consequently contribute to neural function is the testimony of potential future application of hES cells in restoring neurological deficits. We have transplanted hES-cell-derived neuroepithelial cells into the striatum of adult rats that have been subjected to 6-OH dopamine treatment to create a parkinsonian state. In the adult brain environment, hES-cell-derived neuroepithelial cells largely remain as immature, nestin-expressing progenitors for at least 3 mo. A small population of the grafted cells differentiate into neurons that express β_{III}-tubulin. Yet, a smaller number of cells mature and express tyrosine hydroxylase, the enzyme that is required for dopamine synthesis *(41)*. Thus, hES-cell-generated neuroepithelial cells are capable of generating mature neurons in adult brains but are largely dependent on their intrinsic maturation program in the absence of developmental cues. Studies are ongoing to analyze the temporal maturation process and functional integration of the hES-derived neuroepithelia in adult brains.

5. OUTSTANDING QUESTIONS

5.1. How Is the Neural Fate Specified from hES Cells?

Neural induction in vertebrates, particularly in *Xenopus* and chicks, is achieved through FGF and/or anti-BMP signalings. Whether the same signaling mechanism is used in human neuroectoderm induction needs confirmation. Human ES cells offer the only alternative to the inaccessible experimental paradigm. Development of neuroectoderm in humans takes about 3 wk, whereas it takes place in 1 wk in mice and within 1 d in chicks. Understanding how individual signaling molecules at a particular developmental stage of this protracted period orchestrate the neural fate specification from hES cells will still be a challenge to developmental biologists. Yet, this is essential to devising strategies that will ultimately lead to the application of hES cells in repairing neurological disorders.

5.2. What Is the Nature of hES-Cell-Derived Neural Precursor Cells?

"Neural precursor cells" have been designated for neural derivatives generated from both mouse and human ES cells. Are they neuroectodermal cells that are able to generate both CNS and PNS cells? Or are they neural stem cells or neural progenitors that are committed to the CNS fate? Literature to date indicates that ES-cell-derived neural precursors are of CNS characteristics. Does it mean that CNS neural differentiation from ES cells is a default pathway? Or perhaps it is simply because we have not looked into the PNS phenotypes of these in vitro-generated neural precursors. Neural stem/progenitor cells derived from embryonic mouse and human brain tissues are generally regionally specified *(42,43)*. That explains at least in part why human fetal brain-derived neural stem/progenitor cells are difficult to be directed to specialized neuronal and glial lineages. Are the ES-derived neural precursors, using various approaches, regionally specified and/or biased to certain lineage fate as brain-derived neural precursors? Answers to these questions are important both from a biological standpoint and application perspectives.

5.3. Can the In Vitro-Generated Human Neural Cells Integrate into the Brain and Spinal Cord and Contribute to Neural Functions After Transplantation?

Transplantation studies to date clearly indicate that the in vitro-generated human neural precursors differentiate into neurons and glial cells in response to local cues. The key to the application of hES cells in replacement therapy in neurological diseases is the demonstration that these mature neurons and glial cells can incorporate into the neural circuitry in a functional manner and consequently contribute to the recovery of functional deficits.

ACKNOWLEDGMENTS

Studies in my laboratory have been supported by the National Center for Research Resources (NCRR) of NIH, the Michael J. Fox Foundation, the National ALS Association, and the Myelin Project.

REFERENCES

1. O'Rahilly, R. and Muller, F. (ed.) (1994) *The Embryonic Human Brain*, Wiley-Liss, New York.
2. Thomson, J. A., Itskovitz-Eldor, J., Shapiro, S. S., et al. (1998) Embryonic stem cell lines derived from human blastocysts, *Science* **282**, 1145–1147.
3. Reubinoff, B. E., Pera, M. F., Fong, C. Y., Trounson, A., and Bongso, A. (2000) Embryonic stem cell lines from human blastocysts: somatic differentiation in vitro, *Nature Biotechnol.* **18**, 399–404.
4. Bain, G., Kitchens, D., Yao, M., Huettner, J. E., and Gottlieb, D. I. (1995) Embryonic stem cells express neuronal properties in vitro, *Dev. Biol.* **168**, 342–357.
5. Jones-Villeneuve, E. M. V., McBurney M. W., Rogers, K. A., and Kalnins, V. I. (1982) Retinoic acid induces embryonic carcinoma cells to differentiate into neurons and glial cells, *J. Cell Biol.* **94**, 253–262.
6. Wobus, A. M., Grosse, R., and Schoneich, J. (1988) Specific effects of nerve growth factor on the differentiation pattern of mouse embryonic stem cells in vitro, *Biomed. Biochim. Acta* **47**, 965–973.
7. Fraichard, A., Chassande, O., Bilbaut, G., Dehay, C., Savatier, P., and Samarut, J. (1995) In vitro differentiation of embryonic stem cells into glial cells and functional neurons, *J. Cell Sci.* **108**, 3181–3188.
8. Strubing, C., Ahnert-Hlger, G., Shan, J., Wiedenmann, B., Hescheler, J., and Wobus, A. M. (1995) Differentiation of pluripotent embryonic stem cells into the neuronal lineage in vitro gives rise to mature inhibitory and excitatory neurons, *Mech. Dev.* **53**, 275–287.
9. Dinsmore, J., Ratliff, J., Deacon, T., et al. (1996) Embryonic stem cells differentiated in vitro as a novel source of cells for transplantation, *Cell Transplant.* **5**, 131–143.
10. Renoncourt, Y., Carroll, P., Filippi, P., Arce, V., and Alonso, S. (1998) Neurons derived in vitro from ES cells express homeoproteins characteristic of motoneurons and interneurons, *Mech. Dev.* **79**, 185–197.
11. Maden, M. (2002) Retinoid signaling in the development of the central nervous system, *Nat. Rev. Neurosci.* **3**, 843–853.
12. Gottlieb, D. I. and Heuttner, J. E. (1999) An in vitro pathway from embryonic stem cells to neurons and glia, *Cells Tissues Organs* **165**, 165–172.
13. Muhr, J., Graziano, E., Wilson, S., Jessell, T. M., and Edlund, T. (1999) Convergent inductive signals specify midbrain, hindbrain, and spinal cord identity in gastrula stage chick embryos, *Neuron* **23**, 689–702.
14. Okabe, S., Forsberg-Nilsson, K., Spiro, A. C., Segal, M., and McKay, R. D. G. (1996) Development of neuronal precursor cells and functional postmitotic neurons from embryonic stem cells in vitro, *Mech. Dev.* **59**, 89–102.

15. Streit, A., Berliner, A. J., Papanayotou, C., Sirulnik, A., and Stern, C. D. (2000) Initiation of neural induction by FGF signaling before gastrulation, *Nature* **406,** 74–78.
16. Wilson, S. I., Graziano, E., Harland, R., Jessell, T. M., and Edlund, T. (2000) An early requirement for FGF signaling in the acquisition of neural cell fate in the chick embryo, *Curr. Biol.* **10,** 421–429.
17. Brustle, O. Jones, K. N., Learish, R. D., et al. (1999) Embryonic stem cell-derived glial precursors: a source of myelinating transplants, *Science* **285,** 754–756.
18. Lee, S.-H., Lumelsky, N., Studer, L., Auerbach, J. M., and McKay, R. D. (2000) Efficient generation of midbrain and hindbrain neurons from mouse embryonic stem cells, *Nat. Biotechnol.* **18,** 675–679.
19. Kim, J.-H., Auerbach, J. M., Rodriguez-Gomez, J. A., et al. (2002) Dopamine neurons derived from embryonic stem cells function in an animal model of Parkinson's disease, *Nature* **418,** 50–56.
20. Kawasaki, H., Mizuseki, K., Nishikawa, S., et al. (2000) Induction of midbrain dopaminergic neurons from ES cells by stromal cell-derived inducing activity, *Neuron* **28,** 31–40.
21. Kawasaki, H., Suemori, H., Mizuseki, K., et al. (2002) Generation of dopaminergic neurons and pigmented epithelia from primate ES cells by stromal cell-derived inducing activity, *Proc. Natl. Acad. Sci. USA* **99,** 1580–1585.
22. Rathjen, J., Haines, B. P., Hudson, K. M., Nesci, A., Dunn, S., and Rathjen, P. D. (2002) Directed differentiation of pluripotent cells to neural lineages: homogeneous formation and differentiation of a neuroectoderm population, *Development* **129,** 2649–2661.
23. Tropepe, V., Hitoshi, S., Sirard, C., Mak, T. W., Rossant, J., and van der Kooy, D. (2001) Direct neural fate specification from embryonic stem cells: a primitive mammalian neural stem cell stage acquired through a default mechanism, *Neuron* **30,** 65–78.
24. Finley, M. F., Devata, S., and Huettner, J. E. (1999) BMP-4 inhibits neural differentiation of murine embryonic stem cells, *J. Neurobiol.* **40,** 271–287.
25. Wichterle, H., Lieberam, I., Porter, J. A., and Jessell, T. M. (2002) Directed differentiation of embryonic stem cells into motor neurons, *Cell* **110,** 385–397.
26. Mujtaba, T., Piper, D. R., Kalyani, A., Groves, A. K., Lucero, M. T., and Rao, M. S. (1999) Lineage-restricted neural precursors can be isolated from both the mouse neural tube and cultured ES cells, *Dev. Biol.* **214,** 113–127.
27. Li, M., Pevny, L., Lovell-Badge, R., and Smith, A. (1998) Generation of purified neural precursors from embryonic stem cells by lineage selection, *Curr. Biol.* **8,** 971–977.
28. Smith, A. G. (2001) Embryo-derived stem cells: of mice and men, *Annu. Rev. Cell Dev. Biol.* **17,** 435–462.
29. Reubinoff, B. E., Itsykson, P., Turetsky, T., et al. (2001) Neural progenitors from human embryonic stem cells, *Nature Biotechnol.* **19,** 1134–1140.
30. Carpenter, M. K., Inokuma, M. S., Denham, J., Mujtaba, T., Chiu, C. P., and Rao, M. S. (2001) Enrichment of neurons and neural precursors from human embryonic stem cells, *Exp. Neurol.* **172,** 383–397.
31. Schuldiner, M., Eiges, R., Eden, A., et al. (2001) Induced neuronal differentiation of human embryonic stem cells, *Brain Res.* **913,** 201–205.

32. Wilson, S. I. and Edlund, T. (2001) Neural induction: toward a unifying mechanism, *Nature Neurosci.* **4**, 1161–1168.
33. Edlund, T. and Jessell, T. M. (1999) Progression from extrinsic to intrinsic signaling in cell fate specification: a view from the nervous system, *Cell* **96**, 211–224.
34. Wiles, M. V. (1995) Embryonic stem cell differentiation in vitro, *Methods Enzymol.* **225**, 900–918.
35. O'Shea, K. S. (1999) Embryonic stem cell models of development, *Anat. Rec.* **257**, 32–41.
36. Lake, J.-A., Rathjen, J., Remiszewski, J., and Rathjen, P. D. (2000) Reversible programming of pluripotent cell differentiation, *J. Cell Sci.* **113**, 555–566.
37. Zhang, S.-C., Wernig, M., Duncan, I. D., Brüstle, O., and Thomson, J. A. (2001) In vitro differentiation of transplantable neural precursors from human embryonic stem cells, *Nature Biotechnol.* **19**, 1129–1133.
38. Piscitelli, G. M. and Zhang, S.-C. (2002) Differentiation of neural precursors from rhesus monkey embryonic stem cells, *Soc. Neurosci. Abstr.* **7**, 5.
39. Qian, X., Shen, Q., Goderie, S. K., et al. (2000) Timing of CNS cell generation: a programmed sequence of neuron and glial cell production from isolated murine cortical stem cells, *Neuron* **28**, 69–80.
40. McDonald, J. W., Liu, X. Z., Qu, Y., et al. (1999) Transplanted embryonic stem cells survive, differentiate and promote recovery in injured rat spinal cord, *Nature Med.* **5**, 1410–1412.
41. Yan, Y. P., Lyons, E., Moreno, P., and Zhang, S.-C. (2002) Survival and differentiation of human embryonic stem cell-derived neural precursors in a rat model of Parkinson's disease. *Soc. Neurosci. Abstr.* **429**, 8.
42. Hitoshi, S., Tropepe, V., Ekker, M., and van der Kooy, D. (2002) Neural stem cell lineages are regionally specified, but not committed, within distinct compartments of the developing brain, *Development* **129**, 233–244.
43. Ostenfeld, T., Joly, E., Tai, Y. T., et al. (2002) Regional specification of rodent and human neurospheres, *Dev. Brain Res.* **134**, 43–55.

9

Pancreatic Differentiation of Pluripotent Stem Cells

Nadya Lumelsky

1. DIABETES AS A TARGET FOR STEM CELL THERAPY

Since their derivation in 1999, human pluripotent embryonic stem (ES) and embryonic germ (EG) cells have attracted considerable attention *(1–3)*. Mainly this is the result of the promise of application of these cells to the treatment of a variety of diseases that involve tissue and organ degeneration *(4,5)*. There is no doubt that even with the best organ procurement programs in place, the supply of cadaveric organs needed for transplantation would never satisfy the demand. Type I and type II diabetes mellitus, which affect 16 million people in the United States alone, are among the most promising targets for pluripotent-stem-cell-based applications *(6,7)*. Although they are different diseases, type I and type II diabetes both involve inadequate mass of insulin-producing β-cells, a condition resulting in elevated blood glucose level (hyperglycemia). Insulin injections alleviate hyperglycemia in most patients, but they do not provide a stable finely tuned control of glucose homeostasis that is provided by the healthy pancreas. This lack of fine control in turn leads to numerous complications, including cardiovascular and kidney disease, neuropathy, and blindness.

Until several years ago, whole-pancreas transplantation was the only treatment capable of alleviating hyperglycemia in diabetic patients not controlled by insulin injections. However, following recent improvement in human pancreatic islet isolation *(8)* and development of a new glucocorticoid-free immunosuppressive regimen *(9)*, islet transplantation began to be viewed as an ultimate treatment of diabetes *(7)*. There are two major advantages to islet transplantation over the whole-organ transplantation. First, the surgical procedure of islet grafting, which involves injecting islets into the portal vein of the liver, is significantly simpler compared to transplantation of the whole pancreas. It results in fewer complications, requires shorter

From: *Human Embryonic Stem Cells*
Edited by: A. Chiu and M. S. Rao © Humana Press Inc., Totowa, NJ

hospitalization, and allows for additional islet reinjection if deemed necessary. Second, when islets are prepared, the quality requirements for the donated pancreas are not as stringent as for the pancreas to be used for whole-organ transplantation. Unfortunately, because of the severe shortage of tissue, islet transplantation cannot become a widespread therapeutic modality. However, if a safe and abundant source of functional islets can be obtained from stem cells, injection of islets will most likely replace pancreas transplantation as a treatment of diabetes.

2. STEM CELLS: DEFINITIONS

All definitions of stem cells include a notion of self-renewal, coupled with a potential to generate one or more differentiated descendant cell types as well as functional reconstitution of an organ or a whole organism *(10,11)*. It is important to emphasize that differentiation of stem cells is a multistage process. When a stem cell commits to differentiation, it generates an intermediate population of progenitor cells with limited proliferative capacity and restricted differentiation potential. The progenitors then undergo further development to produce fully differentiated progeny. In many adult tissues, including endocrine pancreas, fully differentiated cells are generally postmitotic *(12)*.

Stem cells are classified based on the extent of their differentiation potential. At the top of the hierarchy are totipotent stem cells, which, when introduced in vivo, have the potential to contribute to every tissue of an embryo, including the germ line and extraembryonic tissues. An example of a totipotent stem cell is a fertilized egg. One level lower on the hierarchy are pluripotent ES and EG cells, which have a more limited differentiation potential—they contribute to embryonic, but not to extraembryonic tissues. Mouse ES cells have been first isolated more than 20 yr ago *(13)* and have since been used widely for the introduction of mutations into the mouse germ line. Mouse and, more recently, human ES cells are also widely used for modeling embryonic development in vitro and for developing new approaches for future use in cell-based therapies for treatment of a variety of diseases resulting from cell-type-specific degeneration *(14)*.

Many adult tissues are thought to contain stem cells; these are referred to as multipotent or tissue-specific stem cells *(11)*. Multipotent cells fall lower still on the stem cell hierarchy. They are generally considered to be restricted to producing only the cell types of the tissue in which they reside. This lineage restriction has recently become a center of a controversy, with some works suggesting that adult stem cells can give rise to heterologous cell types, and still others challenging this notion *(15–17)*. Several adult stem cell types, including hematopoietic, neural, and epidermal, have been extensively characterized *(18–20)*. Not so with the adult pancreatic stem cells—despite extensive effort, their nature still remains relatively obscure *(21)*.

3. PROLIFERATION OF STEM CELLS IN VITRO

Our ability to expand adult stem and progenitor cells in vitro without differentiation is limited. With one notable exception described recently *(22)*, long-term expansion of adult stem and progenitor cells in vitro normally leads to loss of proliferation and commitment to differentiation *(23)*. Alternatively, the cells may continue to proliferate, but become resistant to differentiation. Because in vivo adult stem cells must survive and function throughout the life of an organism, it is probable that the difficulty of manipulating these cells in vitro is the result of imperfect culture conditions. As understanding of the mechanisms controlling the balance among proliferation, differentiation, and apoptosis of stem cells develops, this difficulty might be overcome. Unlike adult stem cells, ES cells can easily be expanded in vitro in an undifferentiated state *(1,3,13)*. Thus, until the problem of expansion of adult stem cells in vitro is solved, ES cells are expected to remain a preferred source for obtaining large quantities of undifferentiated cell mass.

4. HETEROGENEITY OF ES CELL DIFFERENTIATION

In addition to their unlimited expansion capacity, the ES cells system has another important advantage, namely a shift from the proliferation to the differentiation mode induced by a simple change in culturing conditions. Despite these advantages, practical applications of ES cells to generation of desired cell types have been have not been achieved so far in any system. The difficulty in applying ES cell technology to clinical use stems from our poor knowledge of how to control normal multilineage differentiation, which takes place in ES cell cultures in directing this differentiation toward specific cell lineage. Although significant enrichment for a desired cell type has been achieved for several lineages, the ES-cell-derived progeny always retains heterogeneity *(24–26)*. Also, a fraction of potentially tumorogenic, undifferentiated cells usually remains in the cultures. Several strategies to overcome these problems have been proposed. First, specific growth factors and stable gene transfer can increase the efficiency of ES cell differentiation *(25,27)*. Second, cell-type-specific selection can be used to purify cell populations of interest or to remove the undesired cells from the heterogeneous mixtures of cells *(28,29)*. Third, suicide genes can be introduced into the cells before transplantation in order to render them sensitive to specific drugs, thereby permitting selective ablation of tumorogenic cells *(30)*. It is probable that a combination of all these approaches will be incorporated into the ES-cell-based therapeutic protocols in the future.

Although recognizing the need for enriching the ES-cell-derived progeny for the cell type of interest, it is important to keep in mind that cell heterogeneity could be a potentially valuable feature of these cultures. It is widely recognized that cell–cell and cell–matrix interactions are essential for normal development. This microenvironment is often collectively referred to as the developmental niche *(31,32)*. It may be possible to mimic some components of the niches by including purified growth factors and extracellular matrix components into the culture medium. However, in most systems, our knowledge of the structure of the niches is far from complete. Because it has been established that during differentiation in vitro ES cell cultures recapitulate many steps of embryonic development *(14)*, it is reasonable to suggest that certain features of the developmental niches may also be recreated in these cultures.

5. PANCREATIC DEVELOPMENT: AN EXAMPLE TO FOLLOW

Achieving the goal of generating functional pancreatic islets in vitro from stem cells will require in-depth knowledge of the mechanisms of pancreatic organogenesis. Although our understanding of this process is still imperfect, many advances have been made in recent years. What follows is a brief review of the current knowledge in the field. Several excellent recent reviews provide additional details on the subject *(21,33,34)*.

5.1. Anatomy of the Pancreas; Pancreatic Stem Cells

The mammalian pancreas is composed of the exocrine acini, endocrine islets of Langerhans, and pancreatic ducts. The exocrine tissue produces digestive enzymes, which are released into the digestive tract through the pancreatic duct system. The islets, which are embedded in the exocrine tissue, are necessary for glucose utilization. They are complex miniorgans containing different cell types. Through a concerted action of these cell types, the islets respond to extracellular environment by releasing pancreatic endocrine hormones, which allow maintenance of glucose homeostasis. The hormone-producing islet cells are α, β, δ, and PP cells, they secrete glucagon, insulin, somatostatin, and pancreatic polypeptide, respectively. Diabetes results from a selective loss (type I), or malfunction and long-term loss (type II), of insulin-producing β-cells. It has been proposed that adult pancreatic ducts contain a population of islet stem and progenitor cells which, in certain situations in vivo or in vitro, can be induced to generate new islets *(35–37)*. However, these putative stem and progenitor cells have not yet been conclusively identified or isolated. There have been numerous attempts to use the genetically engineered transformed cells lines to generate glucose-responsive insulin-secreting cells in vitro *(38)*. Many of the derived cell lines demonstrated glucose-mediated insulin secretion. Nevertheless, these cell lines lack a fine kinetic control of insulin release characteristic of normal pancreatic islets, which is hardly surprising given the complex miniorgan structure of an islet.

5.2. Systems to Study Pancreatic Development

The development of the pancreas in vivo has been studied primarily through the use of mutant mouse model systems. These studies have allowed the identification of many crucial steps in early pancreatic development and have greatly contributed to our knowledge of pancreatic transcription factors and signaling pathways important for development of endocrine and exocrine pancreas *(39,40)*. In other works, pancreatic explant cultures have been used as reconstitution systems for testing effects of various growth factors, cell–cell and cell–matrix interactions on pancreatic differentiation and organogenesis *(41–43)*. Although useful, in vitro explant culture models have serious limitations: They do not provide access to substantial numbers of progenitor cells at different stages of development. An optimal in vitro pancreatic development system should allow significant expansion of pancreatic stem and progenitor cells coupled with generation of functional islets. If ES cells can be induced to efficiently adopt pancreatic fate and undergo islet morphogenesis, they should provide such an optimal in vitro system.

5.3. Mechanisms of Pancreatic Development

When discussing developmental control mechanisms, one should make a distinction between instructive and permissive signals. For a given cell and tissue type, instructive signals are responsible for committing a cell to a specific fate at the expense of other cell fates. Permissive signals allow survival and differentiation of already specified cell types *(16)*. Although a number of permissive signals responsible for pancreatic development have been identified, the available knowledge of instructive signals is still limited *(33)*. This complicates designing rational strategies for differentiation of ES cells toward pancreatic pathway. Consequently, as discussed below, the existing protocols have relied on selection of cells, which spontaneously adopt pancreatic fate, or on adopting strategies developed for generation of cell types related to pancreas.

During embryogenesis, the pancreas develops from a foregut portion of the embryonic endoderm (*see* Fig. 1), which is specified to become pancreatic endoderm through the action of still unidentified instructive factors. A variety of signaling molecules acting alone and in combination, such as the fibroblast growth factor family (FGFs), activins, the transforming growth factor family (TGFs), and a hedgehog family, are thought to play permissive role in early stages of pancreatic development *(33)*. Early in embryogenesis, the pancreas exists in separate dorsal and ventral portions (*see* Fig. 1), which develop in a close proximity to a notochord and a cardiac mesoderm, respectively. It has been shown that these nonpancreatic tissues produce signaling molecules that control development of dorsal and ventral buds *(41,44)*. Another tissue that contacts the notochord during the same developmental window is the neural tube. Hence, it was of little surprise that pancreatic and neural development were found to be controlled by a similar set of signaling molecules *(45)*. In the mouse, the dorsal pancreatic bud loses its contact with the notochord between the 14- and 16-somite stage (roughly embryonic day 8.5 [E8.5]) when the fusion of paired dorsal aorta takes place. At this time, the dorsal aorta is positioned in close proximity to the dorsal pancreatic bud. It has recently been shown *(46)* that after separation of the dorsal bud from the notochord, the dorsal aorta becomes a source of developmental signals for the dorsal pancreas. The first signs of insulin expression in the dorsal pancreatic bud appear around E9.5. The ventral bud reaches this stage about 1 d later. As the development progresses, the dorsal and ventral pancreatic buds fuse and the pancreatic epithelium becomes embedded in the mesenchyme, which then serves as the source of signals for growth, differentiation, and morphogenesis of the exocrine and endocrine pancreas *(45)*.

Transcription factors involved in pancreatic development are outlined in Fig. 2. As with the extracellular ligands, there exists an extensive overlap between transcription factors controlling the development of neural and pancreatic cells *(40)*. PDX-1, a member of ParaHox group of homeodomain transcription factors, and neurogenin3 (ngn3) a helix–loop–helix transcription factor are the two earliest markers defining pancreatic cells. Whereas *PDX-1* is expressed in the pancreas throughout embryogenesis and in the postnatal β-cells, *ngn3* is expressed transiently; it is silenced late in embryogenesis and in the adult islets.

5.4. Similarities Between Pancreatic and Neural Development

Considering their different embryological origins, the common developmental control mechanisms and the gene expression overlap of neural

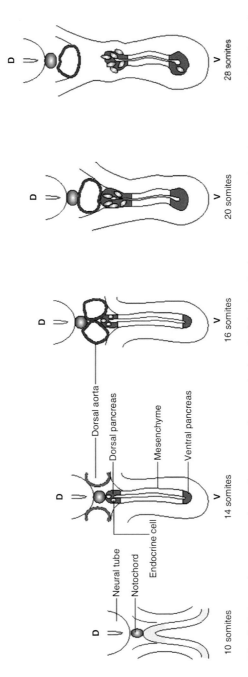

Fig. 1. Schematic cross-sections of mouse embryo at the level of the developing pancreas. D, dorsal; V, ventral. The 10-somite stage roughly corresponds to embryonic day 8 (E8) of development. The 28-somite stage roughly corresponds to E10. (Reprinted with permission from ref. *33*.)

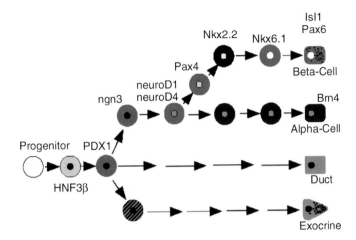

Fig. 2. Transcription factors expressed at different stages of pancreatic development. *HNF-3β* is expressed in the pancreatic endoderm. *PDX-1* marks the progenitor, which gives rise to all pancreatic lineages, whereas *neurogenin 3* (*ngn3*) has a more restricted expression pattern. Some factors function at several steps, but a single step is shown for simplicity. (Reprinted with permission from ref. *40*.)

and pancreatic cells are quite intriguing. In addition, functional overlap exists between these cell types. For example, the release of insulin by β-cells and the release of neurotransmitters by neurons is accomplished through similar mechanisms *(47)*. This suggests a common ancestral origin of insulin-producing cells and neurons. Along the same line of evidence, in *Drosophila*, ablation of insulin-producing cell clusters located in the brain leads to a phenotype reminiscent of diabetes *(48)*. Moreover, a recently described patient with a brain tumor of a neuroectodermal origin developed severe hypoglycemia resulting from insulin production by the tumor *(49)*.

Another common feature of pancreatic and neural cells is expression of a marker of neural progenitor cells, nestin *(50)*. It has been suggested that nestin might also be a marker of pancreatic progenitor cells. This question has recently been explored in several works in the context of the adult *(51–54)* and ES-cell-based *(55,56)* in vitro pancreatic differentiation systems. At this time, conclusive evidence for nestin as a marker of pancreatic progenitor cells is still lacking. In fact, this hypothesis has been challenged in several works, which claim the mesenchymal rather than epithelial nature of nestin-expressing cells found in the pancreas *(33,57,58)*. This debate must be resolved through careful cell lineage tracing approaches.

6. PANCREATIC DIFFERENTIATION
OF HUMAN AND MOUSE ES CELLS

Today there are only few published works in the area of pancreatic differentiation of human ES cells. The results pertaining to both human and mouse ES cells are discussed in the following sections.

6.1. Effect of Transfected Genes and Exogenous Growth Factors on Endodermal and Pancreatic Differentiation of ES Cells

Because in vivo the pancreas develops from embryonic endoderm, a logical strategy for enhancing pancreatic differentiation of ES cells would be to induce them first to adopt endodermal fate. At present, induction of the endoderm in human ES cell cultures has not been achieved. Still, it has been shown that during spontaneous differentiation of human ES cells in vitro, all three embryonic germ layers, including the endoderm, are generated *(59)*. In the mouse ES cell system, enrichment of endodermal fate has been accomplished by the transfer, into the cells, of genes coding for two transcription factors of the *hepatocyte nuclear factor 3* (*HNF3*) gene family known to play a role in endoderm formation *(60)*. Among the induced endodermal genes were *albumin, α_1-antitrypsin, transthyretin*, and *phosphoenolpyruvate carboxykinase* (*PEPCK*).

To investigate the potential of human ES cells for directed differentiation into different cell lineages, Benvenisty and co-workers studied the effect of eight growth factors on the expression of a wide panel of genes characteristic of the ectoderm, mesoderm, and endoderm *(61)*. Specifically, nerve growth factor (NGF), basic FGF, activin-A, transforming growth factor-β_1 (TGF-β_1), hepatocyte growth factor (HGF), epidermal growth factor (EGF), BMP4, and retioic acid (RA) were tested. Among the endodermal genes, expression of *albumin, α_1-antitrypsin, amylase, PDX-1*, and *insulin* was examined. To induce differentiation, human ES cells were removed from embryonic fibroblast feeder layers (these maintain ES cells in an undifferentiated state) and cultured in suspension, where they formed simple-cell aggregates called embryoid bodies (EBs). The EBs were dissociated after 5 d and cultured in adherent cell monolayers for another 10 d with or without growth factors. In a parallel experiment, the EBs were cultured for 20 d in suspension without added growth factors. Gene expression was examined using reverse transcriptase–polymerase chain reaction (RT-PCR). The authors report that insulin, PDX-1, and amylase mRNAs were strongly expressed in the 20-d-old EBs. These results imply that the pancreatic program is activated in spontaneously differentiating human ES cells. When the cell monolayers cultured without growth factors were analyzed for insulin, PDX-1, and amylase expression, only the amylase message was present. In the growth-factor-treated cells, PDX-1 and amylase, but not insulin, were expressed in NGF-treated cultures. Liver-specific genes, *albumin* and *α₁-antitrypsin* were also induced by NGF. None of the other growth factors had any effect on pancreatic gene expression of human ES cells.

6.2. Insulin Production and Secretion During Spontaneous Differentiation of ES Cells

In agreement with the results of Benvenisty et al., Assady and co-workers have found that during spontaneous differentiation of human ES cells in the absence of any added growth factors, several pancreatic genes are induced *(62)*. These experiments were carried out in two different culture formats: EBs in suspension and adherent cultures grown at high cell density. Insulin expression was examined by immunohistochemistry in the 19-d-old EBs. The authors found cells immunoreactive to insulin either scattered throughout EBs or in small clusters. They also found a progressive increase in the number of insulin-positive cells as EBs matured. To characterize these cells further, insulin secretion from 20- to 22-d-old EBs and 22- and 31-d-old high-density monolayers was determined in the presence of 5.5 mM and 25 mM glucose. The authors detected insulin secreted into the medium from both types of culture. However, this insulin secretion was not sensitive to glucose concentration. RT-PCR analysis of a panel of pancreatic endocrine genes was also conducted. The results showed that insulin, PDX-1, ngn3, glucokinase, and the β-cell-specific glucose transporter, Glut2, are induced in EB and in cell monolayers. Interestingly, the expression of PDX-1 and ngn3 preceded expression of insulin, Glut2, and glucokinase during the course of the culture, suggesting that, similarly to the situation in vivo, PDX-1 and ngn3 may control expression of these markers in the in vitro cultures *(40)*.

Shiroi et al. used the zinc-chelating agent dithizone (DTZ), which selectively stains β-cells, to observe the emergence of insulin-positive cell clusters in cultures of spontaneously differentiating mouse ES cells *(63)*. After 21 d in culture, cells faintly stained with DTZ became visible; the intensity of staining became more apparent by d 28 in culture. DTZ-positive cells were selectively extracted from the culture dishes and subjected to RT-PCR analysis for expression of several pancreatic markers. Insulin, glucogon, pancreatic polypeptide, but not somatostatin mRNAs were detected. Also, expression of Glut2, PDX-1, and HNF-3β mRNA was recorded.

6.3. Selection of Insulin-Secreting Cells from Differentiating ES Cell Cultures

A possible approach to obtaining a homogeneous differentiated cell progeny from in vitro ES cell cultures is to use methods that selectively elimi-

Fig. 3. Structure of plasmid used for selection of the insulin-producing mouse ES cell clone. Hygromycin β resistance gene is under the control of the ubiquitous 3-phosphoglycerate kinase (PGK) promoter. The neomycin-resistance gene is under the control of the human insulin promoter. (Adapted from ref. *64*.)

nate the heterologous cell types. This approach has been used in the past to obtain highly enriched populations of cardiomyocytes and neural precursors from mouse ES cells *(28,29)*. Soria and his co-workers used a similar strategy to select insulin-secreting cell clones from spontaneously differentiating mouse ES cells *(64)*. The authors introduced into the ES cells a plasmid conferring resistance to two antibiotics, hygromycin β and neomycin (*see* Fig. 3). A hygromycin β-resistance gene was under the control of a constitutive 3-phosphoglycerate kinase (PGK) promoter, which is expressed in undifferentiated ES cells. A neomycin-resistance gene was under the control of a human insulin promoter. In the first step of the protocol, the ES cells maintained in the undifferentiated state were selected for resistance to hygromycin β. This allowed the generation of a stable cell line where every cell carried the plasmid. Following this step, the ES cells were induced to differentiate in the presence of the second antibiotic, neomycin. Because the neomycin gene was under the control of the insulin promoter, only β-like cells survived the selection. The authors report that the insulin content of the ES-cell-derived progeny obtained using this strategy was approx 90% of the insulin content of normal mouse islets. When the insulin released in response to glucose and other agonists was measured in vitro, the cells showed a good stimulated response. Moreover, when implanted into diabetic mice, the insulin-producing cells normalized hyperglycemia. This normalization was, however, reversed after about 12 wk in about 40% of transplanted animals. The comparison of the glucose tolerance of the transplanted animals with that of the nondiabetic controls showed that the plasma glucose levels was significantly elevated and the recovery to normal glucose levels after meal challenge test was delayed in transplanted animals.

6.4. Generation of Insulin-Secreting Cell Clusters from ES Cells Under Conditions Favoring Neuronal Differentiation

As noted earlier, despite their different embryological origin, neurons and pancreatic cells share multiple developmental and functional similarities. It is thus reasonable to suggest that signals that promote in vitro differentiation of ES cells toward neural pathway might also enhance pancreatic differentiation of ES cells. This hypothesis can be tested, because protocols for efficient neural differentiation of mouse and human ES are available *(24,25,65,66)*. We have recently applied a modified neural differentiation protocol to cultures of mouse ES cells and examined pancreatic phenotype of the ES-cell-derived progeny *(55)*. As outlined in Fig. 4, the protocol consists of five stages: expansion of undifferentiated ES cells (stage 1), EB formation (stage 2), selective enrichment of cultures for nestin-expressing cells in a fetal calf serum (FCS)-free medium (stage 3), expansion of a nestin-enriched cell population in the presence of bFGF (stage 4), and, finally, differentiation by bFGF withdrawal (stage 5). We have found that, in addition to neurons, insulin-, glucagon-, and somatostatin-expressing cells were generated under these conditions. It is noteworthy that hormone-producing cells aggregated to form cell clusters where insulin-positive cells were in the interior and glucagon- and somatostain-positive cells were at the periphery of the clusters. This type of cell sorting is characteristic of normal rodent islets. As expected, neurons were also abundantly generated in these cultures. These neurons were primarily localized to the periphery of the isletlike clusters and were found in close association with hormone-expressing cells. When tested in vitro for response to glucose and other insulin-release agonists and antagonists, the cell clusters secreted insulin in the medium in a regulated fashion, suggesting that they utilize physiological mechanisms of insulin release.

This work indicates that neural differentiating protocols may be useful for enhancing pancreatic differentiation of ES cells. However, these ES-cell-derived cell clusters are still lacking important features of normal islets. For example, as the cells matured in the culture, expression of PDX-1 diminished rather than increased, possibly suggesting that a portion of insulin-producing cells are unstable and were undergoing apoptosis. Also, the content of insulin in the clusters was only about 2% of the normal insulin content of mouse islets. Consequently, when injected into diabetic mice, the clusters, although surviving and continuing to express insulin, were not able to correct hyperglycemia.

This ES cell differentiation protocol has been recently improved by Hori et al. *(56)*. The authors found that the addition of inhibitors of phosphoinositide 3-kinase (PI-3K) during stage 5 (*see* Fig. 4) increases the endocrine cell number and insulin content of mouse ES-cell-derived isletlike clusters. The insulin content of β-like cells obtained in the presence of inhibitor was 30-fold greater than of those in the absence of the inhibitor. They also established that the PI-3K inhibitor arrested cell proliferation in these cultures and thus eliminated tumorogenic cells. Moreover, the islet

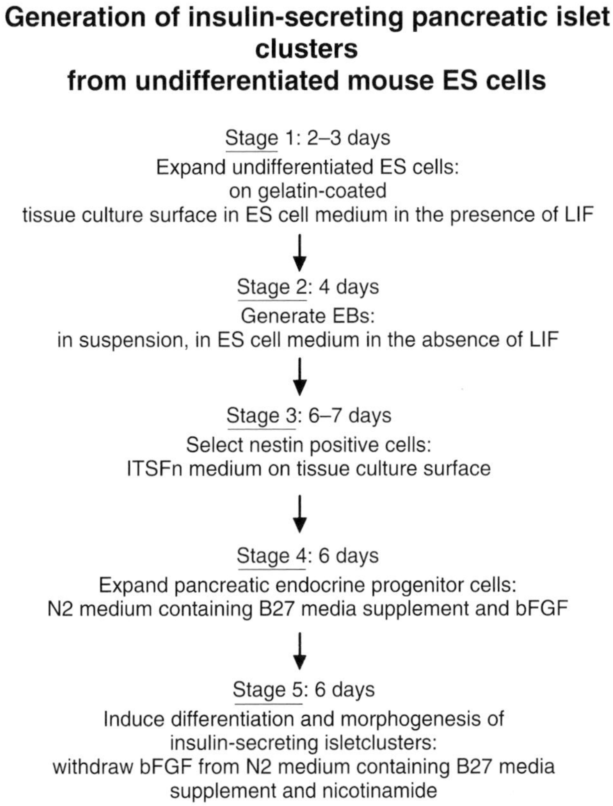

Generation of insulin-secreting pancreatic islet clusters from undifferentiated mouse ES cells

Stage 1: 2–3 days
Expand undifferentiated ES cells:
on gelatin-coated
tissue culture surface in ES cell medium in the presence of LIF

↓

Stage 2: 4 days
Generate EBs:
in suspension, in ES cell medium in the absence of LIF

↓

Stage 3: 6–7 days
Select nestin positive cells:
ITSFn medium on tissue culture surface

↓

Stage 4: 6 days
Expand pancreatic endocrine progenitor cells:
N2 medium containing B27 media supplement and bFGF

↓

Stage 5: 6 days
Induce differentiation and morphogenesis of
insulin-secreting isletclusters:
withdraw bFGF from N2 medium containing B27 media
supplement and nicotinamide

Fig. 4. Scheme of the differentiation protocol used for the generation insulin-producing cells from mouse ES cells. (Reprinted with permission from ref. *55*. Copyright 2001; American Association for the Advancement of Science.)

cultures generated under these conditions did not contain as many neural cells as did the cultures without the added PI3K inhibitor. Accordingly, when transplanted into diabetic mice, these cells, unlike the cells obtained in the absence of inhibitor *(55)*, were able to significantly reduce hyperglycemia, which re-emerged after removal of the graft. Despite significant improvement, the insulin content of the isletlike clusters generated using this protocol is still lower than that of the normal islets. Several other differences with normal islets such as less efficient processing of proinsulin and the absence from cultures of somatostatin- and pancreatic-polypeptide-expressing cells were also noted. Notwithstanding the fact that ES-cell-derived isletlike clusters generated by this protocol remain inferior to normal islets, to date this strategy represents the best available ES-cell-based pancreatic differentiation system. It combines technical simplicity with a relatively high efficiency. Another important advantage of this system is that it allows generation of functional isletlike structures rather than merely insulin-producing cells. Undoubtedly, in the future, attempts will be made to further improve this system and to apply it to the available human ES cell lines.

Two manuscripts of direct relevance to pancreatic differentiation of mouse ES cells appeared in print since this chapter was submitted for publication. In the first work Rajagopal et al. *(67)* reported that their attempt to reproduuce nestin differentiation protocol *(55)* with human and mouse ES cells resulted in significant insulin signal, as detected by immunochemistry, but only in minimal insulin gene expression signal, as determined by RT/PCR. On the basis of their results the authors argue that uptake of insulin from the growth medium significantly contributed to the overall level of insulin, which they detected by immunochemistry. The authors emphasize that in order to avoid possible artifacts of insulin detection, several different methods should be combined for analysis of pancreatic differentiation of ES cell cultures. In the second work, Blyszczuk et al. *(68)* showed that constitutive over-expression of β-cell specific transcription factor Pax4 combined with selection for nestin-positive cells and histotypic maturation significantly promotes pancreatic differentiation of mouse ES cells. This extension of the original nestin ES cell differentiation protocol *(55)* allowed generation of islet-like spheroid structures containing insulin secretory granules. Moreover, these spheroid structures were able to normalize glucose levels in diabetic mice. It is clear that the present debate on the role of nestin in pancreatic differentiation is a reflection of a rapid growth of this still young field. It will undoubtedlyh be resolved by further refinement of the existing differentiation protocols driven by the continued progress in our understanding of the mechanisms of pancreatic differentiation and morphogenesis.

7. CONCLUDING REMARKS

Human ES cells have a potential to provide an abundant supply of functional pancreatic islets for the treatment of diabetes. However, improvements in the efficiency of directed differentiation of ES cells toward the pancreatic pathway are required before this potential can be fully realized. A unique advantage of systems derived from ES cells is the possibility to generate unlimited amounts of undifferentiated cells. Moreover, upon differentiation, these cells may generate organlike structures composed of different cell types. Progress in the development of efficient ES-cell-based pancreatic differentiation systems will be highly dependent on advances in our understanding of normal pancreatic development. It is necessary to elucidate signals, particularly instructive signals, controlling pancreatic commitment. Also, we need to learn more about the genes governing proliferation, differentiation, and survival of pancreatic stem and progenitor cells. Several recent works have significantly advanced this knowledge *(69–75)*, but much still needs to be learned. For example, we still have only limited knowledge on the exact function of genes controlling pancreatic development. The results of gene inactivation studies in mouse models provide important information regarding involvement of specific genes in pancreatic development, however, they do not reveal the exact roles of these genes. In this regard, the approach used in several recent works holds high promise: It combines specific gene inactivation with lineage analysis, thus allowing tracing the fates of the cells carrying the inactivated gene *(76,77)*. Finally, we need to learn more about specific markers of pancreatic progenitor cells. Cell surface markers, are particularly important: They provide a simple means for progenitor cell isolation. This area remains practically unexplored. As in vitro models of pancreatic development, including the ES-cell-based models, are perfected, they will serve as useful assay systems to help to obtain the needed knowledge. Thus, we can expect progress in this area coming from two directions: Information obtained in vivo will help to develop in vitro assay systems, whereas more sophisticated in vitro systems will help elucidate the processes occurring in vivo. These advances will, in turn, lead to the development of clinically useful approaches for the generation of functional pancreatic islets from pluripotent ES cells.

REFERENCES

1. Thomson, J. A., Itskovitz-Eldor, J., Shapiro, S. S., et al. (1998) Embryonic stem cell lines derived from human blastocysts, *Science* **282**, 1145–1147.
2. Shamblott, M. J., Axelman, J., Wang, S., et al. (1998) Derivation of pluripotent stem cells from cultured human primordial germ cells, *Proc. Natl. Acad. Sci. USA* **95**, 13,726–13,731.
3. Reubinoff, B. E., Pera, M. F., Fong, C. Y., Trounson, A., and Bongso, A. (2000) Embryonic stem cell lines from human blastocysts: somatic differentiation in vitro, *Nat. Biotechnol.* **18**, 399–404.
4. Pera, M. F. (2001) Human pluripotent stem cells: a progress report, *Curr. Opin. Genet. Dev.* **11**, 595–599.
5. Park, K. I., Ourednik, J., Ourednik, V., et al. (2002) Global gene and cell replacement strategies via stem cells, *Gene Ther.* **9**, 613–624.
6. Halban, P. A., Kahn, S. E., Lernmark, A., and Rhodes, C. J. (2001) Gene and cell-replacement therapy in the treatment of type 1 diabetes: how high must the standards be set?, *Diabetes* **50**, 2181–2191.
7. Efrat, S. (2002) Cell replacement therapy for type 1 diabetes, *Trends Mol. Med.* **8**, 334–340.

8. Ricordi, C., Lacy, P. E., Finke, E. H., Olack, B. J., and Scharp, D. W. (1988) Automated method for isolation of human pancreatic islets, *Diabetes* **37**, 413–420.

9. Shapiro, A. M., Lakey, J. R., Ryan, E. A., et al. (2000) Islet transplantation in seven patients with type 1 diabetes mellitus using a glucocorticoid-free immunosuppressive regimen, *N. Engl. J. Med.* **343**, 230–238.

10. Morrison, S. J., Shah, N. M., and Anderson, D. J. (1997) Regulatory mechanisms in stem cell biology, *Cell* **88**, 287–298.

11. Clarke, D. and Frisen, J. (2001) Differentiation potential of adult stem cells, *Curr. Opin. Genet. Dev.* **11**, 575–580.

12. Bonner-Weir, S. (2000) Life and death of the pancreatic beta cells, *Trends Endocrinol. Metab.* **11**, 375–378.

13. Evans, M. J. and Kaufman, M. H. (1981) Establishment in culture of pluripotential cells from mouse embryos, *Nature* **292**, 154–156.

14. Rathjen, J. and Rathjen, P. D. (2001) Mouse ES cells: experimental exploitation of pluripotent differentiation potential, *Curr. Opin. Genet. Dev.* **11**, 587–594.

15. Slack, J. M. (2000) Stem cells in epithelial tissues, *Science* **287**, 1431–1433.

16. Weissman, I. L., Anderson, D. J., and Gage, F. (2001) Stem and progenitor cells: origins, phenotypes, lineage commitments, and transdifferentiations, *Annu. Rev. Cell Dev. Biol.* **17**, 387–403.

17. Ying, Q. L., Nichols, J., Evans, E. P., and Smith, A. G. (2002) Changing potency by spontaneous fusion, *Nature* **416**, 545–548.

18. Wagers, A. J., Christensen, J. L., and Weissman, I. L. (2002) Cell fate determination from stem cells, *Gene Ther.* **9**, 606–612.

19. Gage, F. H. (2000) Mammalian neural stem cells, *Science* **287**, 1433–1438.

20. Watt, F. M. (1998) Epidermal stem cells: markers, patterning and the control of stem cell fate, *Phil. Trans. R. Soc. Lond. B: Biol. Sci.* **353**, 831–837.

21. Bonner-Weir, S. and Sharma, A. (2002) Pancreatic stem cells, *J. Pathol.* **197**, 519–526.

22. Jiang, Y., Jahagirdar, B. N., Reinhardt, R. L., et al. (2002) Pluripotency of mesenchymal stem cells derived from adult marrow, *Nature* **418**, 41–49.

23. McKay, R. (2000) Stem cells—hype and hope, *Nature* **406**, 361–364.

24. Lee, S. H., Lumelsky, N., Studer, L., Auerbach, J. M., and McKay, R. D. (2000) Efficient generation of midbrain and hindbrain neurons from mouse embryonic stem cells, *Nat. Biotechnol.* **18**, 675–679.

25. Kim, J. H., Auerbach, J. M., Rodriguez-Gomez, J. A., et al. (2002) Dopamine neurons derived from embryonic stem cells function in an animal model of Parkinson's disease, *Nature* **418**, 50–56.

26. Kabrun, N., Buhring, H. J., Choi, K., Ullrich, A., Risau, W., and Keller, G. (1997) Flk-1 expression defines a population of early embryonic hematopoietic precursors, *Development* **124**, 2039–2048.

27. Czyz, J. and Wobus, A. (2001) Embryonic stem cell differentiation: the role of extracellular factors, *Differentiation* **68**, 167–174.

28. Klug, M. G., Soonpaa, M. H., Koh, G. Y., and Field, L. J. (1996) Genetically selected cardiomyocytes from differentiating embryonic stem cells form stable intracardiac grafts, *J. Clin. Invest.* **98**, 216–224.

29. Li, M., Pevny, L., Lovell-Badge, R., and Smith, A. (1998) Generation of purified neural precursors from embryonic stem cells by lineage selection, *Curr. Biol.* **8,** 971–974.

30. Fareed, M. U. and Moolten, F. L. (2002) Suicide gene transduction sensitizes murine embryonic and human mesenchymal stem cells to ablation on demand—a fail-safe protection against cellular misbehavior, *Gene Ther.* **9,** 955–962.

31. Watt, F. M. and Hogan, B. L. (2000) Out of Eden: stem cells and their niches, *Science* **287,** 1427–1430.

32. Tsai, R. Y., Kittappa, R., and McKay, R. D. (2002) Plasticity, niches, and the use of stem cells, *Dev. Cell* **2,** 707–712.

33. Edlund, H. (2002) Pancreatic organogenesis—developmental mechanisms and implications for therapy, *Nat. Rev. Genet.* **3,** 524–532.

34. Soria, B. (2001) In-vitro differentiation of pancreatic beta-cells, *Differentiation* **68,** 205–219.

35. Bonner-Weir, S., Taneja, M., Weir, G. C., et al. (2000) In vitro cultivation of human islets from expanded ductal tissue, *Proc. Natl. Acad. Sci. USA* **97,** 7999–8004.

36. Ramiya, V. K., Maraist, M., Arfors, K. E., Schatz, D. A., Peck, A. B., and Cornelius, J. G. (2000) Reversal of insulin-dependent diabetes using islets generated in vitro from pancreatic stem cells, *Nature Med.* **6,** 278–282.

37. Bouwens, L. and Kloppel, G. (1996) Islet cell neogenesis in the pancreas, *Virchows Arch.* **427,** 553–560.

38. Newgard, C. B., Clark, S., BeltrandelRio, H., Hohmeier, H. E., Quaade, C., and Normington, K. (1997) Engineered cell lines for insulin replacement in diabetes: current status and future prospects, *Diabetologia* **40(Suppl 2),** S42–S47.

39. Edlund, H. (2001) Factors controlling pancreatic cell differentiation and function, *Diabetologia* **44,** 1071–1079.

40. Schwitzgebel, V. M., Scheel, D. W., Conners, J. R., et al. (2000) Expression of neurogenin3 reveals an islet cell precursor population in the pancreas, *Development* **127,** 3533–3542.

41. Kim, S. K., Hebrok, M., and Melton, D. A. (1997) Notochord to endoderm signaling is required for pancreas development, *Development* **124,** 4243–4252.

42. Cras-Meneur, C., Elghazi, L., Czernichow, P., and Scharfmann, R. (2001) Epidermal growth factor increases undifferentiated pancreatic embryonic cells in vitro: a balance between proliferation and differentiation, *Diabetes* **50,** 1571–1579.

43. Gittes, G. K., Galante, P. E., Hanahan, D., Rutter, W. J., and Debase, H. T. (1996) Lineage-specific morphogenesis in the developing pancreas: role of mesenchymal factors, *Development* **122,** 439–447.

44. Zaret, K. S. (2001) Hepatocyte differentiation: from the endoderm and beyond, *Curr. Opin. Genet. Dev.* **11,** 568–574.

45. Slack, J. M. (1995) Developmental biology of the pancreas, *Development* **121,** 1569–1580.

46. Lammert, E., Cleaver, O., and Melton, D. (2001) Induction of pancreatic differentiation by signals from blood vessels, *Science* **294,** 564–567.

47. Komatsu, M., Yokokawa, N., Takeda, T., Nagasawa, Y., Aizawa, T., and Yamada, T. (1989) Pharmacological characterization of the voltage-dependent calcium channel of pancreatic B-cell, *Endocrinology* **125,** 2008–2014.

48. Rulifson, E. J., Kim, S. K., and Nusse, R. (2002) Ablation of insulin-producing neurons in flies: growth and diabetic phenotypes, *Science* **296,** 1118–1120.

49. Nakamura, T., Kishi, A., Nishio, Y., et al. (2001) Insulin production in a neuroectodermal tumor that expresses islet factor-1, but not pancreatic-duodenal homeobox 1, *J. Clin. Endocrinol. Metab.* **86,** 1795–1800.

50. Lendahl, U., Zimmerman, L. B., and McKay, R. D. (1990) CNS stem cells express a new class of intermediate filament protein, *Cell* **60,** 585–595.

51. Abraham, E. J., Leech, C. A., Lin, J. C., Zulewski, H., and Habener, J. F. (2002) Insulinotropic hormone glucagon-like peptide-1 differentiation of human pancreatic islet-derived progenitor cells into insulin-producing cells, *Endocrinology* **143,** 3152–3161.

52. Hunziker, E. and Stein, M. (2000) Nestin-expressing cells in the pancreatic islets of Langerhans, *Biochem. Biophys. Res. Commun.* **271,** 116–119.

53. Lechner, A., Leech, C. A., Abraham, E. J., Nolan, A. L., and Habener, J. F. (2002) Nestin-positive progenitor cells derived from adult human pancreatic islets of Langerhans contain side population (SP) cells defined by expression of the ABCG2 (BCRP1) ATP-binding cassette transporter, *Biochem. Biophys. Res. Commun.* **293,** 670–674.

54. Zulewski, H., Abraham, E. J., Gerlach, M. J., et al. (2001) Multipotential nestin-positive stem cells isolated from adult pancreatic islets differentiate ex vivo into pancreatic endocrine, exocrine, and hepatic phenotypes, *Diabetes* **50,** 521–533.

55. Lumelsky, N., Blondel, O., Laeng, P., Velasco, I., Ravin, R., and McKay, R. (2001) Differentiation of embryonic stem cells to insulin-secreting structures similar to pancreatic islets, *Science* **292,** 1389–1394.

56. Hori, Y., Rulifson, I. C., Tsai, B. C., Heit, J. J., Cahoy, J. D., and Kim, S. K. (2002) Growth inhibitors promote differentiation of insulin-producing tissue from embryonic stem cells, *Proc. Natl. Acad. Sci. USA* **99,** 16,105–16,110.

57. Selander, L. and Edlund, H. (2002) Nestin is expressed in mesenchymal and not epithelial cells of the developing mouse pancreas, *Mech. Dev.* **113,** 189–192.

58. Lardon, J., Rooman, I., and Bouwens, L. (2002) Nestin expression in pancreatic stellate cells and angiogenic endothelial cells, *Histochem. Cell Biol.* **117,** 535–540.

59. Drukker, M., Katz, G., Urbach, A., et al. (2002) Characterization of the expression of MHC proteins in human embryonic stem cells, *Proc. Natl. Acad. Sci. USA* **99,** 9864–9869.

60. Levinson-Dushnik, M. and Benvenisty, N. (1997) Involvement of hepatocyte nuclear factor 3 in endoderm differentiation of embryonic stem cells, *Mol. Cell Biol.* **17,** 3817–3822.

61. Schuldiner, M., Yanuka, O., Itskovitz-Eldor, J., Melton, D. A., and Benvenisty, N. (2000) From the cover: effects of eight growth factors on the differentiation

of cells derived from human embryonic stem cells, *Proc. Natl. Acad. Sci. USA* **97,** 11,307–11,312.

62. Assady, S., Maor, G., Amit, M., et al. (2001) Insulin production by human embryonic stem cells, *Diabetes* **50,** 1691–1697.

63. Shiroi, A., Yoshikawa, M., Yokota, H., et al. (2002) Identification of insulin-producing cells derived from embryonic stem cells by zinc-chelating dithizone, *Stem Cells* **20,** 284–292.

64. Soria, B., Roche, E., Berna, G., Leon-Quinto, T., Reig, J. A., and Martin, F. (2000) Insulin-secreting cells derived from embryonic stem cells normalize glycemia in streptozotocin-induced diabetic mice, *Diabetes* **49,** 157–162.

65. Okabe, S., Forsberg-Nilsson, K., Spiro, A. C., Segal, M., and McKay, R. D. (1996) Development of neuronal precursor cells and functional postmitotic neurons from embryonic stem cells in vitro, *Mech. Dev.* **59,** 89–102.

66. Bain, G., Kitchens, D., Yao, M., Huettner, J. E., and Gottlieb, D. I. (1995) Embryonic stem cells express neuronal properties in vitro, *Dev. Biol.* **168,** 342–357.

67. Rajagopal, J., Anderson, W. J., Kume, S., Martinez, O. I., and Melton, D. A. (2003) Insulin staining of ES cell progeny from insulin uptake, *Science* **299,** 363.

68. Blyszczuk, P., Czyz, J., Kania, G., et al. (2003) Expression of Pax4 in embryonic stem cells promotes differentiation of nestin-positive progenitor and insulin-producing cells, *Proc. Natl. Acad. Sci. USA* **100,** 998–1003.

69. Bhushan, A., Itoh, N., Kato, S., et al. (2001) Fgf10 is essential for maintaining the proliferative capacity of epithelial progenitor cells during early pancreatic organogenesis, *Development* **128,** 5109–5117.

70. Elghazi, L., Cras-Meneur, C., Czernichow, P., and Scharfmann, R. (2002) Role for FGFR2IIIb-mediated signals in controlling pancreatic endocrine progenitor cell proliferation, *Proc. Natl. Acad. Sci. USA* **99,** 3884–3889.

71. Hart, A. W., Baeza, N., Apelqvist, A., and Edlund, H. (2000) Attenuation of FGF signalling in mouse beta-cells leads to diabetes, *Nature* **408,** 864–868.

72. Tuttle, R. L., Gill, N. S., Pugh, W., et al. (2001) Regulation of pancreatic beta-cell growth and survival by the serine/threonine protein kinase Akt1/PKBalpha, *Nature Med.* **7,** 1133–1137.

73. Laybutt, D. R., Weir, G. C., Kaneto, H., et al. (2002) Overexpression of c-Myc in beta-cells of transgenic mice causes proliferation and apoptosis, down-regulation of insulin gene expression, and diabetes, *Diabetes* **51,** 1793–1804.

74. Pelengaris, S., Khan, M., and Evan, G. I. (2002) Suppression of Myc-induced apoptosis in beta cells exposes multiple oncogenic properties of Myc and triggers carcinogenic progression, *Cell* **109,** 321–334.

75. Gu, G., Dubauskaite, J., and Melton, D. A. (2002) Direct evidence for the pancreatic lineage: NGN3+ cells are islet progenitors and are distinct from duct progenitors, *Development* **129,** 2447–2457.

76. Bort, R. and Zaret, K. (2002) Paths to the pancreas, *Nature Genet.* **32,** 85–86.

77. Kawaguchi, Y., Cooper, B., Gannon, M., Ray, M., MacDonald, R. J., and Wright, C. V. (2002) The role of the transcriptional regulator Ptf1a in converting intestinal to pancreatic progenitors, *Nature Genet.* **32,** 128–134.

10

Human Embryonic
Stem Cell-Derived Cardiomyocytes
Derivation and Characterization

Chunhui Xu and Melissa K. Carpenter

1. INTRODUCTION

The generation of functional cardiomyocytes from human embryonic stem (hES) cells has several potential applications. Myocardial diseases resulting from damage of cardiac tissue affect millions of people, who may benefit from transplantation of cardiomyocytes. Such an application has already been demonstrated in animal models using various cell sources *(1–8)*. However, the availability and proliferative capacity of these cell types may be limiting. Because of the apparent unlimited replicative capacity of hES cells *(9)*, we propose that hES cell-derived cardiomyocytes are ideally suited for cell replacement therapies. Differentiation of cardiomyocytes from hES cells will provide a tool for investigating the molecular pathways associated with cardiogenesis. Further, hES cell-derived cardiomyocytes may provide an appropriate cell source for drug screening and toxicity testing. All of these potential applications of hES cell-derived cardiomyocytes are, however, largely dependent on practical aspects of producing a sufficient amount of these cells. Therefore, several laboratories have been working on the differentiation of functional cardiomyocytes from hES cells. In this chapter, we summarize events in early cardiogenic development, which may provide valuable lessons for differentiation of hES cells into cardiomyocytes. We then discuss current data concerning cardiomyocyte differentiation from hES cells.

2. LESSONS FROM EARLY CARDIOGENESIS

2.1. Overview

The heart is one of the first functional organs to be established in the vertebrate embryo. Development of the heart is a complex process consisting

From: *Human Embryonic Stem Cells*
Edited by: A. Chiu and M. S. Rao © Humana Press Inc., Totowa, NJ

of early specification of cells into cardiac lineage, proliferation and differentiation of progenitors into cardiomyocytes and morphogenesis of the heart (10–12). Morphological studies show that the early steps of heart formation are remarkably conserved among vertebrates including humans; however, the timing of these developmental events differs between species (11,12). Soon after gastrulation, cardiac progenitors migrate from the epiblast and are specified in the anterior lateral plate mesoderm field. Gene expression analysis in avian, Xenopus, and mouse embryos has revealed that transcription factors such as Nkx2.5 (13,14), GATA4 (15,16), and MEF2 (17) are detectable in the precardiac mesoderm and persist during heart development. Subsequently, the cells further differentiate and express cardiac muscle contractile proteins such as cardiac α- and β-myosin heavy chains (MHCs), atrial and ventricular myosin light chains (MLCs), and the troponin complex (for review, see ref. 18). These proteins are then organized into functional structures that allow the cells to contract during and after migration from the two bilateral heart tubes that fuse into the primitive linear cardiac tube. The newly formed cardiac tube undergoes extensive morphological changes before forming the mature heart. Through gene-targeting experiments, factors controlling the later stages of cardiogenesis, including morphogenesis of the primitive heart tube to a mature chambered heart, have been well characterized and are reviewed elsewhere (10).

2.2. Factors Influencing Early Stages of Cardiac Development

The generation of functional cardiomyocytes in embryos appears to be influenced by a combination of positive and negative induction signals produced from adjacent tissues, as demonstrated using avian and amphibian systems (reviewed in refs. 18 and 19). In avian culture, the pregastrula hypoblast (extraembryonic endoderm) has the ability to induce cardiogenesis from cells in the epiblast. This induction is thought to be the result of activin or transforming growth factor-β (TGF-β) present in pregastrula hypoblast (20). Although these studies demonstrate that a population of cells in epiblast has the capacity to enter the cardiogenic pathway, a unique cardiac progenitor has not been identified because of the lack of specific molecular markers. More information is available about cardiogenesis after the cardiac progenitors migrate to the mesoderm. In the avian system, cells in the cardiac mesoderm, located in the anterior primitive streak, interact with the underlying anterior endoderm and differentiate into beating cardiomyocytes (reviewed in refs. 18 and 19). In addition to inducing the cardiac mesoderm, the anterior endoderm also induces the formation of cardiac cells from the noncardiac mesoderm in avian embryo cultures (21,22). In Xenopus, signals from the anterior endoderm are also required for heart formation (23).

Although endoderm signaling is sufficient in the avian system, in *Xenopus*, interactions between the precardiac mesoderm and both the Spemann organizer and endoderm are required to promote cardiogenesis from cardiac and noncardiogenic mesoderm *(23)*.

In the avian system, certain growth factors can mimic the cardiac inductive effect of the endoderm tissue. Factors produced by the endoderm, such as activin A (a member of the TGF-β superfamily), insulin, fibroblast growth factor 1 (FGF1), FGF2, or FGF4 induce the formation of contracting cardiac cells from precardiac mesoderm in vitro in a dose-dependent manner *(24–26)*. In addition, activin A, insulin, insulin-like growth factor (IGF)-I, IGF-II, and FGF2 mimic the ability of endoderm to induce proliferation of precardiac mesoderm and to enhance the rate of cardiac development *(25)*.

Bone morphogenetic proteins (BMPs) expressed adjacent to the heart-forming region in avain endoderm and ectoderm may be critical in cardiogenesis *(27)*. For example, specific application of BMP2-soaked beads adjacent to the heart-forming region in stage HH3-5 chick embryos induces ectopic expression of Nkx2.5 and GATA-4 *(27)*. Administration of soluble BMP2 or BMP4 in serum-containing medium to stage HH5-7 embryos induces cardiomyocyte differentiation, as indicated by the expression of cardiac markers and the presence of beating cells *(27)*. In addition, the BMP antagonist noggin completely inhibits differentiation of the precardiac mesoderm *(27,28)*. Although BMPs contribute to induction of Nkx2.5 expression, BMP2 does not induce the expression of the structural protein MHC, suggesting that additional factors are required for differentiation of precardiac mesoderm into cardiomyocytes *(29)*.

It appears that FGFs and BMPs work together to regulate the specification and differentiation of cardiac cells. Expression of FGF8 is upregulated by low doses of BMP2 but repressed by high doses of BMP2 *(30)*. Removal of the endoderm in chick embryos results in a rapid downregulation of Nkx2.5 and MEF2C, which can be rescued by exogenous application of FGF8 *(30)*. Furthermore, a combination of BMP2 and FGF4, but neither factor alone, can induce formation of contracting cardiac cells in the nonprecardiac embryonic mesoderm *(31)*. The effect of these growth factors together on cardiac differentiation appears to be temporally regulated. For example, in the chick explant system, cardiogenic events require a brief treatment of FGF4 (30 min) and continuous treatment of BMP2 (48 h), which together induce Nkx2.5 and serum response factor *(32)*. In addition to BMP2 and FGF4, other members of the BMP and FGF families also induce cardiogenesis in the nonprecardiac mesoderm. BMP4 and BMP7 have similar activities to BMP2, but BMP6 and BMP12 are less active and BMP13 is not cardiogenic *(32)*. FGF2, but not FGF7, has a similar activity to FGF4

(32). However, other factors produced by anterior lateral plate endoderm, such as activin A and insulin, which support terminal differentiation of precardiac mesoderm, cannot replace BMPs or FGFs to induce cardiogenic events in nonprecardiac mesoderm *(32)*. Thus, different growth factor families interact with each other in the early steps of cardiogenesis in a time- and concentration-dependent manner.

In addition to growth factors, Wnt proteins also participate in cardiogenesis. In contrast to the cardiac inductive activity of anterior endoderm, the neural tube overlying the heart-forming region represses cardiogenesis *(11,33–36)*. The inhibitory effect of the neural tube is thought to be mediated by the Wnt signaling pathway *(36)*. It has been demonstrated that ectopic expression of Wnt-1 and Wnt-3A in chicks mimics the cardiac inhibition effect of the neural tube *(36)*, whereas inhibition of Wnt signals by the Wnt antagonist *crescent* promotes cardiogenesis from the posterior noncardiac mesoderm *(37)*. In *Xenopus*, it has been proposed that organizer-derived Wnt antagonists act directly on the adjacent mesoderm to create a zone of reduced Wnt signaling *(38)*. In addition, the Wnt antagonists *Dkk-1* and *crescent*, but not *frzb*, induce cardiac-specific gene expression and produce beating cardiac cells in noncardiogenic tissue *(38)*. Wnt proteins signal through several pathways, including the canonical β-catenin-mediated pathway and the noncanonical pathway, which involves stimulation of protein kinase C and Jun amino-terminal kinase *(39)*. Ectopic expression of GSK3β, which inhibits β-catenin-mediated Wnt signaling, also induces cardiogenesis in the noncardiogenic ventral mesoderm, whereas overexpression of Wnt3A and Wnt8, but not Wnt5A and Wnt11, blocks cardiogenesis in the cardiac mesoderm *(38)*. Through loss-of-function and gain-of-function experiments it has been demonstrated that cardiogenesis in *Xenopus* requires the activation of Wnt11, which mediates the noncanonical Wnt signaling pathway *(40)*. In addition, Wnt11-conditioned medium stimulates cardiomyocyte differentiation of mEC P19 cells *(40)*. These results indicate that canonical β-catenin-mediated Wnt signaling inhibits cardiogenesis, whereas the noncanonical Wnt signaling pathway promotes cardiac differentiation.

Signals that influence specification and differentiation of cardiomyocytes in vivo are summarized in Fig. 1. Embryo culture experiments also show that the noncardiac mesoderm, which normally generates other mesoderm cell types, can be converted into cardiac cells when placed in certain environments, indicating that these cells may have an extended differentiation potential at particular developmental stages. Although the molecular signaling pathways involved in cardiogenesis need to be further investigated, these findings will, nevertheless, provide some strategies for cardiac differentiation from hES cells.

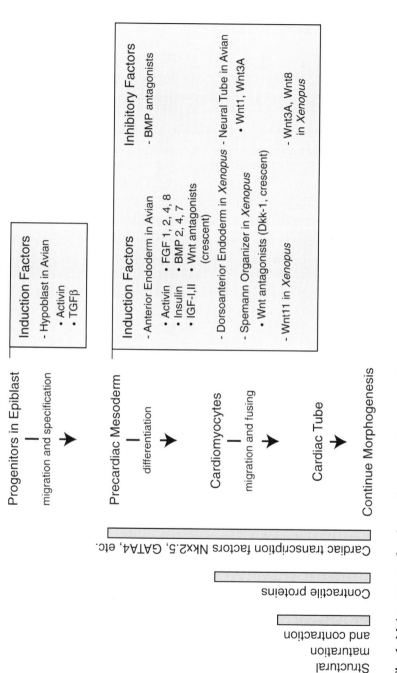

Fig. 1. Major events of early cardiogenesis. The precardiac mesoderm arises from progenitors in the epiblast, differentiates, expresses cardiac-associated markers, and then forms a contracting cardiac tube. Avian or *Xenopus* studies indicate that certain tissues and factors (indicated in boxes) influence early cardiogenesis events.

3. mES CELL-DERIVED CARDIOMYOCYTES

Mouse embryonic stem (mES) cells have the capacity to differentiate into cell types of all primary germ layers, including functional cardiomyocytes (reviewed in refs. *41–45*). The in vitro differentiation process mimics the events that occur in vivo during early embryonic development, including cardiogenesis, as demonstrated by the presence of beating cardiomyocytes as well as developmentally controlled expression of cardiac specific genes *(46)*. The culture system includes an initial differentiation induction stage in which mES cells are cultured in suspension for a few days to form embryolike aggregates called embryoid bodies (EBs). These EBs are then transferred to gelatin-coated plates or continuously cultured in suspension. In both cases, many types of cells representing derivatives of the ectoderm, mesoderm, and endoderm can be generated. The arrangement of these cells allows the interaction among endodermal, mesodermal, and ectodermal cell types and results in complex differentiation. Usually, contracting cardiac cells can be found in certain EBs or EB outgrowth after 1 wk in culture. The efficiency of cardiac differentiation has been significantly enhanced through the generation of uniform EBs using the hanging-drop method *(47)*. Scale-up production of EBs from mES cells has also been achieved by encapsulation of cells into alginate microbeads using various microencapsulation techniques *(48)*. The contracting cardiac cells derived from mES cells have been carefully characterized for their gene expression patterns *(47,49)* and pharmacological responses *(50)* and have been identified as functional cardiomyocytes representing ventricle-like, atrial-like, and pacemaker-like cells by electrophysiological analysis *(51)*.

Because EB formation represents a heterogeneous process, a number of studies have examined the direct application of growth factors to the culture with the goal of specifically inducing cardiomyocyte differentiation. For example, growth factors known to be produced by the endoderm to induce cardiomyocyte formation in embryos were evaluated for their capacity to induce mesoderm and cardiac differentiation of mES cells in chemically defined medium *(52)*. Enhancement of mesoderm and cardiac differentiation was observed in cells treated with specific concentrations of activin A or BMP4 *(52)*.

These results demonstrate that functional cardiomyocytes can be generated from mES cells. Cardiac differentiation of mES cells mimics in vivo cardiogenesis and can be regulated in vitro through different approaches, such as the addition of growth factors to cultures.

4. CARDIOMYOCYTES DERIVED FROM hES CELLS

4.1. Differentiation and Identification of Cardiac Cells

Two different basic approaches have been described for differentiation of hES cells into cardiac cells: (1) EB induction similar to that used for mES cells or (2) prolonged cell culture. Both approaches rely on the apparently heterogeneous differentiation of the hES cells. Cardiac cells are identified by the presence of contracting cells and the expression of cardiac markers. Different efficiencies of cardiomyocyte induction have been reported, which may reflect differences in cell lines, culture systems for maintaining undifferentiated hES cells (such as feeders and media), and/or protocols for the dissociation and the induction of differentiation (summarized in Table 1). We and others have previously reported that undifferentiated hES cells have different properties than undifferentiated mES cells *(9,53–55)*. Consistent with this observation, hES cell cardiomyocyte differentiation is, indeed, quite different from cardiomyocyte differentiation from mES cells. For instance, methods for generating EBs from mES cells cannot be simply applied to hES cells without modification. Reubinoff et al. reported difficulty in the formation of EBs using the hanging-drop method or simple suspension from hES HES-1 and HES-2 cells, although these cells have the capacity to generate contracting cardiomyocytes after prolonged culture on MEF feeders *(54)* or after coculture with an endoderm cell line END-2 cells *(45)*. Using H9 cells maintained on feeders, Itskovitz-Eldor et al. described the formation of EBs containing derivatives of the three embryonic germ layers *(56)*. After culture in medium containing 20% knockout serum replacement (KO-SR), simple EBs formed and some of them further developed into cystic EBs, including spontaneously contracting cells. Using the EB approach, Kehat et al. recently reported that spontaneous contracting cardiomyocytes were derived from approx 8% of EBs after cultured in serum-containing medium *(57)*. These cells express appropriate markers and display cardiac electrophysiological features and pharmacological responses *(57)*.

We have developed methods for the differentiation of cardiomyocytes from hES cells maintained on feeders or in feeder-free conditions *(55,58)*. Differentiation is initiated by culturing hES cell in suspension to form EBs, followed by culturing the EBs on gelatin-coated plates (*see* Fig. 2). The cells differentiate into a heterogeneous cell population that induces spontaneously contracting cells that then can be characterized or further enriched as described in Sections 4.2–4.4. Methods for dissociation of undifferentiated cells appear to influence the efficiency of EB formation and survival of EBs

Table 1
Cardiomyocytes or Cardiomyocytelike Cells Detected in hES-Cell-Derived Populations

	Xu et al. (55,58)	Mummery et al. (45)	Kehat et al. (57)	Itskovitz-Eldor et al. (56)	Schuldiner et al. (59)	Reubinoff et al. (54)
Cell lines	H1, H7, H9, H9.1, H9.2	HES2	H9.2	H9	H9.2	HES-1, HES-2
Yield of beating structures	approx 70% of EBs	15–20% of wells	approx 8% of EBs	Beating cells in minority of EBs	Not reported	Rare event
Maintenance of hES cells	MEF feeders or feeder-free	MEF feeders	MEF feeders	MEF feeders	MEF feeders	MEF feeders
Differentiation conditions:						
A. EB formation • Dissociation	Short collagenase dissociation into large cell clusters		Long collagenase dissociation into small cell clusters	Collagenase or trypsin/EDTA digestion	Trypsin/EDTA digestion	Unable to generate EBs
• Seeding density	High		Medium	Low	Low	
• Time in suspension	4 d		10 d	20 d	5 d	
B. Other differentiation methods		Coculture with END-2 cells				Prolonged culture of hES cells on MEF feeders
Cardiac markers detected	Immunocytochemistry: α/β MHC, βMHC, sMHC, cTnI, cTnT, desmin, α-actinin, tropomyosin, MEF-2, GATA-4, SMA, N-cadherin, myoglobin, CK-MB, β1-AR, RT-PCR: ANF, αMHC. Nkx2.5	Immunocytochemistry: α-actinin	Immunocytochemistry: α/β MHC, α-actinin, cTnI, desmin, ANP, RT-PCR: α-MHC, cTnI, cTnT, GATA-4, ANF, MLC-2A, MLC-2V. Nkx2.5	In situ hybridization: cardiac α-actinin	RT-PCR: cardiac α-actinin	Immunocytochemistry: muscle-specific actin desmin

Abbreviations: α/βMHC, cardiac α/β-myosin heavy chain; αMHC, cardiac α-myosin heavy chain; βMHC, cardiac β-myosin heavy chain; sMHC, sarcomeric myosin heavy chain; MLC2A, myosin light chain 2A; MLC2V, myosin light chain 2V; cTnI, cardiac troponin I; cTnT, cardiac troponin T; ANF, atrial natriuretic factor; CK-MB, creatine kinase-MB; β1-AR, β1-adrenoceptor; RT-PCR, reverse transcription–polymerase chain reaction.

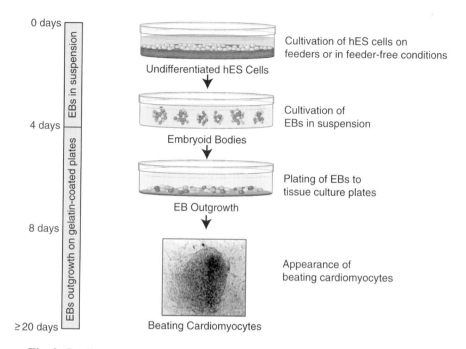

Fig. 2. Cardiomyocyte differentiation from hES cells. Differentiation is initiated by culturing hES cells in suspension to form EBs. After 4 d in suspension culture, EBs are transferred to gelatin-coated plates for 1–9 wk. The EBs attach and differentiate into a heterogeneous cell population, including spontaneously contracting cells, after 8 d in culture (4 d after plating).

in suspension cultures. We use collagenase IV to dissociate the undifferentiated hES cells and culture the clumps of cells in suspension for 4 d before plating (hES cells do not survive well when dissociated into single cells). Typically, the contracting cells are first identified in approx 5–25% of the individual EBs at differentiation d 8 (4 d after plating) and increase to as many as 50–70% of the EBs by d 16. Cardiomyocyte formation from such EB cultures has been observed in three hES cell lines as well as two hES clonal lines (H1, H7, H9, H9.1 and H9.2 *[53]*), even after being maintained for 50 passages (approx 260 population doublings) *(58)*.

4.2. Expression of Cardiac Markers
in hES Cell-Derived Cardiomyocytes

Several laboratories have reported that hES cell-derived cardiomyocytes express cardiac-associated markers. Cardiac α-actin mRNA has been detected by RT-PCR analysis *(56,59)* and *in situ* hybridization *(56)*. Cardiac α/β MHC, cardiac troponin I (cTnI), α-actinin, and desmin have been demonstrated using immunocytochemistry *(57,58)*. Other contractile proteins such as cTnT, tropomyosin, and sarcomeric myosin heavy chain (sMHC) have been detected by immunocytochemistry *(58)* and the production of myosin light chains (MLC2A, MLC2V) has been demonstrated by reverse transcription–polymerase chain reaction (RT-PCR) *(57)*. We have developed quantitative real-time RT-PCR assays for expression of some of these markers to monitor the status of differentiation. For example, we found that cardiac specific α MHC transcripts are undetectable in undifferentiated hES cell cultures or differentiated cultures at early stages, but increase significantly after d 7 of differentiation, correlating with the presence of beating cells *(58)*. Although the hES cell-derived cardiomyocytes express appropriate structural proteins, ultrastructural analysis of myofibrillar and sarcomeric organization indicate an immature phenotype *(57)*.

Cardiomyocytes derived from our cultures also express *N*-cadherin, an adherens junction protein *(60)*, and connexin 43, a gap-junction protein *(61)*, suggesting the presence of electromechanical coupling between hES cell-derived cardiomyocytes (*see* Fig. 3). This is consistent with the observation that cells in the beating EBs contract synchronously. In addition to the structural proteins, the hES cell-derived cardiomyocytes specifically express cardiac transcription factors. RT-PCR showed that Nkx2.5 is present in differentiated cultures containing beating cardiomyocytes, but undetectable in undifferentiated cultures *(57,58)*. In addition, immunocytochemistry showed expression of GATA-4 and MEF2 in nuclei of all cTnI-positive cells *(58)* (*see* Fig. 3), the former of which can also be detected by RT-PCR *(57)* or Western blot analysis (our unpublished data).

Fig. 3. *(opposite page)* Characterization of hES cell-derived cardiomyocytes. H1 cell-derived cardiomyocytes (passage 34) at differentiation d 22 were assayed for immunoreactivity with antibodies against cTnI (green) and *N*-cadherin (red). H1 cell-derived cardiomyocytes (passage 30) at differentiation d 15 were dissociated, maintained for an additional 11 d, and stained with antibodies against sMHC (green) and connexin 43 (red). H9 cells (passage 23) at differentiation d 14 were dissociated, further cultured for an additional 10 d, and stained with antibodies against GATA-4 (red) or MEF2 (red) and cTnI (green) as indicated. Scale bar = 50 mm for A–C and 25 μm for **D**.

Fig. 4. *(opposite page)* Characterization of Percoll-separated cells. Percoll-separated cells were analyzed by immunocytochemistry using various antibodies. Cardiac cells were identified as single cells or groups of cells based on cTnI or sMHC staining, as shown in the representative images. Cumulative data from six experiments using H7 or H1 cell lines at passage 30–47 is summarized in the table. +: positive signal detected in all cardiac cells examined; –: no signal detected; ±: a few positive cells detected. Scale bar = 50 μm.

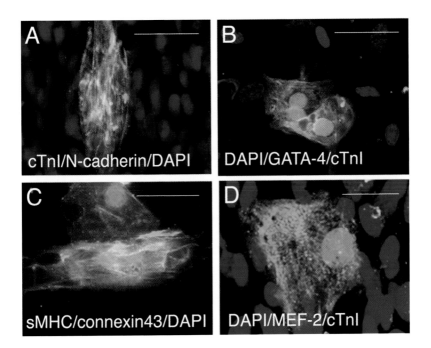

Fig. 3.

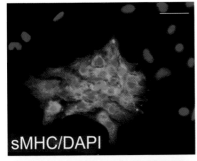

smHC/DAPI

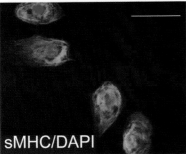

smHC/DAPI

Characterization of Enriched hES Cell Derived Cardiomyocytes

Epitopes	Cardiac Cells	Non-cardiac Cells
cTnI	+	-
α/βMHC	+	-
βMHC	+	-
smHC	+	-
N-cadherin	+	±
Myogenin	-	-
Ki67	subset	subset
BrdU	subset	subset
SSEA-4	-	-
TRA-1-81	-	-
β-tubulin III	-	-
AFP	-	-
SMA	+	subset

Fig. 4.

191

Atrial natriuretic factor (ANF), a hormone expressed in developing atrial and ventricular cardiomyocytes, but downregulated in adult ventricular cells, has been detected in hES differentiated cells by immunocytochemistry *(57)* or RT-PCR *(58)*. We have also found that creatine kinase-MB (CK-MB) and myoglobin are expressed by hES cell-derived cardiomyocytes *(58)*. CK-MB is involved in high-energy phosphate transfer and facilitates diffusion of high-energy phosphate from mitochondria to myofibril in myocytes. Myoglobin is a cytosolic oxygen-binding protein responsible for the storage and diffusion of oxygen within myocytes. The expression of these proteins in hES cell-derived cardiomyocytes indicates that they have appropriate metabolic activity.

Our results and those of others show that hES cells can be differentiated into cardiomyocytes that have appropriate cardiomyocyte-associated markers.

4.3. In Vitro Functional Characterization

The in vitro function of hES cell-derived cardiomyocytes has been examined using electrophysiological and pharmacological techniques. hES cell-derived cardiomyocytes show an average spontaneous beating rate of 94 ± 33 beats/min for H9.2 cells *(57)* or approx 30 pulses/min for H9 cells *(56)*. Mummery et al. reported that cardiac cells derived from cocultures of END-2 and hES HES-2 cells have approx 60 beats/min *(45)*. We found that the beating frequency varies in different contracting areas in cultures. Average contraction rates of H7 cells increase over time during differentiation, with approx 40 beats/min at differentiation d 15–20, which increases to approx 70 beats/min at differentiation d 60–70 *(58)*.

Two laboratories have reported electrophysiological studies on hES cell-derived cardiomyocytes *(45–57)*. Kehat et al. measured transient calcium currents of the cardiomyocytes, which are synchronous with the contraction rate *(57)*. These investigators also measured extracellular electrograms, which had a sharp component lasting 30 ± 25 ms and a slow component of 347 ± 120 ms, representing the depolarization and repolarization processes, respectively. Mummery et al. showed that action potentials in aggregates are detectable by current-clamp electrophysiology *(45)*. Further research is required to distinguish ventricular, atrial, and pacemaker cell types in the cultures using dissociated single cells.

Pharmacological responses can be evaluated in cells after treatment with a number of pharmacological agents that influence the contraction of cardiac cells, as summarized in Table 2. We detected the L-type calcium channel in hES differentiated cells using RT-PCR analysis (our unpublished data). Consistent with this observation, these cells showed a significant dose-dependent and reversible decrease in beating frequency upon incubation with diltiazem, an ion channel blocker *(58)*.

Table 2
Pharmacological Responses of hES Cell-Derived Cardiomyocytes

Agents	Function	Effect on contraction rate after treatment
Diltiazem	Ion channel blocker	Decrease
Isoprenaline	β_1-AR agonist	Increase
Isoproterenol	β_1-AR agonist	Increase
Phenylephrine	α_1-AR agonst	Increase
Clenbuterol	β_2-AR agonist	Increase
Forskolin	Adenylate cyclase activator	Increase
IBMX	Inhibitor of phosphodiesterase	Increase
Carbamylcholine	Muscarinic receptor agonist	Decrease

Note: Studies from Xu et al. *(58)* and Kehat et al. *(57)* on pharmacological responses of hES-derived cardiomyocytes are summarized.

Abbreviations: α_1-AR, α_1-adrenoceptor; β_1-AR, β_1-adrenoceptor; β_2-AR, β_2-adrenoceptor.

Cardiomyocyte contraction is also known to be influenced by the interaction of adrenoceptors (ARs) with their ligands, which affects cytosolic calcium levels *(62)*. As expected, the hES cell-derived cardiomyocytes show positive immunoreactivity using antibodies against α_1-AR and β_1-AR. Furthermore, these receptors appear to be functioning properly because the cells are responsive to treatment with the β_1-AR agonists isoprenaline *(58)* and isoproterenol *(57)* and the α_1-AR agonist phenylephrine *(58)*. We found that isoprenaline and phenylephrine both enhance the contraction rate of the hES cell-derived cardiomyocytes in a dose-dependent manner. In contrast, cells at early stages (differentiation d 22 and 39) did not respond to clenbuterol, a β_2-AR agonist. However, cultures allowed to differentiate for longer periods of time (d 61–72) are responsive to clenbuterol *(58)*. These results suggest that the sensitivity to agents stimulating adrenoreceptors may change over time, similar to observations of mES cell-derived cardiomyocytes *(50)*.

Cyclic AMP (cAMP) signaling appears to be involved in regulation of cell contraction. hES cell-derived cardiomyocytes increase their contraction rate in response to isobutylmethylxanthine (IBMX), an inhibitor of phosphodiesterase *(57,58)* or forskolin, an adenylate cyclase activator *(57)*. In addition, negative chronotropic responses were observed after treatment of cells with carbamylcholine, a muscarinic agonist, which could be reversed by administration of atropine, a muscarinic antagonist *(57)*.

Therefore, the hES cell-derived cardiomyocytes are functional in that they have appropriate electrophysiological features and have appropriate responses to a number of pharmacological reagents.

4.4. Enrichment of hES Cell-Derived Cardiomyocytes

We have examined the effect of differentiation induction reagents on the efficiency of cardiomyocyte differentiation. Dimethyl sulfoxide (DMSO) and all–*trans* retinoic acid (RA) have been shown to induce cardiomyocyte differentiation in murine embryonic carcinoma (mEC) P19 cells *(63)* and mES cells *(64)*, respectively. However, neither DMSO nor RA enhanced hES cell cardiomyocyte differentiation when procedures were similarly applied *(58)*. With different culture methods for H9.1 cells, Schuldiner et al. reported that cardiac α-actin was detected in both undifferentiated and differentiated cultures and increased in cultures treated with RA *(59)*. However, it should be noted that beating cells were not described in this study.

5-Aza-2'-deoxycytidine (5-aza-dC) induces the differentiation of mesenchymal stem cells, presumably via demethylation of DNA *(65)*. We have found that treatment of H9 or H1 cells with 5-aza-dC at d 6–8 has significantly increased the expression of cardiac αMHC, which correlates with increased beating areas in the culture *(58)*.

In order to use hES cell-derived cardiomyocytes in therapeutic or other applications, it would be beneficial to produce a population of cells highly enriched for cardiomyocytes. One possible way of enriching for the desired cell type is by genetic selection, as reported using mES cells *(66)*. These cells were transfected with plasmids containing a cardiac-specific promoter linked to selection markers *(66)* or green fluorescent protein (GFP) *(67)*. Stable clones were selected and induced for cardiac differentiation. Cardiomyocytes are then selected from the mixed population by drug resistance or flow cytometry sorting, resulting in highly purified populations.

We have shown that mechanical separation techniques can be used to enrich hES cell-derived cardiomyocytes *(58)*. Using density gradient centrifugation, we have consistently obtained populations containing 50–70% smHC or cTnI-positive cells *(58)*. Real-time RT-PCR analysis shows higher levels of cardiac αMHC in these cells compared to unpurified cells *(58)*. We have characterized these enriched cell populations by immunocytochemistry analysis with various antibodies *(58)*. As summarized in Fig. 4, these enriched cells express appropriate cardiomyocyte-associated proteins including cTnI and cardiac α/β MHC but are absent of markers for other cell types, such as myogenin for skeletal muscle, AFP for endoderm cell types, β-tubulin III for neurons, and SSEA-4 and TRA-1-81 for undifferentiated hES cells.

The enriched population of cardiomyocytes may contain immature cardiomyocytes. We found that all cTnI-positive cells and a subset of cTnI-negative cells express smooth muscle α-actin (SMA), a marker for embryonic or fetal but not adult cardiomyocytes *(58,68,69)*. The sMHC or cTnI-positive cells are actively proliferating, as demonstrated by expression of Ki-67, a protein that is present in active phases of the cell cycle *(70)*, and by their ability to incorporate BrdU *(58)*.

5. SUMMARY

In conclusion, hES cells can be induced to differentiate into cardiomyocytes and can then be enriched. The hES cell-derived cardiomyocytes are functional cells as demonstrated by the following evidence: (1) spontaneous contraction of the differentiated cultures *(45,54,56–58)*, (2) specific expression of multiple cardiac-associated markers by the differentiated cells *(45,54,56–58)*, (3) appropriate electrophysiological properties *(45,57)*, and (4) appropriate pharmacological responses of the differentiated cells to cardioactive drugs *(57,58)*. We have used mechanical separation to produce the population of cells enriched in cardiomyocytes. The extended replicative life-span and phenotypic and karyotypic stability of hES cells may allow the production of large quantities of cells for research, pharmacology/drug screening, and the development of cell replacement therapies.

ACKNOWLEDGMENTS

We thank Shailaja Police and Namitha Rao for skillful technical assistance, Dr. Joseph Gold, Dr. Jane Lebkowski, Dr. Calvin Harley, and Dr. Michael Schiff for insightful discussions and critical review of the manuscript.

REFERENCES

1. Li, R. K., Weisel, R. D., Mickle, D. A., et al. (2000) Autologous porcine heart cell transplantation improved heart function after a myocardial infarction, *J. Thorac. Cardiovasc. Surg.* **119,** 62–68.
2. Li, R. K., Yau, T. M., Weisel, R. D., et al. (2000) Construction of a bio-engineered cardiac graft, *J. Thorac. Cardiovasc. Surg.* **119,** 368–375.
3. Soonpaa, M. H., Koh, G. Y., Klug, M. G., and Field, L. J. (1994) Formation of nascent intercalated disks between grafted fetal cardiomyocytes and host myocardium, *Science* **264,** 98–101.
4. Jackson, K. A., Majka, S. M., Wang, H., et al. (2001) Regeneration of ischemic cardiac muscle and vascular endothelium by adult stem cells, *J. Clin. Invest.* **107,** 1395–1402.
5. Kocher, A. A., Schuster, M. D., Szabolcs, M. J., et al. (2001) Neovascularization of ischemic myocardium by human bone-marrow-derived angioblasts prevents cardiomyocyte apoptosis, reduces remodeling and improves cardiac function, *Nature Med.* **7,** 430–436.

6. Orlic, D., Kajstura, J., Chimenti, S., et al. (2001) Bone marrow cells regenerate infarcted myocardium, *Nature* **410,** 701–705.

7. Kessler, P. D. and Byrne, B. J. (1999) Myoblast cell grafting into heart muscle: cellular biology and potential applications, *Annu. Rev. Physiol.* **61,** 219–242.

8. Liechty, K. W., MacKenzie, T. C., Shaaban, A. F., et al. (2000) Human mesenchymal stem cells engraft and demonstrate site-specific differentiation after in utero transplantation in sheep, *Nature Med.* **6,** 1282–1286.

9. Amit, M., Carpenter, M. K., Inokuma, M. S., et al. (2000) Clonally derived human embryonic stem cell lines maintain pluripotency and proliferative potential for prolonged periods of culture, *Dev. Biol.* **227,** 271–278.

10. Olson, E. N. and Srivastava, D. (1996) Molecular pathways controlling heart development, *Science* **272,** 671–676.

11. Fishman, M. C. and Chien, K. R. (1997) Fashioning the vertebrate heart: earliest embryonic decisions, *Development* **124,** 2099–2117.

12. Srivastava, D. and Olson, E. N. (2000) A genetic blueprint for cardiac development, *Nature* **407,** 221–226.

13. Fu, Y., Yan, W., Mohun, T. J., and Evans, S. M. (1998) Vertebrate tinman homologues XNkx2-3 and XNkx2-5 are required for heart formation in a functionally redundant manner, *Development* **125,** 4439–4449.

14. Lyons, I., Parsons, L. M., Hartley, L., et al. (1995) Myogenic and morphogenetic defects in the heart tubes of murine embryos lacking the homeo box gene Nkx2-5, *Genes Dev.* **9,** 1654–1666.

15. Kuo, C. T., Morrisey, E. E., Anandappa, R., et al. (1997) GATA4 transcription factor is required for ventral morphogenesis and heart tube formation, *Genes Dev.* **11,** 1048–1060.

16. Molkentin, J. D., Lin, Q., Duncan, S. A., and Olson, E. N. (1997) Requirement of the transcription factor GATA4 for heart tube formation and ventral morphogenesis, *Genes Dev.* **11,** 1061–1072.

17. Lin, Q., Schwarz, J., Bucana, C., and Olson, E. N. (1997) Control of mouse cardiac morphogenesis and myogenesis by transcription factor MEF2C, *Science* **276,** 1404–1407.

18. Nascone, N. and Marcola, M. (1996) Endoderm and cardiogenesis, *Trends Cardiovasc. Med.* **6,** 211–216.

19. Lough, J. and Sugi, Y. (2000) Endoderm and heart development, *Dev. Dyn.* **217,** 327–342.

20. Ladd, A. N., Yatskievych, T. A., and Antin, P. B. (1998) Regulation of avian cardiac myogenesis by activin/TGFbeta and bone morphogenetic proteins, *Dev. Biol.* **204,** 407–419.

21. Sugi, Y. and Lough, J. (1994) Anterior endoderm is a specific effector of terminal cardiac myocyte differentiation of cells from the embryonic heart forming region, *Dev. Dyn.* **200,** 155–162.

22. Schultheiss, T. M., Xydas, S., and Lassar, A. B. (1995) Induction of avian cardiac myogenesis by anterior endoderm, *Development* **121,** 4203–4214.

23. Nascone, N. and Mercola, M. (1995) An inductive role for the endoderm in *Xenopus* cardiogenesis, *Development* **121,** 515–523.

24. Sugi, Y. and Lough, J. (1995) Activin-A and FGF-2 mimic the inductive effects of anterior endoderm on terminal cardiac myogenesis in vitro, *Dev. Biol.* **168,** 567–574.

25. Antin, P. B., Yatskievych, T., Dominguez, J. L., and Chieffi, P. (1996) Regulation of avian precardiac mesoderm development by insulin and insulin-like growth factors, *J. Cell Physiol.* **168,** 42–50.

26. Zhu, X., Sasse, J., McAllister, D., and Lough, J. (1996) Evidence that fibroblast growth factors 1 and 4 participate in regulation of cardiogenesis, *Dev. Dyn.* **207,** 429–438.

27. Schultheiss, T. M., Burch, J. B., and Lassar, A. B. (1997) A role for bone morphogenetic proteins in the induction of cardiac myogenesis, *Genes Dev.* **11,** 451–462.

28. Schlange, T., Andree, B., Arnold, H. H., and Brand, T. (2000) BMP2 is required for early heart development during a distinct time period, *Mech. Dev.* **91,** 259–270.

29. Andree, B., Duprez, D., Vorbusch, B., Arnold, H. H., and Brand, T. (1998) BMP-2 induces ectopic expression of cardiac lineage markers and interferes with somite formation in chicken embryos, *Mech. Dev.* **70,** 119–131.

30. Alsan, B. H. and Schultheiss, T. M. (2002) Regulation of avian cardiogenesis by Fgf8 signaling, *Development* **129,** 1935–1943.

31. Lough, J., Barron, M., Brogley, M., et al. (1996) Combined BMP-2 and FGF-4, but neither factor alone, induces cardiogenesis in non-precardiac embryonic mesoderm, *Dev. Biol.* **178,** 198–202.

32. Barron, M., Gao, M., and Lough, J. (2000) Requirement for BMP and FGF signaling during cardiogenic induction in non-precardiac mesoderm is specific, transient, and cooperative, *Dev. Dyn.* **218,** 383–393.

33. Climent, S., Sarasa, M., Villar, J. M., and Murillo-Ferrol, N. L. (1995) Neurogenic cells inhibit the differentiation of cardiogenic cells, *Dev. Biol.* 171, 130–148.

34. Antin, P. B., Taylor, R. G., and Yatskievych, T. (1994) Precardiac mesoderm is specified during gastrulation in quail, *Dev. Dyn.* **200,** 144–154.

35. Schultheiss, T. M. and Lassar, A. B. (1997) Induction of chick cardiac myogenesis by bone morphogenetic proteins, *Cold Spring Harb. Symp. Quant. Biol.* **62,** 413–419.

36. Tzahor, E. and Lassar, A. B. (2001) Wnt signals from the neural tube block ectopic cardiogenesis, *Genes Dev.* **15,** 255–260.

37. Marvin, M. J., Di Rocco, G., Gardiner, A., Bush, S. M., and Lassar, A. B. (2001) Inhibition of Wnt activity induces heart formation from posterior mesoderm, *Genes Dev.* **15,** 316–327.

38. Schneider, V. A. and Mercola, M. (2001) Wnt antagonism initiates cardiogenesis in Xenopus laevis, *Genes Dev.* **15,** 304–315.

39. Kuhl, M., Sheldahl, L. C., Park, M., Miller, J. R., and Moon, R. T. (2000) The Wnt/Ca2+ pathway: a new vertebrate Wnt signaling pathway takes shape, *Trends Genet.* **16,** 279–283.

40. Pandur, P., Lasche, M., Eisenberg, L. M., and Kuhl, M. (2002) Wnt-11 activation of a non-canonical Wnt signalling pathway is required for cardiogenesis, *Nature* **418,** 636–641.

41. Hescheler, J., Fleischmann, B. K., Lentini, S., et al. (1997) Embryonic stem cells: a model to study structural and functional properties in cardiomyogenesis, *Cardiovasc. Res.* **36,** 149–162.

42. Pedersen, R. A. (1994) Studies of in vitro differentiation with embryonic stem cells, *Reprod. Fertil. Dev.* **6,** 543–552.

43. O'Shea, K. S. (1999) Embryonic stem cell models of development, *Anat. Rec.* **257,** 32–41.

44. Smith, A. G. (2001) Embryo-derived stem cells: of mice and men, *Annu. Rev. Cell Dev. Biol.* **17,** 435–462.

45. Mummery, C., Ward, D., van den Brink, C. E., et al. (2002) Cardiomyocyte differentiation of mouse and human embryonic stem cells, *J. Anat.* **200,** 233–242.

46. Baker, R. K. and Lyons, G. E. (1996) Embryonic stem cells and in vitro muscle development, *Curr. Topics Dev. Biol.* **33,** 263–279.

47. Maltsev, V. A., Wobus, A. M., Rohwedel, J., Bader, M., and Hescheler, J. (1994) Cardiomyocytes differentiated in vitro from embryonic stem cells developmentally express cardiac-specific genes and ionic currents, *Circ. Res.* **75,** 233–244.

48. Magyar, J. P., Nemir, M., Ehler, E., Suter, N., Perriard, J. C., and Eppenberger, H. M. (2001) Mass production of embryoid bodies in microbeads, *Ann. NY Acad. Sci.* **944,** 135–143.

49. Miller-Hance, W. C., LaCorbiere, M., Fuller, S. J., et al. (1993) In vitro chamber specification during embryonic stem cell cardiogenesis. Expression of the ventricular myosin light chain-2 gene is independent of heart tube formation, *J. Biol. Chem.* **268,** 25,244–25,252.

50. Wobus, A. M., Wallukat, G., and Hescheler, J. (1991) Pluripotent mouse embryonic stem cells are able to differentiate into cardiomyocytes expressing chronotropic responses to adrenergic and cholinergic agents and Ca2+ channel blockers, *Differentiation* **48,** 173–182.

51. Maltsev, V. A., Rohwedel, J., Hescheler, J., and Wobus, A. M. (1993) Embryonic stem cells differentiate in vitro into cardiomyocytes representing sinusnodal, atrial and ventricular cell types, *Mech. Dev.* **44,** 41–50.

52. Johansson, B. M. and Wiles, M. V. (1995) Evidence for involvement of activin A and bone morphogenetic protein 4 in mammalian mesoderm and hematopoietic development, *Mol. Cell Biol.* **15,** 141–151.

53. Thomson, J. A., Itskovitz-Eldor, J., Shapiro, S. S., et al. (1998) Embryonic stem cell lines derived from human blastocysts, *Science* **282,** 1145–1147.

54. Reubinoff, B. E., Pera, M. F., Fong, C. Y., Trounson, A., and Bongso, A. (2000) Embryonic stem cell lines from human blastocysts: somatic differentiation in vitro, *Nature Biotechnol.* **18,** 399–404.

55. Xu, C., Inokuma, M. S., Denham, J., et al. (2001) Feeder-free growth of undifferentiated human embryonic stem cells, *Nature Biotechnol.* **19,** 971–974.

56. Itskovitz-Eldor, J., Schuldiner, M., Karsenti, D., et al. (2000) Differentiation of human embryonic stem cells into embryoid bodies compromising the three embryonic germ layers, *Mol. Med.* **6,** 88–95.

57. Kehat, I., Kenyagin-Karsenti, D., Snir, M., et al. (2001) Human embryonic stem cells can differentiate into myocytes with structural and functional properties of cardiomyocytes, *J. Clin. Invest.* **108,** 407–414.
58. Xu, C., Police, S., Rao, N., and Carpenter, M. K. (2002) Characterization and enrichment of cardiomyocytes derived from human embryonic stem cells, *Circ. Res.* **91,** 501–508.
59. Schuldiner, M., Yanuka, O., Itskovitz-Eldor, J., Melton, D. A., and Benvenisty, N. (2000) Effects of eight growth factors on the differentiation of cells derived from human embryonic stem cells, *Proc. Natl. Acad. Sci. USA* **92,** 11,307–11,312.
60. Volk, T. and Geiger, B. (1984) A 135-kd membrane protein of intercellular adherens junctions, *EMBO J.* **3,** 2249–2260.
61. Manjunath, C. K., Goings, G. E., and Page, E. (1987) Human cardiac gap junctions: isolation, ultrastructure, and protein composition, *J. Mol. Cell. Cardiol.* **19,** 131–134.
62. Frey, N., McKinsey, T. A., and Olson, E. N. (2000) Decoding calcium signals involved in cardiac growth and function, *Nature Med.* **6,** 1221–1227.
63. Edwards, M. K., Harris, J. F., and McBurney, M. W. (1983) Induced muscle differentiation in an embryonal carcinoma cell line, *Mol. Cell. Biol.* **3,** 2280–2286.
64. Wobus, A. M., Kaomei, G., Shan, J., et al. (1997) Retinoic acid accelerates embryonic stem cell-derived cardiac differentiation and enhances development of ventricular cardiomyocytes, *J. Mol. Cell. Cardiol.* **29,** 1525–1539.
65. Fukuda, K. (2001) Development of regenerative cardiomyocytes from mesenchymal stem cells for cardiovascular tissue engineering, *Artif. Organs* **25,** 187–193.
66. Klug, M. G., Soonpaa, M. H., Koh, G. Y., and Field, L. J. (1996) Genetically selected cardiomyocytes from differentiating embryonic stem cells form stable intracardiac grafts, *J. Clin. Invest.* **98,** 216–224.
67. Muller, M., Fleischmann, B. K., Selbert, S., et al. (2000) Selection of ventricular-like cardiomyocytes from ES cells in vitro, *FASEB J.* **14,** 2540–2548.
68. Leor, J., Patterson, M., Quinones, M. J., Kedes, L. H., and Klomer, R. A. (1996) Transplantation of fetal myocardial tissue into the infarcted myocardium of rat: a potential method for repair of infarcted myocardium? *Circulation* **94(2),** 332–336.
69. Etzion, S., Battler, A., Barbash, I. M., et al. (2001) Influence of embryonic cardiomyocyte transplantation on the progression of heart failure in a rat model of extensive myocardial infarction, *J. Mol. Cell. Cardiol.* **33,** 1321–1330.
70. Scholzen, T. and Gerdes, J. (2000) The Ki-67 protein: from the known and the unknown, *J. Cell Physiol.* **182,** 311–322.

Vascular Lineage Differentiation from Human Embryonic Stem Cells

Sharon Gerecht-Nir and Joseph Itskovitz-Eldor

1. VASCULOGENESIS AND ES CELLS

1.1. Introduction

In the early stages of embryonic development, vessel formation occurs by a process referred to as vasculogenesis, in which mesodermally derived endothelial cell progenitors undergo *de novo* differentiation, expand, and coalesce to form a network of primitive tubules *(1)*. These blood vessels are generally composed of two cell lineages: internal endothelial cells that form the channels for blood conduction, but alone cannot complete vasculo-genesis; and periendothelial smooth muscle cells that protect and stabilize the fragile channels from rupture and provide hemostatic control *(2)*. The cells that surround the endothelial channels vary throughout the blood vessel system *(3)*. For instance, several layers of smooth muscle cells sur-round the large vessels close to the heart, whereas single cells (which are not joined into layers) called pericytes cover smaller, more distant vessels. In the vertebrate embryo, vasculogenesis occurs in the paraxial and lateral me-soderm, giving rise to the primordia of the heart, the dorsal aorta, and large vessels of the head, lung, and gastrointestinal system. Angiogenesis is the process that involves the maturation and remodeling of the primitive vascu-lar plexus into a complex network of large and small vessels. Angiogenesis also leads to vascularization of initially avascular organs such as the kidney, brain, and limb buds. Angiogenesis is also required for postnatal tissue growth and throughout the adult life (e.g., during neovascularization of the endometrium during normal female cycle, during pregnancy in the placenta, and during wound healing) *(4)*. Therefore, human vascular cells are impor-tant for developing engineered vessels for the treatment of vascular diseases *(5,6)*, for enhancing vessel growth to areas of ischemic tissue or following

From: *Human Embryonic Stem Cells*
Edited by: A. Chiu and M. S. Rao © Humana Press Inc., Totowa, NJ

cell transplantation *(7)*, and for regulating angiogenesis in rheumatoid arthritis, retinopathies, hemangiomas, and cancer *(8,9)*. Moreover, endothelial cells were found to have the main role in organogenesis of the mouse embryo. Endothelial cells promoted liver organogenesis *(10)*, induced pancreas differentiation *(11)*, and were found to transdifferentiate into cardiac muscle under specific conditions *(12)*.

1.2. Vascular Potential in Spontaneous EBs

With the attractive therapeutic potential uses of vasculogenic cells in medicine, much attention has been focused on the differentiation of embryonic stem (ES) cells into mature vasculogenic lineages. In 1992, Wang et al. showed endothelial cells and a primitive vasculature in cystic mouse embryoid bodies (EBs) *(13)*. A number of markers were identified for the early mouse vasculature such as vascular endothelial growth factor-A (VEGF) receptors: the VEGFR2, also known as fetal liver kinase1 (flk-1), and VEGFR1, vascular endothelial cadherin (VE-cad), tie-2, CD34, and SCL/ TAL *(14–16)*. However, platelet endothelial cell adhesion molecule-1 (PECAM1/ CD31) has proven particularly useful because of its abundant early expression in vascular development *(17)*. As human ES (hES) cells also form EBs, some of which develop large lumen (i.e., cystic EBs), Levenberg et al. characterized the vasculogenic potential of hES cells *(18)*. Kinetic analysis for the expression of specific endothelial markers during spontaneous EB formation was preformed using reverse transcriptase–polymerase chain reaction (RT-PCR). The levels of endothelial markers PECAM1, VE-cad, and CD34 increased during the first week of EB differentiation, reaching a maximum at d 13–15 and indicating a differentiation process toward endothelial cells. Furthermore, at d 13, all EBs had defined cell areas expressing PECAM1 arranged in vessellike structures in correlation with VE-cad and von Willibrand factor (vWF) expression. As PECAM1 antibodies have been useful in mouse ES (mES) cells *(17)*, PECAM1-positive cells (2%) were sorted from 13-d-old EBs. The isolated PECAM1+ cells were shown to possess embryonic endothelial cell (EEC) features, such as acetylated low-density lipoprotein (ac-LDL) incorporation, formation of cordlike structures on Matrigel in vitro, and formation of microvessels containing mouse blood cells upon transplantation into severe combined immunodeficient (SCID) mice.

1.3. Induced Vascular Differentiation of hES

In the search for manipulated vascular differentiation, we first explored several matrix substrates that were shown to support endothelial and smooth muscle cell (SMC) cultures *(19)*. Undifferentiated hES cells were removed

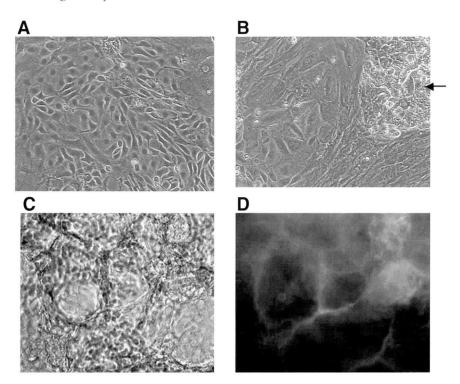

Fig. 1. Vascular manipulated differentiation of hES cells. Small hES cell aggregates grown on collagen-coated dishes showed a mixed population of different cell types along with embryonic colonies (arrow) (×100) **(A,B)**. Longer culture periods resulted in network formation **(C)** visualized by smooth muscle actin (SMA) antibody staining **(D)** (×200). (Figure 1E,F is continued on the next page.)

(H9.2 subclone) *(20)* from their feeder layer and cultured as aggregates on collagen-coated dishes. Under these culture conditions, undifferentiated colonies could be observed along with multiple cell types differentiated from them (*see* Fig. 1A,B). Furthermore, this culture yielded a tubelike structure formation (*see* Fig. 1C), which was found to express smooth muscle α-actin (SMA), a known marker for early smooth muscle cells (SMCs) *(13)* (*see* Fig. 1D). RT-PCR analysis confirmed the expression of vascular endothelial growth factor (VEGF) isomers (*see* Fig. 1E) and other endothelial markers (*see* Fig. 1F). Further experiments, which included isolation of Flk1+ cells from these cultures showed typical endothelial morphology, as previously reported in mouse ES cells *(21)*. These studies led to the conclusion that in order to induce matrix-based differentiation, hES cells should be well digested and seeded as individual cell cultures.

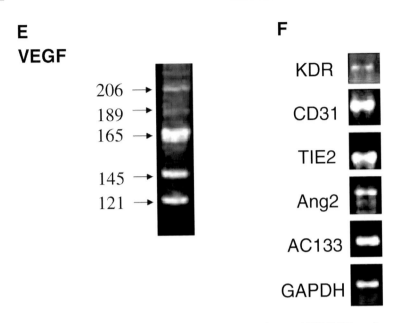

Fig. 1. *(continued from previous page)* RNA isolation and RT-PCR performance revealed the expression of VEGF isomers (**E**) and other endothelial receptors and cadherin genes (**F**).

2. VASCULOGENIC PROGENITOR

2.1. Introduction

In recent years, it has become evident that hematopoietic development and the generation of vascular SMCs are tightly linked to vascular development. In the developing embryo, endothelial cells arise either from precursors that can produce endothelial cells (angioblasts) or from progenitors that give rise to both endothelial and blood cells (hemangioblasts) *(22)*. During mouse embryogenesis, blood cells appear first within the yolk sac and later within the embryo. The contribution of yolk sac hematopoietic progenitors to adult hematopoiesis is still unclear *(22,23)*. The second hematopoietic emergence takes place separately within the embryo in the para-aorta-splanchnopleura region, which later contributes to the future aorta, gonads and mesonephros (i.e., the AGM region). Blood-forming activity in the AGM region has been shown and assessed functionally in bird, mouse, and human embryos *(24)*. Hematopoietic progenitors emerge in early development in proximity of endothelial cells either in the extraembryonic yolk sac or inside the embryo. Both hematopoietic and endothelial cells were reported to exist in blast cell colonies generated from mouse ES-cell-derived EBs

(25). In contrast, Nishikawa and colleagues showed that mouse embryonic-originating VE-cadherin$^+$ cells (i.e., endothelial cells) could produce hematopoietic cells *(26–29)*. Therefore, and congruent to the coexpression of various markers such as Flk-1, SCL/Tal1, AC133, GATA2, CD34, and LMO2, common to both hematopoietic and endothelial cells, two major concepts were suggested (*see* Fig. 2): the bipotential hemangioblast, which produces the primitive erythroid and endothelial progenitor cells, and the hemogenic endothelium, which gives rise to hematopoietic stem cells and endothelial progenitors *(23)*.

Early periendothelial SMCs associated with embryonic endothelial tubes have been shown to transdifferentiate from the endothelium, upregulating markers of SMC phenotype (both surface markers and morphology) *(30)*. Recently, mature vascular endothelium has been shown to give rise to SMCs via "transitional" cells, coexpressing both endothelial and SMC-specific markers *(31)*. Yamashita et al. discovered common embryonic vascular progenitors that differentiate into endothelial cells and SMCs. The SMCs arising from these progenitors were not simply transdifferentiated endothelial cells that expressed the atypical α-SMA marker, but ones that expressed an entire set of SMC markers and surrounded endothelial channels when injected into chick embryos *(21)*.

In conclusion, evidence for the hemangioblast and the hemogenic endothelium has been shown to exist in the mES cell system, whereas only the hemogenic endotheliun has been shown to possess the capability to differentiate into SMCs *(21)*. In human embryos, homogenic endothelium could be observed and isolated in both the extraembryonic and intraembryonic regions *(32,33)*.

2.2. Candidate Markers for Human Vascular Progenitors

Except for the developmental coexpression of common markers for endothelial and hematopoietic cells, a variety of markers were suggested to define endothelial progenitors. Angioblast precursors isolated from human adult mesanchemal stem cells expressed AC133$^+$ and Flk-1$^+$ and did not express CD34$^-$ and VE-cad$^-$ *(34)*. Endothelial progenitors derived from human cord blood *(35)* were isolated based on their CD34$^+$, Flk-1$^+$, and AC133$^+$ expression, whereas endothelial outgrowth from blood was accomplished by the isolation of cells expressing MUC18 *(36)*. Isolation of endothelial cell progenitors from human embryonal aorta was based on their CD34$^+$ expression and not PECAM1$^-$ *(37)*. The latter specifically indicated that the process of endothelial cell (EC) differentiation in humans is different from using mES cells in the acquisition of PECAM1 antigen, which, in humans, occurs after maturation of CD34$^+$ cells in culture. Furthermore,

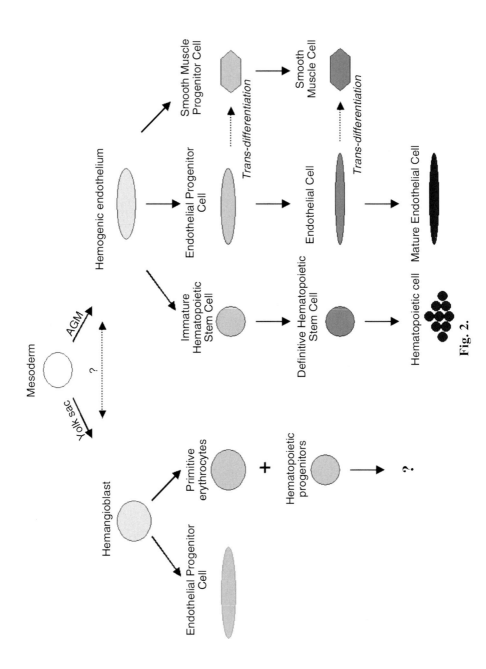

Fig. 2.

these endothelial progenitors are localized along the external layer of the aorta mesenchyme and are most likely to be in close physical contact with the AGM region *(37)*. Differences in expression of embryonic endothelial markers between mES cells and hES cells have also been reported. Flk-1 expression takes an important part in the vascular development of mouse embryo. It has been shown that the expression pattern in the mouse embryo is consistent with Flk-1, marking very early progenitors of the endothelial cell lineage as well as the differentiating endothelial cells themselves *(38)*. Furthermore, exploring the kinetics of a set of endothelial-specific markers in mouse ES-derived EBs revealed that the Flk-1 could be detected in 3-d-old EBs, PECAM1, and Tie2 in 4-d-old EBs, and VE-cad in EBs at least 5 d old. In contrast to mouse ES cells, Flk1 was shown to be expressed in undifferentiated hES cells *(18,39)* and increase very slightly during differentiation *(18)*. As chick embryo Flk-1 cells do not contain multipotent progenitors, it seems that the timing of Flk-1 may vary among vertebrates (also reviewed in ref. *40*). In addition, it appears that other endothelial markers, namely VE-cad and PECAM1, which increased during the first week of the hES cell differentiation *(18)*, play a role in human vascular development. Thus, although hES cells are a potent tool for studying human vasculogenesis, they require respecification of endothelial differentiation markers.

3. VASCULAR PROGENITORS DERIVED FROM hES CELLS

3.1. Motivation

Hematopoietic and endothelial differentiation of hES cells have been shown to require either coculturing with nonhuman cells *(39)* or EB formation prior to the appearance of endothelial phenotypes *(18)*. Levenberg et al. used spontaneous EB formation, followed by immunofluorescent cell sorting of PECAM1-positive endothelial cells. Thus, it would be advantageous to provide a simplified method of culturing, selecting, and directing the differentiation of hES cells, without the limitations of aggregation into EBs or immunofluorescent selection. Hence, we attempted to develop a method

---

Fig. 2. *(previous page)* Pathways of vascular lineage development. Two major pathways have been established for vascular development. The first is the hemangioblast of extraembryonic mesoderm origin, whose contribution to adult hematopoiesis is unclear. The second is the hemogenic endothelium of intraembryonic mesoderm origin, which gives rise to the three lineages that constitute the blood vessels. Transdifferentiation of endothelial-originated cells to smooth-muscle cells have been reported in both embryonic and adult environments. (From refs. *22* and *23*.)

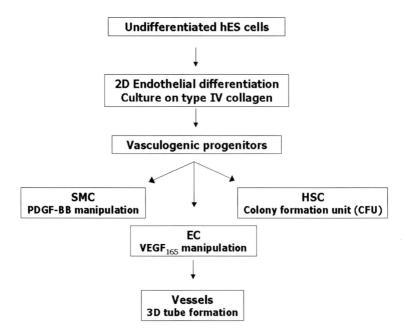

Fig. 3. In vitro differentiation of vasculogenic progenitors derived from hES cells. Outline of the manipulated differentiation method for the enhancement of a vascular progenitor population from hES cells.

for the enrichment of a progenitor cell population common to the three major cells that constitute the human vasculature (i.e., endothelium, smooth muscle, and hematopoietic [ESH] cells). Such a method and the progenitor cells isolated thereby can be used for in vitro vascular engineering, treatment of congenital and acquired vascular and hematological abnormalities, evaluation and development of drugs affecting vasculogenic and angiogenic processes, and further investigation into tissue differentiation and development.

3.2. Methodology

We developed and optimized a method for the enrichment of a vascular progenitor population using a two-dimensional (2D) model. Figure 3 illustrates the in vitro differentiation system used to generate vasculogenic progenitors and, consequently, endothelial, smooth muscle, and hematopoietic cells.

Our approach is based on the following factors:

- Individual undifferentiated hES cell culture, to avoid any aggregation or EB formation
- Specific cell seeding concentration and differentiation medium
- Collagen matrix-based adherent selection

These factors enabled us to enrich a vascular progenitor population without using any feeder layers or exogenous agents. For further manipulation, we used specific growth factors and an environment suitable for the differentiation of the three lineages. VEGF and platelet-derived growth factor BB (PDGF-BB) were used to induce endothelial and SMC differentiation, respectively, whereas semisolid media supplemented with known colony-formation inducers were used for hematopoietic colony formation.

3.3. Procedure

Using our culture conditions, a mixed differentiated populations was observed and could be separated based on cell size. Specific endothelial markers could then be detected in the size-based filtrated cells. RT-PCR analysis revealed upregulation of markers such as specific endothelial PECAM1 and Tie2, as well as AC133/CD133, Gata2, and Tal1, which are the early endothelial/hematopoietic progenitor cell markers *(35,39)*. We have also considered how to induce differentiation along specific lineages. Most reported isolated endothelial precursors divide and differentiate in response to VEGF, which is produced in proximity of forming vessels *(1,41)*. Therefore, in an effort to augment endothelial differentiation, hVEGF was added to the cultures. This resulted in cells continuously expressing VE-cad and vWF; some cells also incorporated Dil-Ac-LDL, and in some cells, stress fibers were observed. To favor SMC differentiation, a known 'recruitment factor' for pericytes—PDGF-BB, which is effective in SMC differentiation of mouse ES cells was used *(21)*. PDGF-BB led to upregulation of SMC-specific markers and fibers such as smooth muscle actin, calponin, smooth muscle-myosin heavy chain (SM-MHC), and smoothelin. Hematopoietic development was induced by the colony-formation unit assay that was previously used in the human ES-cell system *(39)*. The progenitor population yielded hematopoietic colonies under these specific conditions. As in the mouse system *(39)*, we were unsuccessful in committing the population into one specific lineage.

3.4. Vasculogenic Features

Previous studies showed that endothelial cells have the ability to form 3D vessellike structures when cultured in collagen gel *(42)* or Matrigel *(18)* and to further penetrate into the substrate while forming these networks. We therefore tested the vasculogenic progenitors' vessellike formation and their matrix invasion capabilities. Seeding 24-h aggregates of the vasculogenic progenitor cells within collagen gel or Matrigel resulted in aggregate sprouting and assembling vessellike structures while simultaneously penetrating the gels as network and sprouting progressed. Furthermore, few

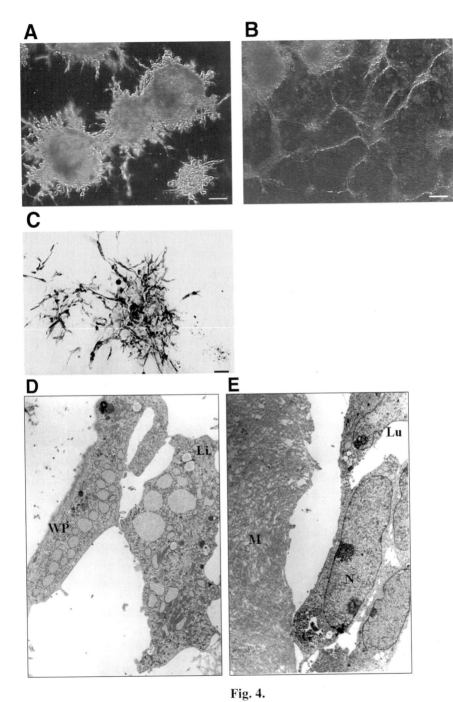

Fig. 4.

blood cells could be observed inside the forming tubes. Electron microscope examination confirmed a typical arrangement of the endothelial cells within the Matrigel. Lipoprotein capsules and Weibel–Palade bodies in the cytoplasm of the cells, previously described as typical for endothelial cells *(43)*, along with the presence of lumen in the cords *(see* Fig. 4) point to the ability of the progenitor cells to differentiate into functional endothelial cells. The vascular progenitors were further analyzed for their angiogenesis potential through their partaking in tumor angiogenesis in vivo. Injecting the vascular progenitors with teratocarcinoma cells (F9 line) subcutaneously to nonobese diabetic severe combined immunodeficient (NOD-SCID) mice resulted in functional involvement in tumor angiogenesis.

3.5. Application Feasibility

3.5.1. Angiogenesis Inhibition Model

In the adult, angiogenesis occurs in physiological conditions, such as the female reproductive tract, but more commonly in various pathological conditions. Most appreciated are the formation of new blood vessels to nurture growing tumors (tumor angiogenesis) and the formation of collaterals in an ischemic heart or limb *(44,45)*. Although the complex occurrences leading to angiogenesis are not completely understood yet, considerable interest is presently centered on the inhibition of new vascular growth to treat the spread of cancer *(46)*. There is a great need for a suitable in vitro model system to study antiangiogenesis agents. Watenberg et al. compared the kinetics of endothelial differentiation within mouse EBs, which were grown with different antiangiogenesis factors *(47)*. Hence, hES cells may serve as an in vitro research model to examine the molecular mechanisms involved in human angiogenesis. Evidence of the key role of VE-cad in tube formation both in mouse *(48)* and human systems *(49)* led to the examination of the angiogenic blocking potential of VE-cad using a 3D model of vascular progenitor cells derived from hES cells. A specific clone of monoclonal anti-hVE-cad, which was found to inhibit in vitro tube formation of human

---

Fig. 4. *(previous page)* Formation of vessellike structures in 3D collagen/Matrigel and sprouting. **(A)** Aggregates sprouting after 7 d in Matrigel and **(B)** network formation in type I collagen gel **(C)** A histological section in the cell-seeded Matrigel stained with toluidine blue, showing endothelial cells forming a tubelike arrangement structure in the matrigel (bar: 100 μm). **(D)** Electron microscope examination revealed typical lipoprotein capsules (Li) and Weibel–Palade bodies (WP) in endothelial cells (magnification: ×5000). **(E)** Electron microscopy of endothelial cell (N, nucleus) arrangement within the matrigel (M) containing lumen (Lu) and glycogen (G), the available energy source (magnification: ×5000). (Reprinted with permission from Gerecht-Nir, S., David, R., Zabura, M., et al. (2003) Human embryonic stem cells for cardiovascular repair. *Cardiovasc. Res.* **58(2)**, 313–323.

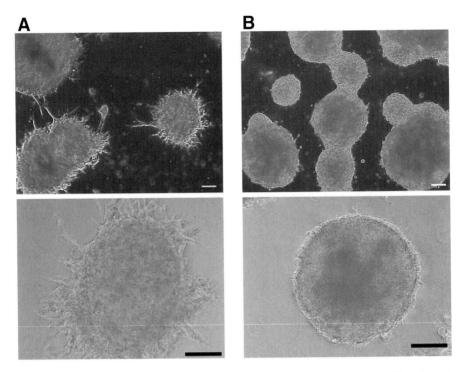

Fig. 5. In vitro human antiangiogenesis model. Low- and high-magnification of vascular progenitor population aggregates seeded for 7 d in Matrigel in a differentiation medium supplemented with hVEGF$_{165}$ and hPDGF-BB **(A)** and with monoclonal VE-cad clone BV6 that blocked sprouting **(B)**. Bar = 100 μm.

endothelial cells *(50)*, was applied to the 3D in vitro vessel-formation model of vascular progenitor cells. Vascular progenitor cells blocked with anti-VE-cad hardly sprouted and failed to form tubes and networks on Matrigel, indicating the important role of VE-cad in the process of capillary tube formation during human vasculogenesis *(see* Fig. 5).

3.5.2. Cell Transplantation and Tissue Engineering

Vasculogenic-derived progenitors, which can be produced in almost unlimited quantities and which can easily become highly available, hold the ability to vascularize regenerating ischemic and engineered tissues. To analyze the therapeutic potential of hES-derived endothelial cells, Levenberg et al. studied their behavior in vivo. PECAM1[+] cells were seeded on highly porous biodegradable polymer scaffolds, commonly used as scaffolds for tissue engineering *(18)*, and were implanted in the subcutaneous tissue of SCID mice. After 14 d of transplantation, human origin microvessels could be detected, and in some of them, mouse blood cells could be observed. Vascularization in vitro is important for the enhancement of cell viability during tissue growth, for the induction of structural organization, and for the encouragement of integration upon implantation. Therefore, an in vitro tissue-engineering model of the therapeutic potential of the vascular progenitor cells, namely seeding these cells within porous natural scaffolds should be examined *(51,52)*.

4. EPILOGUE

Because the means to study human vasculogenesis and embryo development are limited, the mechanisms involved in their regulation remain unclear. Hence, it is very important to explore genetic and molecular events taking part in human vascular development. hES cells can be cultured for prolonged periods (at least 300 population-doubling times) without observed senescence while continuing to preserve normal karyotypes, telomere length, and pluripotency. These cells provide a unique, homogeneous, unlimited preliminary population of cells for exploring and understanding human vasculogenesis and lineage commitment. Furthermore, hES cells can serve as a useful tool for vascular gene and drug discovery, angiogenesis cell therapy, ex vivo blood vessel growth, and vascularization of engineered tissues.

Another aspect worth addressing is the immunological properties of human endothelial cells, which have been suggested to perform an essential role in the acute and chronic rejection following solid organ transplantation *(53)*. As the major objective of hES cells is cell-based therapy, vascular cells derived from them may be useful in studying the various possibilities and difficulties involved in hES-cell-based therapy.

ACKNOWLEDGMENT

We would like to thank Hadas O'Neill for editing this chapter. The fund for Medical Research and Development of Infrastructure and Health Services, Rambam Medical Center, and the Technion Research and Development Foundation Ltd. supported this hES cells work.

REFERENCES

1. Yancopoulos, G., Davis, S., Gale, N., Rudge, J., Wiegand, S., and Holash, J. (2000) Vascular-specific growth factors and blood vessel formation, *Nature* **407,** 242–248.
2. Carmeliet, P. (2000) Mechanisms of angiogenesis and arteriogenesis, *Nature Med.* **6,** 389–395.

3. Gittenberger-de Groot, A. C., DeRuiter, M. C., Bergwerff, M., and Poelmann, R. E. (1999) Smooth muscle cell origin and its relation to heterogeneity in development and disease, *Arterioscler. Thromb. Vasc. Biol.* **19**, 1589–1594.

4. Risau, W. (1997) Mechanisms of angiogenesis, *Nature* **386**, 671–674.

5. Niklason, L. E., Abbott, W. M., Hirschi, K. K., Houser, S., Marini, R., and Langer, R. (1999) Functional arteries grown in vitro, *Science* **284**, 489–493.

6. Kaushal, S., Amiel, G. E., Guleserian, K. J., et al. (2001) Functional small-diameter neovessels created using endothelial progenitor cells expanded ex vivo, *Nature Med.* **7**, 1035–1040.

7. Kawamoto, A., Gwon, H. C., Iwaguro, H., et al. (2001) Therapeutic potential of ex vivo expanded endothelial progenitor cells for myocardial ischemia, *Circulation* **103**, 634–637.

8. Folkman, J. (1995) Angiogenesis in cancer, vascular, rheumatoid and other disease, *Nature Med.* **1**, 27–31.

9. Hanahan, D. and Folkman, J. (1996) Patterns and emerging mechanisms of the angiogenic switch during tumorigenesis, *Cell* **86**, 353–364.

10. Matsumoto, K., Yoshitomi, H., Rossant, J., and Zaret, K. S. (2001) Liver organogenesis promoted by endothelial cells prior to vascular function, *Science* **294**, 559–563.

11. Lammert, E., Cleaver, O., and Melton, D. (2001) Induction of pancreatic differentiation by signals from blood vessels, *Science* **294**, 564–567.

12. Condorelli, G., Borello, U., De Angelis, L., et al. (2001) Cardiomyocytes induce endothelial cells to transdifferentiate into cardiac muscle: implication for myocardium regeneration, *Proc. Natl. Acad. Sci. USA* **98**, 10,733–10,738.

13. Wang, R., Clarck, R., and Batch, V. L. (1992) Embryonic stem cell-derived cystic embryoid bodies form vascular channels: an in vitro model of blood vessel development, *Development* **114**, 303–316.

14. Vittet, D., Prandini, M. H., Schweitzer, A., Mertin-Sisteron, H., Uzan, G., and Dejana, E. (1996) Embryonic stem cells differentiate in vitro to endothelial cells through successive maturation steps, *Blood* **88**, 3424–3431.

15. Drake, C. J., Brnadt, S. J., Trusk, T. C., and Liile, C. D. (1997) TAL/SCL is expressed in endothelial progenitor cells/angioblast and defines a dorsal-to-ventral gradient of vasculogenesis, *Dev. Biol.* **192**, 17–30.

16. Drake, C. J. and Fleming, P. A. (2000) Vasculogenesis in the day 6.5 to 9.5 mouse embryo, *Blood* **95**, 1671–1679.

17. Vecchi, A., Garlanda, C., Lampugani, M. G., et al. (1994) Monoclonal antibodies specific for endothelial cells of mouse blood vessels. Their application in the identification of adult and embryonic endothelium, *Eur. J. Cell. Biol.* **63**, 247–254.

18. Levenberg, S., Golub, J. S., Amit, M., Itskovits-Eldor, J., and Langer, R. (2002) Endothelial cells derived from human embryonic stem cells, *PNAS* **99**, 4391–4396.

19. Kirkpatrick, C. J., Kampe, M., Fischer, E. G., Rixen, H., Richter, H., and Mittermayer, C. (1990) Differential expansion of human endothelial monolayers on basement membrane and interstitial collagen, laminin and fibronectin in vitro, *Pathobiology* **58**, 221–225.

20. Amit, M., Carpenter, M. K., Inokuma, M. S., et al. (2000) Clonally derived human embryonic stem cell lines maintain pluripotency and proliferative potential for prolong periods of culture, *Dev. Biol.* **227,** 271–278.
21. Yamashita, J., Itoh, H., Hirashima, M., et al. (2000) Flk1-positive cells derived from embryonic stem cells serve as vascular progenitors, *Nature* **408,** 92–96.
22. Choi, K. (2002) The hemangioblast: a common progenitor of hematopoietic and endothelial cells, *J. Hematother. Stem Cell Res.* **11,** 91–101.
23. Orkin, S. H. and Zon, L. I. (2002) Hematopoiesis and stem cells: plasticity versus development heterogeneity, *Nat. Immunol.* **3,** 323–328.
24. Mikkola, H. K. A. and Orkin, S. H. (2002) The search for the hemangioblast, *J. Hematother. Stem Cell Res.* **11,** 9–17.
25. Choi, K., Kennedy, M., Kazarov, A., Papadimitriou, J. C., and Keller, G. (1998) A common precursor for hematopoietic and endothelial cells, *Development* **125,** 725–732.
26. Nishikawa, S. I., Nishikawa, S., Hirashima, M., Matsuyoshi, N., and Kodama, H. (1998) Progressive lineage analysis by cell sorting and culture identifies FLK⁺VE-cadherin⁺ cells at a diverging point of endothelial and hematopoietic lineages, *Development* **125,** 1747–1757.
27. Nishikawa, S. I., Nishikawa, S., Kewamoto, H., et al. (1998) In vitro generation of lymphohematopoietic cells from endothelial cells purified from murine embryos, *Immunity* **8,** 761–769.
28. Fujimoto, T., Ogawa, M., Minegish, N., et al. (2001) Step-wise divergence of primitive and definitive hematopoietic and endothelial cell lineage during embryonic stem cell differentiation, *Genes Cells* **6,** 1113–1127.
29. Fraser, S., Ogawa, M., Yu, R. T., Nishikawa, S., Yoder, M. C., and Nishikawa, S.-I. (2002) Definitive hematopoietic commitment within the embryonic endothelial-cadherin+ population, *Exp. Hematol.* **30,** 1070–1078.
30. DeRuiter, M. C., Polemann, R. E., VanMunsteren, J. C. Mironoc, V., Markwald, R. R., and Gittenbereger de Groot, A. C. (1997) Embryonic endothelial cells transdifferentiate into mesenchymal cells expressing smooth muscle actins in vivo and in vitro, *Circ. Res.* **80,** 444–451.
31. Frid, M. G., Kale, V. A., and Stenmark, K. R. (2002) Mature vascular endothelium can give rise to smooth muscle cells via endothelial-mesenchymal transdifferentiation: in vitro analysis, *Circ. Res.* **90,** 1189–1196.
32. Tavian, M., Coulombel, L., Luton, D., Clemente, H. S., Dieterlen-Lievre, F., and Peault, B. (1996) Aorta-associated CD34+ hematopoietic cells in the early embryo, *Blood* **87,** 67–72.
33. Oberlin, E., Tavian, M., Balzsek, I., and Peault, B. (2002). Blood-forming potential of vascular endothelium in the human embryo, *Development* **129,** 4147–4157.
34. Reyes, M., Dudek, A., Balkrishma, J., Koodie, L., Marker, P. H., and Verfaillie, C. M. (2002) Origin of endothelial progenitors in human postnatal bone marrow. *J. Clin. Invest.* **109,** 337–346.
35. Peichev, M., Naiyer, A. J., Pereira, D., et al. (2000) Expression of VEGFR-2 and AC133 by circulating human CD34⁺ cells identifies a population of functional endothelial precursors, *Blood* **95,** 952–958.

36. Lin, Y., Weisdorf, D. J., Solovey, A., and Hebbel, R. P. (2000) Origins of circulating endothelial cells and endothelial outgrowth from blood, *JCI* **105,** 71–77.
37. Alessandri, G., Girelli, M., Taccagni, G., et al. (2001) Human vasculogenesis ex vivo: embryonal aorta as a tool for isolation of endothelial cell progenitors, *Lab. Invest.* **81,** 875–885.
38. Yamaguchi, T. P., Dumont, D. J., Roland, A. C., Breitman, M. L., and Rossant, J. (1993) flk-1, an flt-related receptor tyrosine kinase is an early marker for endothelial cell precursor, *Development* **118,** 489–498.
39. Kaufman, D. S., Hanson, E. T., Lewis, R. L., Auerbach, R., and Thomson, J. A. (2001) Hematopoietic colony-forming cells derived from human embryonic stem cells, *PNAS* **98,** 10,716–10,721.
40. Nishikawa, S. I. (2001) A complex linkage in the development pathway of endothelial and hematopoietic cells, *Curr. Opin. Cell Biol.* **13,** 673–678.
41. Gale, N. and Yancopoulos, G. (1999) Growth factor acting via endothelial cell-specific receptors tyrosine kinase: VEGFs, angiopoietins, and ephrins in vascular development, *Gene Dev.* **13,** 1055–1077.
42. Madri, J. A., Pratt, B. M., and Tucker, A. (1988) Phenotypic modulation of endothelial cells by transforming growth factor beta depends upon the composition and organization of the extracellular matrix, *J. Cell. Biol.* **106,** 1375–1384.
43. Hatzopoulos, A. K., Folkman, J., Vasile, E., Eiselen, G. K., and Rosenberg, R. D. (1998) Isolation and characterization of endothelial progenitor cells from mouse embryos, *Development* **125,** 1457–1468.
44. Schapper, W. and Ito, W. D. (1996) Molecular mechanisms of coronary collateral vessel growth, *Circ. Res.* **79,** 911–919.
45. Ware, J. A. and Simons, M. (1997) Angiogenesis in ischemic heart disease, *Nature Med.* **3,** 158–164.
46. Tabibiazar, R. and Rockson, S. G. (2001) Angiogenesis and the ischaemic heart, *Eur Heart J.* **22,** 903–918.
47. Wartenberg, M., Gunther, J., Hescheler, J., and Sauer, H. (1998) The embryoid body as a novel in vitro assay system for antiangiogenesis agents, *Lab. Invest.* **78,** 1301–1314.
48. Feraud, O., Cao, Y., and Vittet, D. (2001) Embryonic stem cell-derived embryoid bodies development in collagen gels recapitulates sprouting angiogenesis, *Lab Invest.* **81,** 1669–1681.
49. Bach, T. L., Barsigian, C., Chalupowicz, D. G., et al. (1998) VE-cadherin mediates endothelial cell capillary tube formation in fibrin and collagen gels, *Exp. Cell Res.* **238,** 324–334.
50. Corada, M., Liao, F., Lindgren, M., et al. (2001) Monoclonal antibodies directed to different regions of vascular endothelial cadherin extracellular domain affect adhesion and clustering of the protein and modulate endothelial permeability, *Blood* **97,** 1679–1684.
51. Shapiro, L. and Cohen, S. (1997) Novel alginate sponges for cell culture and transplantation, *Biomaterials* **18,** 583–590.

52. Glicklis, R., Shapiro, L., Agbaria, R., Merchuk, J. C., and Cohen, S. (2000) Hepatocyte behavior within three-dimensional porous alginate scaffolds, *Biotechnol. Bioeng.* **67,** 344–353 (2000).
53. Rose, M. L. (1998) Endothelial cells as antigen-presenting cells: role in human transplant rejection, *Cell. Mol. Life Sci.* **54,** 965–978.

12

Hematopoietic Progenitors Derived from Human Embryonic Stem Cells

Dan S. Kaufman

1. HEMATOLOGY AND STEM CELLS

The field of hematology has pioneered both basic research and clinical applications of stem cell biology. Over 40 yr ago, studies to demonstrate clonal cells within bone marrow could be transferred between animals and give rise to multiple blood cell lineages within the spleen were the first experiments to define basic stem cell principles—the ability for a single cell to both self-renew and differentiate into two or more cell types (1,2). Subsequent clinical studies showed the effectiveness of bone marrow transplantation to effectively treat and cure a variety of hematologic disorders, including (but not limited to) aplastic anemia, immunodeficienies, leukemia, lymphoma, and multiple myeloma. Indeed, the National Marrow Donor Program now identifies over 50 diseases that can be treated by bone marrow transplantation. Therefore, although "stem cell therapy" is often described as futuristic medicine, it should be remembered that bone marrow transplantation is, indeed, a form of stem cell therapy and has been successfully used for over three decades (3).

Hematopoietic stem cells were initially identified by surface antigen phenotype as a relative small percentage (less the 0.1%) of cells that reside within mouse bone marrow (4). Similar studies characterized human hematopoietic stem cells (HSCs) initially based on CD34 surface antigen expression, then as a subpopulation of CD34$^+$ cells within the bone marrow (5). HSCs can also be identified in a relatively high number within human umbilical cord blood, and as a much smaller percentage of cells within circulating peripheral blood. Importantly, HSCs can be "mobilized" from the bone marrow into peripheral blood by stimulation with cytokines such as granulocyte colony-stimulating factor (G-CSF) or stem cell factor (SCF). These circulating CD34$^+$ HSCs can be collected by apheresis and used

From: *Human Embryonic Stem Cells*
Edited by: A. Chiu and M. S. Rao © Humana Press Inc., Totowa, NJ

clinically for autologous or allogeneic HSC transplantation. Indeed, the relative ease and utility of the mobilization and collection process has lead to this being the most common way of collecting HSCs for transplantation. For this reason, the field of bone marrow transplantation is now more commonly and correctly termed hematopoietic stem cell transplantation (HSCT). Although some studies have purported to identify CD34$^-$ HSCs within mouse bone marrow *(6)*, the exact properties of these cells and the fluctuation of CD34 surface expression based on cell cycle and activation status remains to be better defined *(7)*. However, for clinical purposes, CD34 remains obviously the most relevant marker, because time to engraftment after transplantation clearly correlates to CD34$^+$ cell dose *(8,9)*. More recent work in the mouse system has begun to identify genetic pathways that regulate the development of HSCs into defined, mature hematopoietic cell types such as lymphocyte, erythrocytes, granulocytes, and megakaryocytes *(10,11)*.

With all of this information known about HSCs from mice and human (and other model organisms such as zebrafish), it could be reasonably asked why studies involving embryonic stem (ES) cells are needed within the field of hematology? Although there are many answers to this rhetorical question, the two most pertinent are as follows: (1) Isolation and subsequent studies of HSCs and their progeny provide no information regarding how human HSCs are derived in the first place, an area now more amenable to study. (2) Continued advances are needed to make HSCT a viable therapy for more patients who would benefit. Although HSCT is used for over 10,000 patients a year, many other patients do not receive a potentially life-saving HSCT because of the lack of a suitable (typically human leukocyte antigen [HLA]-matched) donor. Therefore, as with most every prospective area of study with human ES cells, there are two mutually beneficial aspects of this research: basic developmental biology in a human system and the potential derivation of a novel source of cells for transplantation therapies to treat a host of degenerative or malignant diseases. Because of the known current clinical uses of HSCs, the potential utility of human ES-cell-based research extends even further within hematology compared to less "cell-based" areas of medicine. For example, many groups have furtively sought to define methods to "ex vivo" expand small quantities of HSCs into greater quantities that can be used for transplantation, thus making limited sources of HSCs such as found in individual units of umbilical cord blood more applicable to a greater number of patients *(12)*. However, despite intensive efforts, ex vivo expansion of cord blood or other sources of HSCs has not become routinely feasible for clinical transplantation. Obviously, continued basic and translational research is needed, as success in this field would benefit thousands of patients every year. By better understanding how

Table 1
Potential Clinical Therapies
from Human ES-Cell-Derived Hematopoietic Cells

Hematopoietic stem cell transplantation for hematopoietic malignancies or related disorders
Leukemias
Multiple myeloma
Lymphomas
Myelodysplastic syndromes
Myeloproliferative syndromes
Hematopoietic stem cell transplantation for nonmalignant hematopoietic diseases
Hematopoietic stem cell disorders
Aplastic anemia
Fanconi anemia
Paroxysmal noctural hemoglobinuria
Disorders of other hematopoietic lineages (may not require HSCs)
Pure red cell aplasia
Phagocytic disorders
Histiocytic disorders
Sickle cell disease
Thalasemias
Hemoglobinopathies
Congenital immunodeficiencies
Hemolytic anemias
Liposomal storage disorders
Hematopoietic chimerism for solid organ (or other ES-cell-derived lineage) transplantation without prolonged immunosuppression
Transfusion medicine
Red blood cells
Platelets
Clotting factors

ES cells differentiate into HSCs, specific genes or proteins may be more readily identified that could be used to stimulate HSCs to divide without undergoing further differentiation into more mature blood cells. These findings could be directly applied to ongoing studies of HSC expansion. Additionally, the clinical need for red blood cell, platelet, and other blood components is ever expanding. The potential for human ES cells to provide a source of material for transfusion medicine, especially a well-defined product, free of viruses or other infectious agents, should not be overlooked. Table 1 lists a variety of clinical disorders that could potentially be addressed by human ES-cell-derived hematopoietic cells.

2. HEMATOPOIESIS FROM MOUSE ES CELLS

Mouse ES cells, first derived by two groups over 20 yr ago *(13,14)*, have served as an excellent model for mammalian hematopoietic development. Multiple excellent reviews are available to outline these studies *(15–17)*, which will be only briefly addressed here. In vitro differentiation of mouse ES cells has been achieved using at least three basic strategies. With the first method, undifferentiated ES cells are removed from the feeder cells or soluble factors that maintain them in an undifferentiated state. Individual cells or clusters of cells are then allowed to proliferate in suspension or in semisolid medium to form tight colonies of cells known as embryoid bodies (EBs) *(18)*. Under appropriate conditions, the various hematopoietic lineages develop within the EB in a defined temporal pattern, similar to development of hematopoiesis in the intact embryo *(19)*. Hematopoietic development in the EB can be determined by the detection of lineage-specific mRNAs *(19–21)* or by the detection of hematopoietic precursors based on their capacity to form colonies of mature hematopoietic cells in the colony-forming cell (CFC) assay *(19)*. Erythroid, myeloid, megakaryocytic, lymphoid, and multilineage precursor cells can all be identified within EBs at varying time-points. Endothelial cells and putative hemangioblast cells (a common precursor to blood and endothelial cell lineages) have also been isolated from mouse EBs *(17,22)*.

The second strategy for in vitro differentiation of mouse ES cells into hematopoietic cells involves the co-culture of undifferentiated ES cells with specific feeder cells that are permissive for differentiation. Although ES cells induced to differentiate in adherent cultures look different than EBs, temporal regulation of hematopoietic differentiation is maintained *(23)*. Precursors of erythroid, myeloid, and lymphoid lineages have all been identified during the differentiation of ES cells in this manner. Others have developed a multistep differentiation methods that combines these two methods. Here, culturing EB cells with adherent cells that support hematopoietic differentiation can be used to enrich hematopoietic cells from the whole population of EB cells *(24)*.

A third strategy directly cultures ES cells on collagen-coated plates, thereby eliminating the unknown effects of feeder cells. Growth of ES cells on type IV collagen-coated plates in the absence of leukemia inhibitory factor (LIF) leads to differentiation into a FLK1$^+$ cell population that can then be induced to form blood or endothelial cells, depending on the specific culture conditions *(25)*. Notably, this collagen-based method has also been used to demonstrate that a single flk-1$^+$ cell can form either endothelial cells (by culture with vascular endothelial growth factor [VEGF]) or smooth muscle cells (by culture with platelet-derived growth factor-BB [PDGF-BB]). Most remarkably, these cells can be combined to form vascular-type structures *(26)*.

The utility of these varying methods to characterize hematopoietic differentiation of mouse ES cells has led to more specific applications to understand better the mechanisms that regulate this developmental pathway. For example, specific cytokines or growth factors that modify the derivation of HSCs or mature lineages have been identified *(19,27–30)*. Specific genes that regulate early hematopoiesis have been characterized either by overexpression within mouse ES cells or use of homologous recombination to delete specific gene(s) of interest. Indeed, many "knockout" mice have an embryonic lethal phenotype because of the lack of normal hematopoietic development *(31–36)*.

Until very recently, although many groups were able to use mouse ES cells to analyze hematopoiesis in vitro, it was considerably more difficult to demonstrate normal in vivo function of these cells by long-term engraftment after transplantation into congenic mice. This engraftment potential is crucial to identify HSCs, which, by definition, must be able to sustain long-term self-renewal. Although some groups have been able to demonstrate transplantation of ES-cell-derived blood cells, typically the engraftment has been short-lived and/or limited to a subset of hematopoietic lineages *(37–39)*. Clearly, mouse ES cells have the potential to sustain long-term multilineage engraftment because chimeric mice derived entirely from mouse ES cells have normal hematopoietic development *(40)*.

Recent studies have begun to reveal the barriers to engraftment of ES-cell-derived cells. Expression of the bcr/abl fusion protein in ES cells led to deriviation of clonal HSCs that sustained multilineage engraftment, although, not surprisingly, these cells had a leukemogenic phenotype *(41)*. Subsequently, ES cells that expressed the homeobox protein HOXB4 known to be important for hematopoietic development were established *(42)*. HSCs capable of long-term, multilineage engraftment could be established from these cells. Moreover, bone marrow cells containing these ES-cell-derived HSCs could be serially transplanted into secondary recipients. Notably, hematopoiesis is not completely normal in these mice because granulocytes are produced to a much greater extent than lymphocytes. However, this model was successfully combined with a nuclear reprogramming strategy to repair a Rag2 genetic defect to demonstrate the potential of combined cell and gene therapy to treat and cure immunodeficiencies *(43)*.

3. HUMAN ES CELL AND HEMATOPOIESIS

The derivation and characterization of human ES cells *(44)* set the stage to examine the earliest stages of hematopoietic development in a human system. As indicated, the mouse and other model systems have been incredibly useful in delineating the basic genetic mechanisms of hematopoiesis and other areas of developmental biology. However, there are fundamental differences between mouse and human embryogenesis. Most relevant to this chapter, yolk sac development, the site of the initial stages of blood cell production, is quite different between the two species *(45)*. Therefore, although lessons obtained from studies of the murine system provides a useful guide for analysis of early human hematopoiesis, it cannot be assumed that there is a perfect correlation between the systems. Because human ES cells do not require LIF for maintenance of the undifferentiated state *(44,46)*, removal of LIF from the culture media is not a reasonable strategy for inducing differentiation, as is often done for studies using mouse ES cells *(16)*.

Initial intriguing studies used reverse transcriptase–polymerase chain reaction (RT-PCR) analysis to look for hints of specific lineage development in human ES cells induced to differentiate by the formation of EBs, with or without addition of exogenous cytokines *(47,48)*. Although no specific developmental pathways could be defined by this method, hematopoietic development was suggested by expression of globin chain transcripts *(47)*.

However, at this time, studies with human ES cells induced to form EBs is a somewhat limited strategy because of modest difficulties in obtaining stable genetically modified human ES cells. The most valuable studies using mouse ES cells have come from using homologous recombination to delete individual genes, a technology not yet available for human ES cell studies. In the mouse, the phenotype of the homozygous deletion can then be analyzed to better define the role of the specific gene in particular developmental pathways. This method has been extremely useful in hematopoietic studies because a hierarchical developmental schema has long been conceptualized. Therefore, lack of development of a particular lineage as a result of a specific genetic deletion often corroborates other studies of gene expression that use postnatal tissue (bone marrow or cord blood) as the starting material for hematopoietic studies. For example, flk-1$^{-/-}$ mice die at approximately embryonic day 9 because of the lack of hematopoietic and endothelial cell development *(36)*. Specific transcription factors such as scl and runx1 are likely vital to hemangioblast development based on studies of ES cells derived from these knockout mice *(49,50)*.

Although recent work has found that lentiviral vectors can be used for genetic transduction of human ES cells *(51)*, the long-term, stable expression of these constructs in differentiated ES cells remains to be defined. Progress in this arena should be rapidly advanced as more groups work with human ES cells. However, there has not yet been published reports of using homologous recombination to delete specific genes in human ES cells.

To induce or support differentiation of human ES cells into hematopoietic cells, our group has tested several mouse hematopoietic stromal cell lines such as S17 (mouse bone marrow), C166 (mouse yolk sac), and AFT024 (mouse fetal liver) *(52–54)*. Bone marrow, yolk sac, and fetal liver are all sites of active hematopoiesis. Multiple studies demonstrate that stromal (adherent) cell lines derived from these sites can support maintenance of HSCs and hematopoietic development in vitro *(52–54)*. Somewhat remarkably, mouse stromal cells are quiet adept at supporting human hematopoiesis, suggesting at least some conservation and crossreactivity of agents that promote survival and expansion of blood cells. As described later in this section, S17 and C166 proved most capable of supporting hematopoiesis in this setting. Other human bone-marrow-derived stromal cells, either primary human stroma *(55)* or immortalized cell lines *(56)*, have also been tested and were found to be no better than the murine cell lines. Also, because the human stroma cell lines would raise the possibility of cross-contamination when testing for human ES-cell-derived blood cells, the use of mouse stromal cell lines provides a "cleaner" system.

In our studies, the undifferentiated human ES cells were taken from their normal culture conditions (grown with irradiated mouse embryonic fibroblast in serum-free media) and replated onto irradiated S17 cells (or other cell line of choice) in media containing 20% fetal bovine serum (FBS). Under these conditions, the human ES cells can been seen to differentiate within approx 3–4 d. Multiple morphologies are seen within the cell cultures induced to differentiate in this manner, including endothelial-appearing cells and clusters that have a cobblestone appearance commonly found in cultures of hematopoietic precursor cells. As an initial test for potential hematopoietic cells within this differentiated culture, flow cytometry was done to evaluate cells with surface antigens typical of hematopoietic cells. CD34$^+$ and CD31$^+$ cells are routinely seen to make up 2–5% of cells. CD34 is the best phenotypic marker of HSCs and CD31 can also be found on HSCs. Notably, both these markers can also be found on endothelial cells *(57,58)*. Also, few, if any, CD45$^+$ cells are seen in these cultures. CD45 is found on most mature hematopoietic cells, but its appearance on HSCs is unclear. Intriguing studies using cells isolated from embryonic mice clearly show that CD45$^-$ and CD45$^+$ cells have a similar HSC capacity, as shown by the ability to engraft adult mice.

Next, the presence of functional hematopoietic precursor cells was tested by the methylcellulose-based hematopoietic CFC assay. This well-validated assay is used routinely in clinical bone marrow processing labs. Here, a single CFC will form a colony of defined hematopoietic lineage(s) in the semisolid media supplemented with multiple cytokines and growth factors known to support hematopoietic cell growth and expansion. The human ES cells allowed to differentiate on S17 (or other stromal cell) formed hematopoietic colonies in this assay that appeared identical to those derived from human bone marrow. Multiple lineages could be readily identified including erythroid (red blood cells), myeloid (white blood cells), and megakaryocytic (platelet precursors). Sorting the ES-cell-derived cells for CD34+ cells dramatically increased the yield of CFCs, as expected for hematopoietic precursors isolated from other sites such as bone marrow or umbilical cord blood. Transcription factors known to play an important role in early stages of hematopoiesis are also expressed by these cells, based on RT-PCR analysis. Therefore, human ES cells have the potential to form hematopoietic cells with many of the same characteristics of normal human blood cells.

There are several notable issues in regard to this method of hematopoietic differentiation of human ES cells. First, stromal cells from a hematopoietic microenvironment are required because use of mouse embryonic fibroblast (MEFs) did not lead to identifiable hematopoietic cells. Second, FBS was required because use of serum-free media did not support hematopoietic development. Third, exogenous cytokines, growth factors, or other soluble proteins beyond those found in FBS or secreted by the stromal cells are not required. Indeed, it was originally thought that cytokines such as stem cell factor (SCF) or interleukins that are known to support hematopoietic expansion would be required for successful hematopoietic differentiation of human ES cells. However, this is obviously not the case, and the addition of soluble factors to the FBS-based media has not yet been found to significantly enhance hematopoietic development. These conditions leave many unknowns in clearly defining what "factors" are needed to support hematopoiesis in this system.

The next set of studies (underway) will break down this system further to analyze what variables affect hematopoietic differentiation of human ES cells in this coculture model. For example, is direct stoma–ES cell contact needed or will conditioned media (S17 cell supernatant) suffice? Can serum-free media be used to induce differentiation if appropriate cytokines are added? What is the relative hematopoietic potential of varying subpopulations of cell derived in this system (i.e., CD34+CD31+ cells versus CD34+CD31- cells, or CD133+ versus CD133- cells). Can human EB-derived cells with the same hematopoietic potential as stromal-derived cells be evaluated? What genes are expresses in human ES-cell-derived CD34+ cells compared to those isolated from bone marrow or cord blood? The accessibility and reproducibility of human ES cells makes this an ideal model to address these issues.

Several surrogate culture assays for human HSCs have been developed. In these systems, cell populations, such as the long-term culture-initiating cell (LT-CIC) or the myeloid–lymphoid initiating cell (ML-IC) that can be grown for several weeks in culture while retaining the ability to differentiate into multiple mature hematopoietic cell lineages, are evaluated *(59,60)*. Although the phenotype and other characteristics of LT-CICs and ML-ICs are similar to HSCs, direct correlation is difficult.

Therefore, in vivo transplantation studies are needed to definitively address the characteristics of putative human ES-cell-derived HSCs. As described earlier regarding mouse ES-cell-derived HSCs, HSC populations are evaluated based on their ability to sustain long-term engraftment while retaining the ability to produce multiple mature hematopoietic lineages. In this way, the HSCs demonstrate the hallmarks of stem cells: self-renewal and differentiation. Serial transplantation experiments in which the bone marrow of one animal that has received transplanted HSCs is subsequently injected into secondary, or even tertiary recipients, are even more definitive in identifying HSC populations. The ability to sustain self-renewal through these rounds of transplantation likely requires a quite "primitive" long-term repopulating HSC population.

Two xenogeneic transplantation models have been commonly used to evaluate the long-term engraftment and multilineage potential of prospective human HSCs isolated from bone marrow or cord blood. The first model involves transplantation into immunodeficient mice that do not reject human blood cells. While these studies were first done with severe combined immunodeficient (SCID) mice *(61,62)*, it was subsequently found that nonobese diabetic (NOD)-SCID mice have an even greater degree of immunodeficiency that make them easier to engraft with human cells *(63)*. Other strains of immunodeficient mice such as NOD/SCID/β_2-microglobulin-deficient and Rag-2-deficient mice have also been recently evaluated and may have advantages for these xenogeneic transplantation experiments *(64,65)*. The SCID-repopulating cell (SRC) characterized in these transplant systems likely has a high degree of correlation to human HSCs *(63)*.

The second xenogeneic model uses preimmune fetal sheep as recipients for putative human HSCs. The sheep are injected approximately day 60 of gestation, and growth of human blood cells can be assayed at varying time both before and after birth of the animals *(66)*. In experiments of human bone-marrow- or cord-blood-derived HSCs, long-term engraftment and serial transplantation can be demonstrated *(66,67)*. This system is quite robust and has been used to evaluate the potential of other cells, such as stromal cells, to aid engraftment of human HSCs *(68)*. Preliminary studies of human ES-cell-derived blood cells (created via human ES cell coculture on S17 cells as described earlier) have the potential to engraft in this model *(69)*.

4. FUTURE DIRECTIONS
FOR HEMATOPOIESIS AND HUMAN ES CELLS

The most obvious therapeutic use for human ES-cell-derived HSCs would be as a novel source of transplantable cells. As previously mentioned, HSCT typically uses bone marrow, mobilized peripheral blood, or umbilical cord blood cells as a source of HSCs. Allogeneic transplantation of these cells offers the best chance to cure many hematologic malignancies such as leukemia or multiple myeloma. However, the potential morbidity and mortality associated with allo-HSCT limits use to only a fraction of patients who could potentially benefit. One of the major limiting factors is the need for an HLA-matched donor of HSCs in order to minimize risk of graft-versus-host-disease (GVHD). This potentially lethal disorder results when lymphocytes in the transplanted cell population initiate an immune response against the host tissue. Interestingly, HSCT using cord blood as the source of donor cells seems to allow a lesser degree of HLA-matching without inducing GVHD *(70–72)*. Donor–host matching at only three or four of six HLA molecules is well tolerated in the cord blood setting, whereas severe GVHD almost certainly develops when this degree of mismatch is done with bone marrow or peripheral blood as the source of transplanted HSCs. The better compatibility of cord blood transplants is likely because of at least two facts: (1) Fewer allo-reactive lymphocytes and a greater proportion of HSCs are contained in each cord blood unit; (2) the transplanted cells are more immunologically naïve in the cord blood. The lymphocytes contained within the cord blood unit and lymphocytes derived from transplanted HSCs will become immunologically "educated" in the new host. Therefore, these cells will be tolerant of the new environment and less likely to initiate an allo-reactive immune response. Because human ES-cell-derived HSCs can be created free of any contaminating lymphocytes, GVHD should be a minimal problem in this setting. What degree of HLA mismatch will be tolerated in this setting remains speculative. However, based on the lessons from cord blood transplantation, relatively few donor ES cell lines may be needed to find a population suitable for individual patients.

Many other hurdles will need to be overcome before human ES-cell-derived HSCs become a routine clinical therapy. These barriers such as creation of sufficient number of cells, testing for normal function of the cells, and elimination of residual undifferentiated ES cells within the transplanted population have been described in greater detail elsewhere *(73,74)*. Although it would be marvelous to use human ES cells as a source of HSCs suitable for transplantation to treat those patients who otherwise do not have a suitable donor, studies of hematopoietic development from human ES cells may lead to other suitable therapies. As described, ex vivo expansion of HSCs has long been a "holy grail" for those in the HSCT field. For example, it would be reasonable to expand a small population of highly purified HSCs to produce an amount of cells sufficient for transplantation. This would prevent the collection of large amounts of bone marrow or peripheral blood cells. Importantly, expansion of cord blood units could make this a viable source of cells for larger adults who currently cannot benefit from cord blood transplantation because of the limited number of HSCs within each unit. Indeed, recent clinical trials have been done using two cord-blood-units transplanted into an individual adult. Although successful, this method obviously requires more cord blood units to be used. Based on results of basic studies of hematopoiesis from human ES cells, it is likely that the genes and proteins that regulate early human hematopoiesis will be better understood. Under these circumstances, novel stimuli that would allow HSC expansion without differentiation can be elucidated and applied to the ongoing studies to maximize growth of other HSC populations. In this way, whereas the application of human ES cells to clinical transplantation would be indirect, the number of hurdles to overcome would be fewer.

Transplantation and engraftment of human ES-cell-derived HSCs may have therapeutic benefit in other clinical settings. Over the past 20 yr, multiple anecdotal clinical cases demonstrated the ability of HSCT to induce tolerance for transplanted solid organs (typically kidney) that come from the same donor as the HSCs *(75)*. More recently, prospective clinical trials suggest that cotransplantation of HSCs and kidney from the same donor can lead to graft tolerance that permits immunosuppressive drugs to be weaned of without graft rejection *(76,77)*. In this setting, the donor HSCs are not required to completely replace the host hematopoietic system (considered complete chimerism). Only mixed chimerism, whereby the host has hematopoietic elements of both donor and self, is required to induce immunological tolerance. Although the immune mechanisms are not precisely understood, it is likely that HSC-derived lymphocytes derived in the same setting as the HLA-matched kidney become "educated" and tolerant to the transplanted organ. Application of this principle to transplantation of ES-cell-derived cells and tissues may be one method to avoid immune-mediated rejection of these cells.

Because human ES cells can be induced to differentiate into any tissue, the potential to derive both HSCs and a second cell type (e.g., pancreatic β-cell, hepatocyte, cardiomyocyte) from the same parental cell line is particularly attractive. Here, human ES-cell-derived HSCs would be transplanted, possibly using a minimally myeloablative condition regimen because full donor cell chimerism is not required. Subsequently, a second cell type derived from the same parental ES cell line could be transplanted without need for further immune suppression because the ES-cell-derived HLA-identical lymphocytes should render the host tolerant to the second tissue type. One study using rat ES-cell-like cells finds that intraportal injection of these cells leads to hematopoietic mixed chimerism and prolonged graft survival of cardiac allografts matched to the ES cells *(78)*. This strategy is particularly attractive for treatment of autoimmune diseases such as type 1 diabetes mellitus. If pancreatic islets are derived from ES cells that are precisely matched to the patient (e.g., by substitution of self-HLA genes or via nuclear transfer technology), it would be expected that the donor β-cells would be rejected by the host because of the underlying autoimmune process. This is unfortunately seen in cases of diabetics who receive and subsequently reject pancreas transplants from identical twins *(79)*. Multiple other means to modify human ES cells to permit transplantation tolerance have also been proposed *(73,74)*. Obviously, considerably more research is need before clinical trials involving cellular therapies based on human ES cells are contemplated.

Transfusion medicine can also likely benefit from future advances in hematopoietic differentiation of human ES cells. Red blood cells, platelets, and other blood products are frequently needed to treat patients who undergo surgery, suffer severe trauma, or receive drugs that suppress normal bone marrow function (such as chemotherapy to treat malignancies). Despite millions of people who generously donate blood and the considerable infrastructure to collect and process blood donations, shortages often exist. Moreover, although donated blood is screened for multiple viral infections and is extremely safe, emerging infections continue to be a problem. This hole in the screening process was recently seen with evidence that West Nile Virus was transferred from blood transfusions *(80)*. If (perhaps when) human ES cells can be induced to differentiate into specific lineages of mature hematopoietic cells (e.g., erythrocytes and megakaryocytes/platelet), it should be possible scale production to provide a stable source of blood products to better meet the ever-expanding patient needs. Because well-characterized ES cells would be used as the starting point, these blood products could be guaranteed to be free of infectious agents. Recent studies have defined methods to produce large numbers of red blood cells and megakaryocytes from precursors in human umbilical cord blood *(81,82)*. Similar strategies should be applicable for ES-cell-based production. Because there are only four major blood groups and blood group type O blood serves as a universal donor, tissue matching does not becomes of a barrier in transfusion medicine.

5. CONCLUSION

Currently, research and clinical practice of hematology provides expertise in the understanding and use of "adult" stems cells in the form of HSCs. Because hematology is such a "cell-based" field, the ability of human ES cells to dramatically advance both the basic research and clinical therapies within this field is readily apparent. As described in this chapter, the current successes and limits of HSCT as treatment of malignant and nonmalignant diseases highlight the need for continued progress within this field. Human ES cells will allow us to understand better the general genetic and cellular mechanisms that regulate stem cell self-renewal. Research into the basic biology of human ES cells, in addition to more focused research on hematopoiesis using human ES cells as the starting point, will have broad implications for future advances within hematology. The known therapeutic potential of adult hematopoietic stem cells and the unknown potential of embryonic stem cells provide for mutually beneficial advantages. The close relationship between these stem cell populations will to continue to make hematology a pioneering field for stem cell studies.

Recent studies by Zwaka and Thomson demonstrate the ability to use homologous recombination to achieve stable and permanent deletion (knock-out), or addition (knock-in), of genes in human ES cells *(83)*. These results will further expand the repertoire of studies to be done with human ES cells in regards to hematopoiesis and other areas of research.

REFERENCES

1. Till, J. E. and McCullough, E. A. (1961) A direct measurement of the radiation sensitivity of normal mouse bone marrow cells, *Radiat. Res.* **14**, 213–222.
2. Becker, A. J., McCulloch, E. A., and Till, J. E. (1963) Cytological demonstration of the clonal nature of spleen colonies derived from transplanted mouse marrow cells, *Nature* **197**, 452–454.
3. Thomas, E. D. (1999) Bone marrow transplantation: a review, *Semin. Hematol.* **36**, 95–103.
4. Spangrude, G. J., Heimfeld, S., and Weissman, I. L. (1988) Purification and characterization of mouse hematopoietic stem cells, *Science* **241**, 58–62.
5. Baum, C. M., Weissman, I. L., Tsukamoto, A. S., Buckle, A. M., and Peault, B. (1992) Isolation of a candidate human hematopoietic stem-cell population, *Proc. Natl. Acad. Sci. USA* **89**, 2804–2808.
6. Bhatia, M., Bonnet, D., Murdoch, B., Gan, O. I., and Dick, J. E. (1998) A newly discovered class of human hematopoietic cells with SCID-repopulating activity, *Nature Med.* **4**, 1038–1045.

7. Sato, T., Laver, J. H., and Ogawa, M. (1999) Reversible expression of CD34 by murine hematopoietic stem cells, *Blood* **94**, 2548–2554.

8. Weaver, C. H., Hazelton, B., Birch, R., et al. (1995) An analysis of engraftment kinetics as a function of the CD34 content of peripheral blood progenitor cell collections in 692 patients after the administration of myeloablative chemotherapy, *Blood* **86**, 3961–3969.

9. Wagner, J. E., Barker, J. N., DeFor, T. E., et al. (2002) Transplantation of unrelated donor umbilical cord blood in 102 patients with malignant and nonmalignant diseases: influence of CD34 cell dose and HLA disparity on treatment-related mortality and survival, *Blood* **100**, 1611–1618.

10. Orkin, S. H. (2000) Diversification of haematopoietic stem cells to specific lineages, *Nat. Rev. Genet.* **1**, 57–64.

11. Akashi, K., Traver, D., Miyamoto, T., and Weissman, I. L. (2000) A clonogenic common myeloid progenitor that gives rise to all myeloid lineages, *Nature* **404**, 193–197.

12. Verfaillie, C. M. (2000) Meeting report on an NHLBI workshop on ex vivo expansion of stem cells, July 29, 1999, Washington, D.C. National Heart Lung and Blood Institute, *Exp. Hematol.* **28**, 361–364.

13. Evans, M. J. and Kaufman, M. H. (1981) Establishment in culture of pluripotential cells from mouse embryos, *Nature* **292**, 154–156.

14. Martin, G. R. (1981) Isolation of a pluripotent cell line from early mouse embryos cultured in medium conditioned by teratocarcinoma stem cells, *Proc. Natl. Acad. Sci. USA* **78**, 7634–7638.

15. Smith, A. G. (2001) Embryo-derived stem cells: of mice and men, *Annu. Rev. Cell Dev. Biol.* **17**, 435–462.

16. Keller, G. M. (1995) In vitro differentiation of embryonic stem cells, *Curr. Opin. Cell Biol.* **7**, 862–869.

17. Lacaud, G., Robertson, S., Palis, J., Kennedy, M., and Keller, G. (2001) Regulation of hemangioblast development, *Ann. NY Acad. Sci.* **938**, 96–107; discussion, 8.

18. Doetschman, T. C., Eistetter, H., Katz, M., Schmidt, W., and Kemler, R. (1985) The in vitro development of blastocyst-derived embryonic stem cell lines: formation of visceral yolk sac, blood islands and myocardium. *J. Embryol. Exp. Morphol.* **87**, 27–45.

19. Keller, G., Kennedy, M., Papayannopoulou, T., and Wiles, M. V. (1993) Hematopoietic commitment during embryonic stem cell differentiation in culture, *Mol. Cell. Biol.* **13**, 473–486.

20. Lindenbaum, M. H. and Grosveld, F. (1990) An in vitro globin gene switching model based on differentiated embryonic stem cells, *Genes Dev.* **4**, 2075–2085.

21. Schmitt, R. M., Bruyns, E., and Snodgrass, H. R. (1991) Hematopoietic development of embryonic stem cells in vitro: cytokine and receptor gene expression, *Genes Dev.* **5**, 728–740.

22. Choi, K., Kennedy, M., Kazarov, A., Papadimitriou, J. C., and Keller, G. (1998) A common precursor for hematopoietic and endothelial cells, *Development* **125**, 725–732.

23. Nakano, T., Kodama, H., and Honjo, T. (1994) Generation of lymphohematopoietic cells from embryonic stem cells in culture, *Science* **265**, 1098–1101.

24. Bigas, A., Martin, D. I., and Bernstein, I. D. (1995) Generation of hematopoietic colony-forming cells from embryonic stem cells: synergy between a soluble factor from NIH-3T3 cells and hematopoietic growth factors, *Blood* **85,** 3127–3133.

25. Nishikawa, S. I., Nishikawa, S., Hirashima, M., Matsuyoshi, N., and Kodama, H. (1998) Progressive lineage analysis by cell sorting and culture identifies FLK1+VE-cadherin+ cells at a diverging point of endothelial and hemopoietic lineages, *Development* **125,** 1747–1757.

26. Yamashita, J., Itoh, H., Hirashima, M., et al. (2000) Flk1-positive cells derived from embryonic stem cells serve as vascular progenitors, *Nature* **408,** 92–96.

27. Johansson, B. M. and Wiles, M. V. (1995) Evidence for involvement of activin A and bone morphogenetic protein 4 in mammalian mesoderm and hematopoietic development, *Mol. Cell. Biol.* **15,** 141–151.

28. Adelman, C. A., Chattopadhyay, S., and Bieker, J. J. (2002) The BMP/BMPR/Smad pathway directs expression of the erythroid-specific EKLF and GATA1 transcription factors during embryoid body differentiation in serum-free media, *Development* **129,** 539–549.

29. Faloon, P., Arentson, E., Kazarov, A., et al. (2000) Basic fibroblast growth factor positively regulates hematopoietic development, *Development* **127,** 1931–1941.

30. Nakayama, N., Lee, J., and Chiu, L. (2000) Vascular endothelial growth factor synergistically enhances bone morphogenetic protein-4-dependent lympho-hematopoietic cell generation from embryonic stem cells in vitro, *Blood* **95,** 2275–2283.

31. Pevny, L., Simon, M. C., Robertson, E., et al. (1991) Erythroid differentiation in chimaeric mice blocked by a targeted mutation in the gene for transcription factor GATA-1, *Nature* **349,** 257–260.

32. Robb, L., Lyons, I., Li, R., et al. (1995) Absence of yolk sac hematopoiesis from mice with a targeted disruption of the scl gene, *Proc. Natl. Acad. Sci. USA* **92,** 7075–7079.

33. Tsai, F. Y., Keller, G., Kuo, F. C., et al. (1994) An early haematopoietic defect in mice lacking the transcription factor GATA-2, *Nature* **371,** 221–226.

34. Warren, A. J., Colledge, W. H., Carlton, M. B., Evans, M. J., Smith, A. J., and Rabbitts, T. H. (1994) The oncogenic cysteine-rich LIM domain protein rbtn2 is essential for erythroid development, *Cell* **78,** 45–57.

35. Shivdasani, R. A., Mayer, E. L., and Orkin, S. H. (1995) Absence of blood formation in mice lacking the T-cell leukaemia oncoprotein tal-1/SCL, *Nature* **373,** 432–434.

36. Shalaby, F., Rossant, J., Yamaguchi, T. P., et al. (1995) Failure of blood-island formation and vasculogenesis in Flk-1-deficient mice, *Nature* **376,** 62–66.

37. Palacios, R., Golunski, E., and Samaridis, J. (1995) In vitro generation of hematopoietic stem cells from an embryonic stem cell line, *Proc. Natl. Acad. Sci. USA* **92,** 7530–7534.

38. Hole, N., Graham, G. J., Menzel, U., and Ansell, J. D. (1996) A limited temporal window for the derivation of multilineage repopulating hematopoietic progenitors during embryonal stem cell differentiation in vitro, *Blood* **88,** 1266–1276.

39. Muller, A. M. and Dzierzak, E. A. (1993) ES cells have only a limited lymphopoietic potential after adoptive transfer into mouse recipients, *Development* **118**, 1343–1351.
40. Nagy, A., Rossant, J., Nagy, R., Abramow-Newerly, W., and Roder, J. C. (1993) Derivation of completely cell culture-derived mice from early-passage embryonic stem cells, *Proc. Natl. Acad. Sci. USA* **90**, 8424–8428.
41. Perlingeiro, R. C., Kyba, M., and Daley, G. Q. (2001) Clonal analysis of differentiating embryonic stem cells reveals a hematopoietic progenitor with primitive erythroid and adult lymphoid-myeloid potential, *Development* **128**, 4597–4604.
42. Kyba, M., Perlingeiro, R. C., and Daley, G. Q. (2002) HoxB4 confers definitive lymphoid-myeloid engraftment potential on embryonic stem cell and yolk sac hematopoietic progenitors, *Cell* **109**, 29–37.
43. Rideout, W. M., 3rd, Hochedlinger, K., Kyba, M., Daley, G. Q., and Jaenisch, R. (2002) Correction of a genetic defect by nuclear transplantation and combined cell and gene therapy, *Cell* **109**, 17–27.
44. Thomson, J. A., Itskovitz-Eldor, J., Shapiro, S. S., et al. (1998) Embryonic stem cell lines derived from human blastocysts, *Science* **282**, 1145–1147.
45. Palis, J. and Yoder, M. C. (2001) Yolk-sac hematopoiesis: the first blood cells of mouse and man, *Exp. Hematol.* **29**, 927–936.
46. Reubinoff, B. E., Pera, M. F., Fong, C. Y., Trounson, A., and Bongso, A. (2000) Embryonic stem cell lines from human blastocysts: somatic differentiation in vitro, *Nature Biotechnol.* **18**, 399–404.
47. Itskovitz-Eldor, J., Schuldiner, M., Karsenti, D., et al. (2000) Differentiation of human embryonic stem cells into embryoid bodies compromising the three embryonic germ layers, *Mol. Med.* **6**, 88–95.
48. Schuldiner, M., Yanuka, O., Itskovitz-Eldor, J., Melton, D. A., and Benvenisty, N. (2000) Effects of eight growth factors on the differentiation of cells derived from human embryonic stem cells, *Proc. Natl. Acad. Sci. USA* **97**, 11,307–11,312.
49. Lacaud, G., Gore, L., Kennedy, M., et al. (2002) Runx1 is essential for hematopoietic commitment at the hemangioblast stage of development in vitro, *Blood* **100**, 458–466.
50. Robertson, S. M., Kennedy, M., Shannon, J. M., and Keller, G. (2000) A transitional stage in the commitment of mesoderm to hematopoiesis requiring the transcription factor SCL/tal-1, *Development* **127**, 2447–2459.
51. Pfeifer, A., Ikawa, M., Dayn, Y., and Verma, I. M. (2002) Transgenesis by lentiviral vectors: lack of gene silencing in mammalian embryonic stem cells and preimplantation embryos, *PNAS* **99**, 2140–2145.
52. Collins, L. S. and Dorshkind, K. (1987) A stromal cell line from myeloid long-term bone marrow cultures can support myelopoiesis and B lymphopoiesis, *J. Immunol.* **138**, 1082–1087.
53. Lu, L. S., Wang, S. J., and Auerbach, R. (1996) In vitro and in vivo differentiation into B cells, T cells, and myeloid cells of primitive yolk sac hematopoietic precursor cells expanded > 100-fold by coculture with a clonal yolk sac endothelial cell line, *Proc. Natl. Acad. Sci. USA* **93**, 14,782–14,787.

54. Moore, K. A., Ema, H., and Lemischka, I. R. (1997) In vitro maintenance of highly purified, transplantable hematopoietic stem cells, *Blood* **89,** 4337–4347.

55. Simmons, P. J. and Torok-Storb, B. (1991) Identification of stromal cell precursors in human bone marrow by a novel monoclonal antibody, STRO-1, *Blood* **78,** 55–62.

56. Roecklein, B. A. and Torok-Storb, B. (1995) Functionally distinct human marrow stromal cell lines immortalized by transduction with the human papilloma virus E6/E7 genes, *Blood* **85,** 997–1005.

57. Watt, S. M., Williamson, J., Genevier, H., et al. (1993) The heparin binding PECAM-1 adhesion molecule is expressed by CD34+ hematopoietic precursor cells with early myeloid and B-lymphoid cell phenotypes, *Blood* **82,** 2649–2663.

58. Fina, L., Molgaard, H. V., Robertson, D., et al. (1990) Expression of the CD34 gene in vascular endothelial cells, *Blood* **75,** 2417–2426.

59. Sutherland, H. J., Lansdorp, P. M., Henkelman, D. H., Eaves, A. C., and Eaves, C. J. (1990) Functional characterization of individual human hematopoietic stem cells cultured at limiting dilution on supportive marrow stromal layers, *Proc. Natl. Acad. Sci. USA* **87,** 3584–3588.

60. Punzel, M., Wissink, S. D., Miller, J. S., Moore, K. A., Lemischka, I. R., and Verfaillie, C. M. (1999) The myeloid-lymphoid initiating cell (ML-IC) assay assesses the fate of multipotent human progenitors in vitro, *Blood* **93,** 3750–3756.

61. Lapidot, T., Fajerman, Y., and Kollet, O. (1997) Immune-deficient SCID and NOD/SCID mice models as functional assays for studying normal and malignant human hematopoiesis, *J. Mol. Med.* **75,** 664–673.

62. Dao, M. A. and Nolta, J. A. (1999) Immunodeficient mice as models of human hematopoietic stem cell engraftment, *Curr. Opin. Immunol.* **11,** 532–537.

63. Larochelle, A., Vormoor, J., Hanenberg, H., et al. (1996) Identification of primitive human hematopoietic cells capable of repopulating NOD/SCID mouse bone marrow: implications for gene therapy, *Nature Med.* **2,** 1329–1337.

64. Glimm, H., Eisterer, W., Lee, K., et al. (2001) Previously undetected human hematopoietic cell populations with short-term repopulating activity selectively engraft NOD/SCID-beta2 microglobulin-null mice, *J. Clin. Invest.* **107,** 199–206.

65. Kollet, O., Peled, A., Byk, T., et al. (2000) β2 microglobulin-deficient (B2m^{null}) NOD/SCID mice are excellent recipients for studying human stem cell function, *Blood* **95,** 3102–3105.

66. Civin, C. I., Almeida-Porada, G., Lee, M. J., Olweus, J., Terstappen, L. W., and Zanjani, E. D. (1996) Sustained, retransplantable, multilineage engraftment of highly purified adult human bone marrow stem cells in vivo, *Blood* **88,** 4102–4109.

67. Lewis, I. D., Almeida-Porada, G., Du, J., et al. (2001) Umbilical cord blood cells capable of engrafting in primary, secondary, and tertiary xenogeneic hosts are preserved after ex vivo culture in a noncontact system, *Blood* **97,** 3441–3449.

68. Almeida-Porada, G., Porada, C. D., Tran, N., and Zanjani, E. D. (2000) Cotransplantation of human stromal cell progenitors into preimmune fetal sheep results in early appearance of human donor cells in circulation and boosts

cell levels in bone marrow at later time points after transplantation, *Blood* **95,** 3620–3627.

69. Narayan, A. D., Thomson, J. A., Lewis, R. L., et al. (2002) In vitro and in vivo potential of human embryonic stem cells, *Blood* **100,** 1546 (abstract).

70. Gluckman, E., Rocha, V., Boyer-Chammard, A., et al. (1997) Outcome of cordblood transplantation from related and unrelated donors. Eurocord Transplant Group and the European Blood and Marrow Transplantation Group, *N. Engl. J. Med.* **337,** 373–381.

71. Rubinstein, P., Carrier, C., Scaradavou, A., et al. (1998) Outcomes among 562 recipients of placental-blood transplants from unrelated donors, *N. Engl. J. Med.* **339,** 1565–1577.

72. Laughlin, M. J., Barker, J., Bambach, B., et al. (2001) Hematopoietic engraftment and survival in adult recipients of umbilical-cord blood from unrelated donors, *N. Engl. J. Med.* **344,** 1815–1822.

73. Odorico, J. A., Kaufman, D. S., and Thomson, J. A. (2001) Multilineage differentiation from human embryonic stem cell lines, *Stem Cells* **19,** 193–204.

74. Kaufman, D. S., Odorico, J. S., and Thomson, J. A. (2000) Transplantation therapies from human embryonic stem cells—circumventing immune rejection, *e-biomed: J. Regener. Med.* **1,** 11–15.

75. Dey, B., Sykes, M., and Spitzer, T. R. (1998) Outcomes of recipients of both bone marrow and solid organ transplants. A review, *Medicine (Balt.)* **77,** 355–369.

76. Millan, M. T., Shizuru, J. A., Hoffmann, P., et al. (2002) Mixed chimerism and immunosuppressive drug withdrawal after HLA-mismatched kidney and hematopoietic progenitor transplantation, *Transplantation* **73,** 1386–1391.

77. Spitzer, T. R., Delmonico, F., Tolkoff-Rubin, N., et al. (1999) Combined histocompatibility leukocyte antigen-matched donor bone marrow and renal transplantation for multiple myeloma with end stage renal disease: the induction of allograft tolerance through mixed lymphohematopoietic chimerism, *Transplantation* **68,** 480–484.

78. Fandrich, F., Lin, X., Chai, G. X., et al. (2002) Preimplantation-stage stem cells induce long-term allogeneic graft acceptance without supplementary host conditioning, *Nature Med.* **8,** 171–178.

79. Sutherland, D. E., Goetz, F. C., and Sibley, R. K. (1989) Recurrence of disease in pancreas transplants, *Diabetes* **38,** 85–87.

80. Centers for Disease Control and Prevention (2002) Investigation of blood transfusion recipients with West Nile virus infections, *MMWR* **51,** 823.

81. Neildez-Nguyen, T. M., Wajcman, H., Marden, M. C., et al. (2002) Human erythroid cells produced ex vivo at large scale differentiate into red blood cells in vivo, *Nat. Biotechnol.* **20,** 467–472.

82. Pick, M., Eldor, A., Grisaru, D., Zander, A., Shenhav, M., and Deutsch, V. (2002) Ex vivo expansion of megakaryocyte progenitors from cryopreserved umbilical cord blood. A potential source of megakaryocytes for transplantation, *Exp. Hematol.* **30,** 1079.

83. Zwaka, T. P. and Thomson, J. A. (2003) Homologous recombination in human embryonic stem cells, *Nat. Biotechnol.* **21,** 319–321.

IV
THERAPEUTICS

13

Human Embryonic vs Adult Stem Cells for Transplantation Therapies

Calvin B. Harley and Mahendra S. Rao

1. INTRODUCTION

Cells are the fundamental units of life, the building blocks of all tissue, and the source of all extracellular matrix and interstitial fluids. It is clear that irreversible loss of cells as a result of intrinsic or extrinsic causes can be permanently debilitating or lethal. Examples include loss of islet β-cells (diabetes), cardiomyocytes (heart failure), neurons (Parkinson's, dementias, ataxia, stroke, paralysis), hepatocytes (cirrhosis), chondrocytes (joint diseases), renal cells (kidney failure), and hematopoietic cells (bone marrow failure, anemias). Although traumatic cell loss can occur at any age, as demographics in developed countries shift gradually to older individuals, the most significant unmet clinical need and health care burden will likely be chronic diseases in the elderly in which the underlying etiology is cell aging or degeneration.

1.1. Regenerative Medicine and the Inadequacy of Small-Molecule Drugs

Unfortunately, conventional drug therapies (small-molecule chemicals), cannot currently and may never stimulate fully functional cell or tissue replacement for degenerative diseases. There are several reasons for this. First, in many aging adult tissues, the appropriate stem or progenitor cells that might respond to drugs either do not exist, or they exist in insufficient numbers or with insufficient renewal capacity to restore organ function, even when the tissue is not further compromised by disease. Second, even for tissues in which there are ample stem or progenitor cells that respond to drugs when individuals are young, the inherent ability to renew tissues declines with time or chronic stress because of progressive, degenerative, age-related changes. Finally, if the hope is to stimulate division of remaining,

From: *Human Embryonic Stem Cells*
Edited by: A. Chiu and M. S. Rao © Humana Press Inc., Totowa, NJ

differentiated cells of a similar phenotype with in vivo drug therapy in order to restore function, there is no precedent for this in animal or human studies. Postmitotic terminally differentiated cells do not respond to any known mitogens and, hence, it may be impossible to discover small molecules that reprogram these cells into proliferative progenitor cells of the desired phenotype. It may be that, some day, we will have small-molecule, protein, or gene therapy drugs that will reverse terminal differentiation or cellular aging of the target tissue or will cause transdifferentiation, migration, and functional integration of cells from other compartments. However, as a generalized strategy for the broad spectrum of degenerative diseases, that day is not foreseeable and certainly not on the horizon for patients suffering from most chronic disorders today. The best hope in the near future is arguably cell transplantation therapy.

1.2. Limitations of Lineage-Restricted Somatic Stem Cells

The potential of cell or tissue transplantation therapy for acute trauma was demonstrated with the success of blood transfusions in the 19th century. By the mid-20th century, surgeons recognized that for successful, long-term cell replacement therapies in the hematopoietic system, it was necessary to transplant stem cells resident in the bone marrow. Autologous and allogeneic bone marrow transplants and hematopoietic stem cell transplants from bone marrow or cord blood, with or without ex vivo expansion, are now routinely conducted. However, despite the clinical successes in this area and decades of research on human hematopoietic stem cells, we still cannot culture these cells indefinitely in the laboratory, their isolation is expensive and idiosyncratic, and there are periodic shortages of all cell types, from the most elusive stem cell to terminally differentiated platelets and red blood cells. Thus, whereas bone marrow transplants and cord blood transfusions demonstrate the potential success of stem cell therapies, they clearly are not drugs in the classical sense of being batch manufactured on a large scale under good manufacturing practice (GMP) license. Rather, they are more a boutique solution with individual sites harvesting cells as single-shot therapies with consequent variations in product structure and quality.

The state of basic research knowledge and clinical success with somatic cell therapies for other human tissues, with the possible exception of skin *(1–3)*, is worse than that for the hematopoietic system. Although lineage-restricted and multipotent stem or progenitor cells have been identified with varying levels of rigor in the mammalian central nervous system *(4–7)*, in the stromal or mesenchymal compartment of the bone marrow *(8–11)*, and in muscle *(12,13)*, liver *(14–16)*, pancreas *(17,18)*, and fat *(19,20)*, relatively few laboratories have had success with long-term culture and manipulation of the human equivalents, and cells for therapeutic use are certainly not widely available. In fact, biopharmaceutical companies that are on the market today with cell therapy products based on the best characterized human somatic stem or precursor cells are struggling with the challenges of scale-up, quality control, function, and cost. If these are the "easy" targets, it is clear that if we are to develop cell-based therapies for a wide range of tissues, including those for which adult stem cells have yet to be identified, we need a more reliable source of material.

The critical need for a reliable, large-scale source of material is, in our minds, the single largest factor holding back routine use of cells for replacement therapies. For example, pancreatic islet or dopaminergic precursor transplants have resulted in some patients being symptom-free for 5 yr or longer. However, the lack of adequate numbers of reproducible high-quality islets or specific neural populations has prevented physicians from replicating these sporadic successes on a large scale.

1.3. Human Embryonic Stem Cells, Nuclear Transfer, and Renewed Interest in Multipotent and Transdifferentiating Adult Stem Cells

The reason for optimism that great strides will be made in cell transplantation therapies came with the discovery of human embryonic stem cells (hESCs) capable of indefinite replicative capacity and the ability to differentiate into all somatic cells of the body *(21–24)*. Theoretically, hESCs can be used to generate any desired cell or tissue for the treatment of virtually any degenerative disease *(25)*. Moreover, if hESC derivation and differentiation could be coupled with nuclear transfer *(26,27)*, the mammalian cloning technology in which adult somatic cells can be reprogrammed to the embryonic state by fusing them with enucleated oocytes, it is conceivable that a patient could be treated with genetically identical "cloned" cells of any desired type, thus potentially eliminating histocompatibility problems associated with most allogeneic cell therapies *(28,29)*.

Interestingly, the dramatic potential of hESCs and "therapeutic" cloning has also created fresh interest and optimism for the potential of multipotent and "transdifferentiating" adult stem cells and the directed reprogramming of adult cells for medical uses. Part of this interest was fostered by ethical and political concerns about the derivation and use of embryonic stem cells, including those from nuclear transfer embryos. Nevertheless, it is reasonable to question whether hESC derivatives are necessary, or even preferred, for cell transplantation given the potential of adult stem cells, which could be derived from a patient's body and hence generate autologous therapies without the complexities of nuclear transfer.

This chapter focuses on practical issues of developing cell therapies and argues that if we as a society want to generate cost-effective and broadly available cells for treating the millions of patients with degenerative diseases, we need to more aggressively pursue research and development of hESCs. We first encourage the reader to recognize the differences in cell aging and stem cell biology between standard laboratory animals and humans. Much of the promise of adult stem cell therapies has been generated with rodent data, and there are reasons to believe that results may not easily translate to humans. Second, we argue that stem cell therapies for millions of patients, regardless of the original source of cells, will need to be allogeneic, not autologous, at least for the foreseeable future. Finally, we argue that despite the apparent existence of certain scalable human multipotent stem or progenitor cells from the prenatal or postnatal body, these cells currently suffer from a number of disadvantages relative to hESCs and, thus, our best hope for breakthrough cell transplantation therapies demands expanded research and development of hESCs in addition to that with somatic cells.

2. SPECIES DIFFERENCES AND CELL AGING IN STEM CELL BIOLOGY

Although many somatic stem cells have been well characterized in rodents, the human equivalents have often been difficult to identify or difficult to expand for a significant number of divisions in tissue culture (*see* Table 1). In addition to the trivial explanation that each species is unique and hence special culture methods may need to be discovered, there is at least one important fundamental difference between somatic cells from humans versus rodents: Human somatic cells display telomere-dependent replicative senescence, whereas rodent cells generally do not *(50,51)*.

Telomeres are essential genetic elements that protect chromosome ends from degradation or recombination events associated with broken chromosomes. Because they cannot be fully replicated on each cell division, immortal cells require telomerase, a specialized RNA-dependent DNA polymerase, for *de novo* synthesis of telomeric DNA *(52,53)*. In humans, telomerase is repressed in all somatic tissues before birth *(54,55)*, and telomere length subsequently declines as a function of in vivo donor age and in vitro replicative age, reaching a threshold length that triggers cell cycle arrest and an altered pattern of gene expression *(56–59)*. However, in most laboratory strains of rodents, low levels of telomerase can be detected in most somatic tissues, telomeres are much longer than those seen in humans, and there is little evidence for a telomere-dependent mitotic clock with functional significance in vivo.

When telomerase is genetically knocked out in mice and animals are inbred until telomeres become as short as those seen in aging humans, there is a dramatic, age-related increase in degenerative conditions and susceptibility to stress in proliferative tissues *(51,60,61)*. It is also now clear that the presence of telomerase or long telomeres is associated with increased resistance to stress *(61–65)*. Thus, telomerase-positive rodent cells might have a relatively high frequency of escaping from suboptimal tissue culture conditions. These same tissue culture conditions could limit the proliferative capacity of telomerase-negative human cells in culture.

Overall, the telomere dynamics and phenotype of wild-type versus telomerase knockout mice underscore the differences between rodents and humans and the importance of telomere-dependent replicative senescence in human cell biology and disease. Unfortunately, it is not always clear from a superficial review of some publications when human or rodent cells were used. Thus, it is important to be critical in reading the stem cell literature and, second, to be cautious in extrapolating data from rodent stem or progenitor cell studies to applications with human cells.

Highly proliferative tissues in humans such as the bone marrow, skin, and gut contain cells that are "telomerase competent," or inducible. Telomerase induction in these tissues is a regulated process that may slow the rate of telomere loss. However, even in isolated human somatic stem and progenitor cells, telomere length declines with age and there is a limited proliferative capacity in culture *(66–70)*, unlike the general situation in rodents. A possible exception to this is the multipotent adult progenitor cell (MAPC) recently identified by Verfaille and colleagues *(11,37,71)*. MAPCs are reported to be telomerase positive and immortal in culture. However, most of the tissue culture data presented with human MAPCs have been with cells that have undergone fewer than 50 doublings, and most of the functional data have been collected with the rodent counterparts. Further work in multiple laboratories will be required to fully understand the potential of this interesting human cell lineage.

3. CHALLENGES OF AUTOLOGOUS CELL THERAPIES, INCLUDING TRANSDIFFERENTIATION AND THERAPEUTIC CLONING

Autologous cell therapies for the treatment of degenerative diseases have the major advantage that histocompatability is essentially guaranteed because the transplanted cells should have the same genetic composition as the donor. In conventional autologous therapies, cells from the patient are removed, manipulated ex vivo for enrichment, expansion, or differentiation, and transplanted back into the patient. Provided no errors are made during

Table 1
Representative Comparisons of Somatic Stem Cells in Rodents and Humans

Somatic cell type	Rodent (Relatively long telomeres, generally constitutive telomerase expression, immortal)	Human (Relatively short telomeres, telomerase inducible but not generally constitutive expression, limited life-span)
Hematopoietic	Immortal, clonable, scalable, telomerase positive, limited or no transdifferentiation (30).	Limited life-span in culture, not clonable, questionable whether true stem cell identified (31–33).
Liver	Putative stem or progenitor cells isolated from fetal or newborn rats; adult cells difficult to identify except for rare BM-derived precursors (14,34,35).	No hepatocyte or biliary stem cell unambiguously identified; candidate biliary progenitors found with viral oncoprotein transformation; otherwise proliferative capacity very low (15).
Skin	Epidermal stem cells and epithelial equivalents in multiple tissues isolated and expanded, but cell immortality generally not maintained in culture (1). Clonable, multipotent dermal cells identified, passaged at least 1-yr culture, but growth kinetics and stability not established (36).	No skin epidermal or dermal stem cell capable of indefinite growth in culture characterized; candidate dermal multipotent stem cells identified, but not characterized (36).
Stromal/mesenchymal	Mesenchymal stem cells (MSCs) and multipotent adult progenitor cells (MAPCs) isolated from bone marrow using serial selection methods (9,37).	MSCs or marrow stromal cells mortal, but capable of extended or indefinite replicative capacity following telomerase transduction (38,39).

	Cells telomerase positive and capable of indefinite growth under appropriate culture conditions.		Viral oncoproteins may also be necessary for immortalization in some cases (40).
	Connective tissue/mesenchymal differentiation and clonal transdifferentiation reported.		MAPCs reported to be telomerase positive and immortal, but slow growing and not capable of maintenance at high cell densities.
			Most human data reported with cells from relatively low passage.
Neural	Multipotent neural stem cells isolated from multiple brain regions at different developmental ages (6,41–44).		Multipotent stem cells from fetal and adult CNS reported (46–49).
	Cells clonable, but generally have a slow growth rate (up to 10 d/division). Telomerase constitutively high in progenitor cells and telomeres are long (45).		Some evidence of clonality, transdifferentiation, functional transplantability, and long replicative capacity, but growth rate and density relatively low, and ability to isolate from diseased populations not established.
			Telomerase reported in early passage cells only; proliferative capacity diminishes with age (45). Relatively little data on late-passage adult cells.

245

Table 2
Challenges of Autologous Cell Therapies

A. With conventional methods
  1. Appropriate cell type may not be available.
  2. Age or disease state of patient limits abundance or function of desired cell.
  3. Ex vivo expansion is limited because of starting cell numbers and/or replicative senescence.
  4. Phenotype or karyotype of desired cells drift during expansion.
  5. Transplanted cells may be inherently sensitive to original insult/stress.
  6. Patient-specific factors introduce variables in starting material.
  7. Ex vivo manipulations introduce variables in final product.
  8. Overall procedure is expensive.
  9. Surgery or cell extraction may introduce complications for patient.
B. With transdifferentiation
  1. Unknown conditions for regulated, stable transdifferentiation in culture.
  2. No in vivo proof of concept with animal cells.
  3. No proof of concept with human cells in any disease model.
  4. All of the issues with conventional methods.
C. With therapeutic cloning (nuclear transfer)
  1. Sourcing oocytes.
  2. Low efficiency.
  3. Lengthy product development time.
  4. Ex vivo manipulations introduce variables in final product.
  5. High complexity and cost.
  6. Genetic or epigenetic variability of clones.
  7. Potential of teratomas.
  8. Ethics.
  9. Issues 2 and 5–9 with conventional methods.

processing and handling, the risk of rejection or graft-versus-host disease is minimal. However, even in the simplest scenario, there are numerous challenges for this type of therapy to become widely used for degenerative diseases.

3.1. Conventional Autologous Cell Therapies

Some of the challenges of developing autologous cell therapies are listed in Table 2. For many degenerative diseases in humans, we simply do not have access to the appropriate cell type for therapeutic application. For example, even though there is some evidence for cardiomyocyte regeneration in humans *(72)*, no one today has identified a population of human adult cells that can be reproducibly isolated from a human for expansion and functional replacement of damaged heart muscle. A similar situation exists for most, if not all, neurodegenerative diseases. Even for tissue like the skin, in which autologous stem and progenitor cells appear abundant and have been greatly expanded and successfully used for burn patients, autologous skin grafts do not work well for chronic diseases such as diabetic or venous stasis skin ulcers. The reason for this may relate to the fact that the underlying cause of chronic ulcers impacts the local vascular tissue or the cells that form granulation tissue, and autologous cells for these tissues have not been developed. Another issue with autologous transplants is that in some cases of chronic diseases, the genotype of the patient's cells may have contributed to disease initiation or progression, in which case, the transplant could be inherently sensitive to the stress or insult at the site of injury and succumb with time, as did the original cells.

However, even if the appropriate cell type for the treatment of a particular disease has been identified, there are scientific, clinical, and commercial challenges for autologous cell therapies. The ability to isolate and expand somatic cells, including stem and progenitor populations, declines with donor age *(31,67,73–75)*, and if cells are expanded, their phenotype and/or karyotype can drift, raising the risk of reduced efficacy and safety. Replicative senescence from telomere loss in autologous cells in vivo and ex vivo can contribute to lack of scalability, changes in phenotype, and genomic instability, but other factors associated with aging such as intracellular and extracellular protein modification, apoptosis, somatic mutation, and cell signaling aberrations could also be involved. Moreover, disease or chronic stress in a patient may further compromise isolation of a suitable population of somatic stem cells.

Because most patients in need of regenerative medicine are elderly, these factors alone could severely limit the utility of generalized autologous somatic stem cell therapies. However, there are numerous other problematic issues related to the idiosyncratic nature of autologous therapies. First, the surgical extraction of tissue for isolation of stem cells could create complications for the patient. Second, there are a number of issues related to quality control of the product. Because every patient is unique and different doctors and technical staff would be involved in isolating, expanding, and retransplanting cells, there are multiple points where inherent variables could lead to unpredictable outcomes. Even when a simple cell such as a connective tissue fibroblast is isolated on separate occasions or from healthy versus diseased individuals, the phenotype of the cell population in culture can vary significantly *(76)*. In addition, the purity of starting material and the authenticity and quality of the final product will depend on contaminating cell types present at the time of isolation (donor or nondonor in origin), whether unwanted bacterial, viral, or chemical agents are inadvertently introduced postisolation, and whether bookkeeping, labeling, shipping, and handling are all error-free. Finally, there is an issue of economics. It is difficult to obtain precise information on costs related to the decentralized infrastructure and personnel necessary for autologous therapies, but industry representatives have stated that isolation, expansion, and qualification of cells for every single patient can require up to 10–20 technicians working full time for several weeks in GLP-regulated, specialized laboratories between the initial and final surgical procedures. These extraordinary costs, coupled with all of the previously described challenges with autologous transplants, have made it extremely difficult for companies focusing on this opportunity to survive.

Thus, with each patient receiving a unique product that is infinitely more complex and orders of magnitude more expensive than standard pharmaceutical products, it is difficult to imagine that autologous cell therapies for the wide range of degenerative diseases afflicting millions of patients will become generally available in the foreseeable future. This does not mean that, for specific diseases, autologous cell therapies in specialized settings will never have a major medical impact; it simply argues against the concept that autologous adult stem cell therapies will provide *the* answer for regenerative medicine.

3.2. Transdifferentiation for Autologous Transplants

Some somatic stem or progenitor cells may not be as restricted in their differentiation potential as previously supposed, but may be able to transdifferentiate into essentially any cell type (reviewed in ref. *77*). Proponents of transdifferentiation have thus argued that hESCs are not required, as we can simply harvest readily available cells that are present in sufficient numbers in an adult patient and manipulate them in culture to direct their differentiation into the desired cell type. For example, if we could direct hematopoietic or mesenchymal stem cells isolated from a Parkinson's disease patient to generate dopaminergic neurons, we would not need to utilize embryonic or fetal material and we would not need to worry about immune rejection.

The appeal of such an approach is seductive, but recent data suggest that transdifferentiation is not as straightforward as previously thought and that many of the examples in the literature may not represent true transdifferentiation *(30,78,79)*. Even if specific examples of transdifferentiation were authentic, to date no transdifferentiated cell has been obtained in sufficient numbers for use in autologous cell replacement therapy, and even if it were, it would still suffer the limitations of conventional autologous therapies outlined earlier (*see* Table 2A). Moreover, there are additional concerns with transdifferentiation (*see* Table 2B). We do not yet have any molecular understanding of the induction of transdifferentiation. It is likely to be tightly regulated and highly specific, such that uncontrolled or ectopic transdifferentiation in vivo may create cells of inappropriate phenotype. The timing of when we discover conditions that reproducibly generate stable transdifferentiation in culture is uncertain, and even if we achieve this in the not-to-distant future, it could still be sometime before we know whether the cells are stable in vivo and function in disease models. Finally, if conditions and factors for transdifferentiation are worked out in rodents, this success may not readily translate to human cell biology, as noted earlier (*see* Section 2). We do not doubt that transdifferentiation is possible, but simply emphasize that assuming transdifferentiation of somatic stem cells is a viable alternative to hESCs is unfounded based on current data.

3.3. Therapeutic Cloning for Autologous Transplants

Although no one has yet achieved human therapeutic cloning (i.e., the creation of patient-specific embryonic stem cells from a nuclear transfer embryo for autologous cell therapy applications), the concept has been proven in animal models *(80)*, and in a sense, it is the ultimate form of somatic cell transdifferentiation. Therapeutic cloning technology provides a potential solution to two of the autologous cell therapy limitations: Sourcing and scaling the appropriate somatic cell type without phenotypic or karyotypic drift. By creating a cloned embryo from an enucleated donor oocyte and a patient's skin cell, for example, embryonic stem cells could be isolated, expanded, genetically manipulated if necessary, and differentiated to the desired cell type(s). Assuming that mitochondria or other factors contributed by the oocyte cytoplasm are not an issue, and if all else goes well, the differentiated cells could be further enriched and transplanted back into the patient with minimal risk of immunologic rejection or graft-versus-host disease. Because undifferentiated human embryonic cells are pluripotent and can be expanded without replicative senescence or significant phenotypic or genetic drift (*see* Section 4.3.), the production of sufficient numbers of functional, nonaged cells of the appropriate type is theoretically assured. However, therapeutic cloning suffers from all of the other limitations of patient-specific autologous therapies and has additional challenges of its own (Table 2C).

Successful reprogramming of adult somatic cells to generate embryonic blastocysts and the subsequent derivation of embryonic stem cell lines requires considerable technical expertise and is currently an inefficient process. In fact, even the first step in this process, the creation of a viable nuclear transfer embryo, has not yet been unambiguously accomplished in humans, despite reports to the contrary *(81)*. It is expected that human nuclear transfer will be feasible and accomplished in the near future, but to treat millions of patients with therapeutic cloning, many millions of high-quality oocytes would need to be sourced, raising financial, ethical, technical, and regulatory issues for governments, hospitals, and commercial entities, as well as medical risks for the oocyte donors. In the future, it should be possible to reprogram somatic cells with molecular techniques rather than with enucleated oocytes, eliminating most of these issues, but it is uncertain if this can be achieved in the near future.

Regardless of how or when autologous embryonic or pluripotent stem cells are created, it would subsequently take many weeks to isolate, characterize, and expand the undifferentiated population, and then a number of additional weeks to differentiate, enrich, characterize, and qualify the final transplantable cells. Doing this for each and every patient and ensuring quality control at all steps creates an almost unthinkable effort and cost given today's technologies and resources. Finally, if all of these challenges were met, the time element alone for creating a product from therapeutic cloning means that it would be unsuitable for patients with immediate transplant needs.

4. ADVANTAGES AND CHALLENGES OF ALLOGENEIC STRATEGIES

Allogeneic cell therapies are not patient-specific and, hence, can be centrally produced. This provides a number of advantages over autologous cell therapies relating primarily to cost (no extraction of cells from individual patients and efficiencies of mass production) and quality control (products that can be selected for optimal properties and well characterized at all stages of production). However, for allogeneic therapy to be practical, appropriate starting materials must be available, they must be stable and predicable under modification and scale-up, and they must not be too expensive nor create prohibitive histocompatibility problems.

The issues of starting materials, scale-up, function, stability, and cost will be discussed in this section in considering the differences between somatic and hESC therapies. The histocompatability problem with allogeneic cell therapies is shared by somatic and hESC derivatives, but this medical issue may not be as much of a challenge as originally believed. Even under the worst-case scenario in which tissue matching and immune suppression are required to reduce graft-versus-host disease and rejection, these problems have been addressed for decades with kidney, heart, liver, lung, corneal, skin, bone, and hematopoietic transplants. There are currently more than 300,000 tissue and organ allogeneic transplantations per year in the United States alone, so physicians receiving characterized and qualified allogeneic stem cell derivatives for transplantation should not encounter new histocompatability challenges. To the contrary, because stem cell derivatives for transplantation will be much better characterized than tissue or organ donor material and will typically lack mature immune cells, the complications of graft-versus-host disease common to most organ and tissue transplants should not be an issue. Moreover, there are a variety of strategies being considered to reduce graft rejection from stem cell transplants, one of which, hematopoietic chimerism with hESC-derived hematopoietic precursors, may hold tremendous potential for histocompatible allogeneic hESC products *(82,83)*.

4.1. Challenges of Allogeneic Somatic Stem Cell Therapies

When considering allogeneic cell therapies, it is not necessary to restrict somatic approaches to adult stem cells: Young, newborn, cord blood, or even fetal somatic stem cells should also be considered. Thus, issues related to cellular aging inherent to the donor can be minimized. However, many of the challenges of autologous somatic stem cell therapies remain (Table 3A). First, although almost every specific cell type imaginable has been claimed to be derivable from multipotent somatic stem cells, there have been concerns that many examples of transdifferentiation were artifacts of somatic cell fusions or represent such rare events as to be of uncertain biomedical significance, as described earlier *(30,78,79)*. In addition, for many lineage-specific or multipotent somatic stem cells, the majority of the data has been collected with rodent models (Table 1). Thus, it is possible that for many human diseases such as diabetes or congestive heart failure, the appropriate cell type may not be available from existing somatic stem cells. Where human somatic stem or progenitor cells have been demonstrated to be medically useful for certain lineages (e.g., hematopoietic and mesenchymal lineages), multiple different primary stem cell banks or donor derivations are required, compared to hESCs derivatives for which a single or relatively small number of master cell banks is plausible.

For most existing, well-characterized human somatic stem cells, it appears that ex vivo expansion is limited because of one or more of the following: spontaneous differentiation, slow growth rates at low density, drifting phenotype or karyotype, or replicative senescence. When primary cells need to be sourced on multiple occasions, as is the case for all current somatic stem cell therapies, this introduces inherent donor variability, quality control issues, and increased costs at all stages of production, as described previously for autologous therapies (*see* Section 3). In addition, the limited or difficult growth of undifferentiated somatic stem cells creates another problem. For certain stem cell therapies, genetic modification of cells may be advantageous (e.g., to help correct genetic disorders in the recipient patient or to improve cell isolation, function, transplantation, or integration). Even the simplest genetic modification typically requires dozens of cell divisions to create, confirm or isolate, and qualify the altered cell product prior to transplantation, which may not be possible with most somatic stem cell populations.

Finally, in addition to the histocompatibility issue addressed earlier (*see* Sections 3 and 4), somatic stem cells share a potential liability with hESCs: the ability to form undesired growths in vivo. The key advantage of certain fairly well-characterized somatic stem or progenitor cells, such as those in the hematopoietic and mesenchymal lineage, over hESCs as a source for cell transplantation resides in the fact that they are "closer" to the desired differentiated state and, hence, may more readily generate the relevant therapeutic cell type free of unwanted cells. They also lack the ethical concern that exists for some individuals with cell therapies derived from a human embryonic source. However, the challenges of developing scalable somatic stem cell therapies for multiple chronic diseases are still enormous, arguing that research and development of hESCs for widespread therapeutic use must continue.

4.2. Challenges of Allogeneic hESC Therapies

The issues that face biopharmaceutical companies developing hESCs for cell and tissue transplantation (*see* Table 3B) are not unique or insurmountable. The potential of hESCs to form teratomas (i.e., benign growths of undifferentiated and differentiated cells) is higher than that of somatic stem or progenitor cells, but, as indicated earlier, transplanted somatic cell populations may also form undesired growths (*see* Section 4.3.). It will be necessary for any cell therapy company to ensure that the final population of cells transplanted into patients is sufficiently characterized to have an acceptable risk of complications as a result of the presence or growth of either desired or undesired cells. The advantages of hESCs enumerated in Section 4.3. make it possible that the risk of undesired cells or undesired properties of the transplanted cells is actually lower with hESCs derivatives than with somatic stem cells. For example, in many cases of somatic stem cell therapies, spontaneous or viral oncoprotein transformation of cells is considered or being used in clinical studies in order to overcome growth limitations of the normal human equivalent. Examples include SV40-T antigen transformed islet precursor cells and embryocarcinoma-derived neural cells.

Table 3
Challenges of Allogeneic Cell Therapies

A. With adult stem cells
  1. Appropriate cell type may not be available.
  2. Different stem cell banks required for different lineages.
  3. Ex vivo expansion may be limited or very costly because of starting cell densities, growth kinetics, spontaneous differentiation, or replicative senescence.
  4. Phenotype or karyotype of desired cells may drift during expansion.
  5. Multiple sourcing of primary tissue may be required.
  6. Ability to engineer cells may be limited because of phenotype or replicative senescence.
  7. Potential for undesired growths in vivo.
  8. Histocompatability.
B. With hESCs
  1. Potential for undesired growths in vivo may be higher than that with adult stem cells.
  2. Differentiation methods may be difficult and/or expensive.
  3. Histocompatability.
  4. Ethics.

These cells may have a greater propensity than qualified hESC derivatives to create unwanted cell products posttransplantation.

Although analyses of embryoid bodies and teratomas from hESCs indicate that essentially all somatic lineages are derivable from these cells *(21,24,84)*, the number of well-characterized populations of different somatic cells derived in culture is still relatively small. Among the best characterized are neural cells *(85,86)*, hematopoietic cells *(82)*, cardiomyocytes *(87)*, and hepatocyte-like cells *(88)*. It may be very costly today to isolate some of these cells in quantities needed for transplantation, and for other cell types, efficient derivation may be some years away. However, it has been only a few years since hESCs were discovered, and only a small number of laboratories have been working with the cells. We would argue that the pace of progress has been extraordinary and that improvements in culture conditions and additional research will lead to cost-effective production of essentially any desired somatic cell type from hESCs in the near future.

The issue of histocompatability with allogeneic cell transplants was raised earlier in this section. The present ability of the medical community to deal with the more complex challenges of tissue and organ transplants suggests that the issue should not be a major hurdle for the development of either

Table 4
Advantages of Allogeneic hESC Therapies

1. Apparent source for *all* somatic cells or tissues (published examples listed)
   a. Neural lineages (neurons, astrocytes, oligodendrocytes)
   b. Hematopoietic lineage (erythroid, myeloid, and lymphoid)
   c. Cardiomyocytes
   d. Hepatocyte-like cells
2. Self-organizing potential
3. Scalable
   a. Immortal (telomerase positive)
   b. Reasonable growth rate (20–30 h)
   c. Reasonable growth density (approx 0.5×10^5 to 1×10^5 cells/cm^2)
4. Stable
   a. Maintenance of pluripotency over hundreds of population doublings
   b. No evidence of gross phenotypic or karyotypic drift
   c. Ability to cryopreserve for storage and shipment
5. Modifiable
   a. Gene transduction demonstrated
   b. Immortal and stable single-cell clones achievable
6. May have low histocompatability barriers

somatic or embryonic stem cell therapies. The one remaining challenge for hESC therapies from Table 3B is that of ethics. This issue is complex, largely outside the realm of scientific argument, and has been dealt with elsewhere (*see* Chapter 1). Thus, we will not attempt to summarize the ethical issue or propose resolutions here, but, instead, move to why we believe continued research and development of hESC therapies is vital.

4.3. Advantages of Allogeneic hESC Therapies

Table 4 summarizes the advantages of pursuing hESC research for the development of allogeneic cell therapies. Analysis of teratomas and embryoid bodies shows that hESCs are capable of being a single source for essentially all somatic cells and tissues. Although to date only neural, hematopoietic, cardiomyocyte, and hepatocyte cells as enriched populations from hESCs have been described in the published scientific literature, data for osteoblasts, insulin-producing islet cells, and several mesenchymal derivatives have been presented at scientific meetings. In addition, a variety of neuronal and glial cell types are being found and enriched from within neural lineage cells derived from hESCs. The potential to work from a single master cell bank of undifferentiated hESCs for the production of any transplantable cell type clearly has advantages of scale and quality control over working with a variety of multipotent somatic cell banks derived from multiple donors at multiple times.

In addition to showing that hESCs are pluripotent, teratomas and embryoid bodies also show that hESCs have a self-organizing potential reminiscent of organogenesis (*see* Fig. 1 and refs. *21* and *24*). Thus, when hESCs spontaneously differentiate, nuclei of self-organizing tissues form such that cartilage-containing sacs form, neural cells stratify surrounding ventricle-like structures, kidney epithelial cells arrange in glumerular-like clusters, columnar and glandular gut epithelial cells can form a lumen and be surrounded by smooth muscle cells, and myoblasts can fuse in long strands of skeletal muscle fiber or differentiate into heart tissue and spontaneously pulsate. This inherent, genetic program of hESCs to differentiate along multiple coordinated lineages provides a potential benefit to current or future cell or tissue transplantation applications. It is rarely the case that degenerative diseases involve a single cell type, and thus it may be advantageous to transplant not just a single cell type, but a population of cells that includes support cells, or a progenitor population that has the ability to proliferate and differentiate into a set of desired cell types.

In contrast to essentially all somatic stem or progenitor cells, hESCs are scalable and stable in a commercially practical sense. They are telomerase positive and immortal and have been cultured for hundreds of population doublings while maintaining their pluripotency and showing no signs of gross phenotypic or karyotypic drift *(22)*, with or without cryopreservation (*see also* Chapter 6). The latter property is especially important for establishing working and master cell banks. Although not all hESC derivatives have been tested, it appears that cryopreservation of cell populations suitable for transplantation will be possible, which will facilitate distribution to research centers and hospitals.

The growth rate of hESCs (<30 h/doubling) and cell density (approx 10^5 cells/cm^2) could be improved with further research, but it is suitable for scalable production of cells even with the current state of the art. To appreciate how significant this is compared to MAPCs (arguably the most competitive somatic stem cell currently in terms of cell immortality and scalability), consider generating 1 kg of cells (about 1 L, or 10^{12} cells) from an ampoule of 10^6 cells. Assuming exponential growth and no loss on passaging, this requires about 20 doublings. For hESCs, this translates to <25 d of culture and about 1000 m^2 of tissue culture surface. In contrast, for MAPCs, for example, about 40 d of culture and 100,000 m^2 of tissue culture surface is required with currently described growth rates and cell densities *(11,37,71)*, clearly an impractical situation with today's technologies.

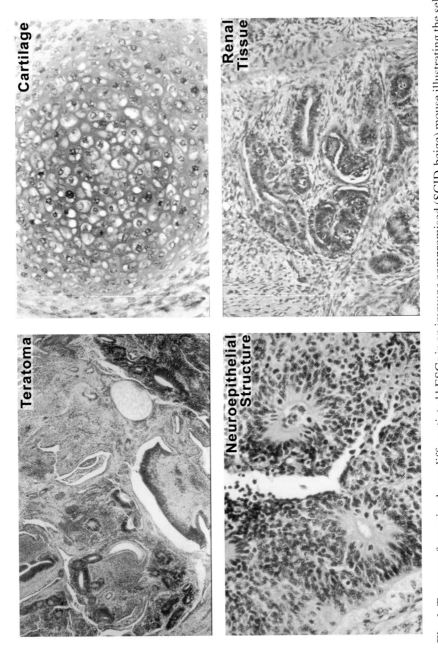

Fig. 1. Teratoma formation by undifferentiated hESCs in an immune compromised (SCID-beige) mouse illustrating the self-organizing potential of hESCs for complex tissue formation and organogenesis. The low-power teratoma panel illustrates

256

Human ESCs have been shown to be genetically modifiable *(24)* and clonable without loss of cellular immortality or pluripotency *(22)*. These properties permit genetic transduction and/or selection of clones with improved function. As mentioned earlier, gene modification may be desirable to help correct genetic disorders in the recipient patient or to improve cell isolation, function, transplantation, integration, or histocompatability. Genetic modification of human somatic stem cells is currently limited by their finite replicative capacity, drift in phenotype, senescence, or loss of "stemness" upon expansion, or simply being refractory to transduction.

Finally, hESCs may offer some unique advantages over adult somatic cells solutions regarding histocompatability: The ability to isolate hematopoietic stem cells from hESCs enables hematopoietic chimerism as a potential solution *(82)*; the apparent ease of genetic transduction enables strategies to create universal donor cells; linking nuclear transfer with hESC derivation enables generation of a cell banks representing major histocompatibility types; and because of the embryonic nature of hESCs and the ability to select and qualify highly pure populations of derivatives, the proportion of cells presenting antigens that might trigger graft rejection could be reduced relative to adult cell transplants *(84,89)*. It will take further research and ultimately human clinical studies to determine whether hESC-based transplantation therapies have fewer graft rejection problems than allogeneic somatic cell therapies. However, the presence of acceptable solutions today (e.g., immunosuppression) and the presence of promising second-generation strategies for graft acceptance warrants the continued development of allogeneic hESC-based cell therapies.

5. SUMMARY AND CONCLUSIONS

There has been an ongoing debate regarding hESCs and somatic stem cells for research and therapeutic use as though this was an "either/or" situation. However, as this chapter indicates, there are multiple complementary approaches to a common goal. In some areas, such as transplants for

---

Fig. 1. *(continued)* the variety of stratified, differentiated cell structures representing multiple germ lineages that are commonly present in all teratoma growths. The remaining higher-power panels illustrate cartilage, neuroepithelial, and renal tissues. The cells used in this study were the H1 line at passage 70. Five million undifferentiated cells were injected intramuscularly into SCID-beige mice. Teratomas were harvested for histological analysis 42 d postinjection.

hematopoietic and certain connective tissue cells, existing somatic stem or progenitor cells, whether autologous or allogeneic, clearly provide a clinical benefit and are part of medical practice today. These and other somatic stem or progenitor cell therapies may get better and in the future provide a safe and cost-effective means to treat many diseases. However, today, all somatic cell therapies, whether autologous or allogeneic, have significant commercial or clinical limitations, and there are many chronic diseases for which no safe and effective somatic cell therapy appears imminent. Thus, we should be fostering research and development on hESCs as they may be the best current option, and perhaps for some diseases, the only long-term option.

We have argued that scalable, reproducible allogeneic cells that can be mass produced and quality assured in centralized facilities are the only practical solution today for the development of cell therapies as drug-like biopharmaceuticals. Such an approach enables the development of cost-effective products and avoids the idiosyncrasies and in vivo and ex vivo variability and risks inherent with all autologous strategies, including transdifferentiation and therapeutic cloning. Today, the only cells with features that enable widespread allogeneic applications (unlimited growth potential, reasonable growth and cell density characteristics, phenotypic and karyotypic stability, and proven pluripotency) are hESCs. These cells also have additional potential advantages over some existing adult somatic stem cells in being easily modified by genetic transduction and offering novel approaches to graft rejection. This argues for strong support for continued research and development of hESC-based cell therapies to complement ongoing work with somatic cells.

At any given point in time in the future, physicians should have the opportunity to weigh the merits of each available cell therapy to determine which one has the most acceptable safety and efficacy profile for any given patient. In tissues and organs where a stem cells are relatively abundant or can be created and where there is ongoing cell replacement, it may be that therapies along the line of the hematopoietic model with transplants of somatic stem cells will be best on balance. In other systems, alternative strategies, including hESC-based products may be the only viable solution.

Human ESCs were isolated only 4 years ago and progress in that short a period has been rapid. Over two decades of experience working with murine ESCs has also generated a wealth of data on ES pluripotency and their safety and efficacy in disease models. As there appears to be no conceptual barrier to hESC therapy and these cells have certain advantages compared to existing somatic cell therapies, we conclude that to ensure the best medical care for patients now and into the future, research and development of hESCs must be broadly promoted and the relative merits of different cell therapies directly compared in clinical studies and ultimately in medical practice.

ACKNOWLEDGMENTS

The authors thank their many colleagues and co-workers at Geron Corporation, the University of Utah, and the NIH for their experimental work on embryonic and somatic stem cells, which contributed to the foundation of this chapter. We also thank David Earp, Tom Okarma, and Jane Lebkowski for critically reading of the manuscript, and Choy-Pik Chiu, Melissa Carpenter, and Ellen Rosler for providing Fig. 1.

REFERENCES

1. Lavker, R. M. and Sun, T.-T. (2000) Epidermal stem cells: properties, markers, and location, *PNAS* **97(25)**, 13,473–13,475.
2. Parenteau, N. (1999) Skin: the first tissue-engineered products, *Sci. Am.* **280**, 83–84.
3. Naughton, G. K. (2002) From lab bench to market: critical issues in tissue engineering, *Ann. NY Acad. Sci.* **961**, 372–385.
4. Sommer, L. and Rao, M. (2002) Neural stem cells and regulation of cell number, *Prog. Neurobiol.* **66**, 1–18.
5. Shihabuddin, L., Horner, P. J., Ray, J., et al. (2000) Adult spinal cord stem cells generate neurons after transplantation in the adult dentate gyrus, *J. Neurosci.* **20(23)**, 8727–8735.
6. Johansson, C. B., Momma, S., Clarke, D. L. et al. (1999) Identification of a neural stem cell in the adult mammalian central nervous system, *Cell* **96**, 25–34.
7. Rietze, R., Valcanis, H., Brooker, G. F., et al. (2001) Purification of a pluripotent neural stem cell from the adult mouse brain, *Nature* **412**, 736–739.
8. Caplan, A. I. and Bruder, S. P. (2001) Mesenchymal stem cells: building blocks for molecular medicine in the 21st century, *Trends. Mol. Med.* **7**, 259–264.
9. Bianco, P., Rimunucci, M., Gronthos, S., et al. (2001) Bone marrow stromal stem cells: nature, biology, and potential applications, *Stem Cells* **19**, 180–192.
10. Liechty, K., MacKenzie, T. C., Shaaban, A. F., et al. (2000) Human mesenchymal stem cells engraft and demonstrate site-specific differentiation after in utero transplantation in sheep, *Nature Med.* **6(11)**, 1282–1286.
11. Reyes, M., Lund, T., Lenvik, T., et al. (2001) Purification and ex vivo expansion of postnatal human marrow mesodermal progenitor cells, *Blood* **98(9)**, 2615–2625.
12. Seale, P., Asakura, A., and Rudnicki, M. A. (2001) The potential of muscle stem cells, *Dev. Cell.* **1**, 333–342.
13. Deasy, B. M. and Huard, J. (2002) Gene therapy and tissue engineering based on muscle-derived stem cells, *Curr. Opin. Mol. Ther.* **4**, 382–389.
14. Agelli, M., Dello Sbarba, P., Halay, E. D., et al. (1997) Putative liver progenitor cells: conditions for long-term survival in culture, *Histochem. J.* **29**, 205–217.

15. Allain, J., Dagher, I., Mahieu-Caputo, D., et al. (2002) Immortalization of a primate bipotent epithelial liver stem cell, *PNAS* **99(6),** 3639–3644.
16. Alison, M. (1998) Liver stem cells: a two compartment system, *Curr. Opin. Cell Biol.* **10,** 710–715.
17. Bonner-Weir, S. and Sharma, A. (2002) Pancreatic stem cells, *J. Pathol.* **197,** 519–526.
18. Gmyr, V., Kerr-Conte, J., Belaich, S., et al. (2000) Adult human cytokeratin 19-positive cells reexpress insulin promoter factor 1 in vitro, *Diabetes* **49,** 1671–1680.
19. Zuk, P., Zhu, M., Mizuno, H., et al. (2001) Multilineage cells from human adipose tissue: implications from cell-based therapies, *Tissue Eng.* **7,** 211–228.
20. Hauner, H., Skurk, T., and Wabitsch, M. (2001) Cultures of human adipose precursor cells, *Methods Mol. Biol.* **155,** 239–247.
21. Thomson, J. A., Itskovitz-Eldor, J., Shapiro, S. S., et al. (1998) Embryonic stem cell lines derived from human blastocysts, *Science* **282,** 1145–1147.
22. Amit, M., Carpenter, M. K., Inokuma, M. S., et al. (2000) Clonally derived human embryonic stem cell lines maintain pluripotency and proliferative potential for prolonged periods of culture, *Dev. Biol.* **227,** 271–278.
23. Trounson, A. O. (2001) The derivation and potential use of human embryonic stem cells, *Reprod. Fertil. Dev.* **13,** 523–532.
24. Lebkowski, J., Gold, J., Xu, C., et al. (2001) Human embryonic stem cells: culture, differentiation, and genetic modification for regenerative medicine applications, *Cancer J.* **7,** S83–S93.
25. Pedersen, R. A. (1999) Embryonic stem cells for medicine, *Sci. Am.* **280(4),** 68–73.
26. Campbell, K. H. S., McWhir, J., Ritchie, W. A., et al. (1996) Sheep cloned by nuclear transfer from a cultured cell line, *Nature* **380,** 64–66.
27. Lanza, R. P., Cibelli, J. B., and West, M. B. (1999) Prospects for the use of nuclear transfer in human transplantation, *Nature Biotechnol.* **17,** 1171–1174.
28. Munsie, M. J., Michalska, A. E., O'Brien, C. M., et al. (2000) Isolation of pluripotent embryonic stem cells from reprogrammed adult mouse somatic cell nuclei, *Curr. Biol.* **10,** 989–992.
29. Lanza, R., Chung, H. Y., Yoo, J. J., et al. (2002) Generation of histocompatible tissues using nuclear transplantation, *Nature Biotechnol.* **20,** 689–696.
30. Wagers, A. J., Sherwood, R. I., Christensen, J. L., et al. (2002) Little evidence for developmental plasticity of adult hematopoietic stem cells, *Science* **297,** 2256–2259.
31. Thornley, I. and Freedman, M. (2002) Telomeres, X-inactivation ratios, and hematopoietic stem cell transplantation in humans: a review, *Stem Cells* **20,** 198–204.
32. Feugier, P., Jo, D. Y., Shieh, J. H., et al. (2002) Ex vivo expansion of stem and progenitor cells in co-culture of mobilized peripheral blood CD34+ cells on human endothelium transfected with adenovectors expressing thrombopoietin, c-kit ligand, and Flt-3 ligand, *J. Hematother. Stem Cell Res.* **11,** 127–138.
33. Bonnet, D. (2002) Haematopoietic stem cells, *J. Pathol.* **197,** 430–440.

34. Forbes, S., Vig, P., Poulsom, R., et al. (2002) Hepatic stem cells, *J. Pathol.* **197,** 510–518.
35. Susick, R., Moss, N., Kubota, H., et al. (2001) Hepatic progenitors and strategies for liver cell therapies, *Ann. NY Acad. Sci.* **944,** 398–419.
36. Toma, J., Akhavan, M., Fernandez, K. J., et al. (2001) Isolation of multipotent adult stem cells from the dermis of mammalian skin, *Nature Cell Biol.* **3,** 778–784.
37. Jiang, Y., Jahagirdar, B. N., Reinhardt, R. L., et al. (2002) Pluripotency of mesenchymal stem cells derived from adult marrow, *Nature* **418,** 41–49.
38. Simonsen, J., Rosada, C., Serakinci, N., et al. (2002) Telomerase expression extends the prolifrative life-span and maintains the osteogenic potential of human bone marrow stromal cells, *Nature Biotechnol.* **20,** 592–595.
39. Shi, S., Gronthos, S., Chen, S., et al. (2002) Bone formation by human postnatal bone marrow stromal stem cells is enhanced by telomerase expression, *Nature Biotechnol.* **20,** 587–591.
40. Darimont, C., Avanti, O., Tromvoukis, V., et al. (2002) SV40 T antigen and telomerase are required to obtain immortalized human adult bone cells without loss of the differentiated phenotype, *Cell Growth Differ.* **13,** 59–67.
41. Laywell, E. D., Rakic, P., Kukekov, V. G., et al. (2000) Identification of a multipotent astrocytic stem cell in the immature and adult mouse brain, *PNAS* **97(25),** 13,883–13,888.
42. Uchida, N., Buck, D. W., He, D., et al. (2000) Direct Isolation of Human central nervous system stem cells, *PNAS* **97(26),** 14,720–14,725.
43. Reynolds, B. A. and Weiss, S. (1992) Generation of neurons and astrocytes from isolated cells of the adult mammalian central nervous system, *Science* **255,** 1707–1710.
44. Morshead, C. M., Reynolds, B. A., Craig, C. G., et al. (1994) Neural stem cells in the adult mammalian forebrain: a relatively quiescent subpopulation of subependymal cells, *Neuron* **13,** 1071–1082.
45. Ostenfeld, T., Caldwell, M. A., Prowse, K. R., et al. (2000) Human neural precursor cells express low levels of telomerase in vitro and show diminishing cell proliferation with extensive axonal outgrowth following transplantation, *Exp. Neurol.* **164,** 215–226.
46. Pagano, S., Impagnatiello, F., Girelli, M., et al. (2000) Isolation and characterization of neural stem cells from the adult human olfactory bulb, *Stem Cells* **18,** 295–300.
47. Gage, F. H. (2000) Mammalian neural stem cells, *Science* **287,** 1433–1438.
48. Vescovi, A. L., Parati, E. A., Gritti, A., et al. (1999) Isolation and cloning of multipotential stem cells from the embryonic human CNS and establishment of transplantable human neural stem cell lines by epigenetic stimulation, *Exp. Neurol.* **156,** 71–83.
49. Wu, P., Tarasenko, Y. I., Gu, Y., et al. (2002) Region-specific generation of cholinergic neurons from fetal human neural stem cells grafted in adult rat, *Nature Neurosci.* **5,** 1271–1278.
50. Harley, C. B., Kim, N. W., Prowse, K. R., et al. (1994) Telomerase, cell immortality, and cancer, *Cold Spring Harbor Symp. Quant. Biol.* **59,** 307–315.

51. Wright, W. E. and Shay, J. W. (2000) Fundamental differences in human and mouse telomere biology, *Nature Med.* **6,** 849–851

52. Greider, C. W. and Blackburn, E. H. (1985) Identification of a specific telomere terminal transferase enzyme with two kinds of primer specificity, *Cell* **51,** 405–413.

53. Morin, G. B. (1989) The human telomere terminal transferase enzyme is a ribonucleoprotein that synthesizes TTAGGG repeats, *Cell* **59,** 521–529.

54. Ulaner, G. A., Hu, J. F., Vu, T. H., et al. (1998) Telomerase activity in human development is regulated both by hTERT transcription and by alternate splicing of hTERT transcripts, *Cancer Res.* **58(18),** 4168–4172.

55. Wright, W. E., Piatyszck, M. A., Rainey, W. E., et al. (1996) Telomerase activity in human germline and embryonic tissues and cells, *Dev. Genet.* **18,** 173–179.

56. Harley, C. B., Futcher, A. B., and Greider, C. W. (1990) Telomeres shorten during ageing of human fibroblasts, *Nature* **345(6274),** 458–460.

57. Allsopp, R. C., Vaziri, H., Patterson, C., et al. (1992) Telomere length predicts replicative capacity of human fibroblasts, *Proc. Natl. Acad. Sci. USA* **89,** 10,114–10,118.

58. Allsopp, R. C., Chang, E., Kashefi-Aazam, M., et al. (1995) Telomere shortening is associated with cell division in vitro and in vivo, *Exp. Cell Res.* **220,** 194–200.

59. Shelton, D. N., Chang, E., Whittier, P. S., et al. (1999) Microarray analysis of replicative senescence, *Curr. Biol.* **9,** 939–45.

60. Blasco, M. A., Lee, H. W., Hande, M. P., et al. (1997) Telomere shortening and tumor formation by mouse cells lacking telomerase RNA, *Cell* **91,** 25–34.

61. Rudolph, K. L., Chang, S., Lee, H. W., et al. (1999) Longevity, stress response, and cancer in aging telomerase-deficient mice, *Cell* **96(5),** 701–712.

62. Goytisolo, F. A., Samper, E., Martin-Caballero, J., et al. (2000) Short telomeres result in organismal hypersensitivity to ionizing radiation in mammals, *J. Exp. Med.* **192(11),** 1625–1636.

63. Wong, K.-K., Chang, S., Weiler, S. R., et al. (2000) Telomere dysfunction impairs DNA repair and enhances sensitivity to ionizing radiation, *Nature Genet.* **26,** 85–88.

64. Oh, H., Taffet, G. E., Youker, K. A., et al. (2001) Telomerase reverse transcriptase promotes cardiac muscle cell proliferation, hypertrophy, and survival, *PNAS* **98(18),** 1–22.

65. Fu, W., Begley, J. G., Killen, M. W., et al. (1999) Anti-apoptotoc role of telomerase in pheochromocytoma cells, *J. Biol. Chem.* **274(11),** 7264–7271.

66. Chiu, C.-P., Dragowska, W., Kim, N. W., et al. (1996) Differential expression of telomerase activity in hematopoietic progenitors from adult human bone marrow, *Stem Cells* **14,** 239–248.

67. Vaziri, H., Dragowska, W., Allsopp, R. C., et al. (1994) Evidence for a mitotic clock in human hematopoietic stem cells: Loss of telomeric DNA with age, *Proc. Natl. Acad. Sci. USA* **91,** 9857–9860.

68. Taylor, R. S., Ramirez, R. D., Ogoshi, M., et al. (1996) Detection of telomerase activity in malignant and nonmalignant skin conditions, *J. Invest. Dermatol.* **106,** 759–765.

69. Weng, N.-P., Levine, B. L., June, C. H., et al. (1996) Regulated expression of telomerase activity in human T lymphocyte development and activation, *J. Exp. Med.* **183,** 2471–2479.
70. Forsyth, N. R., Wright, W. E., and Shay, J. W. (2002) Telomerase and differentiation in multicellular organisms: turn it off, turn it on, and turn it off again, *Differentiation* **69,** 188–197.
71. Schwartz, R. E., Reyes, M., Koodie, L., et al. (2002) Multipotent adult progenitor cells from bone marrow differentiate into functional hepatocyte-like cells, *J. Clin. Invest.* **109,** 1291–1302.
72. Beltrami, A. P., Urbanek, K., Kajstura, J., et al. (2001) Evidence that human cardiac myocytes divide after myocardial infarction, *N. Engl. J. Med.* **344(23),** 1750–1757.
73. Allsopp, R. C., Morin, G. B., DePinho, R., et al. (2003) Telomerase is required to slow telomere shortening and extend replicative lifespan of HSC during serial transplantation, *Blood* **101,** in press.
74. Schlessinger, D. and Van Zant, G. (2001) Does functional depletion of stem cells drive aging?, *Mech. Ageing Dev.* **122,** 1537–1553.
75. Effros, R. B., Boucher, N., Porter, V., et al. (1994) Decline in CD28⁺ T cells in centenarians and in long-term T cell cultures: a possible cause for both in vivo and in vitro immunosenescence, *Exp. Gerontol.* **29(6),** 601–609.
76. McCulloch, C. A. and Bordin, S. (1991) Role of fibroblast subpopulations in periodontal physiology and pathology, *J. Periodont. Res.* **26,** 144–154.
77. Liu, Y. and Rao, M. (2003) Tansdifferentiation—fact or artifact, *J. Cell. Biochem.* **88,** 29–40.
78. Terada, N., Hamazak, N., Oka, M., et al. (2002) Bone marrow cells adopt the phenotype of other cells by spontaneous cell fusion [letter], *Nature* **416,** 1–3.
79. Holden, C. and Vogel, G. (2002) Stem cells. Plasticity: time for a reappraisal?, *Science* **296,** 2126–2169.
80. Rideout, W., Hochedlinger, K., Kyba, M., et al. (2002) Correction of genetic defect by nuclear transplantation and combined cell and gene therapy, *Cell* **109,** 17–27.
81. Cibelli, J., Lanza, R. P., West, M. D., et al. (2002) The first human cloned embryo, *Sci. Am.* **286(1),** 44–51.
82. Kaufman, D. S., Hanson, E. T., Lewis, R. L., et al. (2001) Hematopoietic colony-forming cells derived from human embryonic stem cells, *Proc. Natl. Acad. Sci. USA* **98,** 10,716–10,721.
83. Kaufman, D. S. and Thomson, J. A. (2002) Human ES cells—haematopoiesis and transplantation strategies, *J. Anat.* **200,** 243–248.
84. Itskovitz-Eldor, J., Schuldiner, M., Karsenti, D., et al. (2000) Differentiation of human embryonic stem cells into embryoid bodies comprising the three embryonic germ layers, *Mol. Med.* **6(2),** 88–95.
85. Carpenter, M., Inokuma, M. S., Denham, J., et al. (2001) Enrichment of neurons and neural precursors from human embryonic stem cells, *Exp. Neurol.* **172,** 383–397.
86. Zhang, S., Wernig, M., Duncan, I. D., et al. (2001) In vitro differentiation of transplantable neural precursors from human embryonic stem cells, *Nature Biotechnol.* **19,** 1129–1133.

87. Xu, C., Police, S., Rao, N., et al. (2002) Characterization and enrichment of cardiomyocytes derived from human embryonic stem cells, *Circ. Res.* **91,** 501–508.
88. Rambhatla, L., Chiu, C. P., Kundu, P., et al. (2003) Generation of hepatocyte-like from human embryonic stem cells, *Cell Transplant.* **12,** 1–11.
89. Draper, J., Pigott, C., Thomson, J. A., et al. (2002) Surface antigens of human embryonic stem cells: changes upon differentiation in culture, *J. Anat.* **200,** 249–258.

14

Genetic Manipulation
of Human Embryonic Stem Cells

Micha Drukker and Nissim Benvenisty

1. INTRODUCTION

Human embryonic stem (ES) cells are pluripotent cell lines derived from the inner cell mass of blastocyst stage embryos obtained by in vitro fertilization *(1,2)*. These unique cell lines may develop into cells from the three embryonic germ layers and they can be propagated continuously in culture while remaining undifferentiated. Their wide developmental potential can be demonstrated in vitro and in vivo. In vitro, the cells can undergo spontaneous differentiation into various cell types, by forming cell aggregates called embryoid bodies (EBs) *(3)*. In vivo, following injection to immunodeficient mice, they develop into teratomas comprising ectodermal, mesodermal, and endodermal derivatives *(1,2)*. Their pluripotency and unlimited growth in culture could make them extremely advantageous for the production of pure populations of specific cell types required for transplantation therapeutics. The addition of growth factors to human ES cells was shown to direct cell fate toward specific lineages *(4)*. To date, various protocols were developed for in vitro production of enriched populations of specific cell types such as neurons *(5–8)*, pancreatic β-cells *(9)*, cardiomyocytes *(10)*, and hematopoietic cells *(11)*. However, isolation of a pure differentiated cell type still remains a challenge. It was also shown that a single human ES cell could proliferate to form pluripotent ES culture *(12,13)*. This is important because it proves that human ES cells are indeed pluripotent and not a mixture of stem cells, each capable to differentiate into specific tissues.

Extensive genetic modifications of mouse ES cells have enabled the isolation of specific differentiated derivatives and greatly increased our knowledge of genes involved in developmental processes. Genetic modifications of human ES cells has also become feasible recently, with the development of transfection *(12)* and infection techniques *(14)*. The use of these methods

From: *Human Embryonic Stem Cells*
Edited by: A. Chiu and M. S. Rao © Humana Press Inc., Totowa, NJ

should improve our understanding of the biological properties of human ES cells and assist in the isolation of better cell types for cellular therapy. This chapter focuses on the procedures and the experimental designs required to perform genetic modifications in human ES cells. In addition, future applications of genetic manipulation methodologies are discussed.

2. METHODS FOR INTRODUCTION OF FOREIGN DNA INTO ES CELLS

2.1. Introduction

Derivation and propagation in vitro of mouse ES cells have enabled one to genetically modify mammals by initially carrying out the genetic alternations of the cells in culture. Thus, over time, protocols and procedures for the introduction of DNA into ES cells have been developed. These include transfection by chemical reagents, electroporation, and infection by viral vectors. A highly efficient technique for introducing DNA constructs into ES cells employs retrovirus-based vectors. Although very efficient, the virus integrates randomly into the genome, and the expression cassette within the provirus (integrated virus) tends to be silenced. Introduction of DNA by means of transfection into mouse ES cells, however, can be targeted by homologous recombination to a specific locus and the transgene is usually not silenced. These methods allow one to screen individual cell lines for desired expression levels of the transgenes or the modified genes. This section will mainly focus on protocols developed so far for the introduction of foreign DNA into the newly discovered human ES cells. For review on the introduction of foreign DNA into mouse ES cells, *see* ref. *15*.

2.2. Transfection

Transfection is the process by which exogenous DNA is introduced into cultured cells by biochemical or physical methods. Various methods were developed over the years for this purpose, such as chemical reagents or electroporation, but for each experiment, the method of choice is dependent on the cell line to be transfected. Although ES-like cells are now available from various species, transfection protocols were mainly established for mouse ES cells and recently for human ES cells *(12)*. For mouse ES cells, transfection by means of electroporation was found to be a simple technique that gives high success rate *(16)* (for a review, *see* ref. *17*). In contrast, for human ES cells, electroporation is not the method of choice, as the human cells did not survive the voltage shock when electroporation conditions, similar to those applied for mouse ES cells, were tested *(12)*. Other commonly used transfection methods include cationic lipid reagents, multicomponent lipid-based reagents, linear polyethylenimine (PEI), and calcium phosphate/DNA coprecipitation. Cationic lipid reagents form small uni-lamellar liposomes, which harbor a positive charge at the periphery of the liposome. When mixed with DNA, the positive charge is attracted electro-statically to the phosphate backbone of the DNA as well as to the negatively charged surface of the cell membrane. Although very efficient with many eukaryotic cell types, LipofectAMINE PLUS (Life Technologies), a cat-ionic lipid reagent, was found to give only poor transient transfection efficiencies. Similar results were obtained with FuGENE 6 (Roche), a non-liposomal lipid-based reagent. In contrast, Exgen 500 (Fermentas), which consists of linear PEI molecules that have a high cationic-charge density, gave a considerably higher transfection rate by approximately one order of magnitude. The transgene expression was found to be stable during expan-sion of undifferentiated cells in approx $10^{-6}-10^{-5}$ of the transfected cells *(12)*. This expression subsists even during in vitro differentiation of the transfected clones into EBs. Thus, transfection by Exgen 500 remains, to date, the method of choice for the transfection of human ES cells *(12)*.

2.3. Infection

Historically, retroviral-based vectors were the first to be used for the gen-eration of genetically modified chimera from mouse ES cells *(18)*. Although highly efficient as most of the cells become infected, retroviral-based trans-duction has some major limitations that prevent their broad use: (1) The provirus transgene tends to be silenced by either cis (methylation) or trans (protein) factors (*see* references in ref. *19*); (2) elements within the proviral LTR (long terminal repeat) tend to express transgenes constitutively, thus tissue-specific promoters cannot be studied; (3) the size of the transgene is limited to approx 6 kb. Because of these limitations, the use of retroviral-based vectors for genetic manipulation of mouse ES cells was somewhat restricted.

A recently published report by Pfeifer et al. *(14)* demonstrated that vec-tors derived from lentiviruses (complex retroviruses) can efficiently deliver genes to the mouse ES cells. The transgene was actively expressed in undif-ferentiated mouse ES cells and also when the cells were differentiated in vitro into EBs or in vivo to teratomas. Moreover, viable mouse chimeras were produced from the transduced cell line and the transgene expression appeared in many tissues. Further crossings resulted in fully transgenic animals. Human ES cells were also infected and transgene expression was stable during proliferation as undifferentiated cells *(12)*. Yet, expression following differentiation was not examined. As efficiency rates were very high (virtually all of the cells were infected), lentiviral infection has a great potential to serve as an excellent delivery system of transgenes into human ES cells. Nonetheless, lentiviral vectors may still express their transgenes constitutively and integrate into the genome randomly; thus, these vectors may better suit those experiments where random integration and high expression levels of the transgene are desired.

2.4. Transient vs Stable Transfection

When choosing a DNA construct to be introduced into cells, either by transfection or infection, it is important to determine that it serves all of the requirements of the experiment. One of the factors that must be taken into consideration is the duration of the experiment (e.g., to decide whether transient or stable modification is desired). If only a short expression period is sufficient for a given experiment, then transient transfection should be the method of choice. Transient expression of transgenes is achieved ideally by transfection of the ES cells with a circular plasmid by means of one of the methods outlined in Section 2.2. It is important to stress that no matter what technique is used, the plasmids should be supercoiled and not linearized. This is because transcription from a linear plasmid is not as efficient as from a supercoiled plasmid (20). The maximum expression level in transient transfection experiments is typically detected 48 h after transfection.

Stable integration is the process by which a DNA fragment integrates stably into a host chromosome. For this purpose, either transfection of a linear plasmid or retroviral infection can be used. As noted in Section 2.3., infection by retroviruses is highly efficient but cannot be directed to a specific locus. In contrast, integration frequencies following transfection are very poor but allow selection of clones where the DNA was targeted into a specific locus. As infection rates are so high (usually more than 50% of the cells are infected), there is no need to use a selectable marker in order to isolate cell clones.

Other than infection all other stable transfection techniques require selection markers to be used for the identification of cells in which DNA integration has occurred, as these events are very rare. Following selection, only few clones will survive and the resistant colonies are transferred separately for the establishment of individual clones. As each colony represents a different integration event, clones can be screened for the level of transgene expression or integration site. Colony transfer is usually performed by micropipet, either with or without preincubation of the cells in the presence of proteolytic enzymes such as trypsin or dispase. As human ES cells are usually propagated in direct contact with inactivated mouse embryonic fibroblasts (MEFs) *(1)*, it is important that the MEF used be resistant to the selection reagent when applied to the coculture.

3. GENETIC MANIPULATION EXPERIMENTS IN ES CELLS

3.1. Introduction

The relative ease by which mouse ES cells can be genetically manipulated, cultured, and integrated into the germ line of chimeric animals has made the production of genetically altered mammals a routine. Moreover, early embryonic developmental processes may also be studied in vitro using genetically modified ES cells following their differentiation in culture into EBs. These opportunities called for the development of DNA constructs to be exploited in the generation of genetically altered ES cells. The type of DNA construct to be used relies primarily on the nature of the experiment. The experiments employing modification of ES cells may be divided into two major categories: the introduction of foreign genes into ES cells and the alternation of endogenous gene expression. Introduction of foreign genes into ES cells is used mainly for the expression of various reporter genes. Alternation of gene expression is highly important in order to understand gene function. Many forms of constructs were developed for these purposes, such as gene targeting, antisense, and overexpression constructs. In addition, double-stranded RNA (dsRNA) molecules were also shown to alter protein levels. This section is dedicated to the different types of experiments, which are being performed using the introduction of DNA constructs, and, to a lesser extent, RNA molecules, to mouse and human ES cells. As attempts to introduce and alter genes in human ES cells are still scarce, experiments that should be performed on them in the future and had been performed in their mouse counterparts will also be discussed.

3.2. Expression of Foreign Genes and Inducible Systems in ES Cells

The introduction of marker genes into cells allows tracking of the modified cells within a mixed cell population or even in an organ. Also, various cellular processes can be followed directly by means of fusing an endogenous protein to a marker protein. The expression of the marker gene can be either constitutive or under the control of a promoter that is expressed only under certain conditions. In the case of a constitutive promoter, the transgene is introduced into the genome randomly by means of viral infection or transfection. In contrast, a nonconstitutive promoter such as a tissue-specific promoter should not be introduced by viral vectors because its regulation may be affected by the viral LTR (*see* Section 2.3.). The expression of the transgene can be also regulated by a noncellular conditional system such as the tetracycline-responsive transactivator. This section will focus on studies demonstrating the use of reporter genes as traceable markers for monitoring and selecting specific cell lineages in vivo and in vitro and on inducible systems available for regulating gene expression in mammalian cells.

3.2.1. Lineage Selection

Lineage selection of highly purified cell populations is extremely important for the study of specific cell lineages and for cellular transplantation therapeutics. For this purpose, it is advisable to use a marker gene that can be directly viewed in viable cells without fixation. Moreover, the labeled cells may be sorted from the heterogeneous population by means of fluorescence-activated cell sorter (FACS) or other related methods. This principle was demonstrated recently in human ES cells by Eiges et al. *(12)*. The aim of this work was to identify and select undifferentiated cells from heterogeneous cell populations in culture. For this purpose, the gene-encoding enhanced green fluorescent protein (EGFP) was fused to the mouse *Rex-1* promoter, which is specifically expressed by undifferentiated cells. Following transfection, EGFP expression was monitored in undifferentiated ES cells and in both young and mature EBs, which represent early and late differentiation stages, respectively. As expected, EGFP expression was observed in the undifferentiated ES cells. Interestingly, EGFP expression was also evident in cells within the EBs, but their proportion decreased as differentiation was taking place. Next, the cells were sorted according to EGFP expression, allowing the isolation of pure populations of either undifferentiated or fully differentiated cells. This strategy of fusing a cell-specific promoter to a marker gene may also be used to isolate a specific population of differentiated or partially differentiated cells. To this end, we have established human ES cell lines that stably express EGFP under the regulation of the nestin enhancer *(21)*. As this enhancer restricts nestin expression to developing neuronal cells *(22)*, we expect EGFP expression only in neuronal progenitor cells (the experimental paradigm is shown in Fig. 1A). When EGFP expression was monitored in a culture of the nestin–EGFP

Fig. 1. *(opposite page)* Genetic manipulation of human ES cells to identify neuronal progenitors. **(A)** Schematic representation of the transfection protocol. Human ES cells are grown on a feeder layer (MEF) while being transfected with the nestin–EGFP construct. Following transfection, cells undergo neomycin selection. Neomycin-resistant colonies are collected using a micropipet and their ability to form neural progenitor cells is assayed by FACS. **(B)** Spontaneous differentiation of human ES cells into neuronal stem cells. Human ES cells were stably transfected with a construct containing the neural progenitor-specific regulatory sequences

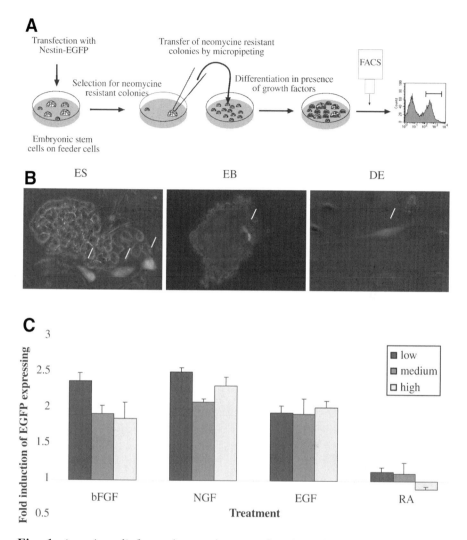

Fig. 1. *(continued)* from the nestin gene fused to the EGFP gene. Three differentiation stages are presented: EGFP positive cells appear in the periphery of an ES cell colony (**left**), within an EB (**middle**) and differentiated ES cells (DE) (**right**). Arrows point to fluorescent cells. (**C**) Growth factors induce in vitro induction of neuronal stem cells from ES cells. In order to amplify the neural progenitors populations in culture, different growth factors were added to human ES cells transfected with the neural progenitor-specific promoter fused to the EGFP gene. Fold induction of EGFP-positive cells was monitored using FACS. Three concentrations were used for each of the growth factors: basic fibroblast growth factor (bFGF) 5 ng/mL, 10 ng/mL, 20 ng/mL; neural growth factor (NGF) 50 ng/mL, 100 ng/mL, 150 ng/mL; epidermal growth factor (EGF) 50 ng/mL, 100 ng/mL, 150 ng/mL; retinoic acid (RA) 10^{-5} M, 10^{-6} M, 10^{-7} M (designated respectively as low, medium, and high concentrations). Each experiment was carried out four times and standard error (SE) values are shown. Note that at least a twofold induction was observed after the addition of bFGF, NGF, and EGF.

cells, only few differentiated cells in the periphery of the undifferentiated colonies were marked by the fluorescent protein (*see* Fig. 1B, left). Small numbers of fluorescent cells were also detected in the EBs (Fig. 1B, center). When the EBs were dissociated and plated on fibronectin-coated plates, EGFP expression was apparent only in a few cells with similar morphology to radial glial cells, which are known to be nestin positive (Fig. 1C, right) *(23)*. In order to induce differentiation of human ES cells into neuronal progenitors, various growth factors were added to the growth media and EGFP levels were monitored. The addition of different concentrations of human basic fibroblast growth factor (bFGF), neural growth factor (NGF), and EGF caused up to 2.5-fold increase in the number of cells expressing EGFP (see Fig. 1C). However, the addition of retinoic acid (RA), which is a very potent inducer of neuronal maturation of human ES cells *(5)*, had no effect, as it may have driven the progenitor cells into a more differentiated state in which nestin is no longer expressed. These results suggest that, similarly to mouse ES cells, the introduction of marker genes expressed under the control of lineage-specific promoters can be used to identify and select specific cell types from a mixed population of cells.

Expression of a dominant selection marker under the control of specific promoters was also shown in the mouse to enrich for cells committed for a specific lineage. Li et al. *(24)* targeted one allele of the *sox2* gene with a βgeo construct (a bicistronic vector expressing *neo* and *lacZ* genes *[25]*); thus, only the differentiated cells that express the transgene are resistant to the neomycin drug G418. As *sox2* is exclusively expressed in early neuronal cells, only these cells can survive G418 selection. Analysis of neuroepithelial markers following G418 selection showed that most surviving cells express these markers.

Taken together, these methodologies offer an easy and efficient means to identify and select specific cell types derived from human ES cells. This may become critical in cellular therapies because the transplanted population of cells must be of a certain type in order to treat a specific pathology. Moreover, undifferentiated ES cells must be removed from the transplanted population because they may form teratomas in vivo. As many genes that encode for fluorescent proteins are available, it may be even possible to select simultaneously for and against different markers.

3.2.2. Overexpression of Cellular Genes in ES Cells

Expressing cellular genes in ES cells can serve to direct their differentiation into particular lineages or to produce a cell type with a new activity. Overexpression of cellular genes was mainly used in the field of in vitro differentiation of mouse ES cells. The general scheme of these experiments is to constitutively express a gene believed to play an important role in the development of a specific cell type and thus direct the ES cells into a specific cell lineage. For example, overexpression of hepatocyte nuclear factor 3α or 3β (HNF3α or HNF3β) during differentiation of mouse ES cells induced expression of the endodermal markers albumin and cystic fibrosis transmembrane conductance regulator (CFTR), demonstrating the importance of HNF3 genes in endodermal differentiation *(26)*. A more complicated and sophisticated system of this kind employs conditional expression in order to express varying levels of the transgene. Niwa et al. *(27)* demonstrated that altered levels of the stem cell factor Oct3/4 caused different lineage commitment of the differentiated cells. A twofold increase in Oct3/4 resulted in increased endoderm and mesoderm induction, whereas a higher ratio of trophoblast cells formed when Oct3/4 was repressed. For more examples of the use of forced differentiation in mouse ES cells, *see* the references in ref. *28*.

3.2.3. Promoter Usage in ES Cells

The introduction of constitutively expressed genes into ES cells can serve to induce their differentiation *(see* Section 3.2.2.). Alternatively, it may be of use as a method for tracking in vivo ES cells expressing a marker gene, as recently demonstrated by Goldstein et al. *(29)*. Selection of an appropriate promoter system in order to constitutively express genes in ES cells is of great importance. Viral promoters were suggested to be silenced in undifferentiated mouse embryonal carcinoma (EC) cells, which resemble ES cells *(30)*. Similarly, in undifferentiated mouse ES cells, the cytomegalovirus (CMV) promoter was unable to drive the expression of transgenes *(31)*. To date, cellular and viral promoter systems were tested for their ability to constitutively express EGFP in human ES cells *(12,14)*. Either cellular or viral promoters were shown to highly express EGFP when introduced by transient transfection, stable transfection, or infection *(12,14)*. Most of the isolated clones expressed the transgene uniformly in undifferentiated human ES cells and in differentiated derivatives. Moreover, the expression did not seem to be silenced even after many passages in vitro (up to 20 passages were tested—authors data). Thus, the differentiation process and prolonged culturing did not alter the promoter's ability to express the transgene. It remains to be determined what the intrinsic differences between mouse and human ES cells that interfere with expression from viral promoters in the mouse cells are.

Controlling gene expression in transgenic systems has particular advantages over constitutive expression. For instance, regulated gene expression allows one to explore the effects of gene products, which promote cell death.

Similarly, constitutive expression of a factor, which drives rapid differentiation of ES cells, will not enable isolation of stable transfected undifferentiated clones. As many forms of regulated expression systems are available today, the choice must rely on experimental design and demands. The factors that should be taken into account are (1) whether high or low levels of expression are necessary, (2) what the response time for maximal induction of expression is, and (3) what the basal expression levels that can be tolerated are.

One of the most commonly used systems is the Tet-On/Tet-Off system that uses elements of the tetracycline derepression mechanism from *Escherichia coli*. This system has a very high-inducible expression level but some basal expression usually exists. It seems that it works well in mouse ES cells, as demonstrated by Niwa et al. *(27)*. Such an expression system and many others available today may aid researchers to better understand the function of genes governing cell death and differentiation of human ES cells.

3.3. Alternation in Expression Levels of Endogenous Genes

Several approaches can be used in order to alter the expression levels of endogenous genes. This section will concentrate on the methods that have been applied to the mouse ES cells and most probably will be performed in the future in human ES cells. Altering gene expression can be achieved through the use of several systems. In order to eliminate endogenous gene expression completely and indefinitely, the best option is to alter its sequence so that no expression of the native transcript will be possible. Other means to reduce the expression level are antisense expression and RNA interference (RNAi) through the action of dsRNA molecules. The rational of the antisense method is to express a gene in its reverse orientation so that it will anneal with the native RNA and, thus, translation will not be initiated. The RNAi system uses RNA molecules that have inhibitory effects on products of endogenous genes, in a sequence-specific manner.

3.3.1. Gene Targeting in Mouse ES Cells

Transfection of mouse ES cells with a construct containing a modified DNA sequence allows the targeting of an endogenous sequence. Thus, if homologous recombination occurs, the modified allele will replace the endogenous one. Various methods were developed over the years for this purpose and were extensively used to manipulate the mouse genome. The most common is the deletion of a large fragment of the gene resulting in a nonfunctional gene or a loss-of-function mutation. For this purpose a "knockout" construct contains a genomic fragment of the targeted gene, from which an important region is substituted by a selection-conferring gene. Upon a successful homologous recombination event, the mutated construct replaces the native one. Yet, the vast majority of integrations occur randomly and not in the homologous gene. A commonly used approach to enrich for the targeted alleles is to add a negative selection marker adjacent to the homologous tract, a method known as positive negative selection (PNS). This methodology was suggested to dramatically enrich for targeted colonies *(32)*.

Another general strategy to inactivate an endogenous gene is to select directly for homologous recombination events with a targeting vector containing a promoterless positive selection marker. Thus, the selection marker is expressed only if it landed downstream to an active promoter *(33)*. Using promoterless constructs, efficiencies as high as 85% of targeted integrations were reported *(34)*. It is important to note that this method is valid only for genes expressed in undifferentiated ES cells, otherwise the selection marker will obviously not be expressed.

Although the gene-targeting technique is highly important for understanding gene function, it requires prior knowledge of the targeted gene. Thus, unknown genes cannot be targeted by homologous recombination. One way of mutating a gene at random is to insert a selection marker into it using a methodology called the gene trap. In this system, the selection marker is expressed depending on the promoter, enhancer, or poly A of the endogenous gene. The gene trap methodology produced a huge amount of information regarding gene function in the mouse (for review, *see* ref. *35*) and is currently tested in human ES cells.

In the past years, numerous "knockout" and other modified mice were developed through gene-targeting experiments in mouse ES cells. In order to create a knockout mouse from ES cells, one usually needs to form a chimera from a blastocyst-stage embryo injected with a modified ES cell line. Thus, when ES cells contribute to the germ cells, heterozygous offsprings can be produced that then may be mated to form a mouse homozygous for the mutation. In human ES cells on the other hand, it is impossible to create a homozygote by way of chimera production. Hence, apart from genes on the X and Y chromosomes in male ES cell lines, methods for the production of knockouts in both alleles of autosomal chromosomes need to be developed. To this aim, cells that were modified in one allele can be retransfected with another vector containing a different selection-conferring gene. Another strategy developed in order to induce homozygosity in mutated ES cell lines is to grow the heterozygous cells in high levels of the selection drug in order to enrich for gene conversion events that will result in two copies of the selection markers disrupting the two endogenous alleles *(36)*. The ability to mutate specific genes in human ES cells should enable the establishment of genetic systems to study human embryonic differentiation in vitro.

3.3.2. Silencing Gene Expression at the Posttranscriptional Level in ES Cells

Gene knockout experiments in ES cells gave rise to an explosion of information regarding gene function in mammals. Although very informative, the procedures involved in this process are very tedious. Thus, alternative methods for silencing genes of interest were developed and gained much attention, especially in somatic cells. Among these methods are antisense expression and dsRNA-mediated inhibition. Although these procedures were not extensively practiced in mouse ES cells and not at all in human ES cells, they most probably will become more prominent as RNAi techniques advance.

3.3.2.1. ANTISENSE

Antisense techniques are relatively simple to design and perform. In a typical antisense experiment, a construct expressing the gene of interest is introduced into ES cells in its reverse order. Later, specific antibodies can assay the effect at the protein level. For instance, antisense RNA of the *vav* gene was shown to inhibit hematopoietic differentiation of mouse ES cells in vitro *(37)*. Although very simple, the main drawbacks of this method are low efficiency and artifacts that arise from questionable specificity. Hence, only a few experiments were performed to date in ES cells using these constructs.

3.3.2.2. RNAi

In the past few years, a new mechanism of gene silencing in eukaryotes was discovered. It had become apparent that dsRNA can direct mRNA degradation in a sequence dependent manner. This effect termed "RNA inhibition" (RNAi) was found to exist in several organisms and was studied extensively in *Caenorhabditis elegance* and in *Drosophila melanogaster*. Posttranscriptional gene silencing (PTGS) in plants and quelling in *Neurospora crassa* were found to be related phenomena to RNAi (for an excellent review on RNAi, *see* ref. *38*) In short, the mechanism by which specific degradation is achieved involves fragmentation of dsRNA to short tracts of approx 21–23 nucleotides termed "short inhibitory RNAs" (siRNAs). These sequences bind a nuclease complex and thus direct the endogenous mRNA towards degradation in a sequence-dependent manner. Two recent works have shown that long dsRNAs are able to induce a specific RNAi response in mouse ES and EC cells *(39,40)*. Extracts prepared from undifferentiated ES or EC cells were able to convert long dsRNA into siRNAs. Interestingly, undifferentiated mouse ES cells exhibited a sequence-specific RNAi effect, but their differentiated derivatives did not *(39)*.

The dsRNAs are either produced *in situ* by transient transfection of a plasmid harboring an inverted repeat of the gene to be silenced *(39)* or by direct transfection of in vitro-prepared long dsRNAs *(39,40)*. A new system for *in situ* synthesis of siRNAs within the targeted cell was found to be highly efficient in mammalian cells and caused specific downregulation of the targeted gene *(41)*. As it involves the simple ligation of a short sequence from the targeted gene into an expression vector and its introduction into the genome, it may serve as a powerful tool for analyzing gene function *(41)*. It remains to be determined whether this system will be efficient in ES cells also.

The hallmark of RNAi is its high specificity. Thus, this method may be ideal for the silencing of endogenous genes expressed in ES cells. This may turn out to be especially convenient for human ES cells that otherwise inhibition of expression of endogenous gene may require independent mutations in both alleles. It remains to be determined whether RNAi will be efficient in silencing of endogenous genes in human ES cells before and after their differentiation.

4. CLINICAL APPLICATIONS OF GENETICALLY MODIFIED HUMAN ES CELLS

4.1. Introduction

Since the establishment of the first human ES cell lines in 1998 *(1)* hopes were raised that they could serve as an unlimited cell source for replacement therapy to treat various diseases and disabilities such as cardiomyopathies, Parkinson's disease, and diabetes. It is estimated that approx 3000 people die every day in the United States alone from diseases that may be treatable in the future with human ES-derived cells *(42)*. Isolation of different types of differentiated derivative from human ES cells may, therefore, be valuable for clinical applications of these cells. The ability to genetically manipulate the cells may aid researchers to achieve this goal, as it may enable one to mark specific cell types during the differentiation process. Moreover, it may serve to eliminate the transplanted cells if they cause deleterious effects. This chapter will present the different genetic modification strategies that could pave the way toward the use of human ES for clinical applications in the future.

4.2. Genetic Modifications of Human ES Cells for Cellular Therapy

One of the main objectives for performing research on human ES cells is their potential use in transplantation therapy. These cells may serve as better reagents for cellular therapy than other cells used to date. For example, fetal-derived dopamine neurons were used to treat Parkinson's disease (for references, *see* review by Freed *[43]*). These cells are hard to obtain, grow poorly in culture, and are difficult to be genetically manipulated. Thus, human ES cells may have some advantage over fetal cells because they can be propagated indefinitely in culture and be genetically modified. In order to obtain an enriched population of a specific cell type, human ES cells may be transfected with a marker gene controlled by a tissue-specific promoter, thus allowing the identification and purification of specific cell types from a mixed population of cells. Enrichment for marker-expressing cells can be performed by various methods, including FACS sorting. Alternatively, cells of a specific type can be selected directly if a dominant selection gene is expressed in a lineage-specific manner (*see* Section 3.2.1.). Another use for genetically manipulated human ES cells is to isolate specific types of differentiated cell through overexpression of a key regulating gene involved in lineage commitment, thus allowing differentiation to specific cell types (*see* Section 3.2.2.).

Moreover, it may become necessary to interfere with the expression of cellular genes in order to enable safe transplantation (*see* Section 3.3.). For example, undifferentiated human ES cells were found to express relatively low levels of class I major histocompatibility complex (MHC-I) proteins. Yet, induction of differentiation was shown to dramatically increase expression levels of MHC-I, making the cells susceptible for tissue rejection (*44*). Thus, deleting genes that encode for MHC-I may be obligatory in order to enable transplantation of differentiated human ES cells. This is because transplantation of allogeneic (nonself) cells into a patient will cause rejection of the transplanted cells by the host immune system.

Similarly, introduction of genes, which will allow negative selection of ES cells, may also produce safer cell types for transplantation medicine. One of the risks of cellular transplantation is overproliferation of the transplanted cells. For example, uncontrolled proliferation of transplanted fetal cells in Parkinson's patients caused worsening of the symptoms in some of the patients (*45*). The risk with human ES cells is even greater, as undifferentiated ES cells are tumorgenic and if they are not eliminated prior to transplantation, teratomas may develop. Thus, in order to be able to eliminate the cells if they overproliferate after their transplantation, it would be beneficial to introduce into them a suicide gene that can be activated if they create tumors. The various types of genetic modification that may serve to improve cellular transplantation are summarized in Fig. 2.

4.3. Genetic Manipulation in Nuclear Transplantation Therapy

The ultimate goal for performing research on ES cells is to generate a homogenous cell population that may serve to treat various human diseases.

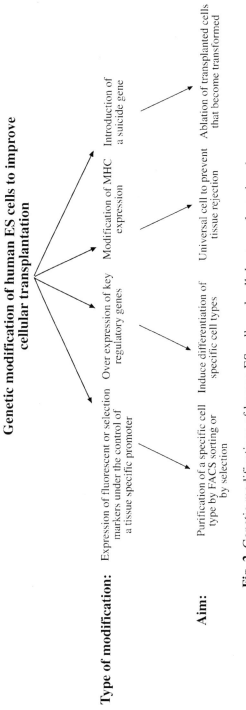

Fig. 2. Genetic modifications of human ES cells and cellular transplantation therapy.

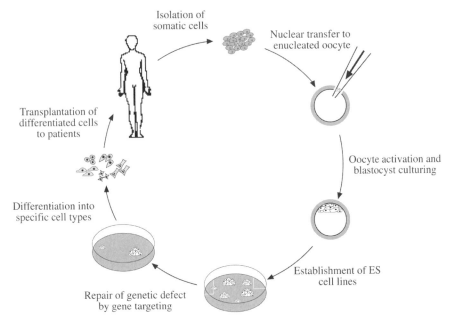

Fig. 3. Schematic outline of the nuclear transplantation therapy for inherited disorders. A nucleus from a somatic cell of a patient with a genetic disease is transplanted into an enucleated oocyte. The oocyte is then activated, and the resulting blastocyst is used to establish a human ES cell line. Repair of the genetic defect is performed by replacement of the mutated allele by a normal one using homologous recombination techniques. Specific cell types are sorted out of the differentiated ES cells and transplanted back into the patient.

To achieve this ambitious task, derivation of a pure cell population is indeed obligatory but not sufficient. As human ES derived cells can express high levels of MHC molecules *(44)*, rejection by the host immune system is expected if the patient and the transplanted cells differ in their MHC alleles. This problem may be circumvented by genetic manipulation, as discussed above or by the transfer of a somatic nucleus into an enucleated oocyte resulting in a reconstructed embryo from which an ES cell line can be established. This procedure, termed "therapeutic cloning," was recently demonstrated in mice *(46,47)*. Not only were the cells genetically identical to the nucleus donor, the derived ES cell lines were shown to differentiate into various cell types, including dopaminergic and serotonergic neurons *(46)*. However, establishment of ES cell lines via nuclear transfer alone cannot by itself take care of genetically inherited disorders because the established cell line harbors the same genetic defect as the patient. A possible solution might be to genetically correct the defect in the ES cells, resulting in a functional and MHC identical cell line. The general scheme of this methodology is presented in Fig. 3. Recently, Rideout et al. *(48)* showed that a genetic defect could be corrected in a mouse ES cell line that was derived by somatic nuclear transfer into an enucleated oocyte. In short, nuclei of somatic cells from immunodeficient mice were transferred into enucleated mouse oocytes and ES cell lines were derived. Next, the gene defect underlying the immunodeficiency was corrected using homologous recombination. The resulting ES clone was differentiated in vitro to produce hematopoietic stem cells and transplanted into the original immunodeficient mice. The transplanted cells were shown to extensively contribute to myeloid and lymphoid lineages. This work represents the first attempt to genetically modify ES cells in order to treat inheritable diseases and shows that gene therapy via nuclear transplantation is possible.

5. CONCLUSIONS

Since the establishment of mouse ES cell lines 20 yr ago, numerous works have enabled the discovery and comprehensive understanding of complicated developmental processes in mammals. In addition, many investigations have demonstrated that specific cell types could be isolated for transplantation purposes. These tasks could not have been accomplished without genetic modification techniques. The isolation of human ES cell lines is a great opportunity to harness the knowledge obtained while working with mouse ES cell for introducing genetic modifications into the genome. Thus far, the creation of specific ES cell clones have enabled one to follow different cell types during differentiation and to isolate the cells by sorting of the marked population. It is likely that direct isolation of specific cell types will be also possible in human cells, as was demonstrated in mouse ES cells. Similarly, the forced expression of genes involved in differentiation will allow to enrich the culture for specific cell types and deepen the knowledge regarding the function of specific genes involved in commitment and differentiation. Homologous recombination will open the way to produce genetically altered ES cell lines, which will help to understand gene function and may create better cell lines for cellular transplantation. It is clear that the methods applied in mouse ES cells cannot be applied without modifications to their human counterparts, as demonstrated with the transfection protocols. Improved procedures for the genetic manipulations of human ES cells are likely to have a great impact on the study of early human development as well as generating unlimited supply of better reagents for cellular therapy in humans.

REFERENCES

1. Thomson, J. A., Itskovitz-Eldor, J., Shapiro, S. S., et al. (1998) Embryonic stem cell lines derived from human blastocysts, *Science* **282**, 1145–1147.
2. Reubinoff, B. E., Pera, M. F., Fong, C. Y., Trounson, A., and Bongso, A. (2000) Embryonic stem cell lines from human blastocysts: somatic differentiation in vitro, *Nature Biotechnol.* **18**, 399–404.
3. Itskovitz-Eldor, J., Schuldiner, M., Karsenti, D., et al. (2000) Differentiation of human embryonic stem cells into embryoid bodies comprising the three embryonic germ layers, *Mol. Med.* **6**, 88–95.
4. Schuldiner, M., Yanuka, O., Itskovitz-Eldor, J., Melton, D. A., and Benvenisty, N. (2000) Effects of eight growth factors on the differentiation of cells derived from human embryonic stem cells, *Proc. Natl. Acad. Sci. USA* **97**, 11,307–11,312.
5. Schuldiner, M., Eiges, R., Eden, A., et al. (2001) Induced neuronal differentiation of human embryonic stem cells, *Brain Res.* **913**, 201–205.
6. Zhang, S. C., Wernig, M., Duncan, I. D., Brustle, O., and Thomson, J. A. (2001) In vitro differentiation of transplantable neural precursors from human embryonic stem cells, *Nature Biotechnol.* **19**, 1129–1133.
7. Reubinoff, B. E., Itsykson, P., Turetsky, T., et al. (2001) Neural progenitors from human embryonic stem cells, *Nature Biotechnol.* **19**, 1134–1140.
8. Carpenter, M. K., Inokuma, M. S., Denham, J., Mujtaba, T., Chiu, C. P., and Rao, M. S. (2001) Enrichment of neurons and neural precursors from human embryonic stem cells, *Exp. Neurol.* **172**, 383–397.
9. Assady, S., Maor, G., Amit, M., Itskovitz-Eldor, J., Skorecki, K. L., and Tzukerman, M. (2001) Insulin production by human embryonic stem cells, *Diabetes* **50**, 1691–1697.
10. Kehat, I., Kenyagin-Karsenti, D., Snir, M., et al. (2001) Human embryonic stem cells can differentiate into myocytes with structural and functional properties of cardiomyocytes, *J. Clin. Invest.* **108**, 407–414.
11. Kaufman, D. S., Hanson, E. T., Lewis, R. L., Auerbach, R., and Thomson, J. A. (2001) Hematopoietic colony-forming cells derived from human embryonic stem cells, *Proc. Natl. Acad. Sci. USA* **98**, 10,716–10,721.
12. Eiges, R., Schuldiner, M., Drukker, M., Yanuka, O., Itskovitz-Eldor, J., and Benvenisty, N. (2001) Establishment of human embryonic stem cell-transfected clones carrying a marker for undifferentiated cells, *Curr. Biol.* **11**, 514–518.
13. Amit, M., Carpenter, M. K., Inokuma, M. S., et al. (2000) Clonally derived human embryonic stem cell lines maintain pluripotency and proliferative potential for prolonged periods of culture, *Dev. Biol.* **227**, 271–278.
14. Pfeifer, A., Ikawa, M., Dayn, Y., and Verma, I. M. (2002) Transgenesis by lentiviral vectors: lack of gene silencing in mammalian embryonic stem cells and preimplantation embryos, *Proc. Natl. Acad. Sci. USA* **99**, 2140–2145.
15. Lovell-Badge, R. (1987) Introduction of DNA into embryonic stem cells, in *Teratocarcinomas and Embryonic Stem Cells* (Robertson, E. J., ed.), IRL, Oxford, pp. 153–182.

16. Thomas, K. R. and Capecchi, M. R. (1987) Site-directed mutagenesis by gene targeting in mouse embryo-derived stem cells, *Cell* **51**, 503–512.

17. Torres, M. (1998) The use of embryonic stem cells for the genetic manipulation of the mouse, *Curr. Topics Dev. Biol.* **36**, 99–114.

18. Robertson, E., Bradley, A., Kuehn, M., and Evans, M. (1986) Germ-line transmission of genes introduced into cultured pluripotential cells by retroviral vector, *Nature* **323**, 445–448.

19. Cherry, S. R., Biniszkiewicz, D., van Parijs, L., Baltimore, D., and Jaenisch, R. (2000) Retroviral expression in embryonic stem cells and hematopoietic stem cells, *Mol. Cell. Biol.* **20**, 7419–7426.

20. Weintraub, H., Cheng, P. F., and Conrad, K. (1986) Expression of transfected DNA depends on DNA topology, *Cell* **46**, 115–122.

21. Kawaguchi, A., Miyata, T., Sawamoto, K., et al. (2001) Nestin-EGFP transgenic mice: visualization of the self-renewal and multipotency of CNS stem cells, *Mol. Cell. Neurosci.* **17**, 259–273.

22. Lothian, C. and Lendahl, U. (1997) An evolutionarily conserved region in the second intron of the human nestin gene directs gene expression to CNS progenitor cells and to early neural crest cells, *Eur. J. Neurosci.* **9**, 452–462.

23. Stemple, D. L. and Mahanthappa, N. K. (1997) Neural stem cells are blasting off, *Neuron* **18**, 1–4.

24. Li, M., Pevny, L., Lovell-Badge, R., and Smith, A. (1998) Generation of purified neural precursors from embryonic stem cells by lineage selection, *Curr. Biol.* **8**, 971–974.

25. Kim, D. G., Kang, H. M., Jang, S. K., and Shin, H. S. (1992) Construction of a bifunctional mRNA in the mouse by using the internal ribosomal entry site of the encephalomyocarditis virus, *Mol. Cell. Biol.* **12**, 3636–3643.

26. Levinson-Dushnik, M. and Benvenisty, N. (1997) Involvement of hepatocyte nuclear factor 3 in endoderm differentiation of embryonic stem cells, *Mol. Cell. Biol.* **17**, 3817–3822.

27. Niwa, H., Miyazaki, J., and Smith, A. G. (2000) Quantitative expression of Oct-3/4 defines differentiation, dedifferentiation or self-renewal of ES cells, *Nature Genet.* **24**, 372–376.

28. O'Shea, K. S. (2001) Directed differentiation of embryonic stem cells: genetic and epigenetic methods, *Wound Repair Regen.* **9**, 443–459.

29. Goldstein, R. S., Drukker, M., Reubinoff, B. E., and Benvenisty, N. (2002) Integration and differentiation of human embryonic stem cells transplanted to the chick embryo, *Dev. Dynam.* **225**, 80–86.

30. Gorman, C. M., Rigby, P. W., and Lane, D. P. (1985) Negative regulation of viral enhancers in undifferentiated embryonic stem cells, *Cell* **42**, 519–526.

31. Chung, S., Andersson, T., Sonntag, K. C., Bjorklund, L., Isacson, O., and Kim, K. S. (2002) Analysis of different promoter systems for efficient transgene expression in mouse embryonic stem cell lines, *Stem Cells* **20**, 139–145.

32. Mansour, S. L., Thomas, K. R., and Capecchi, M. R. (1988) Disruption of the proto-oncogene int-2 in mouse embryo-derived stem cells: a general strategy for targeting mutations to non-selectable genes, *Nature* **336**, 348–352.

33. Doetschman, T., Maeda, N., and Smithies, O. (1988) Targeted mutation of the Hprt gene in mouse embryonic stem cells, *Proc. Natl. Acad. Sci. USA* **85,** 8583–8587.

34. te Riele, H., Maandag, E. R., Clarke, A., Hooper, M., and Berns, A. (1990) Consecutive inactivation of both alleles of the pim-1 proto-oncogene by homologous recombination in embryonic stem cells, *Nature* **348,** 649–651.

35. Stanford, W. L., Cohn, J. B., and Cordes, S. P. (2001) Gene-trap mutagenesis: past, present and beyond, *Nat. Rev. Genet.* **2,** 756–768.

36. Mortensen, R. M., Conner, D. A., Chao, S., Geisterfer-Lowrance, A. A., and Seidman, J. G. (1992) Production of homozygous mutant ES cells with a single targeting construct, *Mol. Cell. Biol.* **12,** 2391–2395.

37. Wulf, G. M., Adra, C. N., and Lim, B. (1993) Inhibition of hematopoietic development from embryonic stem cells by antisense vav RNA, *EMBO J.* **12,** 5065–5074.

38. Hammond, S. M., Caudy, A. A., and Hannon, G. J. (2001) Post-transcriptional gene silencing by double-stranded RNA, *Nat. Rev. Genet.* **2,** 110–119.

39. Yang, S., Tutton, S., Pierce, E., and Yoon, K. (2001) Specific double-stranded RNA interference in undifferentiated mouse embryonic stem cells, *Mol. Cell. Biol.* **21,** 7807–7816.

40. Billy, E., Brondani, V., Zhang, H., Muller, U., and Filipowicz, W. (2001) Specific interference with gene expression induced by long, double-stranded RNA in mouse embryonal teratocarcinoma cell lines, *Proc. Natl. Acad. Sci. USA* **98,** 14,428–14,433.

41. Brummelkamp, T. R., Bernards, R., and Agami, R. (2002) A system for stable expression of short interfering RNAs in mammalian cells, *Science* **296,** 550–553.

42. Lanza, R. P., Cibelli, J. B., West, M. D., Dorff, E., Tauer, C., and Green, R. M. (2001) The ethical reasons for stem cell research, *Science* **292,** 1299.

43. Freed, C. R. (2002) Will embryonic stem cells be a useful source of dopamine neurons for transplant into patients with Parkinson's disease?, *Proc. Natl. Acad. Sci. USA* **99,** 1755–1757.

44. Drukker, M., Katz, G., Urbach, A., et al. (2002) Characterization of the expression of MHC proteins in human embryonic stem cells, *Proc. Natl. Acad. Sci. USA* **99,** 9864–9869.

45. Freed, C. R., Greene, P. E., Breeze, R. E., et al. (2001) Transplantation of embryonic dopamine neurons for severe Parkinson's disease, *N. Engl. J. Med.* **344,** 710–719.

46. Munsie, M. J., Michalska, A. E., O'Brien, C. M., Trounson, A. O., Pera, M. F., and Mountford, P. S. (2000) Isolation of pluripotent embryonic stem cells from reprogrammed adult mouse somatic cell nuclei, *Curr. Biol.* **10,** 989–992.

47. Wakayama, T., Tabar, V., Rodriguez, I., Perry, A. C., Studer, L., and Mombaerts, P. (2001) Differentiation of embryonic stem cell lines generated from adult somatic cells by nuclear transfer, *Science* **292,** 740–743.

48. Rideout, W. M., 3rd, Hochedlinger, K., Kyba, M., Daley, G. Q., and Jaenisch, R. (2002) Correction of a genetic defect by nuclear transplantation and combined cell and gene therapy, *Cell* **109,** 17–27.

15
Human Therapeutic Cloning

Jose B. Cibelli

Although the method of nuclear transplantation should be valuable principally for the study of nuclear differentiation, *it may also have other uses*

—Briggs and King *(1)*

1. INTRODUCTION

Is it possible to transform a skin cell into a neuron? Yes, nuclear transfer
can do it.

Spemann in the 1930s gave credence to the idea that a cell during development is not irreversible committed to a certain lineage but can go back to
early development *(2)*. At the time, he was trying to answer the question of
whether, as cells differentiate, genes are turned off forever during development or they can be reactivated by a specific mechanism. Spemann's early
experiments were rudimentary but groundbreaking. Working with newt
embryos, he allowed them to divide four or five times and then isolated one
cell from the embryo and let it continue developing as two separate entities.
As a result, he obtained two individuals. This simple event showed that even
though a cell was isolated from an embryo already committed to form one
individual, such cell was still capable, when isolated and placed on a different environment, to grow into another being (*see* Fig. 1). At that time,
Spemann did not have the technology available to micromanipulate cells,
but in his mind, he was sure that we might be able, some day in the future, to
take a cell from a developing embryo and transplant it to a different egg; he
called this "the fantastical experiment." Spemann indeed created, although
not quite in practical terms, the nuclear transfer or cloning field. As a side
note, it is interesting that now, almost 70 yr later, scientists are trying to
undo this semantic connection (nuclear transfer and/or cloning) for reasons
that have little to do with science but more with politics.

From: *Human Embryonic Stem Cells*
Edited by: A. Chiu and M. S. Rao © Humana Press Inc., Totowa, NJ

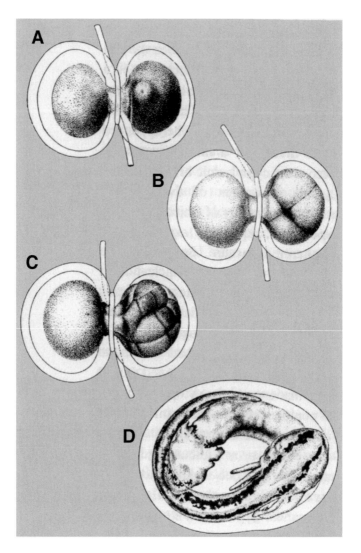

Fig. 1. Spemann experiment in newt embryos published in 1938. This was the first demonstration that a cell can return to early embryonic stages after development has been triggered. (**A**) A recently fertilized embryo is partially split. There is one nucleus in the right side. (**B**) The right side is allowed to divide up to the four-cell stage. (**C**) Only one cell from the 16-cell embryo is allowed to migrate to the left side and, soon after, the embryo is completely separated into two entities. (**D**) Two whole normal embryos will appear, one (upper left) slightly delayed in development in comparison with the one in the bottom right. (From ref. 2.)

Years later, Briggs and King succeeded in performing nuclear transfer experiments in frogs (*Rana pipiens*) *(1)*, but it was not until 1983 in the United Kingdom that the first mammal, a sheep, was cloned for the first time using an embryonic cell as a nuclear donor *(3)*. Efforts to repeat these results in other mammals were soon attempted in the mouse, with controversial results. Ilmensee claimed the successful generation of mice from an inner-cell-mass cell transplanted into an enucleated zygote. These experiments were soon challenged by McGrath and Solter and the capacity to clone from a mammalian cell was again questioned *(4,5)*. Soon after, successful experiments were repeated in the United States in cows, rabbits, and pigs *(6–8)*. During the late 1980s and early 1990s, there was a great deal of excitement about the use of this "new" technology and scientists as well as the investment community tried to use it for agricultural purposes. The reality was different, we knew very little about how this biological process worked and the clear limitation turned out to be its extremely low efficiency. Only a few scientists continued to work relentlessly, and in the winter of 1997, Dolly, the cloned sheep, was introduced to the world as the first mammal ever cloned from a somatic (body) cell, opening the door to many wonderful possibilities in medicine, agriculture, basic science, and any other scenario the human imagination could dream of, including cloning of human beings *(9)*.

In 1998, a report from our laboratory on nuclear transfer in bovines, introduced the notion of generating embryonic stem (ES) cells from a somatic cell *(10)*. By taking a genetically modified fetal fibroblast and fusing it to an enucleated bovine egg, we were capable of isolating pluripotent cells that, when introduced into a host embryo, were capable of differentiating into tissues such as brain, skin, muscle, spleen, and mammary gland. Studies in mice later confirmed these findings *(11–13)*. Of particular interest is the study by Rideout et al. in which somatic cells from a RAG(–/–) mouse were used as nuclear donors, ES cells were generated via nuclear transfer, genetically corrected in the laboratory (ex vivo gene therapy), induced to differentiate into blood progenitors, and transferred intravenously into another RAG(–/–) animal. Remarkably, the host animal was able to have their immune system restored *(14)*.

2. THE PROMISE

The current hypothesis in human therapeutic cloning (HTC) is based on the notion that creating autologous stem cells could some day be possible by isolating a somatic cell from a patient, fusing it with an egg whose DNA is previously removed, artificially triggering development, obtaining a blastocyst from it, and isolating human ES (NT-ES) cells from such blastocyst.

Cells derived in this fashion could later be used to generate any cell type for the patient from whom the original somatic cell was isolated (*see* Fig. 2) *(15)*. Although this scheme has been proven possible in cattle and mice, in man seems to be more difficult than previously thought. At present, there are several ethical and technical difficulties that cast doubts on the validity of such an approach.

3. THE NEED FOR HUMAN EGGS

Unless an alternative source of eggs is discovered, this technology will rely on the donation of human eggs. A woman normally has 400,000 to 500,000 eggs at puberty and by the time she reaches menopause, all of them will be lost via apoptosis (cell death) or ovulation. It would be tempting to assume that a woman may have enough eggs to supply a large number of individuals that need NT-ES cells. Unfortunately, these eggs ought to be isolated from the ovary after the donor has undergone a superovulation procedure, something that should not be taken lightly. This protocol can generate a certain level of discomfort to women, which, in turn, leads us to ask the underlying ethical question: Is it right to seek human eggs for therapeutic cloning? Advanced Cell Technology (ACT), a company pursing this line of research in the private sector, has appointed its own ethic advisory board (EAB) to address some of these issues. ACT's EAB recognized that there are risks associated with the procedure and that these risks should be minimized at all cost *(16,17)*. Donors who agree to participate are subject to a thorough medical examination and the potential risks of the procedure are carefully explained; if they decide to continue with the protocol, they will be subject to a superovulation regime that is designed to be far more conservative than the ones used for conventional in vitro fertilization (IVF). Another important question has to do with the level of compensation offered and the possibility that this is the sole motivator for the donation. Although the answer to this question will vary among individuals, ACT's EAB has set a maximum compensation that shall not exceed (indeed is lower) the amount offered by IVF egg donor programs in the area.

As of today, there is only one report describing the introduction of the nucleus of a somatic cell into a human egg with the purpose of generating autologous stem cells. In this study, 19 eggs were reconstructed with somatic cells, either granulosa cell or skin fibroblasts. Once fibroblasts were used, the embryos developed to the pronuclear stage; cumulus cells, however, were able to sustain development up to the six-cell stage (*see* Fig. 3). In both instances, the nucleus was injected into the egg *(18)*. More research is under way in order to determine if these results could be improved.

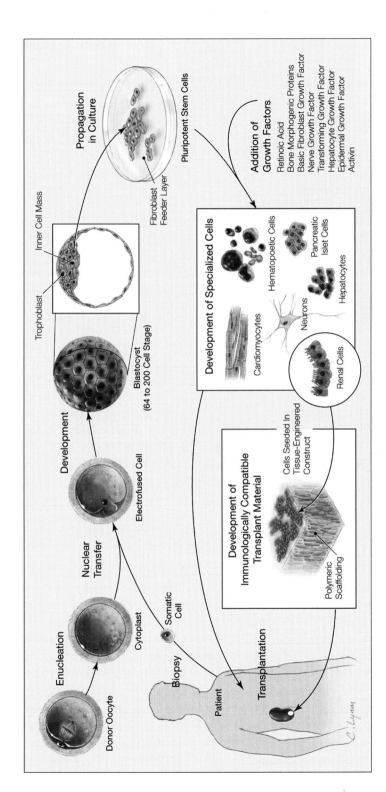

Fig. 2. Proposed scheme for human therapeutic cloning. A cell is taken from the skin of the patient who needs therapy and after being induced to return to an embryonic state (NT-ES), a colony is induced to differentiate into any cell type the patient would need. (Reprinted with permission from ref. *15*.)

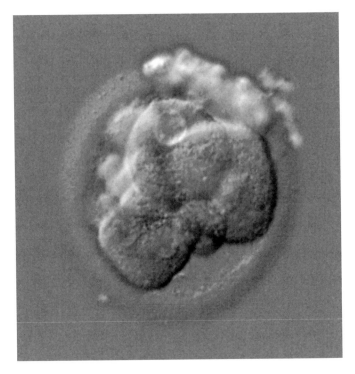

Fig. 3. Six-cell human nuclear transfer embryo. Embryo generated at ACT from the reconstitution of an enucleated human egg and a granulosa cell from the same donor. (Reprinted with permission from ref. *18.*)

4. TECHNICAL VARIABLES THAT INFLUENCE THE OUTCOME OF HTC

Like many other nuclear transfer protocols, HTC has great potential for improvement. There are many variables that need to be considered; among them, the age of the donor of the egg, the degree of maturation of the egg, and the kind of cell that is used as nuclear donor play a major role. The latter one is particularly important because it is clear from studies in mice and cattle that not all cell types offer the same developmental potential. It is also important to identify the proper activation protocol for human eggs. During normal fertilization, the sperm is responsible for activating the egg, but during nuclear transfer, in the absence of sperm, this procedure must be mimicked artificially. Once the embryo is activated, in vitro culture conditions will greatly impact development. It is essential not only to select for embryos that can sustain development to blastocyst but for those embryos that are capable of giving a robust inner cell mass from which NT-ES cells can be isolated.

Human therapeutic cloning is technically demanding and it is composed of a series of steps that carry its own efficiency. The cumulative overall efficiency today is almost as low as 1%. This means that 100 human eggs will be needed to generate 1 HT-ES cell line. As soon as the efficiency reaches 10–15%, we could envision a scenario where donor cells from patients are taken early in life, perhaps as early as birth, transformed into NT-ES cells, and stored frozen until the patient needs their own cells returned for therapy. There are hundreds of in vitro fertilization clinics throughout the United States; these same clinics could perform HTC at an affordable cost. At the same time, there are enough health care providers already performing cell therapy with a high level of sophistication, such as bone marrow stem cells isolation, purification, cryopreservation, and transplantation. Protocols could be standardized and followed consistently in a similar fashion for the generation of other specific cell types.

5. DO WE REALLY NEED HUMAN EGGS?

Claims have been made on the media that NT-ES cells have been derived as the result of fusing human somatic cells with animal eggs. Cows and rabbits have been mentioned as the source of eggs. At the time this manuscript was prepared, no peer-review articles have shown that this was possible. Our personal experience with these two species as cytoplasmic donors indicates that although morula-stage embryos can be easily obtained (*see* Fig. 4), the production of blastocysts or ES cell lines using cross-species nuclear transfer has been unsuccessful. Nevertheless, we believe that this approach is worth pursuing in order to overcome one of the major limitations of HTC—the lack of an unlimited supply of affordable eggs.

Another alternative to the use of human eggs, and therapeutic cloning altogether, could be the possibility of adding the nucleus of a specific differentiated cell into the cytoplasm of another cell that belongs to a different tissue type. It has been recently shown that leukocyte-specific genes can be reactivated by placing the nucleus of a fibroblast into the cytosol of a homogenate of leukocytes *(19)*. Although this is not a new technology, it certainly has its own merits and should be consider as a potentially valuable alternative to HTC. Preliminary studies in mice seem to indicate that a pluripotent state might be achieved by fusing or injecting the nucleus of a differentiated cell into the cytoplasm of undifferentiated cells.

6. MORE REASONS IN FAVOR OF HTC

In addition to the fact that a somatic cell can be potentially transformed into any cell type of the body, two other major advantages are unique to HTC: (1) the capacity to generate ES cells that are 100% compatible with

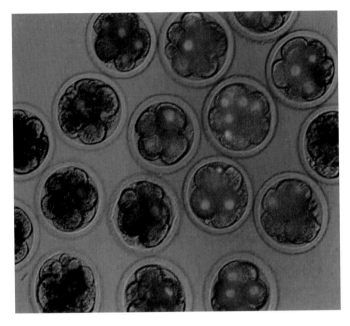

Fig. 4. Cross-species nuclear transfer embryos. Morula-stage embryos derived from the reconstitution of enucleated cow eggs and the nucleus of a human somatic cell.

the patient and (2) the intrinsic property of the egg to regenerate the life-span or resetting the biological clock of a cell. Somatic cells have a finite life-span; this was demonstrated by Hayflick in 1965 and subsequent studies have shown a correlation between in vitro cellular aging and in vivo human age *(20,21)*. It was later discovered that mammalian cells have an internal clock capable of recognizing the number of cell divisions that a cell has undergone. One explanation for this phenomenon has to do with the shortening of the telomeres and the lack the enzyme telomerase in somatic cells. Our laboratory as well as others have demonstrated that during the nuclear transfer procedure, the telomeres can be restored to their normal length (if not larger) most likely the result of the up regulation of telomerase during early embryonic development *(22–24)*. The clinical implications of this particular feature of HTC are far reaching. This means that regardless of the age of the donor, we could generate cells for therapy that have the life-span equivalent to that of cells from a newborn.

7. SUMMARY

Human therapeutic cloning has the potential to transform the lives of many people suffering from degenerative and genetic disorders and injuries.

However, before applying this technology, we, as a democratic society, must consider all of the ethical and moral issues associated with it. Regardless of whether we consider a nuclear transfer embryo a real embryo or not, the fundamental question is when a human being is considered a human being. For some people, it is at the moment of conception, others believe that the embryo is a group of cells with the *potential* to become a human being, and for many, it is somewhere in between or perhaps they have never been challenged to think on these terms. In 2001, President Bush appointed a Bioethics Council 2002 to try to address this issue clearly, and perhaps as a reflection of what happen in our society today, no consensus was reached and, indeed, two different opinions were clearly voiced. One was the opinion of Robert George from Princeton University who said: "The blastocyst is a human being at a certain stage of development. It is a human being." Conversely, Michael Gazzaniga, from Dartmouth College used an interesting analogy: "An embryo is a potential human, not a human itself; destroying an embryo would be like a fire at a Home Depot, the headline isn't '30 houses burn down, it's Home Depot burns down.'"

However, perhaps the loudest voice in this debate belongs to the patients. As Christopher Reeve, a strong supporter of HTC, expressed during a US Senate hearing in the spring of 2002: "Senator Kennedy, Members of the Committee: Thank you for the opportunity to testify this afternoon. For the record, I am a C-2 ventilator-dependent quadriplegic, which means that I am paralyzed from the shoulders down and unable to breathe on my own. For the last 7 years, I have not been able to eat, wash, go to the bathroom, or get dressed by myself. Some people are able to accept living with a severe disability. I am not one of them."

Progress continues to be made in the private sector in order to have HTC available for those patients in need. Society and lawmakers will have to reach a consensus as to whether this is an acceptable therapeutic alternative. We certainly hope that their response is supportive and manifested with not further delay, millions of people in the United States alone will benefit from it *(25)*.

REFERENCES

1. Briggs, R. and King, T. (1952) Transplantation of living nuclei from blastula cells into enucleated frogs' eggs, *Proc. Natl. Acad. Sci. USA* **38**, 455–463.
2. Spemann, H. (1938) *Embryonic Development and Induction*, Yale University Press, New Haven, CT.
3. Willadsen, S. M. (1986) Nuclear transplantation in sheep embryos, *Nature* **320**, 63–65.
4. Illmensee, K. (1978) Transplantation of embryonic nuclei into unfertilized eggs of *Drosophila melanogaster*, *Nature* **219**, 1268, 1269.

5. McGrath, J. and Solter, D. (1983) Nuclear transplantation in mouse embryos, *J. Exp. Zool.* **228,** 355–362.

6. Prather, R. S. and First, N. L. (1990) Cloning of embryos, *J. Reprod. Fertil.* **40(Suppl.),** 227–234.

7. Prather, R. S., Sims, M. M., and First, N. L. (1989) Nuclear transplantation in early pig embryos, *Biol. Reprod.* **41,** 414–418.

8. Stice, S. L. and Robl, J. M. (1988) Nuclear reprogramming in nuclear transplant rabbit embryos, *Biol. Reprod.* **39,** 657–664.

9. Wilmut, I., Schnieke, A. E., McWhir, J., Kind, A. J., and Campbell, K. H. S. (1997) Viable offspring derived from fetal and adult mammalian cells, *Nature* **385,** 810–813.

10. Cibelli, J. B., Stice, S. L., Golueke, P. J., et al. (1998) Transgenic bovine chimeric offspring produced from somatic cell-derived stem-like cells, *Nature Biotechnol.* **16,** 642–646.

11. Kawase, E., Yamazaki, Y., Yagi, T., Yanagimachi, R., and Pedersen, R. A. (2000) Mouse embryonic stem (ES) cell lines established from neuronal cell-derived cloned blastocysts, *Genesis* **28,** 156–163.

12. Munsie, M. J., Michalska, A. E., O'Brien, C. M., Trounson, A. O., Pera, M. F., and Mountford, P. S. (2000) Isolation of pluripotent embryonic stem cells from reprogrammed adult mouse somatic cell nuclei, *Curr. Biol.* **10,** 989–992.

13. Wakayama, T., Tabar, V., Rodriguez, I., Perry, A. C., Studer, L., and Mombaerts, P. (2001) Differentiation of embryonic stem cell lines generated from adult somatic cells by nuclear transfer, *Science* **292,** 740–743.

14. Rideout, W. M., 3rd, Hochedlinger, K., Kyba, M., Daley, G. Q., and Jaenisch, R. (2002) Correction of a genetic defect by nuclear transplantation and combined cell and gene therapy, *Cell* **109,** 17–27.

15. Lanza, R. P., Caplan, A. L., Silver, L. M., Cibelli, J. B., West, M. D., and Green, R. M. (2000) The ethical validity of using nuclear transfer in human transplantation, *JAMA* **284,** 3175–3179.

16. Green, R. (2002) Therapeutic cloning: the ethical considerations, *Sci. Am.* 53–45.

17. DeVries, K. O. and Members of the Ethics Advisory Board, ACT (2002) Overseeing therapeutic cloning research: a private ethics board responds to its critics, *Hastings Center Rep.* **32(3),** 2–7.

18. Cibelli, J., Kiessling, A., Cunniff, K., Richards, C., Lanza, R., and West, M. (2001) Somatic cell nuclear transfer in humans: pronuclear and early embryonic development, *e-biomed: J. Regener. Med.* **2,** 25–31.

19. Hakelien, A. M., Landsverk, H. B., Robl, J. M., Skalhegg, B. S., and Collas, P. (2002) Reprogramming fibroblasts to express T-cell functions using cell extracts, *Nature Biotechnol.* **20,** 460–466.

20. Cristofalo, V. J. and Pignolo, R. J. (1993) Replicative senescence of human fibroblast-like cells in culture, *Physiol. Rev.* **73,** 617–638.

21. Hayflick, L. (1965) The limited in vitro lifetime of human diploid cell strains, *Exp. Cell Res.* **37,** 614–636.

22. Lanza, R. P., Cibelli, J. B., Blackwell, C., et al. (2000) Extension of cell life-span and telomere length in animals cloned from senescent somatic cells [see comments], *Science* **288,** 665–669.

23. Tian, X. C., Xu, J., and Yang, X. (2000) Normal telomere lengths found in cloned cattle, *Nature Genet.* **26,** 272, 273.
24. Wakayama, T., Shinkai, Y., Tamashiro, K. L., et al. (2000) Cloning of mice to six generations, *Nature* **407,** 318, 319.
25. Perry, D. (2000) Patients' voices: the powerful sound in the stem cell debate, *Science* **287,** 1423.

16

Therapeutic Uses of Embryonic Stem Cells

Alexander Kamb, Mani Ramaswami, and Mahendra S. Rao

1. INTRODUCTION

The discovery of human embryonic stem (hES) cells and the ability to propagate and maintain hES cells in an undifferentiated state have sparked the realization that, conceptually, nothing prevents scientists from manipulating the human genome using techniques already verified in mouse experiments. A glance through the literature shows that the promise of ES cell technology has been realized in mice where transgenesis, homologous recombination, and somatic nuclear transfer are common procedures. The possibility that we now have the capacity to clone human beings has incited ethical debate and a call to ban all hES cell research (*see* Chapter 1).

At the other extreme, people have argued that there is more hype than substance to claims of therapeutic benefit. Scientists have made inflated claims about technologies in the past that have failed to live up to their promise. Detractors point out with some justification that, as yet, there is no example of a clinical use of hES cells. Moreover, the risk that hES cells may form teratocarcinomas after transplantation will limit any potential therapeutic applications. The current controversy is much like the debate that surrounded recombinant DNA technology when it was first proposed as a therapeutic tool. As in the heady early days of DNA cloning, it is tempting for visionaries to view the medical potential of hES cell technology as virtually boundless and for opponents to point out the potentially insurmountable problems.

The reality probably lies between these two extremes. Although much has been claimed without direct clinical evidence, it is important to realize that hES cells were identified only 3 yr ago and have for the most part been off limits to basic scientists for research purposes. Even as we write this chapter, only five cell lines are available at a cost of approximately $5000 per vial. The drug-approval process is slow and expensive and it will take

From: *Human Embryonic Stem Cells*
Edited by: A. Chiu and M. S. Rao © Humana Press Inc., Totowa, NJ

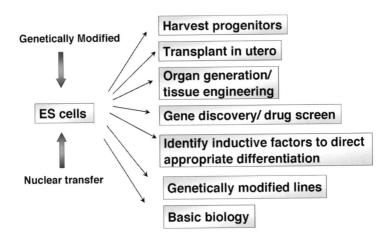

Fig. 1. Potential uses of hES cells are listed. Nuclear transfer and homologous recombination can be used to generate modified cell lines that can expand the range and utility of ES cells.

time before any hES cell therapies reach the clinic. It is therefore understandable that little direct evidence is available.

Embryonic stem cells can differentiate into all major cell types present in the embryo and do so with reliability and precision when transplanted into early-stage embryos, even after maintenance in culture for long periods. Current policy and law allow certain avenues of research with existing cell lines. Funding agencies will encourage permissible avenues of research. The rationale for such support is based on therapeutic opportunities derived from fundamental distinguishing traits of ES cells. They differ from all other cells in their ability to self-renew and confer the ES cell state on somatic nuclei. ES cells are the cellular version of life's blueprint or code. They contain not only the information but also the capacity to create an entire organism when nurtured in the proper environment (e.g., a suitably prepared uterus). These same properties that would play pivotal roles in human cloning efforts of dubious social value also afford unique advantages for ES-cell-based treatments of a variety of disorders and research into fundamental biological questions. As additional human cell lines become available, researchers must evaluate therapeutic and diagnostic possibilities allowed under current guidelines and move to exploit the potential of hES cells.

In this chapter, we address the therapeutic potential of hES cells. We regard certain features of ES cell biology as inherent and unalterable. These traits restrict the types of therapeutic application somewhat. From such considerations, we attempt to define the boundaries to future applications of ES

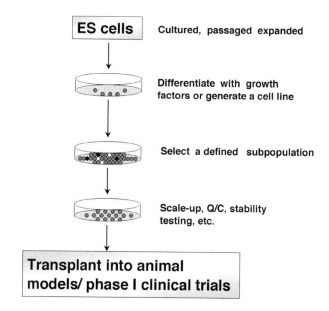

ES cells — Cultured, passaged expanded

Differentiate with growth factors or generate a cell line

Select a defined subpopulation

Scale-up, Q/C, stability testing, etc.

Transplant into animal models/ phase I clinical trials

Fig. 2. Cell transplantation. A strategy that uses hES cells as a cell bank is outlined. This is the approach that has been taken as an initial demonstration of the utility of hES cells.

cells in therapy, foreseeable at present: what ES cells eventually will do in a therapeutic setting and what they will not. The most obvious use of hES cells is as a source of differentiated cells for cell replacement therapies. The relative merits of ES cells and other somatic cells in this regard are discussed elsewhere (*see* Chapter 1). However, the horizon of potential uses for ES cells in therapy is much broader (*see* Fig. 1).

2. TRANSPLANT THERAPY

In its general form, cell replacement therapy involves culturing stem cells, either ES cells or lineage-restricted precursors (*see* Fig. 2). These cultures, or selected subpopulations, may be transplanted into specific regions of the body where they fall under the control of the normal tissue formation and maintenance apparatus. Alternatively, the cultures can be treated with agents that stimulate some change (e.g., a differentiation-promoting factor that sets in motion a particular developmental program, prior to transplantation). Because the cells are cultured in vitro, we have, in principle, a greater degree of control over their growth. We can use native factors or artificial ones to manipulate cellular behavior. In the following subsections, we discuss several types of transplant therapy based on stem cells, from amorphous aggregates and suspensions of cells to fully formed organs.

2.1. Cell Replacement

Cell replacement therapy is the most direct and obviously conceived form of stem-cell-based therapy (*see* Fig. 2). Missing, degenerating, or malfunctioning endogenous cells may be replaced by healthy cells derived in vitro from cultured ES cells. This is feasible if the host environment is "normal" inasmuch as it provides tropic growth factor support required for the survival, appropriate differentiation, and seamless integration of transplanted cells in the host tissue.

The relative advantages of hES cells over other somatic stem cell populations was discussed in detail in Chapter 13. A few important points worth making here are that ES cells are likely to be the single best source of cells that can be maintained in culture for prolonged time periods. In many somatic tissues, stem cell populations do not exist in sufficient numbers for therapy. Tissues for which such limitations are known include neural tissue, cardiac and skeletal muscle, the pancreas, and the liver. Even in tissue types for which somatic cells exist, evidence has accumulated to suggest that adult stem cells age and their replicative potential is reduced. Differentiation of hues cells in culture appears to mimic the differentiation during normal development and, as such, the differentiated cells are likely to have normal properties. ES cells have not yet acquired rostrocaudal and dorsoventral positional markers that may restrict their differentiation potential. ES cells, however, can acquire appropriate positional identity in vitro, suggesting that hES cells may be more appropriate for cell replacement in tissues where positional identity is critical such as in the brain.

The major disadvantage of ES cells is the fear that undifferentiated ES cells will pass through the selection process and, when implanted, may generate teratocarcinomas. This is a valid concern and one under intense study. Preliminary results indicate that once differentiated in culture to beyond the embryoid body stage, few undifferentiated cells remain and that these are insufficient in number to generate tumors. If one adds a stringent selection criteria such as fluorescence-activated cell sorting (FACS) to positively select for a desired population, it appears even less likely that undifferentiated cells may pass through in sufficient numbers to be a major concern. However, detailed experiments still need to be performed before ES-cell-derived cell therapy will be considered routine and safe.

However, it is important to emphasize that no major technical barrier to their use as a source of cells for therapy exists and there are ongoing efforts to begin using ES-cell-derived cells for transplantation in diabetes, Parkinson's disease, hepatic damage, and stroke. Data from hES-cell-derived cell replacement is limited, as these cells have been available to most researchers for less than 1 yr, but data from most studies are quite encouraging (*see* Chapter 18).

Cell replacement strategies, although conceptually appealing, can sometimes fail to consider important limitations in the host environment that specifically preclude the success of simple replacement therapy such as the absence of appropriate cues to differentiate or the altered milieu in a diseased state. In this context, it is attractive to consider alternative strategies that intend to either bypass this major limitation or specifically attempt to alter the tissue environment into one suitable for stem cell therapy. These strategies include *in utero* transplantation at early stages of embryonic development, cell and organ differentiation in vitro prior to transplant, or developing bio-organs in xeno transplant models. These are discussed briefly in section 2.2.

2.2. In Utero *Transplantation*

Complexities and developmental limitations of adult hosts for cell transplants need not preclude the utility of stem-cell-based therapies. A special opportunity for stem cell therapy may exist early in development (*see* Fig. 3). *In utero* therapy is likely to succeed. Migration and connectivity cues are present and appropriate feedback loops exist to regulate the final cell number. The limited data suggest that the incorporation of stem cells is robust and reproducible. In short, the environment permits the support of cell transplants. It is also worth pointing out that identification of antigens as self versus nonself occurs late in development and cells transplanted prior to this stage are likely to be recognized as self and, therefore, will not provoke a immune response. Furthermore, the blood–brain barrier is not yet fully formed and interactions between molecules or cells delivered intravenously are more likely to transit to the brain.

Development in humans is spread over a prolonged period and many techniques for *in utero* manipulation and surgery have been developed. Delivering cells to the embryo, although technically challenging, is feasible with current technology (*1*). The uterus can be readily accessed throughout most embryonic development and specific tissues can be targeted with relative ease. Detailed time tables of organ generation are available and ultrasound or other noninvasive methods of assessing fetal development have been developed (*2–4*), enabling therapy to be targeted to the most appropriate developmental time period (*see* Fig. 6). Surgical intervention *in utero* has been described (*1,5,6*) and the limited data that are available in rodents suggest that incorporation is robust, reliable, and reproducible (*1*). For example,

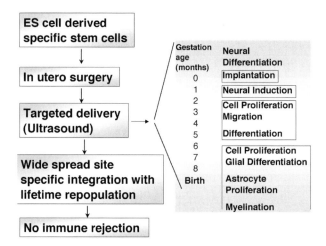

Fig. 3. *In utero* surgery. A possible strategy for *in utero* intervention is outlined. Co-opting existing surgical expertise, our knowledge of normal development and the ability to grow hES cells in large number allow one to approach therapy in ways that were impossible as little as a decade ago.

in utero transplantation into the brain at embryo day 18 (E18) in rodents results in reliable integration *(7)*. Ourednik and colleagues, for example, have shown integration of stem cells into the ventricular zone after intrauterine transplantation in monkeys *(8)*.

Although any cell that has the capability of integrating into the tissue targeted for therapy and responding appropriately to local signals is useful for intrauterine replacement, we suggest that fetal or ES-cell-derived cells will be the most appropriate. It is unlikely, in our opinion, that appropriate cues to direct ES cells to differentiate will be available at most stages of embryonic development. Progenitor or intermediate precursors may not provide lifetime replacement and, thus, may be of limited benefit. Tissues in which self-renewing stem cells exist and show lifetime self-replenishment are the best systems in which *in utero* transplantation should be considered. Perhaps the greatest concern is identifying prospective patients for cell replacement therapy. It is clear that *in utero* transplantation cannot be an option for diseases that manifest late in development or cannot be unambiguously diagnosed early. However, for genetic disorders and childhood diseases, *in utero* stem cell therapy may be a possibility that offers a high potential for success, as it appropriates normal developmental feedback loops. Possible disease targets are osteogenesis imperfecta, glycogen storage disorders, cerebral palsy, and other disorders in which candidate cells are available and the diagnosis can be made early.

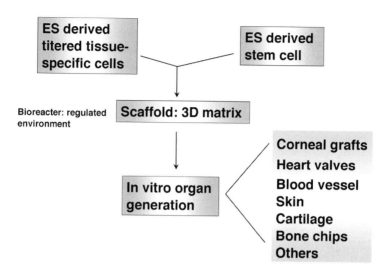

Fig. 4. Organ generation in vitro. A schematic of how hES cells can be combined with scaffolds or molds to generate three-dimensional organs that are difficult to generate otherwise. Note that no additional breakthrough in ES cell biology is required to take advantage of existing breakthroughs in bioengineering. Rather, combining technologies offers the possibility of synergy.

2.3. Organ Replacement

A more involved version of stem cell transplantation is organ replacement. In this application of stem cell biology to therapy, most of the work of producing a repaired or substitute organ is performed in vitro. Stem cells supply the cell types that divide, differentiate, and orient themselves into functional units of tissue that can be inserted into the patient (*see* Fig. 4). To facilitate formation of the three- or two-dimensional (in the case of skin) organ, a mold of some sort is typically used (or at least contemplated). The mold guides the proper alignment and physical disposition of the cells and extracellular matrix to produce a reasonable facsimile of an organ. Organ formation in vitro can be nearly complete or partial. In the latter case, the organ structure is seeded with stem cells and introduced into the body, where the majority of differentiation and tissue integration ensues. In the former case, the organ is manufactured in totality in tissue culture *(9)*. The procedure of organ generation is outlined in Fig. 5. Advances in biomaterial engineering and stem cell technology have made the creation of biohybrid organs feasible *(10,11)*. Indeed, commercial products are available and biotechnology companies that market these products have multimillion dollar revenues. Although a detailed discussion of recent breakthroughs is beyond the scope of this chapter, it is important to point out how some advances may expand

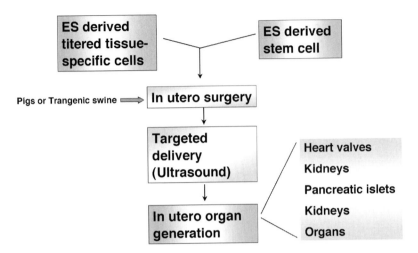

Fig. 5. Organ generation in vivo. A schematic of how hES cells could be used to generate a particular tissue or organ by combining expertise in xenotransplants with the ability to use hES cells as an unlimited source of naive or partially differentiated cells.

the range of stem cell therapy. Skin is perhaps the best example of an organ that is synthesized from precursor cells in vitro and sold as a commercial product *(12–14)*. Apligrap®, Dermagraft®, OrCel®, Alloderm®, Epicell®, and so forth are commercial products routinely used in therapy. Other tissues or organs where bioengineering and specialized culturing techniques have resulted in usable products are cartilage, bone, and smooth muscle (reviewed in refs. *15–18*). Osiris Therapeutics Inc. has demonstrated good bone engraftment using a carbonate scaffold, as has Orthovista *(19)*. Urinary organs have been repaired by combining endothelium with smooth muscle using calcium biphosphate ceramics *(9,20)* and periodontal implants have been developed as well *(21,22)*. The strategy used here is harvesting autologous or heterologous cells that are amplified in culture in a proprietary fashion and then grown into a tissue using specialized matrices. Titered mixtures of cells are then resupplied to the donor for use in rebuilding damaged tissue.

In principle, any stem cell with a sufficient replicative potential that can differentiate into the appropriate cell type can be used for this sort of therapy and in most of the examples described in this subsection, fibroblasts or other mesenchymal cells have been used. ES and ES-derived cells offer several advantages over other cell types. Banks of cells can be maintained rather than having to amplify from a small donor sample, which saves considerable time for therapy. ES cells can serve as a source of cells that are difficult to isolate or propagate from adult or fetal tissue and, finally, ES cells may not express positional cues, which can restrict the repertoire of differentiation of adult cells. It is important to note that most adult-derived stem cells cannot be amplified or maintained in culture for prolonged periods. Indeed, with the possible exception of neural and mesenchymal stem cells, the propagation of stem cells in culture has been extremely modest. Overall, the ability to combine biodegradable matrices that have been assembled into three-dimensional structures with dividing stem and progenitor cells to generate complex forms that closely mimic the contours and functionality of the natural structure provide an additional dimension to the use of ES cells in cell replacement strategies. Advances have also been made in providing hybrid biological/synthetic structures for organs and tissue for which it is still impossible to generate synthetic organs. This approach of mechanical or electronic devices that signal to biological tissues will, in turn, process the information. Examples of such devices are cochlear and retinal implants or nerve-stimulation devices *(23–26)*. This synthetic biological interface is one in which stem cells and progenitor cells may play a useful role as well *(7)*.

2.4. Organ Generation In Vivo

An even more radical possibility is to develop organs in swine by transplantation of human stem cells or progenitor cells into an organ *(27,28)*. Such a transplant could take place *in utero* or in early stages of postnatal development or in cadaveric tissue (*see* Fig. 5). The organ along with the transplanted stem and progenitor cells can be harvested when it has reached an appropriate size and then transplanted as xenoplants (or cadaveric transplants). Experience with xenotransplants is available and mechanisms to inhibit the acute immune response have been developed.

The host immune system will, in time, destroy the xenobiotic cells while sparing the matched human cells, which will gradually expand to replace the destroyed cells, leaving a fully formed human organ that maintains the complex morphology and contours of the normal organ *(29)*. Because we cannot as yet generate complex organs in vitro, this strategy offers a mechanism whereby organs can be generated by co-opting the complex regulatory controls that biological system have evolved to generate complex three-dimensional structures. Heart valves may be an ideal test of such a strategy *(30)*. They are relatively straightforward structures, and cells that would appropriately repopulate the valve can be grown in culture and valve transplants are routinely performed *(30)*.

Embryonic stem cells present their potentially unlimited replicative capacity with their innate ability to differentiate into multiple phenotypes, and their ability to be stored and recovered as undifferentiated or partially differentiated cells make them the ideal cell source. The recent evidence that ES cells and their derivatives may not be as immunogenic as adult cells further suggests that these cells are the cell of choice for such therapy as compared to other cell types overall. We suggest that combining our ability to harvest and grow stem cells and progenitor cells in culture with evolving techniques in bioengineering to provide tissue or organ replacement represents an alternative strategy to harness the power of stem cells.

2.5. Using Cells to Deliver Peptide Molecules, Repair Gene Defects, and Generate Universal Donor Cells

As has been discussed earlier, it is possible to use cells for replacement therapy in a variety of ways, and in many of these replacement strategies, no alternative to cell therapy exists (*see* Fig. 6). These diseases represent direct and obvious target for ES-cell-derived cell therapy. However, we would suggest that this is not the only use of ES cells or stem cells in general. Cells can be used to deliver peptides or growth factors that cannot be delivered easily otherwise. If this approach is used to deliver factors that mobilize endogenous progenitors or stem cells, then it is termed "inductive therapy." If it is used to enhance the survival of cells damaged in a particular disease, then the more generalized term "drug delivery" has been used.

Therapies termed "inductive" attempt to alter the tissue environment in order to induce or stimulate the innate therapeutic potential of endogenous stem cells. The best example of an inductive therapy is the use of "Epo," the hormone erythropoietin that stimulates the production of red blood cells (erythrocytes), for the treatment of anemia. As a therapy, the simple systemic injection of a soluble molecule is clearly preferable to cell replacement by bone marrow transplants or periodic blood transfusions. Although remarkably successful for the treatment of anemia, the use of inductive therapies has yet to be established in other therapeutic contexts. In part, this is because few diseases are the result of the specific loss of a naturally replenished cell type; however, it is also because few hormones and growth factors have been shown to have as highly specific a biological function as erythropoietin.

For complex tissues like the nervous system, where systemic injection of growth factors is unlikely to be successful, such a cell-based drug delivery approach may prove particularly important. Methods for the delivery of these proteins to the brain are currently limited. If a survival factor, for instance, must be continuously present in the environment for therapeutic effects, then the necessary frequent injections of the factor into brain, although feasible, is a major barrier to clinical treatment. If specific factors

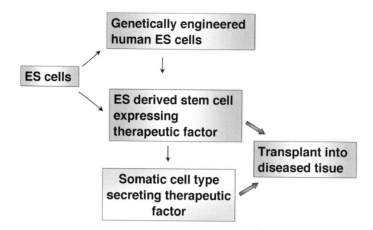

Fig. 6. Embryonic-stem-derived cells as delivery vehicles. Human ES cells can be used to create cell-based delivery vehicles for therapeutic factors. Typically, ES cells may be genetically modified to express the therapeutic proteins under the control of promoters appropriately active in cell types derived from genetically modified cells. In some cases (e.g., when the patient is genetically defective in production of a tissue-specific secreted protein), some the genetic modification steps may be unnecessary. Other modifications, which either reduce the likelihood of an immune response or allow induction of cell death in case of malfunction of transplanted cells, may, nevertheless, be of therapeutic value.

need to be delivered to specific subcellular locations (e.g., the postsynaptic membrane), then concentrations required for injection would need to be high, increasing the consequences of side effects. The use of genetically modified stem cells is one potential route for the efficient delivery of protein therapeutics. In the context of such cell-based inductive and drug delivery therapies, ES cells offer attractive possibilities: ES cells may be engineered to express the appropriate growth factor or peptide therapeutic, differentiated in vitro into the appropriate stem or progenitor cell type, and then transplanted into the diseased tissue where the growth factor or peptide drug is provided. In section 3., we consider different ways that ES cells may be engineered to enable therapeutic goals.

3. GENETIC ENGINEERING

Embryonic stem cells are unique among all cells in their ability to be maintained in an undifferentiated, karyotypically normal state for prolonged periods in culture. This ability has been utilized in mouse ES cells to produce genetic knockouts, transgenics, and conditional alleles for a variety of uses. Although we do not anticipate hES cells being used in the same fashion, these results in mice do suggest that strategies to introduce or remove genes are likely to be successful in hES cell cultures as well. One of the many exciting possibilities in the ES cell therapeutic arena is the combination of stem cell biology with genetic engineering *(31,32)*. This therapeutic mode can take the form of in vivo gene therapy, targeting gene transfer procedures to stem cells in the body, or ex vivo therapy, in which stem cells are manipulated in culture prior to reintroduction. As discussed earlier, standard methods for DNA transfection, including viral infection, lipid delivery and electroporation, work with reasonable fidelity in hES cells. Selection techniques for isolating successfully transfected cells including use of standard drug resistance markers or FACS sorting, highly feasible in mouse ES cells, are currently under test in hES cells as well. Our preliminary results and those of others suggest that efficiencies of cell selection procedures in hES cells are similar to those obtained in mouse cells and, in many cases, little adaptation of the methodology is required. The following are some examples of various strategies that can be used. It is important to emphasize that these strategies are relatively well tested in mouse ES cell cultures, have more recently been adapted to other ES cells, and are likely, although not yet proven, to work in hES cells as well.

1. *Classes of drugs.* The only requirement of "drugs" to be delivered by cells is that they be genetically encodable. The vast majority of such drugs will be protein based and fall into three categories: (1) Secreted growth factors or other secreted extracellular proteins. Examples of growth factors include but are not limited to neuroprotective agents or proteins that induce division of endogenous tissue precursor cells; extracellular molecules include extracellular matrix components that are, for example, either involved in musculoskeletal disease *(33)* or have the ability to induce tissue regeneration *(34)*. (2) Enzymes that if secreted may be taken up by host cells with therapeutic consequences. The best examples of these are enzymes genetically deficient in a large range of lysosomal storage disorders *(35)*. (3) Non-native therapeutic proteins or peptides. Proteins and peptides with important pharmacological activities have been identified from a large number of diverse biological sources (e.g., see ref. *36*). In addition, emerging technologies for the generation of large peptide libraries and rapid isolation of molecules with desired activities make peptides a very promising source of high-affinity, high-specificity therapeutics *(37)*. The impact of these three classes of therapeutic agents may be greatly enabled by cell-based delivery approaches enabled by hES cell technology.

2. *Use of transgenes expressed via viral vectors.* DNA constructs expressing secreted proteins or bioactive peptides in appropriate form can be expressed in a variety of cell types. Viral vectors allow a high efficiency (greater than 10%) of transgenesis in cultured cells and, in principle, should allow the generation of ES cells that may be differentiated into the appropriate stem or progenitor cell type expressing the appropriate therapeutic molecule, prior to transplanta-

tion. In the case of diseases associated with multinucleate muscle fibers (e.g., muscular dystrophy), their fusion with myoblasts derived from engineered ES cells may be sufficient for gene therapy. For other diseases, appropriate cells types derived from engineered ES cells may allow ex vivo gene therapy; for instance, mesenchymal cells derived from ES cells overexpressing appropriate forms of the extracellular matrix protein collagen may be used for the treatment of osteogenesis imperfecta, and glial or neuronal precursor cells overexpressing LINCL1 may be beneficial for treatment of neuronal ceroid lipofucinoses *(35)*. However, despite the ease of DNA virus- or retrovirus-based cell transfection, this approach to gene therapy is associated with two major problems unconnected to the traditional complications associated with transplantation (considered previously) *(38)*. First, the site for genomic integration of the transgene cannot be controlled and, thus, varies substantially between procedures. This could lead not only to a loss of control over therapeutic efficiency but also to varied and hard-to-predict complications if integration occurs within essential sequences. Although this problem may be overcome, a second so far insurmountable problem is that of transgene silencing. Retrovirus-based transgenes allow brief periods of transgene expression, but then, as a result of intrinsic regulatory mechanisms that likely operate at the level of chromatin structure, they are transcriptionally silenced. Several methods to allow long-term persistent expression of therapeutic transgenes are currently in development *(38)*. These include the use of autonomously replicating human minichromosomes as expression vectors that may be transfected into ES cells by electroporation or lipofectin. However, the most promising and advanced approach has to be the use of homologous recombination.

3. *Targeted integration-based transgene expression via native promoters.* Homologous recombination in mammalian ES cells is routinely used to replace endogenous sequences with sequences modified in vitro. This has been used to generate gene disruptions, subtle gene alterations, or gene replacements in cells and in animals generated from modified ES cells *(39)*. In the engineered cells and animals, the redesigned genes are expressed stably at the targeted locus in the manner expected for endogenous genes. If the biology and technologies associated with this procedure work similarly in human cells, then homologous-recombination-based gene replacement or gene correction may be envisioned as a therapeutic procedure in the future.

A particularly attractive notion is that transgenes integrated by homologous recombination downstream of specific endogenous promoters may be resistant to genetic silencing. Thus, sequences encoding protein or peptide therapeutics could be targeted to sequences downstream of a ubiquitous or cell-type-specific promoter. The engineered cell would have one wild-type allele of the targeted locus, usually sufficient for normal health and function of the engineered cell. As discussed previously, homologous recombination would be performed in the hES cell that would subsequently be differentiated in vitro into a cell type suitable for transplantation.

Figure 7 shows the design and mechanism of the simplest construct used for a purpose. The targeted locus in this case is modified to include the therapeutic coding sequence under control of the endogenous promoter as well as a cDNA-conferring antibiotic resistance under control of a constitutive promoter. If desired, the latter sequences can be precisely excised in a second step if in the initial construct they were flanked by C-recombinase (Cre) target sequences. ES cells that have undergone homologous recombination could then be transiently transfected with a Cre-expressing plasmid to generate cells containing the therapeutic transgene but not the selectable marker. Further modifications to the original construct could confer additional properties to the transgene. For instance, conditional expression under regulation of a systemically or orally delivered stimulus may be possible if the endogenous promoter activity is gated by an additional transcriptional control element that would act as a repressor in the absence of the appropriate stimulus *(38)*.

In addition to the possibilities for engineering delivery of therapeutic molecules, two further possibilities opened up by homologous recombination technologies address problems of cell malfunction after transplantation and immune cell rejection. An off switch that allows a physician the option of destroying malfunctioning transplanted cells is of obvious utility. ES cells could be engineered to encode an inducible pro-apoptotic protein that might be activated to trigger cell suicide. To avoid immune rejection, a class of universal donor ES cells may be established by appropriate disruption of major histocompatibility complex (MHC) genes that are involved in tissue recognition and, hence, graft rejection.

3.1. Generating Universal Donors

The MHC was discovered originally as a genetic locus controlling rapid rejection of tissue grafts. Subsequently, the study of antibody responses in vivo and T-cell responses in vitro to MHC antigens identified the presence of a number of closely linked loci within the MHC. Immune response (Ir) genes also mapped to the MHC (reviewed in ref. *40*). Both direct pathways involving donor MHC class I and class II expression activating CD4+ and CD8+ cells as well as indirect pathways involving donor antigen (foreign) presentation by self (recipient) MHC class antigens have been recognized as being important in the rejection process *(41)*. The relative importance of different pathways regulating graft rejection depend on the target organ *(42)*. Similar xenotransplantation experiments involving animals deficient in either class I or class II antigens or both have revealed that donor MHC deficiency offers no protection to the xeno graft, suggesting that strategies to eliminate MHC antigen expression will not be successful in generating "universal xenotransplant donors" *(43,44)*. However, these strategies may be useful in enhancing the survival of transplants in heterologous transplants with limited mismatch. Thus, one possible mechanism of enhancing the util-

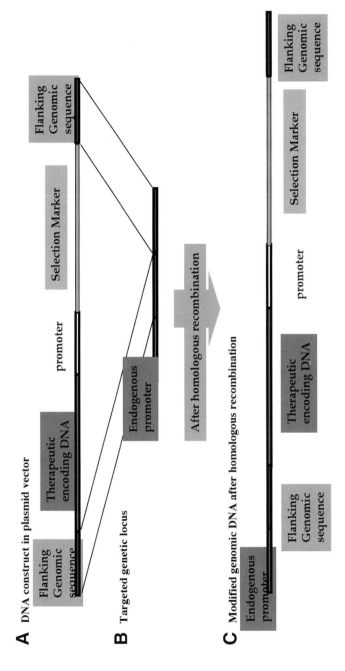

A DNA construct in plasmid vector

Flanking Genomic sequence

Therapeutic encoding DNA

promoter

Selection Marker

Flanking Genomic sequence

B Targeted genetic locus

Endogenous promoter

After homologous recombination

C Modified genomic DNA after homologous recombination

Endogenous promoter

Flanking Genomic sequence

Therapeutic encoding DNA

promoter

Selection Marker

Flanking Genomic sequence

Fig. 7. Placing a gene downstream of a known locus. Outline of a simple homologous recombination-based strategy for placing a therapeutic-encoding transgene under control of a selected endogenous promoter. (**A**) The DNA construct used contains the sequences to be inserted, flanked by genomic DNA that is homologous to the targeted locus shown in (**B**). After homologous recombination, one allele of the targeted locus will be modified, as shown in (**C**). For additional details of varied construct design and selection techniques, the reader is referred to ref. *39*.

311

ity of ES cells and resolving the potential immune rejection issue is to generate hES cell line subclones in which the MHC class 1 and II gene complexes have been deleted by homologous recombination.

3.2. Cloning and Somatic Nuclear Transfer

Nuclear transfer and cloning are perhaps the ultimate manifestation of genetic engineering. Here, an entire genome is transferred into the recipient blastocyst. The media has focused almost entirely on the creation of identical twins, an entire organism that shares genes with the donor. Cloning as in creating a genetically identical copy is theoretically possible in humans based on successful cloning in multiple species. Cloned mice, pigs, sheep, and cows have been generated. It is, however, important to remember that success has come at a high price, as current cloning techniques are relatively inefficient. Even when cloning has resulted in a viable offspring, many of these offspring have had subtle defects and poor viability. Cloning is not only morally and ethically controversial, but also scientifically questionable. The relative inefficiency of the cloning process, the long gestation period, and late sexual maturity are all significant barriers to any even theoretical utility of the cloning process and scientists have been nearly unanimous in their opposition to the idea.

However, more restricted and perhaps less controversial uses of this technique also exist (reviewed in ref. 45). Nuclear transfer might be used in the combination of ES cells and other methods that we have discussed, such as organ replacement, to produce autologous organs for transplantation. Rather than whole organisms, only a single tissue or organ is cloned for therapeutic use. Scientists have strongly supported the idea of therapeutic cloning or the use of nuclear transfer to create cell lines of ES cells that are genetically identical is of therapeutic use. In this process, somatic nuclei from a potential cell transplant recipient are transferred to a donated blastocyst and an ES cell line derived. This ES cell line would be genetically identical to the donor and could serve as a source for matched cells that would not provoke an immune response. Matched cell lines could be generated at any stage and stored much the same way that cord blood stem cells are being stored and used when required for therapy. Studies with somatic nuclear transfer have suggested that different cell nuclei may have different efficiencies for generating ES cell lines. Currently, it appears that gonadal cells may offer the best efficiency. This makes sense when one considers normal biology where imprinting is erased as gonadal cells are prepared for fertilization and are thus primed to accept signals from the blastocyst cytoplasm.

The process of therapeutic cloning, although having a higher success rate than cloning an entire organism, is still by no means an efficient process and much work is required to understand the process *(46,47,47a)*. Understanding the basic biology of how nuclear reprogramming occurs is of great importance and recent breakthroughs highlight the advantage of having ES cells available *(48)*. It appears likely that ES cell cytoplasm may possess the same reprogramming ability *(49)*, and if this is validated, then studying the process of reprogramming will not be hampered by the limited access to blastocyst cytoplasm.

Understanding the process of somatic cell nuclear reprogramming may lead to a scenario in which unfertilized eggs can be used for nuclear transfer or even normal cells can be reprogrammed to generate ES-like cells and thus bypass any possible moral caveats to using discarded blastocysts. These ES like cells would have the virtue of being autologous as well as solving the issue of immune rejection. The huge potential upside of understanding the biology of nuclear reprogramming and the advantage of having a source of material for studies has been overshadowed by the debate on cloning and we would urge readers to consider the benefits of these studies as well.

3.3. Gene Discovery

Human ES cells offer the possibility to study differentiation under favorable experimental conditions. Cells in culture can be exposed to a wide range of environmental stimuli. In addition, the availability of large, homogeneous cell populations enables a variety of biochemical studies that are otherwise difficult or impossible. In the same way that hES cells will contribute to unraveling questions in basic biology, these cells provide a route to discover novel genes that regulate differentiation and growth (*see* Figs. 8 and 9). Such findings have obvious therapeutic implications. For instance, genes that bias the production of one cell type over another, stimulate production of a particular cell type, or block formation of a cell type may prove useful in treating diseases as diverse as neurodegeneration and cancer.

How can genes important in stem cell biology be defined? The traditional path to genes involved in complex cellular behavior is genetics. Somatic cell genetics is a relatively mature area in biology, but it remains challenging. Nonetheless, there are new tools that improve the yield and speed of genetic experiments in somatic cells. These include mRNA expression profiling, RNA interference, homologous recombination, and viral gene expression vectors to generate stage-specific lines. Genetic selections for particular phenotypes, mutants, or mutant phenocopies are often problematic in somatic cells, but a large number of genes have been identified this way. In addition, genomic technologies coupled with molecular genetics provide a powerful combination to fuel hypothesis creation and testing.

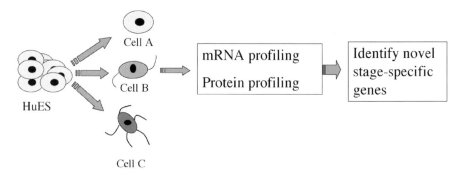

Fig. 8. Gene discovery. Human ES (HuES) cells can be exposed to a variety of signals and/or induced to differentiate into homogeneous populations of cell types (e.g., A, B, or C). These cell types can be used to prepare protein or mRNA for analysis using tools for comparative gene expression or protein expression. Such tools include hybridization-based (e.g., microarrays) and sequence-based (e.g., SAGE and MPSS) methodologies for mRNA, and gel-based or mass-spectrometry-based techniques for proteins. Comparisons among the different samples facilitate discovery of genes whose regulation are important in the induction or maintenance of particular cell fates.

A relatively recent arrival on the technical scene is gene expression profiling (i.e., high-throughput quantitative measurements of mRNA levels in specific cell or tissue samples). Gene expression profiles afford clues to the genes that may play critical roles in regulating cell growth and differentiation. Recently, several groups have begun to explore the identities of genes specifically expressed in hES cells (Carpenter-Geron, personal communication). Such studies are a starting point for functional tests. Overexpression of selected genes or targeted suppression can elucidate the subset of genes with specific effects of possible therapeutic value. Perhaps the most interesting genes are those whose expression and, more importantly, functions are highly restricted to hES cells or specialized offspring.

3.4. Drug Discovery

On a more applied level, stem cells hold promise to facilitate drug research. Genes discovered as outlined earlier might lead directly to new therapies. Growth factors such as erythropoietin and granulocyte colony-stimulating factor (G-CSF) are the backbone of the biopharmaceutical industry. These stimulate specific hematopoietic stem cells. It may be possible to define other factors that have important effects on hES cells. Alternatively, receptors for growth factors and other ligands may provide novel targets for agonists or antagonists. Finally, other traditional drug target

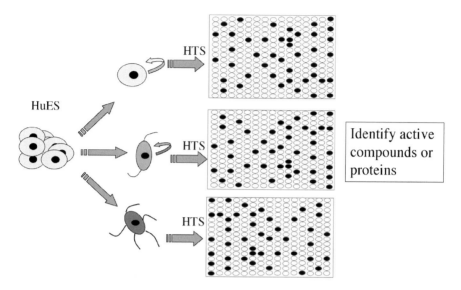

Fig. 9. Cell screening assays. Human ES cells as the starting point for cell-based assays. hES cells are induced to differentiate into specialized cell types or used as is in microplate screens. Such screens may lead to novel inductive, proliferation, or antiproliferation agents. This type of cell-based system can be used to assay specific proteins defined through, for example, comparative gene expression studies, antibodies against specific cell surface markers or secreted proteins, or chemical agents. The potential for miniaturization of such assays affords the possibility of high-throughput screens of, for example, chemical libraries.

classes such as ion channels and G-protein-coupled receptors (GPCRs) whose activities may influence hES cell behaviors are worthy of serious attention.

The process of drug development is at present highly inefficient, often ineffective, and always very expensive. ES cells could improve the overall process by supplying the materials for cell-based screens and assays. Indeed, hES cells could be bioassay tools par excellence. One can envision many areas in which hES cells may prove useful, but we mention one in particular: drug screening. With their characteristics of immortality, self-renewal, and pluripotency, ES cells possess ideal qualities for a cell-based drug screening tool. They can be cultured indefinitely and triggered to create any particular cell type desired. Thus, isogenic lines of any cell type can be obtained. Indeed, those with specific genotypes (allele combinations) can be generated in principle with ease. A panel of appropriate cell types provides a test bed to identify and characterize functional properties of protein factors and chemicals Antibodies also can be validated in the context of such an assay. Pharmaceutical researchers might, for example, seek out molecules that control cell growth and differentiation. Alternatively, they might simply use the panel to examine potential toxicity in vitro, as a surrogate for, and preliminary to, expensive and time-consuming human trials.

These considerations suggest that hES cells may bridge the large gap between assays that involve purified protein and those that use animals. Cell culture has heretofore relied mainly on tumor cell lines. hES cells may yield a large number of new cell lines with important properties for pharmaceutical development.

4. UNDERSTANDING THE BASIC BIOLOGY OF STEM CELLS

As early as 1887, it was postulated that the mature oocyte possesses all of the elements necessary for embryonic development. After fertilization, initial developmental events are regulated primarily by maternal inherited mRNA, and embryonic transcription is not initiated until later in development *(50,51)*. The major male contribution appears to be to provide the active division center, called the centrosome, which consists of two centrioles in a perpendicular arrangement and pericentriolar material and is considered to be responsible for nucleation of microtubules and the formation of the mitotic spindle. There is a paternal pattern of inheritance of the centrosome in humans, and human oocytes lack centrioles, whereas spermatozoa carry two *(52,53)*. Maternal contributions are more major to all other aspects of differentiation. Maternal RNA provides the initial impetus for differentiation and regulates embryonic imprinting *(54)*. Inheritance of mitochondria is maternal, with the paternal mitochondria likely being important only for placental differentiation *(55)*. Unique methylases and demethylases are expressed at this early age and cell cycle regulatory events and telomeric biology are likely to be unique at this stage. Genomic analysis has suggested that the transcriptosome is maintained in an open state with decondensed chromatin. Maintenance of an open transcriptosome in multipotent ES cells likely requires both the presence of positive factors as well as the absence of negative regulators *(see* Fig. 10). Factors that maintain an open transcriptosome include as yet unidentified factors such as demethylases, reprogramming molecules present in blastocyst cytoplasm, and regulators of heterochromatin modeling *(56)*. These positive factors are segregated as early progenitor cells undergo asymmetric cell division. The cell that receives these factors remains undifferentiated while the other daughter either degrades these factors or does not receive them to activate cell-type-specific programs. Global activators, global repressors, and master regulatory genes play important regulatory roles in switching on or off cassettes of genes while methylation, heterochromatin remodeling *(57)*, and

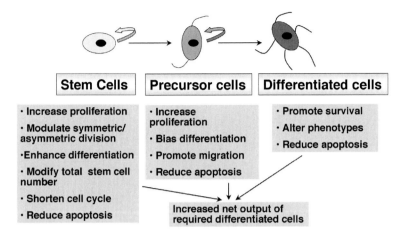

Fig. 10. Possible points and mechanisms of therapeutic intervention in hES cells and their progeny. By targeted manipulation of genetic programs in particular cell types, specific cell populations can be expanded or reduced in size.

perhaps siRNA (reviewed in ref. *58*) maintain a stable phenotype by specifically regulating the overall transcriptional status of a cell. Allelic inactivation and genome shuffling further sculpt the overall genome profile to generate sex, organ, and cell-type specification.

Although much has been learned from studying discarded blastocysts in human eggs and eggs from other species, evidence has begun to accumulate that difference in fundamental biology exist at this early stage and that many of these events can be usefully studied in embryonic stem cells and, as such, hES cell lines offer the unprecedented opportunity to begin to understand fundamental issues of symmetric versus asymmetric cell division, telomerase biology, chromatin condensation, regulation of imprinting, mitochondrial inheritance, cell fusion, and so forth (*see* Fig. 11). For example, mitochondrial segregation is not well understood and the ability to isolate hES cells allows one the possibility of making cybrids (mitochondrial hybrids) at various stages in development with mitochondria from various cell types, and inheritance can be assessed readily. Likewise, aging and the change in telomere length and its regulation by telomerase can be assessed in a normal cell type. Much of what we have learned about telomerase biology has been by in vivo experiments requiring follow-up over many years or studies in transformed cell lines in which the biology of telomerase is unlikely to be normal (*59,60*). This is true for many of the fundamental aspects of development where we lack an appropriate readily accessible model that mimics normal development in vitro. Whereas adult stem cells may be use-

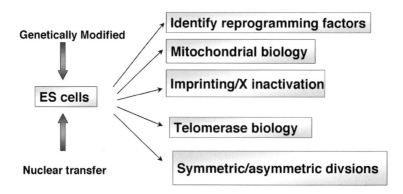

Fig. 11. Potential basic biology issues that can be studied in ES cells. Some aspects of biology that can be readily studied in human ES cells are listed. Note that many of these issues would be difficult to study in other cell lines.

ful for a small subset of these issues, and likewise we may learn something from either blastocysts from other species or embryo carcinoma lines from human samples, it is unlikely that we will be able to test and validate hypothesis without the ready availability of well-characterized normal cell lines that offer sufficient material to perform genomic and biochemical experiments as well as the opportunity to watch events as they unfold in vitro. The ability to test the effect of single-gene perturbations on events in a controlled and reproducible manner, which is currently impossible in human cells, offers a unique ability to understand how a single cell can orchestrate the development of an entire organism in such an error-free fashion.

5. CONCLUSION

The unique properties of embryonic stem cells endow them with exceptional power in the context of therapy—both as tools for discovery and as therapeutic agents themselves. Indeed, the potential of stem cell biology is so revolutionary in this regard that a variety of social and ethical issues will likely need serious attention. Huge benefits will ensue and, as with all paradigm-shifting advances in technology, care must be exerted so that the dangers of irresponsible application of molecular and cellular stem cell biology will be avoided.

REFERENCES

1. Hayashi, S. and Flake, A. W. (2001) In utero hematopoietic stem cell therapy, *Yonsei Med. J.* **42(6)**, 615–629.
2. Hill, L. M., Guzick, D., Fires, J., et al. (1990) The transverse cerebellar diameter in estimating gestational age in the large for gestational age fetus, *Obstet. Gynecol.* **75(6)**, 981–985.

3. Harrington, K. and Campbell, S. (1993) Fetal size and growth, *Curr. Opin. Obstet. Gynecol.* **5(2)**, 186–194.
4. Lysikiewicz, A., Bracero, L. A., and Tejani, N. (2001) Sonographically estimated fetal weight percentile as a predictor of preterm delivery, *J. Matern. Fetal Med.* **10(1)**, 44–47.
5. Agarwal, S. K. and Fisk, N. M. (2001) In utero therapy for lower urinary tract obstruction, *Prenat. Diagn.* **21(11)**, 970–976.
6. Stevens, G. H., Schoot, B. C., Smets, M. J., et al. (2002) The *ex utero* intrapartum treatment (EXIT) procedure in fetal neck masses: a case report and review of the literature, *Eur. J. Obstet. Gynecol. Reprod. Biol.* **100(2)**, 246–250.
7. Teng, Y. D., Lavik, E. B., Qu, X., et al. (2002) Functional recovery following traumatic spinal cord injury mediated by a unique polymer scaffold seeded with neural stem cells, *Proc. Natl. Acad. Sci. USA* **99(5)**, 3024–3029.
8. Ourednik, V., Ourednik, J., Flax, J. D., et al. (2001) Segregation of human neural stem cells in the developing primate forebrain, Science **293(5536)**, 1820–1824.
9. Park, K. D., Kwon, I. K., and Kim, Y. H. (2000) Tissue engineering of urinary organs, *Yonsei Med. J.* **41(6)**, 780–788.
10. Fuchs, J. R., Nasseri, B. A., and Vacanti, J. P. (2001) Tissue engineering: a 21st century solution to surgical reconstruction, *Ann. Thorac. Surg.* **72(2)**, 577–591.
11. Zandstra, P. W. and Nagy, A. (2001) Stem cell bioengineering, *Annu. Rev. Biomed. Eng.* **3**, 275–305.
12. Yang, E. K., Seo, Y. K., Youn, H. H., et al. (2000) Tissue engineered artificial skin composed of dermis and epidermis, *Artif. Organs* **24**, 7–17.
13. Bello, Y. M., Falabella, A. F., and Eaglstein, W. H. (2001) Tissue-engineered skin. Current status in wound healing, *Am. J. Clin. Dermatol.* **2(5)**, 305–313.
14. Kuroyanagi, Y., Yamada, N., Yamashita, R., and Uchinuma, E. (2001) Tissue-engineered product: allogeneic cultured dermal substitute composed of spongy collagen with fibroblasts, *Artif. Organs* **25(3)**, 180–186.
15. Boyan, B. D., Lohmann, C. H., Romero, J., and Schwartz, Z. (1999) Bone and cartilage tissue engineering, *Clin. Plast. Surg.* **26(4)**, 629–645.
16. Freed, C. R., Greene, P. E., Breeze, R. E., et al. (2001) Transplantation of embryonic dopamine neurons for severe Parkinson's disease, *N. Engl. J. Med.* **344(10)**, 710–719.
17. Goble, E. M., Kohn, D., Verdonk, R., and Kane, S. M. (1999) Meniscal substitutes—human experience, *Scand. J. Med. Sci. Sports* **9(3)**, 146–157.
18. Muschler, G. F. and Midura, R. J. (2002) Connective tissue progenitors: practical concepts for clinical applications, *Clin. Orthoped.* **395**, 66–80.
19. Pittenger, M. F., Mosca, J. D., and McIntosh, K. R. (2000) Human mesenchymal stem cells: progenitor cells for cartilage, bone, fat and stroma, *Curr. Topics Microbiol. Immunol.* **251**, 3–11.
20. LeGeros, R. Z. (2002) Properties of osteoconductive biomaterials: calcium phosphates, *Clin. Orthop.* **395**, 81–98.
21. Buckley, M. J., Agarwal, S., and Gassner, R. (1999) Tissue engineering and dentistry, *Clin. Plast. Surg.* **26(4)**, 657–662.

22. Malament, K. A. (2000) Prosthodontics: achieving quality esthetic dentistry and integrated comprehensive care, *J. Am. Dent. Assoc.* **131(12)**, 1742–1749.

23. Humayun, M. S. (2001) Intraocular retinal prosthesis, *Trans. Am. Ophthalmol. Soc.* **99**, 271–300.

24. Raine, C. H. and Martin, J. (2001) Cochlear and middle ear implants: advances for the hearing impaired, *Hosp. Med.* **62(11)**, 664–668.

25. Kerdraon, Y. A., Downie, J. A., Suaning, G. J., et al. (2002) Development and surgical implantation of a vision prosthesis model into the ovine eye, *Clin. Exp. Ophthalmol.* **30(1)**, 36–40.

26. Rauschecker, J. P. and Shannon, R. V. (2002) Sending sound to the brain, *Science* **295(5557)**, 1025–1029.

27. Tuch, B. E. and Beretov, J. (1994) Interaction between xenografted human fetal pancreas and liver, *Transplant Proc.* **26(6)**, 3333.

28. Angioi, K., Hatier, R., Merle, M., and Duprez, A. (2002) Xenografted human whole embryonic and fetal entoblastic organs develop and become functional adult-like micro-organs, *J. Surg. Res.* **102(2)**, 85–94.

29. Macchiarini, P., Candelier, J. J., Coullin, P., et al. (2000) Use of embryonic human trachea grown in nude mice to patch-repair congenital tracheal stenosis, *Transplantation* **70**, 1555–1559.

30. Zeltinger, J., Landeen, L. K., Alexander, H. G., Kido, I. D., and Sibanda, B. (2001) Development and characterization of tissue-engineered aortic valves, *Tissue Eng.* **7(1)**, 9–22.

31. Arbones, M. L., Austin, H. A., Capon, D. J., and Greenburg, G. (1994) Gene targeting in normal somatic cells: inactivation of the interferon-gamma receptor in myoblasts, *Nature Genet.* **6(1)**, 90–97.

32. Yanez, R. J. and Porter, A. C. (1998) Therapeutic gene targeting, *Gene Ther.* **5(2)**, 149–159.

33. Niyibizi, C., Wallach, C. J., Mi, Z., and Robbins, P. D. (2002) Approaches for skeletal gene therapy, *Crit. Rev. Eukaryot. Gene Expr.* **12(3)**, 163–173.

34. Condic, M. L. (2001) Adult neuronal regeneration induced by transgenic integrin expression, *J. Neurosci.* **21(13)**, 4782–4788.

35. Hofmann, S. L., ed. (2002) The expanding spectrum of lysosomal storage disorders, *Curr. Mol. Med.* **2**, 423–437.

36. Adams, M. E. and Olivera, B. M. (1994) Neurotoxins: overview of an emerging research technology, *Trends Neurosci.* **17(4)**, 151–155.

37. Wilson, D. S., Keefe, A. D., and Szostak, J. W. (2001) The use of mRNA display to select high-affinity protein-binding peptides, *Proc. Natl. Acad. Sci. USA* **98(7)**, 3750–3755.

38. Pfeifer, A. and Verma, I. M. (2001) Gene therapy: promises and problems, *Annu. Rev. Genom. Hum. Genet.* **2**, 177–211.

39. Jackson, I. J. and Abbott, C. M. (2001) *Mouse Genetics and Transgenics. A Practical Approach*, Hames, B. D. (ed.), UK, Oxford University Press, Oxford.

40. Simpson, E. (1998) Minor transplantation antigens: mouse models for human host-versus-graft, graft-versus-host and graft-versus-leukemia reactions, *Arch. Immunol. Ther. Exp. (Warsz.)* **46(6)**, 331–339.

41. Chitilian, H. V. and Auchincloss, H., Jr. (1997) Studies of transplantation immunology with major histocompatibility complex knockout mice, *J. Heart Lung Transplant.* **16(2)**, 153–159.
42. Auchincloss, H., Jr. and Sultan, H. (1996) Antigen processing and presentation in transplantation, *Curr. Opin. Immunol.* **8(5)**, 681–687.
43. Markmann, J. F., Campos, L., Bhandoola, A., et al. (1994) Genetically engineered grafts to study xenoimmunity: a role for indirect antigen presentation in the destruction of major histocompatibility complex antigen deficient xenografts, *Surgery* **116(2)**, 242–248; discussion, 248–249.
44. Lee, R. S., Grusby, M. J., Laufer, T. M., et al. (1997) CD8+ effector cells responding to residual class I antigens, with help from CD4+ cells stimulated indirectly, cause rejection of "major histocompatibility complex-deficient" skin grafts, *Transplantation* **63(8)**, 1123–1133.
45. Colman, A. and Kind, A. (2000) Therapeutic cloning: concepts and practicalities, *Trends Biotechnol.* **18(5)**, 192–196.
46. Yanagimachi, R. (2002) Cloning: experience from the mouse and other animals, *Mol. Cell. Endocrinol.* **187(1–2)**, 241–248.
47. Brem, G. and Kuhholzer, B. (2002) The recent history of somatic cloning in mammals, *Cloning Stem Cells* **4(1)**, 57–63.
47a. Dinnyes, A., De Sousa, P., King, T., and Wilmut, I. (2002) Somatic cell nuclear transfer: recent progress and challenges, *Cloning Stem Cells* **4**, 81–90.
48. Wade, P. A. and Kikyo, N. (2002) Chromatin remodeling in nuclear cloning, *Eur. J. Biochem.* **269(9)**, 2284–2287.
49. Serov, O. L., Matveeva, N. M., Serova, I. A., and Borodin, P. M. (2000) Genetic modification of mammalian genome at chromosome level, *An. Acad. Bras. Cienc.* **72(3)**, 389–398.
50. Sathananthan, A. H., Tarin, J. J., Gianaroli, L., et al. (1999) Development of the human dispermic embryo, *Hum. Reprod. Update* **5(5)**, 553–560.
51. Foulk, R. A. (2001) From fertilization to implantation, *Early Pregn.* **5(1)**, 61–62.
52. Palermo, G. D., Colombero, L. T., and Rosenwaks, Z. (1997) The human sperm centrosome is responsible for normal syngamy and early embryonic development, *Rev. Reprod.* **2(1)**, 19–27.
53. Sutovsky, P. and Schatten, G. (2000) Paternal contributions to the mammalian zygote: fertilization after sperm-egg fusion, *Int. Rev. Cytol.* **195**, 1–65.
54. Eichenlaub-Ritter, U. and Peschke, M. (2002) Expression in in-vivo and in-vitro growing and maturing oocytes: focus on regulation of expression at the translational level, *Hum. Reprod. Update* **8(1)**, 21–41.
55. Poulton, J. and Marchington, D. R. (2002) Segregation of mitochondrial DNA (mtDNA) in human oocytes and in animal models of mtDNA disease: clinical implications, *Reproduction* **123(6)**, 751–755.
56. Schultz, R. M., Davis, Jr., W., Stein, P., and Svoboda, P. (1999) Reprogramming of gene expression during preimplantation development, *J. Exp. Zool.* **285(3)**, 276–282.
57. Geiman, T. M. and Robertson, K. D. (2002) Chromatin remodeling, histone modifications, and DNA methylation-how does it all fit together?, *J. Cell Biochem.* **87(2)**, 117–125.

58. Cai, J., Wu, Y., Mirua, T., et al. (2002) Properties of a fetal multipotent neural stem cell (NEP cell), *Dev. Biol.* **251(2),** 221–240.
59. Fajkus, J., Simickova, M., and Malaska, J. (2002) Tiptoeing to chromosome tips: facts, promises and perils of today's human telomere biology, *Phil. Trans. R. Soc. Lond. B. Biol. Sci.* **357(1420),** 545–562.
60. Forsyth, N. R., Wright, W. E., and Shay, J. W. (2002) Telomerase and differentiation in multicellular organisms: turn it off, turn it on, and turn it off again, *Differentiation* **69(4–5),** 188–197.

17
Human Embryonic Stem Cells and the Food and Drug Administration
Assuring the Safety of Novel Cellular Therapies

Donald W. Fink, Jr.

1. INTRODUCTION

As evidenced by the breadth and scope of information detailed in the chapters of this book that describe ongoing scientific advances in the arena of human stem cells, the isolation of these unique cell populations continues to fuel the expectation that an array of promising novel cellular therapeutics will be developed. It is anticipated that biologic therapies either comprised of or derived from human stem cells will be effective in treating a broad spectrum of medical conditions that necessitate replacement, restoration, repair, or regeneration of damaged or diseased tissues and organ systems. In addition to contributing directly to therapeutic efficacy through seeding and repopulation of areas ravaged by trauma and disease, stem cells are being contemplated as an efficient means to deliver functional genes and gene products to target sites where degenerative damage and disease are the result of genetic abnormalities. Only the realization that novel human stem cell biologic therapies will ultimately be subject to the rigors of a carefully conducted clinical investigation appears sufficient to temper the ebullient enthusiasm resulting from the outcomes of basic science research. Efforts to analyze and critically assess issues pertaining to the safe use of human stem cells are of vital importance in assuring that the ultimate goal of generating novel human stem cell therapies is achieved. Tackling this challenge is the responsibility of the Center for Biologics Evaluation and Research (CBER) within the Food and Drug Administration (FDA) *(1)*. The explicit mission of CBER is to ensure the safety, purity, potency, and efficacy of new biologic therapies through a review process that is founded squarely on scientific principles, thus making available to the public innovative new

From: *Human Embryonic Stem Cells*
Edited by: A. Chiu and M. S. Rao © Humana Press Inc., Totowa, NJ

treatments *(2)*. This chapter will describe the approach used by CBER to evaluate investigational new therapies comprised of stem cells—in particular, cellular therapies derived from human embryonic stem (hES) cells. The principal focus will be on issues related to the preparation of the investigational cellular therapy, as a discussion of topics relevant to clinical trials that involve hES cells occurs elsewhere in this book.

2. CBER'S APPROACH TO ASSURING
THE SAFETY OF STEM CELL THERAPIES

Cellular therapies derived from populations of human stem cells represent examples of complex, dynamic biological entities. It is expected that transplanted stem cells will interact intimately with and be influenced by the physiologic milieu of the recipient. Prior to transplantation, cultures of human stem cells may be maintained under conditions that either favor retention of undifferentiated, self-renewing properties or promote acquisition of differentiated properties characteristic of the phenotypes that the undifferentiated cultures of stem cells are destined to assume. Following transplantation of partially differentiated stem cell populations, additional fine-tuning is likely to occur as a consequence of signals received from the physiologic microenvironment of the recipient. It is these intrinsic capabilities of stem cells for self-renewal and phenotypic differentiation that contribute simultaneously to their touted therapeutic potential and the challenge of performing a reliable safety assessment.

Assessing the safety of novel therapeutic candidates derived from cultures of hES cells requires a comprehensive strategy. An initial glimpse into CBER's approach with respect to the identification of issues germane to the safety evaluation of stem cell therapies is provided in the National Institutes of Health report on stem cells *(3)*. The schematic diagram presented in Fig. 1 represents a summary of elements essential to the completion of an adequate safety assessment of cellular therapies derived from hES cells. This paradigm constitutes the basis for the discussion that follows. The information capsules that appear in the perimeter of the diagram are deliberately depicted so as to surround their focus, namely safety assurance. This feature conveys the all-encompassing nature of the review process. Each component of the comprehensive review is interconnected and interrelated. Beginning with the evaluation and selection of eligible egg/sperm donors, all steps in the process of developing a hES-based cellular therapy projected for clinical testing are subject to careful scrutiny. Derivation, expansion, manipulation, and characterization of cell lines established from hES cells are among the items included in this thorough assessment. In conjunction with the analysis of processes, procedures, and analytical tests used to generate an

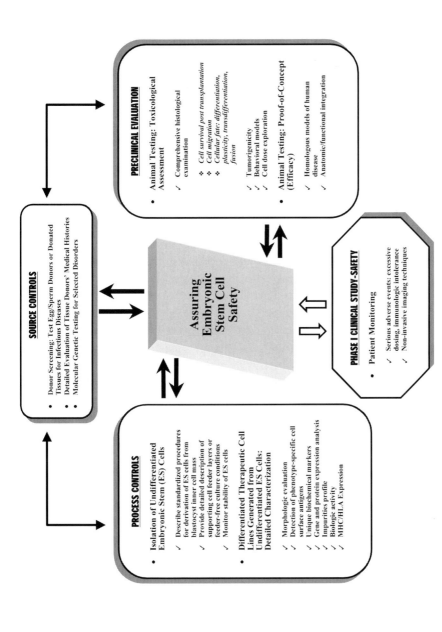

Fig. 1. Schematic representation depicting FDA current thinking regarding elements deemed necessary for assuring the safety of novel cellular therapies derived from human embryonic stem cells.

hES cellular therapy, equal emphasis is placed on preclinical testing conducted in appropriate animal models. Animal testing of "clinical-grade" cellular preparations is performed for the expressed purposes of detecting
potential toxicities and providing evidence indicative of possible efficacious
outcomes that will be explored in human clinical safety studies. A key feature of the review approach is the principle of linkage, or traceability, which
is illustrated by the inward/outward bidirectional arrows shown in Fig. 1. In
those situations when an unanticipated adverse event occurs during the
course of an investigative clinical trial, linkage allows investigators and
study reviewers alike to trace back from the patient through the process of
cellular preparation all the way to the initial acquisition of biologic materials used to generate the founder hES cells. Linkage makes it possible to
establish where significant problems may have occurred that contribute to
the observation of untoward clinical events.

Sharply divided opinion exists among investigators as to the feasibility of
initiating pilot clinical studies that involve human stem cells, especially hES
cells. Some maintain that it is reasonable to expect that within a relatively
short period of time, perhaps as few as 5 yr, transplantation of human stem
cells will be used clinically to replace dead or dying cells within organs such
as the failing or myocardial infarct-damaged heart or that genetically transduced stem cells will be generated for the delivery of therapeutic genes.
Others argue that there is insufficient information pertaining to the basic
biology of human stem cells to justify initiation of clinical trials in the near-
term, citing concerns about inadequate characterization of unique stem cell
populations with respect to identification, demonstration of functional integration, and an incomplete understanding of biological phenomena such as
commitment and plasticity (4). Regardless of one's position with respect to
the issue of readiness, submission of an application to the FDA requesting
permission to conduct a clinical study involving a novel cellular therapy
derived from hES cells will require CBER to perform an evaluation of the
proposal and reach a determination regarding safety based on all pertinent,
available scientific information.

To effect consolidation of a previously fragmented approach to regulatory oversight of human cellular and tissue-based therapies, including stem
cells regardless of their source (embryonic, fetal, or adult), CBER has
developed a framework that is intended to provide a unified paradigm for
evaluation of cell- and tissue-based treatments (5). The intent of this initia-
tive is to provide a tiered structure that is risk-based with respect to the
public health. The specific aims are to (1) prevent the unintended transmission of infectious disease, (2) guard against improper handling or processing that might contaminate or damage cells/tissues intended for therapeutic

use, and (3) ensure demonstration of clinical safety and effectiveness. In addition, CBER has formulated a Tissue Action Plan conceived to promote the timely development of policies, rules, and guidance necessary to assure full implementation of the unified, tiered approach for evaluating the safety of cell- and tissue-based therapies *(6)*.

The evaluation of novel hES-based cellular therapies will occur in a manner that is consistent with the principles outlined in the tiered-approach framework and predicated on previous CBER experience assessing the safety of stem cell preparations. Investigative clinical studies that involve transplantation of hematopoietic stem cells have been underway for a number of years. Reconstitution of the hematopoietic and immune systems is a crucial component of therapeutic procedures designed to treat hematological malignancies such as leukemia and lymphoma. Transplantation of hematopoietic stem cells resident in the bone marrow or isolated from circulating peripheral blood or umbilical cord blood is used to counter the destruction of specialized marrow cells responsible for hematopoiesis caused by the high-intensity chemotherapeutic regimens used to combat hematologic malignancies and various solid tumors. As an extension of experience gained using stem cell transplantation to elicit hematopoietic reconstitution, current ongoing clinical studies explore the utility of coupling hematopoietic stem cell transplantation with lower-intensity chemotherapy conditioning regimens as a means to treat autoimmune diseases such as multiple sclerosis, lupus, and rheumatoid arthritis. Additionally, mesenchymal stem cell transplantation is being investigated as a possible treatment for a variety of indications, including metastatic cancer, high-risk hematologic malignancies, correction of genetic defects, and regeneration of bone. Corneal surface reconstruction achieved through the use of limbal epithelial stem cells represents another example of a clinical investigation that involves stem cells. In their collective, the accumulated experience gained through safety assessments of these investigative clinical studies involving stem cell treatments serves to shape and direct the approach CBER will take in evaluating biological therapies derived from hES cells.

3. DEVELOPING A SAFE EMBRYONIC STEM CELL BASED THERAPY: KEY ELEMENTS

As depicted in Fig. 1, the approach CBER uses currently to assess the safety of a novel cellular therapy derived from hES cells is multidimensional, encompassing acquisition of the starting biological materials, isolation of undifferentiated hES cells, generation of differentiated cell lines from undifferentiated cells, and preclinical animal testing. Reduced to the simplest of terms, demonstrating the safety of a hES-based biological product is all about exerting complete and total control over every facet of the process. Rigorous control over the process used to produce a hES cell therapy ensures predictable consistency from cell preparation to cell preparation thereby maximizing safety assurance from the perspective of the cellular product.

3.1. Assuring Safety Begins at the Source

Demonstrating an acceptable safety profile for a hES cell therapy intended for use in an investigative clinical study is an effort that is initiated at the earliest stages, namely qualification of the donor-derived egg and sperm that will be used to generate embryos from which embryonic stem cells are to be isolated. Acceptance of donor tissues as suitable source material principally depends on screening and testing for infectious diseases *(7)*. In addition, donor medical history evaluations and molecular genetic testing for selected diseases and disorders are being contemplated as assessments of potential importance with respect to the qualification of donor tissue.

3.1.1. Donor Eligibility

One of the FDA's principal public health obligations is to prevent introduction, transmission, and spread of communicable diseases. Therapies derived from hES cells pose a potential risk for transmitting communicable disease from donor to recipient because of their nature as derivatives of the human body. Recognizing that certain diseases are transmissible through the implantation, transplantation, infusion, or transfer of cells and tissues derived from donors infected with disease, it is essential that appropriate measures be taken that guard against the use of cells or tissues obtained from infected donors. To achieve this goal, screening and testing for relevant infectious and communicable diseases is performed to determine the eligibility of donors volunteering to provide eggs and sperm used for in vitro fertilization and subsequent isolation of embryonic stem cells from the inner cell mass of resulting blastocysts. In situations in which freshly derived blastocysts serve as the starting source material for the generation of embryonic stem cells, the donors themselves may be screened for high-risk behaviors and tested serologically for evidence of infectious disease using FDA-licensed test kits. In circumstances when cryopreserved fertilized embryos are to be used and the tissue donors are not available for disease testing, embryonic stem cells themselves may be assessed directly for evidence of infectious disease after expanding the cell populations to numbers that are sufficient to support sampling. When possible, it is encouraged that collection and archiving of additional donor blood samples occurs to provide for retrospective, post-hoc testing. As new information and diagnostic techniques become available, this permits testing for additional disease markers, thus providing an added measure of assurance with respect to the safety of a particular hES cell line.

3.1.2. Donor Medical Histories and Molecular Genetic Testing

Consideration of the intended clinical uses for a hES-cell-derived cellular therapy provides the rationale for positing as reasonable the evaluation of tissue donor medical histories and conducting genetic testing as measures for establishing source material acceptability. Specifically, information obtained from these assessments could serve to establish whether a particular preparation of hES cells is suitable for use in the context of a particular clinical setting. As a hypothetical example, hES cells isolated from a blastocyst for which either the egg or sperm donor had a family medical history tainted by serious cardiovascular disease may not represent the best choice for derivation of myocardial cells intended to repair damaged heart tissue. In a similar fashion, the use of genetic testing could, in theory, detect mutations in genes such as α-synuclein. Evidence from genetic studies has identified two specific mutations in α-synuclein that cause familial Parkinson's disease with autosomal dominant inheritance *(8,9)*. Detection of such genetic anomalies in a population of hES cells could negate their use in generating neuronal progenitor cells for treating a number of neurodegenerative conditions, including Parkinson's disease. Clearly, the total number of genes known to be directly responsible for causing disease or contributing to aberrant physiologic function is relatively small. Undoubtedly, this number will increase, perhaps substantially, as a result of advances in techniques for identifying, isolating, and analyzing genes, coupled with a growing abundance of information certain to become available as one outcome of the human genome sequencing projects. In addition, it is expected that a great deal more will be discovered as to how multiple gene products, each contributing incrementally to an overall outcome, will predispose individuals to developing particular diseases. Ultimately, it will become impossible and thus impractical to screen every donor or preparation of hES cells for the entire panoply of disease-associated genes. In this context, screening would be conducted only for targeted genes that are relevant to a specific clinical indication.

3.2. Controlling Processes Used to Isolate hES Cells and Establish Derivative Cell Lines Enhances Safety

Demonstrating that rigorously controlled standardized practices and procedures are followed during isolation, derivation, and maintenance of cultured hES cells ensures the integrity, uniformity, and reliability of candidate hES-cell-derived therapies generated from hES cells. The initial stage of producing a hES cell therapy may be separated conveniently into two processes: (1) isolation of hES cells from the inner cell mast of the blastocyst and (2) induction of differentiation to push undifferentiated hES cells toward selected phenotypic lineages. Both processes involve culturing of cells for some specified time period in order to obtain numbers that are sufficient for establishing master cell storage banks or testing in a clinical transplantation study. Procedures for culturing hES cells and derivative lines require the use of formulated liquid media supplemented with cytokines, growth factors, and chemical reagents that promote cellular proliferation and govern differentiation. Because hES cells represent a complex, dynamic biological entity, failure to standardize procedures for expansion and maintenance of cultured hES cells could result in unintended alterations in their intrinsic properties. For example, density of the initial cell seeds, the frequency with which culture medium is replenished, the quality of the reagents used to supplement the culture media, and the density cell cultures are permitted to achieve prior to subdividing could all have an impact on the characteristics of cultured hES cells. Inadvertent changes in concentrations of supplemental growth factors and chemical substances added to the media, even simply switching from one supplier to another, may contribute to changes in the properties of cultured hES cells, including their intrinsic growth rate, expression of defining phenotypic markers, and differentiation potential. When available, use of clinical-grade reagents, recombinant human materials, and FDA-licensed products is preferred. It is expected that variances in hES cell properties resulting from the use of nonstandardized culture practices will affect their stereotypic behavior and biological effectiveness once transplanted in the recipient.

3.2.1. Isolation of Undifferentiated hES Cells from the Inner Cell Mass of the Blastocyst

Attention to detail and standardization of procedures used to acquire egg and sperm from volunteer donors, perform in vitro fertilization, and isolate cleavage-stage embryos is crucial to the generation of quality hES cell cultures. In circumstances when hES cells will be derived from cryopreserved embryos, it is essential that critical information be documented and records maintained regarding the reagents and procedures used to perform embryo cryopreservation. Additionally, providing details as to how cryopreserved embryos are stored is important. Perhaps the most significant issue with respect to use of frozen fertilized embryos is the specification of prospective criteria that are used to ascertain the acceptability of a thawed, cryopreserved embryo as source material for generating hES cell cultures.

3.2.1.1. ESTABLISHING HES CELL CULTURES

Historically, undifferentiated hES cells teased from the inner cell mass of cleavage-stage embryos have been propagated on layers of irradiated feeder cells often composed of fibroblasts obtained from fetal mice. Moreover, it is not uncommon that medium used to sustain hES cell cultures is supplemented with bovine serum. It has been suggested that hES cells cultured under either of these conditions would be disqualified for use in creating a clinical-grade cellular preparation. It must be emphasized that this perception is inaccurate.

3.2.1.1.1. Culturing with Bovine Serum. Because of the outbreak of bovine spongiform encephalopathy (BSE) in cattle herds, primarily those raised in the United Kingdom, there is a measure of justifiable apprehension regarding the use of bovine serum as a media supplement when culturing hES cells designated for eventual clinical use. Consumption of beef contaminated with the agent responsible for causing BSE has led to limited emergence of a new variant of Creutzfeldt–Jakob disease in humans. This disease results in the relentless destruction of brain tissue and is invariably fatal. Unquestionably, transplanting neural progenitor cells derived from cultures of hES cells contaminated with BSE infectious agent into the nervous system of a patient with neurological disease represents an act that is both devastating and irresponsible. There are several options in place to safeguard against the inadvertent transmission of BSE to patients through transplantation of contaminated hES-cell-derived therapies. Cultures of hES cells may be maintained in medium containing bovine serum if it is demonstrated that the reagent is produced from cows reared for the entirety of their lives in countries certified to be free of BSE. A more favorable alternative is to replace bovine serum with clinical-grade serum sourced from humans. In this case, testing of the serum donors or final product for infectious human diseases is a crucial element. The optimal solution is development of a serum-free, chemically defined medium, thus obviating risks associated with serum supplementation. Obviously, CBER endorses vigorous efforts in this latter category; however, eliminating serum from culture media is not a requirement.

3.2.1.1.2. Nonhuman Feeder Layers and Xenotransplantation. Transplanting patients with cellular therapies composed of, or derived from, hES cells maintained originally on a murine fetal fibroblast feeder layer constitutes xenotransplantation. As defined in the Public Health Service (PHS) guideline on this topic, xenotransplantation is any procedure that involves the transplantation, implantation, or infusion into human recipients of either

(1) live cells, tissues, or organs from a nonhuman animal source or (2) human body fluids, cells, tissues, or organs that have had ex vivo contact with live nonhuman animal cells, tissues, or body organs *(10)*. The PHS guideline on infectious disease issues in xenotransplantation was jointly developed by five components within the Department of Health and Human Services (DHHS), the Centers for Disease Control and Prevention, the Food and Drug Administration, Health Resources and Services Administration, National Institutes of Health, all parts of the US Public Health Service (PHS), plus the DHHS Office of the Assistant Secretary for Planning and Evaluation. This PHS document serves as a foundation for the Xenotransplantation Action Plan, which details the FDA's comprehensive approach to regulating xenotransplantation therapies *(11)*. The stated purpose of the action plan is to address potential public health safety issues associated with xenotransplantation, including the possible infection of recipients with both recognized and unrecognized infectious agents that could result in subsequent transmission to their close contacts as well as the population at large. Consistent with its stated purpose, the FDA action plan provides guidance for the safe development of xenotransplantation therapies. Statements suggesting that culture methods for hES cells must be devised so as to exclude contact with potentially contaminating mouse or other nonhuman cells in order to permit use of hES cells and their derivatives in an investigative clinical study are misleading *(12)*. Presently, the FDA oversees a number of xenogeneic investigational therapies being evaluated in for safety and efficacy clinical trials. These include hepatocytes, fetal neuronal cells, and pancreatic islet cells, all of porcine origin, for treating a variety of indications such as hepatic failure, Parkinson's disease, Huntington's disease, epilepsy, refractory pain associated with spinal cord injury, and type I diabetes mellitus. To ensure an adequate measure of safety for clinical studies involving xenotransplantation, additional requirements are overlaid on top those associated with the development of investigational human cellular therapies. Admittedly, the challenge of complying with these added measures is not a trivial matter.

For clinical development of cell therapies derived from hES cells that have been cultured on murine fetal fibroblast feeder layers at any point during the derivation process, it is necessary to provide detailed information about sourcing of the nonhuman feeder cells. This includes a description of animal husbandry practices, housing conditions, and routine testing for murine infectious agents. Of particular interest is whether or not feeder cell source animals are obtained from specific pathogen-free colonies, and, if so, what infectious agent testing is performed on the animal colony as a whole and on the individual mice that serve as source animals for generating feeder cells. Additional details about the banking of feeder layer cells, the passage numbers of the cultures used, and testing conducted to verify their purity, phenotype, and viral adventitious agent profile are expected.

Arguably, efforts to develop culture conditions for hES cells that do not rely on the use mouse feeder cells are motivated by the increased level of testing called for in the xenotransplantation guidelines. Also included are requirements for indefinite patient monitoring and lifestyle modification following transplantation of xenotransplantation products. Two reports have been published that represent potentially groundbreaking advances with respect to the challenging technical issue posed by the stringent culture conditions that govern successful expansion and maintenance of undifferentiated hES cells. The first report details development of a feeder-free culture environment in which hES cells are propagated on an extracellular matrix in media preconditioned by cultures of murine embryonic fibroblast cells *(13)*. From CBER's perspective, treating patients with a hES cell therapy comprised of cells cultured under these conditions would not constitute xenotransplantation because intimate, direct contact between the hES cells and nonhuman murine cells is avoided. Importantly, the investigators report that the hES cells maintained in feeder-free culture retained genotypic and phenotypic properties characteristic of undifferentiated embryonic stem cells. It should be noted, however, that the hES cells used to establish the culture conditions described in this study were originally derived on murine embryonic fibroblast feeder layers and then migrated to the feeder-free culture system. Consequently, because of their initial contact with a nonhuman feeder layer, transplantation of these hES cells or their derivatives would still be considered as xenotransplantation.

The contributors of a second report adopted a different strategy because of concerns over the use of conditioned media obtained from cultures of nonhuman feeder cells and poorly characterized extracellular substrates such as Matrigel, which is comprised of a solubilized basement membrane extracted from a mouse sarcoma. These investigators describe the successful prolonged culture of undifferentiated hES cells on feeder layers comprised of human fetal and adult fibroblasts *(15)*. Significantly, this report describes the generation of a nascent hES cell line in conditions completely free of animal-derived feeder layers and preconditioned media. By definition, transplanting a cellular therapy derived from a culture of hES cells created under these conditions would not be considered xenotransplantation. As is the case for donors providing egg and sperm used to establish a hES cell line, it is necessary to screen and test human fibroblast donors or the acquired tissue for a panel of infectious diseases in order to prevent inadvertent disease transmission.

3.2.1.2. MONITORING STABILITY OF hES CELL LINES

It appears likely that during early phases of development, a limited number of banked hES cell lines will serve as source material for generating novel investigational cellular therapies. This is the result of the complexities and technical challenges affiliated with the derivation of hES cell lines from the expanded stage blastocyst inner cell mass. One key to the feasibility of this strategic approach is a periodic, prospectively scheduled assessment of the stability of the founder hES cells. Monitoring stability provides a means for ensuring consistency with respect to the repeated generation of a specific cellular therapy from a single source of well-characterized hES cells. Stability assessments are based on a panel of analytic tests and evaluations comparable to those used to perform the initial characterization of a particular hES cell line. It is expected that stability testing programs for hES cell lines will include, but not be limited to, (1) demonstrating pluripotency, namely the capacity for hES cells to differentiate into tissues representative of all three germ layers, (2) assessing genetic stability involving karyotypic chromosomal analysis, (3) monitoring the rate of proliferation, (4) measuring telomerase activity, and (5) assessing the expression of molecular markers that serve to uniquely identify undifferentiated hES cells.

3.2.2. Detailed Characterization of Differentiated Cell Lines Derived from Undifferentiated hES Cells

Human embryonic stem cells are defined by their capacity for continuous self-renewal and the ability to differentiate into the complete spectrum of tissue types found in the body. At present, most counsel against developing cellular therapies comprised principally of undifferentiated HES cells. In part, this stance is the result of the risk posed by unregulated cellular proliferation, which could occur following transplantation. Buttressing this position is an incomplete understanding of mechanisms that govern the processes of fate specification, differentiation, and plasticity. Consequently, in order to effect an increased measure of control over these biologic complexities, it is envisioned that hES cells will be subject to specified in vitro culture conditions that elicit differentiation of the hES cells along phenotypic lineages best suited for treating a selected disease or medical condition. Detailed characterization of the resulting differentiated cell populations intended for transplantation is critical to their development as cellular therapies for clinical investigation. Testing conducted to characterize unique cell lines derived from undifferentiated hES cells serves to establish their identity. The specific analytical tests used and their results should be of sufficient rigor and accuracy so as to permit unambiguous identification of a cellular preparation. Valid identity testing of undifferentiated, partially differentiated, or fully differentiated stem cell populations poses considerable challenges.

Based on intricate biological properties, including the potential to differentiate along multiple lineages and give rise to a variety of cell types, characterization of tissue-specific stem cell lines derived from undifferentiated hES cells requires a constellation of orthogonal assessments, as is illustrated in Fig. 1. Parameters judged to be useful in establishing identity include (1) cell morphology (visual microscopic inspection of cells to assess their appearance, electron microscopy to detect characteristic structural elements), (2) detection of unique, phenotype-specific cell surface antigens (as is currently the case for use of the CD34 antigen to isolate and identify populations of hematopoietic stem cells), (3) assessment of exclusive biochemical markers, including tissue-specific enzymatic activity (e.g., enzymes involved in the production of neurotransmitters selective for distinct neuronal populations), and (4) evaluation of tissue-specific gene and protein expression patterns. It is expected that the continued development and standardization of DNA microarray (permitting the simultaneous screening for many genes) and proteomics (protein profiling) technologies will significantly enhance the effectiveness and precision of stem cell characterization. Presenting a recommended panel of parameters for consideration when performing detailed characterization of human stem cells should not be construed, by any means, to suggest that the list is either definitive or all encompassing. As information and experience are accumulated, the composition of the list will be re-evaluated and modified as appropriate, consistent with advances being made in the areas of stem cell basic science and clinical research.

It is evident that regardless of the degree of differentiation, cell preparations derived from hES cells will not be homogeneous with respect to the cellular phenotypes expressed. Rigorous and quantitative identification of all cell types within a heterogeneous population of differentiating human stem cells serves as a means to establish the characteristic phenotypic composition or impurities profile for a given cellular preparation. In turn, this permits an evaluation of the extent to which the cell phenotype profile predicts efficacy of a stem cell therapy following transplantation. It is not necessarily the case that homogeneously pure populations comprised of a single cell type will be more efficacious as a cell replacement therapy than cell preparations composed of multiple phenotypes. As an example, it is conceivable that the reason in vitro differentiation of cultured neural stem cells isolated from brain tissue results in formation of all the cell types found within the nervous system (namely neurons, astrocytes, and oligodendrocytes) is the required coincidental expression of each distinct phenotype to ensure maximal survival and optimal functional capability of all the cells in the culture. Elucidating the neighborhood effect of cells with different phenotypes interacting with one another within populations of differentiating stem cells constitutes an area of active basic research.

In addition to serving as an indicator of identity, the phenotypic impurities profile of a stem cell preparation is also useful for detecting changes that have occurred either as a consequence of inadvertent and undocumented errors made during the preparation of the cells or as a function of their intrinsic biologic volatility. Deviations in the impurities profile that fall outside expected biologic variability for a differentiated cell line derived from hES cells using standardized procedures serve as a harbinger that significant, and possibly deleterious, changes may have occurred. Such anomalies could reflect genetic instability that is driven by culture conditions used to promote expansion and trigger differentiation of progenitor cell populations.

In addition to developing procedures/specifications for identity testing of cellular preparations derived from hES cells and establishing an impurities profile, it is essential to demonstrate that investigational human stem cell therapies possess a relevant biologic activity prior to initiating a clinical study. Relying on various types of bioassays to assess biologic activity provides a quantitative measure of the potency of a cellular preparation and ensures that cells destined for transplantation are not inert biologically. Glucose-dependent secretion of insulin from pancreatic-islet-like cells, demonstration of glycogen storage by cells intended for the regeneration of liver tissue, evidence of synchronous contraction in populations of stem-cell-derived cardiomyocytes to be used to repair damaged heart muscle, and depolarization-evoked release of neurotransmitter from putative neuronal cell populations represent examples of bioassays that are based on specific, characteristic biologic activities. Alternatively, it is appropriate to consider the use of surrogate markers that predict the eventual acquisition of an intended biologic activity in circumstances when populations of stem cells are to be transplanted prior to acquiring fully differentiated functionality. In these cases, the expectation is that stem cells will continue to differentiate functionally following transplantation. For instance, quantifying the number of tyrosine hydroxylase-expressing neural progenitors present in a mixed population of cells intended to supply dopaminergic neurons for treating Parkinson's disease represents an acceptable approach for anticipating acquisition of a targeted biologic activity after transplantation.

Histocompatibility constitutes an additional safety issue meriting consideration with respect to the characterization of cell lines established following differentiation of hES cells. Initial reports suggested that hES cells derived from the inner cell mass of blastocyst-stage embryos do not express immune-recognition proteins, raising the possibility that hES cells might be immuno-privileged and, therefore, unrecognized by the immune system of the recipient. This led investigators to the hope that transplanted tissue derived from hES cells would be immunologically silent and thus remain undetected by the recipient immune surveillance system. The low-level expression of major histocompatibility complex (MHC) class I proteins on the surface of hES cells is described in a more recent report, thus providing evidence that it might not be the case that hES cells are immunologically inert *(16,17)*. MHC-I proteins are involved in tissue rejection mediated by cytotoxic T-lymphocytes. In addition to the detection of low levels of MHC-I proteins on the surface of hES cells, moderate increases in their expression are noted in conjunction with in vitro or in vivo differentiation. At this point, it is uncertain whether the level of MHC-I expressed in hES cells is sufficient to elicit a vigorous, let alone even a tepid rejection response, however, it does appear that the characterization of MHC profiles for stem cell therapies derived from hES cells is warranted.

3.3. Enhancing HES Safety Through Preclinical Evaluation: Proof-of-Concept and Toxicity Testing in Animal Models

Preclinical evaluation of cellular therapies derived from hES cells in appropriate animal models is a linchpin of the safety assessment process, as is illustrated in the schematic diagram presented in Fig. 1. Two critical objectives for preclinical testing are appraisal of toxicity and demonstration that the candidate therapy is capable of doing what it is projected to ("proof-of-concept"). With respect to the core biology of hES cells, namely a propensity for continuous self-renewal and broad differentiation potential, it is crucial that experimental animals are inspected carefully for evidence of unregulated growth and genesis of inappropriate cell types following transplantation of cell preparations derived from hES cells.

3.3.1. Preclinical Animal Testing: Toxicological Assessment

Conducting a comprehensive histological examination following transplantation of cellular preparations derived from hES cells in immunosuppressed animals is fundamental to fulfilling requirements for an adequate preclinical toxicological safety study. Important issues pertaining to the fate of stem cells posttransplantation are addressed in these types of preclinical studies. These include assessment of acute and long-term cell survival, the extent and pattern of cell migration, evidence for differentiation and plasticity, indications of hyperplastic growth or tumorigenicity, and anatomical/functional integration. The extent of valid safety information extracted from investigations conducted in animals will be a direct function of the experimental design as well as technical limitations associated with specific tests used to identify and track transplanted populations of stem cells. Continued advancements in noninvasive imaging technologies such as magnetic resonance imaging and positron emission tomography will allow cell fate monitoring to be conducted in real time with reasonable resolution and without having to expend excessively large numbers of animals. Monitoring the posttransplantation disposition of hES cell therapies in preclinical animal models provides considerable challenges; however, the information gleaned will be vitally important to the overall safety assessment for these as yet untested therapies.

The significance of evaluating the survival index of transplanted human stem cell preparations is intuitively obvious; however, there are subtleties that may not be as readily apparent. Unquestionably, cell survival immediately posttransplantation is crucial, but perhaps more important is the persistence of viable cells over time, particularly considering the fact that stem cell therapies are expected to elicit long-term clinical benefit. Assessing cell survival over a prolonged time frame will provide a glimpse of the enduring robustness and durability of transplanted stem cells. To capture this information, longitudinal preclinical studies of sufficient duration must be conducted. Analysis of which specific cellular phenotypes persist and whether or not they continue to contribute to the overall biologic activity of the transplant is equally important.

Beyond an evaluation of posttransplantation survival of stem cell preparations, it is imperative that the propensity for stem cells to migrate and home to specific tissue/organ targets is assessed. Signals emitted by damaged or diseased tissues/organs appear to serve as potent attractants for cognate stem cell populations. On the one hand, migration could be considered a favorable attribute of stem cells, particularly with respect to their potential efficacy as tissue-specific delivery systems of gene products, however, less desirable consequences might occur as an outcome of "misdirected" migration to adjacent or distant tissues. Migration of transplanted human stem cells to a nontarget site and subsequent differentiation into a tissue type that is inappropriate for that specific anatomic location could prove to be problematic.

Precise assessment of the posttransplantation phenotypic fate for cellular therapies offers, perhaps, the single greatest challenge from a technical and conceptual perspective. A recent review on the topic of stem cell plasticity serves to capture the breadth of possibilities that might occur following transplantation of human stem cell preparations. Once situated within the physiologic microenvironments of the recipient, transplanted stem cells will be faced with several cell fate decisions, including self-renewal versus induction of differentiation *(17)*. The prospect of determining with rigor in a preclinical animal model whether final phenotypic expression of transplanted stem cells is the result of activating an intrinsic differentiation program or occurs as a consequence of either transdifferentiation *(18)* or spontaneous cell–cell fusion *(19)* further complicates the picture.

Arguably, tumorigenicity or hyperplastic, unregulated growth represents the single most important safety issue with respect to the toxicological preclinical animal testing of cellular therapies composed of or derived from hES cells. A fundamental feature of true stem cells is their capacity to effect expansion through proliferation. In the case of hES cells, concern is heightened by their pluripotential character and ability to generate teratomas when injected into immunodeficient strains of mice. This property, alone, provides a forceful argument against clinical testing of cellular therapies that are composed exclusively of undifferentiated hES cells. Results from a recent report that describe an animal study involving transplantation of embryonic stem cells in a Parkinson rat model support the position that the proliferative potential of undifferentiated hES cells is prohibitive with respect to their use as a cellular transplant therapy. Although the results demonstrate that transplanted embryonic stem cells develop spontaneously into dopamine-producing neurons capable of supporting behavioral recovery in this model of neurodegenerative disease, they also reveal that teratomalike structures developed in a significant proportion of the animals (25%) after the injection of what were considered low doses of embryonic stem cells *(20)*. A critical question that needs to be addressed is at what point during the differentiation of hES cells do risks attributable to tumorigenic potential and hyperplasia become insignificant, if ever? Identifying a precise stage in the differentiation process at which the risk for tumor formation is minimized will depend on whether or not cellular differentiation proceeds in unidirectional manner, or is reversible. Moreover, it is highly unlikely that inducing differentiation of undifferentiated hES cell cultures into populations composed of more phenotypically mature cells will be a total and completely efficient process. With this in mind, it will be essential to carefully evaluate cell preparations derived from hES cells in order to determine the number of residual undifferentiated hES cells and partially differentiated intermediates that are present. The issue of unregulated growth potential and its relationship to the process of cell differentiation must be thoroughly evaluated prior to initiating clinical trials involving cellular therapies derived from hES cells. This will require completion of careful toxicological studies that are of appropriate duration and involve transplanting undifferentiated or partially differentiated hES cells into immunocompromised animals.

Effective control of the dose-related effects of stem cell preparations observed following transplantation is an issue appropriately addressed during the preclinical animal testing phase. Once delivered, a cellular therapy is not readily retrievable nor is the cell dose level easily amenable to adjustment. Furthermore, it is extremely difficult to predict with accuracy cell dose-dependent events elicited by transplanted stem cell preparations. Each of the above-mentioned items pertaining to cellular fate posttransplantation, namely survival, migration, and plasticity, will have an impact on dose-related effects. The importance of conducting carefully designed preclinical studies using well-characterized cellular preparations to explore issues related to cell dosing is underscored by published results from a clinical trial involving transplantation of minimally characterized fetal neuronal tissue fragments to treat severe Parkinson's disease *(21)*. Profound untoward effects on muscle tone and motor function were observed in approx 15% of the patients receiving transplants. The presumed cause for these events is an "overdose" effect as a result of the survival of too many dopamine-producing neurons, although there are other plausible explanations. Developing strategies for effectively controlling cell-dose levels after stem cell transplantation will improve the safety profile for hES-cell-derived therapies. This includes ongoing efforts to introduce into populations of stem cells genetic switches that are capable of regulating cellular proliferation and that may be turned on or off after the cells have been transplanted.

3.3.2. Preclinical Animal Testing: Proof-of-Concept

Evaluation of studies performed using animal transplant models of human disease is an essential component of the paradigm used for assessing the safety of cellular therapies derived from hES cells. Results from these proof-of-concept investigations serve to substantiate a rationale for conducting a proposed clinical trial with an investigational stem cell therapy. The ideal animal model will recapitulate with fidelity most if not all of the features of human disease. The reality is that animal models, in general, are imperfect because of the fact that most human maladies and medical conditions do not occur spontaneously in animals. Chemical, surgical, and immunologic methods are used to damage neural tissue, induce diabetes, simulate myocardial infarction, stroke and hypertension, or compromise organ function. In selected situations in which focal monogenetic lesions are know to cause disease, the creation of transgenic animals in which the culpable gene is either eliminated or overexpressed results in the generation of animal disease models that are often more faithful in their recapitulation of human disease-specific pathologies.

It is neither essential nor required to demonstrate that a novel investigational stem cell therapy is curative in an animal model of human disease. The principal aim of studies performed in animal disease models is to obtain results that provide a reasonable level of assurance regarding the biologic activity of a novel therapy. For example, transplantation of neuronal cells or progenitors derived from cultures of hES cells should demonstrate restorative activity in animal models that mimic human neurodegenerative diseases or neurological disorders such as Parkinson's disease, Huntington's disease, Alzheimer's disease, amyotrophic lateral sclerosis, spinal cord injury, and stroke. Improved liver function after transplantation of hepatocyte precursors derived from hES cells should be observed in an animal model of hepatic failure, whereas normalization of blood insulin concentrations and amelioration of diabetic disease symptoms should result from the transplantation of pancreatic islet progenitors in a diabetes animal model. In all cases, immunosuppression will be required because of immunologic incompatibility between humans and the species of animal used to generate the disease model. The impact of immunosuppressive regimens on the biologic properties of a cellular transplant will need to be considered. Importantly, demonstration of any corrective effects resulting from transplantation of an investigational cellular therapy in an animal model of disease will provide circumstantial evidence for anatomic/functional integration of transplanted cells within the host physiology that may be corroborated upon histopathological examination.

4. SUMMARY

To summarize, a paradigm is provided that describes FDA's approach to assessing the safety of novel cellular therapies destined to derive from an explosion of ongoing basic research focused on human embryonic stem cells. Demonstrating control over every facet of the development process from acquisition of starting materials through derivation, characterization, and preclinical testing is an overarching tenet of safety evaluation. The topics discussed have been deliberately restricted to issues that pertain to the preparation of cellular therapies derived from human embryonic stem cells, but this is not to suggest that they overshadow, in importance, the role of clinical testing with respect to establishing safety and efficacy. In acknowledgment of the rapidity with which new scientific information is generated resulting in an ever-shifting landscape with regard to what is known about the biology of human embryonic stem cells, it is advisable to consider the framework presented here as a work in progress. Although founded on concrete principles formulated to ensure the safety of biologic therapies, this paradigm is not to be viewed as static or immutable. Rigorous, but flexible is a more apt descriptor. In order to be credible and effective, any approach to evaluating the safety human embryonic-cell-based therapies must be tailored to accommodate their dynamic complexity. In the end, operating within the jurisdiction of its legal authority, the FDA will meet its obligation to ensure the safety of nascent cellular therapies derived from human embryonic stems cells by relying on a review approach that is iterative, collaborative, and based on the best available science.

REFERENCES

1. US Food and Drug Administration, Center for Biologics Evaluation and Research, available at www.cber.fda.gov. Accessed April 4, 2003.
2. Mission statement of the CBER. Available from www.fda. gov/cber/inside/mission.
3. Assessing Human Stem Cell Safety, available from www.nih.gov/news/stemcell/scireport.htm. Accessed July 2001.
4. Holden, C. and Vogel, G. (2002) Plasticity: time for reappraisal?, *Science* **296,** 2126–2129.
5. Proposed Approach to Regulation of Cellular and Tissue-based Products, available from www.fda.gov/cber/gdlns/CELLTISSUE.pdf. Accessed February 28, 1997.
6. Tissue Action Plan, available from www.fda.gov/cber/tissue/tissue.htm. Accessed March 1998, updated: April 4, 2003.
7. USPHS (1999) Suitability determination for donors of human cellular and tissue-based products. *Fed. Reg.* **64(189),** 52,696–52,723.
8. Polymeropoulos, M. H., Lavedan, D., Leroy, E., et al. (1997) Mutation in the α-synuclein gene identified in families with Parkinson's disease, *Science* **276,** 2045–2047.
9. Krüger, R., Kuhn, W., Müller, T., et al. (1998) Ala30Pro mutation in the gene encoding alpha-synuclein in Parkinson's disease, *Nature Genet.* **18,** 106–108.
10. PHS uidline on Infection Disease Issued in Xenotransplantation, available from www.fda.gov/cber/gdlns/xenophs0101.pdf. Accessed January 19, 2001.
11. Xenotransplantation Action Plan: FDA Approach to the Regulation of Xenotransplantation, available from www.fda.gov/xber/xap/xap.htm. Accessed December 17, 2001.
12. Holden, C. (2002) Versatile cells against intractable diseases, *Science* **297,** 500–502.
13. Xu, C., Inokuma, M. S., Denham, J., et al. (2001) Feeder-free growth of undifferentiated human embryonic stem cells, *Nature Biotechnol.* **19,** 971–974.
14. Richards, M., Fong, C.-Y., Chan, W.-K., Wong, P.-C., and Bongso, A. (2002) Human feeders support prolonged undifferentiated growth of human inner cell masses and embryonic stem cells, *Nature Biotechnol.* **20,** 933–936; online at www. nature. com/naturebiotechnology.
15. Vogel, G. (2002) Stem cells not so stealthy after all, *Science* **297,** 175–176.

16. Drukker, M., Katz, G., Urbach, A., et al. (2002) Characterization of the expression of MHC proteins in human embryonic stem cells, *Proc. Natl. Acad. Sci. USA* **99,** 9864–9869.
17. Lemischka, I. (2002) A few thoughts about the plasticity of stem cells, *Exp. Hematol.* **30,** 848–852.
18. Yang, L., Li, S., Hatch, H., et al. (2002) *In vitro* trans-differentiation of adult hepatic stem cells into pancreatic endocrine hormone-producing cells, *Proc. Natl. Acad. Sci. USA* **99,** 8078–8073.
19. Ying, Q-L., Nichols, J., Evans, E. P., and Smith, A. G. (2002) Changing potency by spontaneous fusion, *Nature* **416,** 545–547.
20. Björklund, L. M., Sánchez-Pernaute, R., Chung, S., et al. (2002) Embryonic stem cells develop into functional dopaminergic neurons after transplantation in a Parkinson rat model, *Proc. Natl. Acad. Sci. USA* **99,** 2344–2349.
21. Freed, C. R., Greene, P. E., Breeze, R. E., et al. (2001) Transplantation of embryonic dopamine neurons for severe Parkinson's disease, *N. Engl. J. Med.* **344,** 710–719.

18

Studies of a Human Neuron-Like Cell Line in Stroke and Spinal Cord Injury

Preclinical and Clinical Perspectives

Paul J. Reier, John Q. Trojanowski, Virginia M-Y. Lee, and Margaret J. Velardo

1. INTRODUCTION

Although complex circuit reconstruction in the brain or spinal cord still presents many formidable challenges *(1)*, an extensive body of embryonic central nervous system (CNS) transplantation research has established a compelling proof of principle for cell replacement therapies in animal models of Parkinson's disease (PD) *(2)* and other intractable neurological disorders *(3,4)*. With the advent of stem/progenitor cell biology, focus has shifted from primary fetal CNS tissue to alternative cell sources, and many of these options offer promise for therapeutic neuronal replacement in the brain and spinal cord. The interest this has generated also has contributed to a greater focus on the progression toward clinical trials within the academic and commercial biotechnology sectors, as well as among patient advocacy groups. However, despite exciting recent advances, many intersecting scientific issues and principles of translational bench-to-bedside neurobiology remain to be resolved before the high expectations of stem or progenitor-cell-based therapies can be fully realized.

A primary consideration is that any human trials should be predicated on a comprehensive understanding of the in vivo fates of different neural progenitor cell (NPC) or neuronally restricted progenitor (NRP) cell populations and their anatomical and functional integration within the injured or diseased nervous system *(2,5)*. Likewise, it will be essential to determine how such cells fare in vivo relative to what has been achieved with primary embryonic CNS grafts, which some investigators consider to be the gold standard in both the laboratory and clinical settings. As many studies have

From: *Human Embryonic Stem Cells*
Edited by: A. Chiu and M. S. Rao © Humana Press Inc., Totowa, NJ

shown, fetal CNS tissue grafts can achieve a remarkable level of neuronal differentiation and three-dimensional organotypic organization in the immature and adult brain. That level of development has not yet been obtained with grafts of murine, rat, or human CNS-derived precursors or mouse and human embryonic stem (ES) cells. Rather, in striking contrast with the collective fetal grafting experience in brain and spinal cord, most published reports indicate considerable variability in neuronal yields and phenotypic differentiation with grafts of well-characterized NPC populations from embryonic and adult sources *(6–12)*. Some evidence suggests that the degree and rate of neuronal differentiation exhibited by grafted stem cells is dependent on the region of the CNS into which neural progenitors are grafted. Thus, they appear to show greater adherence to neuronal lineages in the developing brain *(13)* or neurogenic regions of the adult brain *(14,15)* than in either in the mature spinal cord *(8,16)* or other regions of the CNS *(14,17)*. How various forms of CNS pathology and different transplantation parameters will affect the terminal lineage of unmodified NPCs, NRPs, or ES cells remains an open question at this stage of analysis *(6,18–20)*. This underscores a clear need to understand how to lock such cells into a desired neuronal fate prior to transplantation, as well as how to modify intrinsic instructional cues that may be differentially provided by selected graft sites in the injured or diseased CNS *(5,7)*.

Another related issue is that of timely and effective translation of this evolving biology to clinical trials. This will eventually require striking a delicate balance between the urgent need for such therapies and the naturally slower pace of the preclinical scientific process itself. The latter is often dictated by efforts to achieve ideal scientific outcomes. The consensus opinion is that prior to clinical testing, several prerequisites should be satisfied. These include optimal cell handling and transplantation procedures, irrefutable proof of efficacy, absence of toxicity or overgrowth, and thorough delineation of underlying mechanisms in animal models of targeted disorders—including proofs of principle in nonhuman primates *(5,21)*.

Along these lines, the collective fetal CNS cell transplantation experience in neurodegenerative disease and spinal cord injury (SCI) has established an important template for caution and rigor in the development of future stem cell therapies *(2,4)*. In particular, the history of cellular replacement in animal models of PD and subsequent advances to clinical studies has provided valuable lessons. The first clinical trials in this field emerged from equivocal laboratory findings involving adrenal chromaffin cell grafts. These unsubstantiated suggestions of efficacy *(22)* ultimately led to the procedure being done on an inordinate number of patients. It was later observed that such grafts survived poorly and the procedure entailed a high risk-to-benefit ratio leading to unacceptable degrees of mortality and morbidity (e.g., refs. *23–27*).

Subsequent trials with primary human fetal mesencephalic dopaminergic grafts into PD patients have been more carefully orchestrated and solidly based on extensive preclinical data *(2,5)*. However, as judicious as those clinical studies have been in design and execution, controversies continue to arise regarding such issues as surgical approaches, placebo controls, outcome measures, patient sampling, immunosuppression or lack thereof, and duration of functional assessment (e.g., refs. *2, 28,* and *29*). Furthermore, other debates regarding the fidelity of animal models of PD *(30)* illustrate that absolute compliance with the most well-conceived guidelines may not always be possible because of the inherently limited matching of animal models to authentic human disease. However, as the PD experience has demonstrated, initial clinical trials can play an integral role in shaping continuously improving protocols for human applications if appropriately designed in concert with existing animal data. In addition, they can aid in identifying specific biological problems requiring further basic research that otherwise might not be as immediately apparent from the basic science perspective alone. Therefore, although serial progression from the laboratory to the clinic is the most desirable course, the development of cell-based therapies might be more efficient when basic and clinical investigations are conducted in parallel and thus provide feedback on the efficacy or validity of each other.

In view of potential routes for future clinical applications of neural stem/progenitor cells, the following review details basic science and clinical data related to Phase I and Phase II efficacy trials in stroke patients involving transplants of a neuronally committed human cell line derived from a pluripotent embryonal carcinoma. These cells are of particular interest because they exhibit characteristics of differentiated neurons in tissue culture, and after transplantation, many mimic in certain respects properties seen in primary fetal CNS cells. Notably, a significant number of these features are fundamental properties that are likely to be surpassed or further optimized with stem or precursor cell grafts. At the same time, regarding basic-to-clinical translation, the decision to test these cells in stroke patients attracts attention because of the controversy associated with what some consider to be a premature advance into the clinical arena *(5)*. The second part of this chapter discusses recent studies of this neuronally committed cell line in animal models of SCI. The observations made thus far in the spinal cord raise several philosophical and practical considerations, including such issues as the timing of responsible basic science to clinical translations and the complexities of strict adherence to philosophically, ethically, and practically compelling protocols.

2. A HUMAN CLONAL CELL LINE COMMITTED TO NEURONAL DIFFERENTIATION: IN VITRO FEATURES

Many embryonal carcinoma (EC) cell lines have been established from testicular germinal cell tumors *(31,32)*. Unlike their murine counterparts, however, most human EC cells (i.e., pluripotent stem cells of teratocarcinomas) are resistant to manipulations that have been shown to promote cellular differentiation *(33)*. One exception is the Tera-2 cell line that was originally derived from a lung metastasis of a testicular teratocarcinoma *(34)*. Subsequently, a subclone of this cell line was generated by passage through nude mice and these immortalized tumor cells were given the designation NTera-2. This population of human EC-derived cells was further subcloned into another line referred to as NTera2/D1 (NT-2). Under defined tissue culture conditions, NT-2 cells remain as a population of mitotically active, undifferentiated precursors. However, when NT-2 cells are exposed to retinoic acid (RA) and mitotic inhibitors, their rate of proliferation attenuates over the course of several days, and the cells undergo differentiation during a 2- to 5-wk or longer interval *(35–37)*. Differentiation appears to be dependent on the functional integrity of gap junctions *(38)* and entails a reduction in telomere length and telomerase activity *(39)* consistent with what has been observed with human neural precursor cells *(40)*. NT-2 cells subsequently acquire features associated with the expression of a terminally differentiated, postmitotic, neuronallike phenotype *(34,41–43)*. Cells committed to the neuronal lineage via RA induction have been designated as NT2N, hNT, or LBS neurons (Layton Bioscience, Inc.), and these cells can achieve >99% neuronlike purity by sequential replating without the emergence of other cell lineages *(36)*. Once differentiated, NT2N neurons (hereafter referred to as hNT neurons or cells) do not exhibit any proliferative activity in vitro and are capable of long-term survival (e.g., for months in vitro). One report has noted, however, that when mitotic inhibitors are used at lower concentrations and for longer periods (e.g., 8 wk), NT-2 cells will differentiate into both neurons and astroglia *(35)*, but no evidence of oligodendrocyte or non-CNS lineage differentiation by these cells has been reported. Although this suggests a dual neural potential of the NT-2 cell line, all other studies characterizing these cells in vitro and in vivo have used protocols that favor the generation of high-yield hNT neurons. Although still requiring independent confirmation, the fact that one group could induce these cells to express an astrocytic phenotype cautions against referring to the NT-2 line as an exclusively committed neuronal progenitor population, as is often stated in the literature.

Conservatively, hNT neurons should be regarded more as being neuronallike because definitive proof of their capacity to function as bona fide neurons in vivo is still lacking (discussed later in more detail; for related discussion, see ref. *44*). In vitro studies, however, have provided significant evidence in favor of their neuronal character, and for that reason, these cells have served as useful models for such purposes as addressing various issues of neuronal differentiation, as well as neuropathology *(45–48)*. These human cells also have been used as cellular platforms for pharmacological testing and considered potential cellular candidates for ex vivo gene delivery *(49)*.

To summarize, as the RA-treated NT-2 cells begin to differentiate, they form axons and dendrites wherein an extensive complement of neuronal cytoskeletal and other proteins are found *(41–43)*. At the ultrastructural level, hNT neurons are characterized by small cell bodies having large nuclei with minimal surrounding cytoplasm *(37)*. This type of cytological appearance closely resembles (for descriptive purposes only) neuronal profiles seen in the superficial dorsal horn of the spinal cord and cerebellar granule cells. Although immature appearing in that regard, intracellular Lucifer Yellow injections have revealed multipolar cells in vitro with what appear to be single axonal processes and multiple dendrites radiating from the cell soma (*see* Fig. 1 in ref. *50*).

Some gene expression profiling of these cells during RA-induced differentiation in vitro has been reported *(51–54)*, and it was noted that the expression of several Hox and helix–loop–helix domain-containing genes was altered during the process of neuronal specification after RA treatment *(55,56)*. The hNT neurons in tissue culture also express neural cell adhesion molecules, neuronlike calcium and voltage-sensitive Na+ channels, trk receptors, major histocompatibility complex (MHC) class I, and GAP-43 *(43,57–63)*. Subsequently, the differentiated hNT neurons display a wide range of neurotransmitter phenotypes in vitro *(37,50,64–69)*, some of which are coexpressed by individual cells *(37)*. GABAergic, cholinergic, and catecholaminergic immunocytochemical markers have been observed as being the most prominently exhibited neurotransmitter phenotypes of cultured hNT neurons *(37,50)*. Other studies have demonstrated that hNT neurons are capable of ^{3}H GABA uptake and express GAD (glutamic acid decarboxylase) 67, which is involved in GABA synthesis *(37,70,71)*. Molecular and physiological characterizations of terminally differentiated hNT neurons show that they transcribe nine non-*N*-methyl-ᴅ-aspartate (NMDA) glutamate receptor (GluR) genes and express functional NMDA and non-NMDA GluR channels *(60,66,72)*, as well as delta and mu opioid receptor mRNAs and protein *(73)*. Compared with human fetal striatal tissue, these cells also express higher levels of D_1 and D_2 dopamine receptor mRNA *(70)*. In addition, these cells in vitro show intrinsic excitability by virtue of producing action potentials with amplitudes, durations, and after-hyperpolarization characteristics that differ to variable extents based on whether cells are cultured alone or cocultured with astrocytes *(50,66)*. Previous studies of tetrodotoxin-sensitive Na^+ currents raised the possibility that hNT neurons more closely resembled embryonic, rather than adult, neurons *(60,74)*; however, those initial data now appear to have reflected the level of development that was achieved under specific culture conditions. The hNT neurons also acquire glutamate transport activity and are susceptible to glutamate-induced neurotoxicity *(66,75,76)*.

These data are in accordance with other observations showing that functional synapses are formed between hNT neurons in tissue culture. At the ultrastructural level, hNT neurons form classical synaptic relationships with each other and exhibit miniature excitatory and inhibitory postsynaptic currents *(50)*. This is in agreement with the expression of synapsins *(77)* and previously described synaptophysin and chromogranin immunostaining results *(43)*. As Hartley et al. (50) showed, hNT neurons form glutamatergic excitatory and GABAergic inhibitory synapses based on ion selectivity, gating kinetics, and pharmacological sensitivities of postsynaptic receptors. In other studies, Matsuoka et al. *(60)* showed that whole-cell currents could be evoked in hNT neurons using the glutamate receptor agonists NMDA and kainate. They also showed that GABA-evoked currents could be inhibited with bicuculline, a selective $GABA_A$ receptor antagonist. The differentiating hNT neurons thus express at least two forms of major excitatory and inhibitory neurotransmitter receptors found in the CNS. In addition, changes in $GABA_A$ receptor subunit expression and pharmacology occur as NT2 cells differentiate into hNT neurons, which is a pattern similar to that seen during CNS development *(78)*. On the basis of these and other criteria outlined earlier, it is generally agreed that hNT or NT2N neurons in culture share many attributes of primary CNS neurons *(43,66)*, although it cannot be dismissed that the cells also share some attributes of peripheral nervous system (PNS) neurons, particularly those found in autonomic ganglia. Further, the actual level of maturity that the hNT neurons attain in vitro is an important consideration relating to their in vivo behavior following transplantation, as discussed next.

3. IN VIVO CHARACTERISTICS
OF hNT NEURON GRAFTS INTO THE NORMAL CNS

Studies of hNT neurons grafted into the neonatal or mature rodent brain *(49)* and retina *(79)* have demonstrated that these cells are capable of long-term survival (>14 mo in the brain) and acquisition of many of the neuronallike features seen in vitro without reversion to their neoplastic ancestry *(80)*. An impressive biological feature of hNT neurons is their highly reproducible behavior following transplantation. This lends itself to many interesting basic scientific as well as potential therapeutic investigations, that relate to several vital issues based on the derivation of the hNT neurons and their application to CNS repair in general.

3.1. General Maturation of hNT Neuron Grafts

Transplanted hNT neurons can be distinguished from rat brain neurons with human-specific antibodies to several epitopes that are expressed in immature and/or mature neurons. In an initial study, hNT neurons were grafted into the neocortex, subjacent white matter, or hippocampus of immunocompetent Sprague–Dawley rats where they became integrated as neurons *(81)*. Although hNT neurons express an apparently novel immunosuppressive protein *(59)*, these cells are nevertheless immunogenic in vivo *(63)*. Although capable of surviving for as long as 8 wk in the absence of immunosuppression, most hNT grafts into cyclosporine-naive rats are typically rejected within 4 wk posttransplantation *(81)*. However, even in cyclosporine-naive recipients, hNT neurons could be identified with a panel of human-specific markers that demonstrated these cells become exclusively committed to the neuronal phenotype irrespective of the selected graft site. Just like their in vitro counterparts, the hNT neurons in these grafts displayed a characteristic neuronal morphological and molecular polarity in terms of axonal and dendritic profiles, phosphorylated neurofilament protein expression primarily in axons, and microtubule-associated protein 2 (MAP-2) almost exclusively in perikarya and dendrites *(81)*.

From a basic cell biology perspective, it is noteworthy that the preparation of cultured hNT neurons for transplantation entails considerable damage to formed neuritic processes. The fact that these cells re-establish such elaborate molecular and structural polarity following grafting is indicative of their intrinsic growth capacity. Whether or not this represents a regenerative potential, however, is an interesting consideration because the actual in vitro maturational level of these cells at the time of transplantation is more that of an immature neuron under the standard culture conditions employed. As Kleppner et al. *(80)* have extensively discussed, cultured hNT neurons fail to acquire a fully developed or mature neuronal phenotype even after prolonged maintenance in vitro for time intervals of many months *(42)*. For example, unlike terminally differentiated adult human CNS neurons, hNT neurons only express fetal CNS tau and the incompletely phosphorylated isoforms of the heavy neurofilament subunit.

The developmental status of grafted hNT neurons may thus have some impact on the degree or mode of functionality that hNT neurons can exert independently in vivo at various postgraft time intervals without employing other complementary strategies or modified grafting protocols. Specifically, Trojanowski et al. *(81)* reported that even at 3 mo posttransplantation in immunocompetent rats, the grafted hNT neurons did not become fully mature as evidenced by their failure to express the most highly phosphory- lated high-molecular-weight neurofilament (NF-H) proteins. In a subsequent study, hNT neurons were grafted into several different CNS regions of adult and neonatal athymic nude mice to determine whether they would mature into adult neuronlike cells following long-term survival in vivo *(80)*. In this experiment, the grafted hNT neurons showed some evidence of terminal differentiation by 6–8 wk and longer postimplantation depending on the indices of maturity used. Two monoclonal antibodies recognizing the most highly phosphorylated form of NF-H in mature, human CNS neurons stained hNT neurons by approx 8 wk. On the other hand, adult CNS tau isoforms were not detected until 4 mo or later postgrafting.

The fact that cell body staining was observed with NF-H is unusual, but it is known to occur in rare circumstances in vivo *(82)*. However, expression of NF-H in mature neuronal perikarya has been observed in both cells undergoing regeneration *(83)*, as well as in some cells facing imminent axotomy-induced cell death *(84,85)*. The long-term survival of hNT neu- rons favors the notion that persistence of NF-H cell body staining may reflect a maintained growth state of these cells, in which case even the expression of the adult form of NF-H may not be conclusive evidence of their having reached a fully mature state.

The protracted development of hNT cell grafts under certain circum- stances has been confirmed in subsequent studies—most recently in the spi- nal cord (*see* Section 7). Interestingly, these neuronlike cells behave analogously to xenografts of primary fetal human *(86–89)* and other nonrodent/murine CNS tissue, which others have noted to become slowly vascularized *(90)* and progress through prolonged periods of maturation. Whether prior tissue culture protocols have any bearing upon the rate of differentiation *(86)* is unclear.

3.2. Degree of hNT Neuron Survival After Grafting

Frozen, as well as fresh, preparations of hNT neurons have been used for grafting into the neonatal and adult rat brain. High incidences of graft sur- vival (> 90%) have been consistently reported based on the number of recipients showing viable grafts. On the other hand, although initial cell densities (approx 10^4–10^5) and viability (60–80%) data have been reported *(80,81)*, no information is available from those initial studies concerning the actual percentage of cells that had survived the grafting procedure because of technical limitations in methods to assess this reliably. Nonetheless, some more recent data suggest an estimated 12% survival rate in the normal brain using frozen–thawed hNT cell preparations *(65,91)*.

3.3. Potential Pathogenicity of hNT Neurons

Other studies have shown important biological differences between the NT2 and hNT neurons both in vitro and in vivo that also address some important issues related to toxicity and the use of these cells in clinical trials. A primary emphasis has been on the potential for renewed neuroplastic growth; however, even with postgraft survival times of > 1 yr, no evidence of tumorigenesis by hNT neurons has been observed either in the brain *(43,80)* or in the spinal cord (*see* Section 9). In contrast, the undifferentiated NT2 precursor cells (i.e., those that are RA naive or untreated) grafted into the subarachnoid space and superficial neocortex, as well as liver and muscle, established aggressive tumors that became lethal within 70 d postimplantation *(92)*. In striking contrast, however, when placed into the caudoputamen of severe combined immunodeficient (SCID) mice, the non-RA-treated NT2 cells did not form tumors. Instead, they became postmitotic and acquired molecular markers of fully mature neurons *(93)*, as described earlier in this section. It remains to be determined what unique conditions exist in the mouse caudoputamen that promoted differentiation of the NT2 cells. Other studies have likewise demonstrated biological variability between hNT neurons and the NT2 predecessors in terms of selective lysis of NT2-generated tumors in vivo by the avirulent-rendered HSV Type 1 strain 1716, whereas hNT neurons were unaffected *(94)*.

No observations have been reported of unusual or extensive hNT neuron migration following transplantation into the adult CNS. Even in the immature CNS, the migratory behavior of these cells after grafting into the anterior limb of the subventricular zone (SVZa) was in register with that of endogenous SVZa-derived cells and homotopically grafted SVZa cells *(95)*. A similar pattern of migration has been described following transplantation of human neural stem cells (NSCs) or progenitors *(96,97)*. The hNT neurons thus share some migratory attributes exhibited by other human neural progenitors, including an apparent recognition of various tissue guidance and differentiation cues.

Finally, consideration has been given to the possibility of hNT neurons contributing to the initiation of other neuropathological disorders postimplantation in the host CNS. In particular, it has been seen in tissue culture that hNT neurons secrete amyloid-β (Aβ) and other proteins that could induce AD or AD-like lesions in the host brain *(46,48,98)*. Although hNT neurons secrete Aβ (and the NT2 cells do not) in vitro, no neurodegenerativelike disease pathologies were triggered in the rodent brain by hNT neuron grafts that survived up to 46 wk *(41)*. Thus, neurological disease is not an obligatory consequence of grafting and prolonged survival of hNT neurons in the rodent brain.

3.4. Neurotransmitters Exhibited by Grafted hNT Neurons

In contrast to the extensive characterizations of neurotransmitter expression by hNT neurons in vitro *(see* Section 2), only one study thus far has extensively investigated the neurotransmitter phenotypes acquired by these cells following transplantation *(65)*. According to this report, hNT neurons show differential neurotransmitter expression in the normal brain depending on the graft site selected (i.e., striatum, hippocampus, cortex/corpus callosum). Approximately 33% of hNT neurons in striatal grafts were ChAT (choline acetyltransferase)-positive, whereas only 4% of hNT neurons in cortical graft sites expressed ChAT. GAD- or TH (tyrosine hydroxylase)-positive hNT neurons were found at this or the other two graft sites. Postgraft survival time was limited to 30 d in this study. Because the maturation of hNT neurons is very protracted after transplantation *(see* Section 3.2), it is unclear to what degree the findings reflect a pattern of stable differentiation or the full range of neurotransmitter phenotypes that hNT neurons are able to acquire in vitro. This especially applies to the absence of GAD and TH immunostaining. The significance of the apparent graft-site-associated differences in cholinergic phenotype also remains to be resolved and is inconsistent relative to intrinsic cholinergic neuronal populations present in the striatum and cortex.

3.5. Neuritic Outgrowth from Grafted hNT Neurons and Synaptogenesis

One of the more impressive features about hNT neurons is their capacity for achieving long-distance axonal outgrowth *(80,91)*. By 4 wk posttransplantation into the adult rat brain, these cells were already found to have extended axons over several millimeters, and the length and trajectory of these processes appeared to be influenced by the host brain. For example, grafts in the dentate gyrus projected processes along mossy fiber axonal pathways, whereas grafts in deeper layers of neocortex followed subcortical white matter *(49)*. Similarly in the mouse brain, these cells were seen with processes growing along white matter tracts by 3 wk after transplantation *(80)*, and, as in the rat brain *(49)*, the location of the grafts (e.g., septum versus neocortex) appeared to determine the extent to which these processes were elaborated. The time-course for the extension of these processes and the expression of neuronal polypeptides by hNT grafts were similar in neonatal and adult mice. Interestingly, the long-range outgrowth of hNT axons, defined, for example, by immunostaining with the human-specific antibody HO14, which is directed at the highly phosphorylated NF-H subunit, is reminiscent of what has been observed with xenografts of primary human fetal *(99–101)* and human neural precursors *(86)* in the adult rat CNS.

One of the more critical, yet poorly understood features of hNT neuron transplants relates to their ability to form functional neuritic interactions with the host CNS. Although synapses have been described between differentiated hNT neurons grafted into the athymic "nude" mouse brains, no evidence of synaptogenesis between mouse and human cells has been reported *(80,93)*. Synaptophysin staining was observed within hNT neuron grafts as early as 3 d posttransplantation *(80)*, which seems unusual given that hNT neuron-derived process outgrowth is minimal at that time. To date, there is no definitive evidence of host–graft neuronal interactions between hNT and surrounding CNS neurons despite the fact the hNT neurons are capable of establishing unprecedented long projections even in white matter, which is considered to be nonpermissive to axonal elongation (*see* Section 9). Thus, from the perspective of circuit reconstruction, no tangible basis has been established favoring any form of CNS functional repair with hNT neurons involving neuronal replacement and subsequent re-establishment of neural circuits.

4. PRECLINICAL OBSERVATIONS OF hNT NEURONS IN CEREBRAL ISCHEMIA

However, other considerations regarding the CNS neuronallike nature of hNT neurons led to a series of experiments that subsequently formed the basis of a Phase I clinical trial (*see* Section 5). Specifically, three articles were published *(102–104)* in which the potential beneficial effects of grafted hNT neurons were explored in a rodent model of stroke. Different aspects of this paradigm are described in ref. *104*. A point of emphasis is that middle cerebral artery occlusion (MCAo) induces a transient focal cerebral ischemia that largely targets the striatum. This pathology leads to deficits in learning, memory, and motor function that also share some similarities with neuroanatomical and behavioral outcomes associated with excitotoxic damage to the striatum in models of Huntington's disease (HD). Earlier investigations reported improvements in passive avoidance behavior of rats that had received fetal striatal tissue grafts into ischemic regions of the host striatum *(105,106)* (also reviewed in ref. *107*). Other studies of cellular grafts into global ischemia models suggested some improvements in function *(108–112)*. Therefore, it was of interest to explore whether grafts of hNT neurons could have therapeutic value in an animal model of stroke.

The first study *(104)* was carried out to test efficacy in an experimental stroke model using fresh and cryopreserved hNT neuron preparations, both of which were reported to remain viable in the ischemic brain for posttransplantation intervals extending to 6 mo. Eight-week-old rats received ischemic insults, and 1 mo later, baseline passive avoidance and asymmetric motor behavior data were obtained. The animals were then randomly assigned to treatment groups: (1) fresh hNT neurons and cyclosporine-A immunosuppression (CsA), (2) cryopreserved/thawed hNT neurons and CsA, (3) fresh hNT neurons without CsA, (4) primary rat fetal cerebellar tissue grafts without immunosuppression, and (5) control rats receiving either tissue culture medium or CsA alone. The animals were subsequently evaluated on the same behavioral tasks at 1, 2, 3, and 6 mo posttransplantation. These striatally mediated behaviors were significantly impaired in control animals, whereas the hNT cell recipients all were reported to show significant treatment effects that were seen as early as 1 mo after transplantation (see also ref. *102*). Immunosuppression prolonged the apparent therapeutic benefit. Although fresh hNT neurons without concomitant immunosuppression produced functional outcomes at 1 mo postgrafting that were virtually identical to those obtained with the other cell preparations, this beneficial effect diminished over time (*see* Fig. 1A–C in ref. *102*). Interestingly, the nonimmunosuppressed hNT recipients still showed significantly improved behaviors even at 6 mo, although viable grafts were not found in these animals based on staining with a human-specific N-CAM monoclonal antibody (i.e., MOC-1). Also, apart from passive avoidance retention, it is possible that the effect of hNT grafts was not an enduring one because by 2–6 mo, the more dramatic effects seen at 1 mo after transplantation appeared to wane and the behavioral patterns more closely resembled those seen in controls and cerebellar graft recipients. It was concluded that because histological analysis demonstrated surviving hNT neurons in grafted animals showing apparent improvements in ischemia-associated behavioral deficits, these human cells in this xenograft model could exert significant impact on functional repair after stroke. However, the premise of these conclusions is open to debate because anatomical analyses indicated at least five transplant cases in which the grafts were either poorly stained with MOC-1, small in size, or located in the lateral ventricle. In nine animals, the grafts also occupied adjacent brain regions.

Whether such anatomical/survival variability could affect the behavioral data was more directly explored in a subsequent investigation *(102)* in which the viability and survival of hNT neurons were assessed immediately before transplant surgery and then at 3 mo posttransplantation, respectively. Findings were reported showing a correlation between a robust and sustained recovery in asymmetrical motor behavior in rats with induced cerebral ischemia as a function of high cell viability at the time of transplantation and the number of surviving hNT neurons observed at 3 mo posttransplantation. This effect, however, was time dependent in that the most robust effects relative to initial cell viability and postgraft cell survival were seen at 1 mo. The issue of initial "cell dosage" and graft viability and functional efficacy was more extensively investigated in a subsequent study *(113)*. Grafts consisting of $(40–160) \times 10^3$ hNT neurons/recipient resulted in a dose-dependent improvement in passive avoidance and elevated body swing. As before and in subsequent articles noted later in this section, the most robust recovery was seen at 1 mo posttransplantation. From a practical perspective, the results also showed that $(10–20) \times 10^3$ hNT neurons/recipient was the minimum starting dose required for any long-term graft survival, which was 0.5% for that range. At $(80–160) \times 10^3$ hNT neurons/recipient, survival rates of 12% and 15%, respectively, were obtained.

A third set of experiments was designed to further validate the therapeutic efficacy of hNT neurons in the MCAo model by comparing functional improvements achieved with hNT neuron grafts to those obtained with homotopic grafts of fetal rat striatal tissue *(103)*. As in the previous studies *(102,104)*, grafts were made at 1 mo post-MCAo induction of striatal ischemic damage. Dramatic improvements in the retention of passive avoidance behaviors and motor asymmetric performance were promoted by both human and rat tissue. On the latter test, the hNT transplants appeared to have a more robust effect at 1 mo posttransplantation; thereafter, no differences were seen between the effects of hNT neuron and fetal striatal tissue grafts on that task. In the case of the elevated body swing test, the hNT neurons also had a more pronounced effect; however, by 6 mo posttransplantation, no differences were seen. There was a pattern, however, although not mentioned by the authors of that study, that suggested that between 2 and 6 mo posttransplantation, the hNT neuron recipients were beginning to show some progressive worsening of function (see also ref. *103*). In contrast, the trend of the fetal striatal recipients was continuing to show increasingly normalized swing activity. Although the results were interpreted as showing that hNT neurons are as effective, if not more so, than fetal tissue (thus providing a significant alternative source of cells), these subtle trends could be early reflections of different mechanisms whereby fetal and hNT neurons could be having any effect on these behaviors. In light of these interpretations, it is important to note that, at the very least, fetal rat CNS tissue develops at a much faster rate than even fetal human CNS or hNT neurons. There is thus a valid reason to question how much significance can be attached to a comparison of this nature.

5. CLINICAL TRANSLATION
OF hNT GRAFTS INTO STROKE PATIENTS

5.1. Prelude

Collectively, the above studies were considered to have established a pre-clinical basis for advancing the application of hNT neurons to a Phase I clinical trial. As reported, a measure functional efficacy was demonstrated with grafts of this extensively characterized human cell line in an animal model of cerebral ischemia that closely parallels the neuropathological and functional hallmarks of this disorder in humans. In addition, none of the data suggested that these cells, even at high initial cell numbers, had any adverse effects on the behaviors tested. In conjunction with previous investigations, these studies did not reveal any indications of neoplastic growth. The data from those experiments further suggested a transplant dose-dependent index of therapeutic benefit and long-term graft survival. From an additional technical perspective, it was shown that effective transplants could be obtained with freshly thawed, cryopreserved cells. Finally, as these studies were underway, corporate efforts (Layton Bioscience, Inc., Sunnyvale, CA) were being made to generate a clinical-grade preparation of hNT neurons that was subsequently designated as LBS-Neurons.

Having established in principle a method of potential therapeutic use for improving memory, learning, and motor behavior after stroke, the question arises whether the decision to move these cells to the clinic was timely and justified on basic scientific and clinical grounds. There is little question that with advances that have evolved in the treatment of stroke, along with various demographic considerations, there is now an even greater need to develop strategies whereby stroke patients can achieve a significantly improved quality of life. So, there is clearly a major health care imperative. Despite this, some cogent challenges have been raised to this clinical study. These are important to consider because they could be applicable to future clinical translations of stem cell technology.

One of the more compelling issues that has been raised is that the LBS-Neuron stroke trial was not based on a sufficient understanding of the underlying biological mechanisms of the improvements described (5). The data in the first report (104) suggested that any effect produced by the hNT neurons was largely, if not exclusively, trophic in nature. Essentially, no basis for cellular replacement could be inferred from that study. First, although the authors considered the possibility of hNT neurons being induced to become striatal-like neurons, no evidence was obtained either in that or subsequent studies to support that hypothesis. Second, given the very slow maturation of hNT neurons, the fact that the most robust effect was routinely seen within the first month after grafting is inconsistent with any mechanism requiring fully differentiated neurons. A third, closely related issue is that neither early nor prolonged effects of these hNT neuron grafts could be attributed to any form of synaptic circuitry reconstruction. Interestingly, part of the rationale for the original animal study was the similarity between HD and stroke in terms of behavioral outcomes and striatal involvement. The fact that fetal striatal grafts were previously shown to modulate HD outcomes also added impetus to the hNT study in stroke models. However, most of the research on transplantation and HD suggests that functional improvements require some form of connectivity between host and graft tissues *(114,115)*. Notably, there is still no evidence of host–graft functional connectivity in these or other hNT graft paradigms studied to date. Thus, if the functional improvements noted were the result of a trophic effect, one then has to first ask whether this inference alone constitutes "…enough known about the mechanisms of the disease and the proposed intervention to suggest that the risks to the subject are reasonable in light of a possible benefit to the subjects or society from the knowledge acquired" *(21)*. From that perspective, it might be argued that initial efforts should have been made to test the trophic hypothesis in animal models and to identify which trophic agent(s) are involved. This information would have built a stronger rationale for the clinical studies and also provided a more tangible basic science framework for interpreting the clinical trial's outcomes.

Some other criticisms that have been raised about the hNT cell line or the ischemia grafting preclinical studies are less justified. Contrary to the challenges raised *(116)*, there are many reports, as already noted, showing that the hNT neurons are capable of integrating with the host CNS and then extending processes for unprecedented distances. The point also was made *(116)* that purification of hNT cell cultures may limit their utility in the infarcted microenvironment because oligodendrocytes will be required to myelinate new axonal processes and astrocytes could be vital to the long–term survival of these cells. However, long-term survival of hNT grafts has been demonstrated both in the ischemically injured brain as well as other regions of the normal or damaged CNS. Furthermore, studies discussed later *(117)* have provided evidence suggesting that host astrocytes and perhaps other glia migrate into these transplants over time. All these important concerns notwithstanding, the hNT neurons were the first laboratory grown and manufactured cells approved by the FDA for use in clinical trials for stroke therapy. This pioneering trial, which began in 1998, stimulated efforts to formulate more standardized and rationally reproducible criteria for use by the FDA in evaluating other candidate cell therapies for assessment in clinical trials to treat neurological disease.

5.2. Clinical Trials and Reported Findings

A Phase I safety-feasibility, dose-randomized, open-label trial was performed *(118)* on 12 patients with basal ganglia stroke and fixed motor deficits. Patient histories entailed subjects aged 44–75 yr with infarcts dating 6 mo to 6 yr prior to transplant surgery. A dose of 2×10^6 LBS neurons per pass was introduced stereotactically into the region of infarction either as single or multiple passes. The total number of cells delivered on each pass represented approximately one-twentieth of the effective dose defined in the previous focal ischemia rodent studies. All the patients were immunosuppressed with CsA beginning 1 wk prior to transplantation and then continued for 8 wk except for one subject who was taken off the drug because of adverse reactions and declining renal function several days after surgery. Observer-blinded (relative to cell dose) neurological evaluations were then performed at routine intervals from 1 to 24 wk posttransplantation, after which the observer blind was broken and further evaluations were then carried out to 52 wk postsurgery. Neurological evaluations included the NIH Stroke Scale (NIHSS) and European Stroke Scale (ESS) performed at baseline and up to 24 wk after the operation. Positron-emission tomographic (PET) images were obtained at baseline and 24 and 52 wk (data not discussed in the initial report); magnetic resonance imaging (MRI) was done at baseline and then at wk 4 and 24.

In terms of feasibility and safety, implantation was successfully performed in all 12 patients with no procedural or outcome complications that could be directly or indirectly associated with the transplantation procedure itself or presence of LBS-Neurons in the recipient brains. No abnormalities were observed on MRI examination. ESS scores collected at 24 wk posttransplantation revealed that 6 patients had 3- to 10-point improvements, 3 were unchanged, and 3 showed deterioration ranging from −1 to −3 relative to their baseline scores. The NIHSS data were essentially similar. The mean ESS score from baseline was 2.9 points ($p = 0.046$), with most of this difference being attributed to the motor component of the ESS. Because the purpose of this study was to demonstrate safety and feasibility, the neurological outcomes were conservatively treated by the authors. For reasons outlined in their discussion, no inferences about efficacy (or lack thereof) should be made based on the limited power of this study and other design considerations, including questions about the use of these scales for assessing CNS repair.

In terms of safety, it was noted that some patients had worse ESS and NIHSS scores than before surgery, yet almost an equivalent number showed apparent improvements. Independent reviews *(119,120)* of this initial report raised several considerations in that regard. One is that the number of LBS-Neurons implanted into these patients may have been too low for detecting a dose-related toxicity. A corollary to this is that for potential beneficial effects to be seen, a much higher cell dose will be required. These and other issues are now being addressed in an ongoing Phase II trial *(77)*. As with previous animal studies, these early reports did not identify any malignant transformation of the grafted LBS-Neurons, but it has been argued the 1-yr follow-up at the time this first report was published might be inadequate for any long-range predictions about this to be made. In reply to that comment *(119)*, it was noted that no abnormalities were seen even at 2 yr posttransplantation.

Central to any discussion of safety, feasibility, and, most certainly, efficacy is the question of how definitively can cell survival be defined in a human recipient of a cellular-based therapy? In the absence of such information, the only true measure of safety relates to the actual surgical procedure and the introduction of cells that may or may not have survived. In that event, the most that can be concluded is that cell delivery and the potential death of donor cells have no obvious deleterious effects. However, in terms of true biological vs procedual, not to mention efficacy, very little information of a substantive nature can be obtained.

In the initial LBS-Neuron stroke study, MRI did not show any remarkable changes in the targeted infarction regions that would be suggestive of a significant mass of graft tissue. Initial PET scans of 11 subjects at 6 mo after surgery showed a $\geq 15\%$ increase in the relative uptake of ^{18}F-fluoro-deoxyglucose (FDG) at the graft sites and in some cases within surrounding ipsilateral brain regions in six of the patients examined. As the authors and others have appropriately noted *(119)*, this may reflect nothing more than an inflammatory response either in response to the grafting procedure or the death of donor cells. An important consideration in this trial is that CsA immunosuppression was only carried out for 8 wk. No specific rationale was originally discussed for the limited duration of immunosuppression, although it has been more recently acknowledged that decision was based on experience with primary fetal CNS tissue grafts in patients with PD. While those studies suggested that long-term immunosuppression may not be obligatory with human allografts *(121–124)*, this issue is still unresolved.

In a more detailed description of the PET imaging results of this Phase I trial *(125)*, several interesting issues were raised. First, at 6 mo FDG brain uptake in the stroke area targeted by the LBS-Neuron grafts was greater in 7 of the 11 patients studied than the 10% threshold of normal FDG PET variation. However, by 12 mo, this effect was only sustained in three patients. At 6 mo, increased activity also was observed in surrounding tissue in 2 of the

11 patients, but 5 patients showed elevated activity in adjacent regions. Pooled 6- and 12-mo FDG uptake data correlated significantly with performance on the ESS motor subscale. The issue of inflammation was further considered, and the position taken that a persistence of an inflammatory process at 6 mo seemed an unlikely explanation for the increased activity in the region of the cell grafts. Debris removal and other related cellular events associated with inflammation, however, could persist for many months in the CNS. In that event, the grafts or some remnants of those transplants could be exerting a trophiclike or cytokine-mediated effect by acting as an immune cell activator or attractant. This question remains unsettled, and the need for placebo-controlled studies is readily apparent. Interestingly, the initial increase in FDG brain uptake at 6 mo followed by a decline by 1 yr mirrors the early response seen with hNT grafts in the rat focal ischemia model discussed earlier. Although the authors of this study alluded to engraftment of LBS-Neurons, the data may actually reflect more closely on preclinical findings than considered. In that context, as noted earlier, more intensive studies of the trophic impact of hNT grafts in animal models could have facilitated a better interpretation of these initial clinical data. Perhaps an even more pressing issue that has surfaced from this clinical trial is that very limited information exists about the long-term metabolic responses in the brain following stroke *(125)*. Thus, placebo-controlled follow-up studies represent a logical next step, as well as carefully orchestrated longitudinal baseline studies.

Some recent insights about LBS-Neuron survival have been obtained from a postmortem evaluation of brain tissue obtained from one of the graft recipients in the Phase I trial *(126)*. This individual was a 71-yr-old man who exhibited fixed motor deficits as a result of a lacunar stroke of the right putamen. He received stereotaxic implantation of 2×10^6 hNT neurons adjacent to the infarct at 34 mo postinfarction. This patient died of a presumptive myocardial infarction 27 mo postimplantation, but unlike 6 of the 12 patients in the trial who showed motor improvements, this patient demonstrated no motor recovery. Fluorescent *in situ* hybridization (FISH) was used to identify the grafted LBS-Neurons because hNT neurons have been shown to be polyploid for chromosome 21, but not for chromosome 13. No tumor was identified anywhere in the brain, and histological examination demonstrated extensive gliosis with numerous glial fibrillary acidic protein (GFAP) immunoreactive astrocytes and some hemosiderin-laden macrophages surrounding the infarct. How these features related to PET imaging results on this individual was not mentioned and no insights could thus be garnered regarding the question of inflammation and increased FDG uptake noted earlier. A cocktail of FISH probes for multiple chromosomes demonstrated polyploidy in chromosome 21, but not chromosome 13, at the hNT neuron implantation site (*see* Fig. 1) where NF protein immunoreactive cells were present. Thus, the presence of cells with a neuronal phenotype and distinctive chromosomal features of hNT neurons is consistent with survival of a population of grafted hNT neurons in the brain of this patient 27 mo postimplantation. Interpretation of the FISH data in this study hinges on the fact that this patient did not have AD wherein mature neurons duplicate chromosomes prior to cell death and fibroblasts can be trisomic for chromosome 21. That, combined with an absence of FISH staining in a control brain, lends further credence to the conclusion that grafted hNT neurons survived in the brain of this patient for 27 mo after implantation without reverting to neoplastic growth. No counts of the number of putative LBS-Neurons could be obtained, however, and so no estimate of the percent cell survival was made partly because of quantitative stereological limitations.

6. hNT NEURON TRANSPLANTATION AND OTHER NEURODEGENERATIVE DISEASES

Interest in the hNT cell line has extended to other neurological disorders such as Parkinson's disease, Huntington's disease (HD), and amyotrophic lateral sclerosis (ALS). In the case of animal models of PD, the number of tyrosine hydroxylase-positive neurons in these animals was too low to produce significant functional recovery, and further improvements in cell production are required to enhance TH expression (*127*). Using rats with unilateral striatal lesions as a rodent model of HD, comparisons were made of the functional effects exerted by fetal rat striatal grafts relative to hNT transplants (*70*). Both grafts improved methamphetamine-induced circling behavior and partial recovery in skilled use of the paw contralateral to the side of the lesion at 10–12 wk posttransplantation. Whereas complementary in vitro analyses showed that differentiated hNT neurons acquired several neurotransmitter phenotypes of striatal neurons, no evidence was presented of hNT neurons having similar neurotransmitter properties following transplantation. In fact, the hNT neurons did not exhibit any immunoreactivity for the striatal cell marker DARPP-32. Because hNT neurons, like fetal striatal tissue, required 8–10 wk before showing a behavioral effect, it was proposed that the functional data thus reflected the establishment of some form of neuronal connectivity. As noted elsewhere in this chapter, the issue of hNT synaptic integration with surrounding rat tissue must still be regarded with extreme caution. It would appear from published data that significant

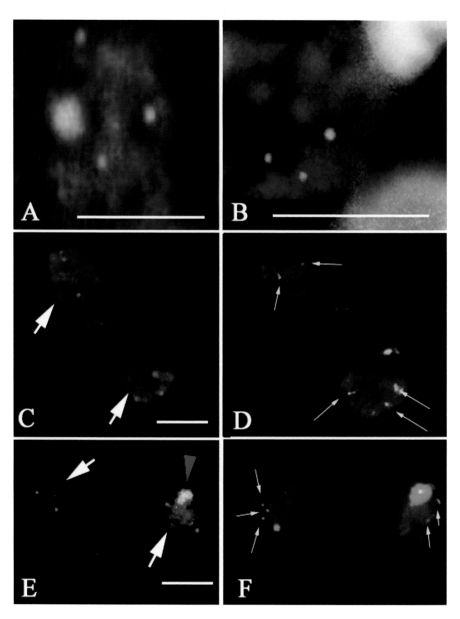

Fig. 1. FISH signals within neuronal nuclei in hNT graft site. **(A)** A section of a rat stroke model brain with immunohistochemically identified hNT neurons at the graft site was probed by FISH to show neurons that are polyploid for chromosome 21 (orange fluorescence; scale bar = 40 μm). **(B)** Using the same chromosome 21 probe as in **(A)**, FISH shows polyploid nuclei in neurons at the graft site of the current patient in sections adjacent to those in **(E)** and **(F)** (scale bar = 40 μm). **(C–F)** FISH staining of sections adjacent to those in **(E)** and **(F)** shows more than

basic research and product development are still required before a compelling case for a clinical trial in either PD or HD can be forthcoming.

Some recent studies have begun to focus on hNT cell transplants in mouse models of ALS. This disorder entails a progressive loss of motoneurons in the spinal cord, brainstem, and cortex, resulting in paralysis and ultimately death. Although it is difficult to envision how neurotransplantation could be applied to this disease based on its widespread nature, there has been reported indication that intraspinal transplantation of hNT neurons can be of some therapeutic benefit *(128,129)*. In a mouse model of familial ALS, long-term (10–11 wk) hNT neuron transplants made into the lumbar spinal cord 6 wk before the overt onset of behavioral symptoms led to improvements on a battery of motor functions and a 3-wk delay in the onset of motor function loss *(128)*. hNT cell recipients also showed prolonged lifespans compared with media-only controls (128 vs 106 d). Some effect of hNT neurons on the rate of motoneuron loss has also been reported when these cells have been transplanted at more advanced stages of declining motor function *(130)*. The mechanism for these findings is unresolved. The notion of motoneuron replacement has been raised *(128)*; however, no supporting evidence has yet been obtained for this. Functional motoneuron replacement would require cellular differentiation leading to a cholinergic phenotype, axonal elongation out to the periphery and targeting of appropriate muscles, intrasegmental connectivity, and appropriate descending inputs leading to an orchestrated regulation of motoneuron excitability, among many other considerations. The most parsimonious explanation, as noted by Garbuzova-Davis et al. *(130)*, is the elaboration of neurotrophic factors *(131,132)*, much like what has already been discussed in relation to other models.

7. SPINAL CORD INJURY

hNT neuron transplantation also has been extended to models of neurotrauma. Two reports noted survival of these cells in the traumatically injured rat brain, but the postgraft duration of those experiments was limited to only 2–4 wk *(133,134)*. A more recent report *(135)* has shown survival of these

---

*(**Fig. 1.** continued)* two signals for chromosome 21 but two signals for chromosome 13 in neurons within the implantation site (scale bars = 40 μm). The panels on the left **(C,E)** show neurons viewed with a rhodamine filter to visualize FISH signals corresponding to chromosome 13 in red, whereas the panels on the right **(D,F)** show neurons viewed with an FITC filter to demonstrate FISH signals corresponding to chromosome 21 in green. Note that in **(D)** and **(F)**, one of two neuronal nuclei (arrow) demonstrates three FISH signals.

cells up to 12 wk post-transplantation following a 1 mo post-lesion delay. No functional outcomes were noted.

Extending the comprehensive early studies of the basic transplantation neurobiology of hNT neurons or their Ntera-2 precursors in rostral regions of the CNS *(80,81,93)*, subsequent studies have provided detailed descriptions of the survival and integration of hNT neurons in the normal *(117)* and injured spinal cord *(136–139)*. Other interesting findings have begun to emerge regarding the functional effects of hNT neuron grafts on hind-limb spasticity in a rat model of spinal ischemia (Marsala et al., unpublished observations). In recent years, there has been increasing emphasis on clinical trials associated with SCI and cell-based interventions, some of which have been proposed (www.alexionpharmaceuticals.com/products/sci), already initiated (www.proneuron.com/ClinicalStudies/index, www.diacrin.com/newpage1), or completed *(4,140,141)*. The following overview identifies issues in SCI that complement several points already raised relative to translational advances with hNT neurons in animal models of stroke to clinical trials. In addition, information already acquired from laboratory studies of intraspinal hNT transplantation pose interesting considerations relative to any potential advance of these or other cell-based preparations to SCI patients.

8. HUMAN NT2N NEURON GRAFTS IN THE UNINJURED SPINAL CORD OF IMMUNOCOMPROMISED NUDE MICE

In the first transplant studies of NT2N cells in the spinal cord *(117;* reviewed in ref. *142)*, hNT neuron grafts were established by injecting cells with glass micropipets into the otherwise intact spinal cords of homozygous athymic mice. Final graft volumes (ranging over a few microliters) after postimplantation periods of up to 15 mo did not correlate with survival times even after postimplantation periods of up to 15 mo. As consistently seen in other studies, there was no evidence of graft rejection or tumorigenicity, and by 7 mo postimplantation, the grafted NT2N cells (i.e., hNT neurons) expressed highly phosphorylated NF-H and adult tau isoforms typical of fully mature human CNS neurons *(143)*. Neither the graft site nor recipient age affected either the phenotype or morphology of the implanted NT2N neurons. Further, NT2N neurons did not migrate from the original graft sites.

These intraspinal hNT grafts also exhibited exuberant neuritic outgrowth patterns similar to what has been observed with intracerebral transplants of these cells *(80)*. Numerous axonlike processes extended from these grafts into host gray and white matter. The trajectory these processes established appeared to be defined by their location. Thus, hNT-derived fibers grew in parallel with host white matter axons, whereas in gray matter, they were more randomly oriented. In some cases, fibers in gray matter coursed in register with commissural axons of the host. Most of the processes of the hNT neurons (approx 82%) extended for distances of >1 cm in the host spinal cord white matter, with some even exceeding 2 cm *(117)*. This also appeared to be independent of the age of the host when grafts were made. The rate of maximal outgrowth was estimated to be approx 1.4 mm/wk for the first 6 wk, followed by a rate of at least 1 mm/mo thereafter. The extent of outgrowth was similar in regions rostral and caudal to the graft site, as well as in both the dorsal and ventral white matter tracts. Of particular interest was the fact that some NT2N axons were ensheathed by oligodendrocytes.

Thus, human hNT neurons implanted into the spinal cords of nude mice demonstrated several fundamental and intriguing features: (1) they integrate and survive for >15 mo in gray and white matter sites with the phenotypic features of mature human CNS neurons; (2) these cells exhibit a neuronal phenotype that is similar at different graft sites regardless of the host age at implantation; (3) hNT neurons extend dendritelike and axonlike processes, and axonal trajectories are differentially influenced and molded by gray and white matter; (4) some hNT-derived axons were ensheathed by oligodendrocytes; these are presumably of host origin because no evidence exists for these cells differentiating along an oligodendroglial lineage; (5) finally, the appearance of some NT2N processes expressing synaptophysin suggests that they may have formed synaptic contacts with host cells.

9. GRAFTS OF hNT NEURONS TO THE INJURED SPINAL CORD

These observations of hNT neurons in the mouse spinal cord were independently extended to a model of acute and chronic cervical spinal cord contusion injury by Velardo et al. *(136–139)* (summarized in ref. *144*). hNT neuron grafts were thus made from 1 h up to 180 d postcontusion injury in adult rats and graft maturation and host–graft integration were subsequently analyzed from 30 d to 14 mo posttransplantation. The results to be summarized here largely derive from studies in which hNT neurons were placed into chronic, 1- to 6-mo-old C_4–C_5 contusions. The animals were maintained on daily administration of CsA (10 mg/kg, s.q.). This challenging and clinically relevant cervical spinal lesion condition allows exploration of basic questions of graft survival and maturation, host–graft integration, cell infiltration of the grafts, and immune responses to these xenografts. In addition to other measures, proximity of the grafts to the phrenic motoneuron pool and brainstem allows for other indices of toxicity. We previously described instances in which fetal neocortical transplants were placed into similar lesions of the cervical spinal cord that resulted in an intentionally designed natural donor tissue overgrowth that became life-threatening *(145)*. Thus, this model offers a very dramatic test environment for potential tissue toxicity.

As with grafts of fetal rat and cat spinal cord tissue into contusion lesions *(146–149)*, the hNT neuron transplants developed into cohesive masses that often filled the cystic cavities that form after these injuries (*see* Fig. 2A). These grafts also showed a high degree of host–graft integration with both white (*see* Fig. 2B) and gray matter that closely paralleled the degree of tissue integration that has been observed with primary fetal CNS transplants to the spinal cord *(150)*. Detailed histological analyses, coupled with nuclear magnetic resonance (NMR) microimaging showed that even 1 yr after grafting, there was no adverse donor tissue overgrowth or cellular migration. No initial neurophysiological or behavioral indications of respiratory compromise were noted in any of the animals, and initial behavioral observations have not revealed general forelimb deficits beyond what would be expected from these lesions.

Similar to what has been observed with human fetal CNS transplants into the rodent *(151,152)*, hNT neurons in grafts that filled or came close to filling areas of cystic cavitation differentiated very slowly, such that at postgraft intervals of < 120 d, the cells were typically still very immature in appearance. Thereafter, more widespread cell differentiation was apparent throughout the grafts (*see* Fig. 2C). Initial immunocytochemical characterization of neurotransmitter phenotypes represented by mature hNT neurons have thus far revealed TH-positive and GABAergic profiles and other evidence suggests that the majority of the hNT neurons may be acquiring a spinal inhibitory interneuronallike phenotype. In the mouse spinal cord, Hartley et al. *(117)* observed that GAD immunoreactivity increasing over time after transplantation but did not find any of the grafted NT2N neurons expressing tyrosine hydroxylase (TH). Some faint ChAT-like neuronal immunoreactivity also was observed in some grafts, suggesting that hNT neurons may have the potential to acquire a cholinergic phenotype. An abundant growth of host 5HT and CGRP fibers extended through the grafts, suggesting possible trophic/tropic effects for some fiber systems. No evidence of corticospinal ingrowth, however, has been obtained to date. Whether this applies to other long-tract descending or ascending systems is currently under investigation.

The robust axonal outgrowth from hNT neuron grafts seen in the normal mouse spinal cord *(117)* also was an exceptionally dramatic feature of these transplants into the chronically injured spinal cord. Thus, the complex molecular and cellular pathology of that injury does not interfere with the intrinsic growth potential of these neuronallike cells (for illustrations, *see*

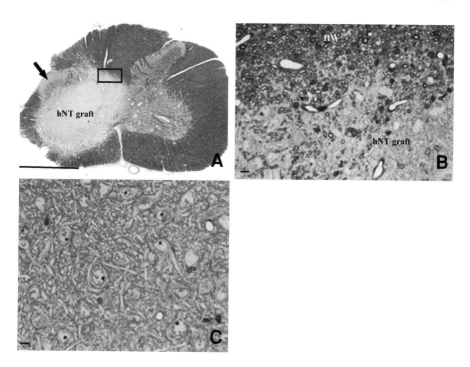

Fig. 2. Transplantation of hNT neurons into a lateralized, chronic (i.e., 6 mo) contusion injury of the rat spinal cord. The 2-μm plastic section shown at low (**A**) and high (**B,C**) magnification was obtained at a level rostral to the lesion epicenter. The hNT neuron graft in (**A**) thus occupies a region that would otherwise be a zone of cavitation among ventromedial, ventral, and ventro- to dorsolateral white matter and preserved superficial dorsal horn (arrow) of the host spinal cord. As seen with fetal CNS transplants (refer to text), these hNT neuron grafts are capable of forming a three-dimensional cohesive transplant that extensively fills regions of tissue degeneration. At higher magnification [**B**, boxed area shown in **A**], the hNT neuron graft can be seen intimately blending with host dorsal column white matter (HW) without any overt intervening partition of fibroglial scarring. By 120 d, hNT neurons are mature process-bearing cells with nuclear profiles that are distinct from any seen in the host spinal cord (**C**). With the exception of some Schwann and oligodendroglial myelinated axons (not shown), the neuropil of the graft is a complex meshwork of interlacing, unmyelinated neuritic processes. As a rule, these grafts also exhibit minimal vascularity. Most neurons are characterized by small cell bodies with distinguishing convoluted nuclei and distinct nucleolar profiles. [Bar = 1 mm (**A**); bar = 10 μm (**B,C**).]

Figs. 9B and 10 in ref. *144*). HO-14-immunostained fibers emerged from the grafts and then extended for long distances (at least 10–14 mm) in both white and gray matter. The distribution of these processes appeared to favor white matter more than gray, and they seemed to have an affinity for regions of white matter degeneration.

Some processes extended into gray matter at varying distances from the graft site. To what extent they actually terminated in gray matter is uncertain. In regions such as the dorsal horn, the hNT axons appeared to favor primary afferent fiber tracking, and Hartley et al. *(117)* also noted hNT-derived fibers in spinal nerves. As also reported by those authors, the establishment of terminal arborizations by the HO-14-positive fibers could not be determined because HO14 immunostaining is undetectable in axon terminal arborizations. With immunocytochemical staining for selected neurotransmitters, however, hNT neurons did exhibit very characteristic neuronallike shapes, and definable terminal arbors were seen with punctuate boutonlike profiles (refer to Fig. 9A in ref. *144)*. This, combined with staining of these transplants with a human-specific synaptophysin antibody (hSYN), suggested that hNT neurons might form synapses with each other *in situ* as they do in vitro *(50)*. Whether the correlation among synaptogenesis, synaptic activity, and synaptophysin expression during normal development (for references, *see* ref. *142)* also applies equally well to the developmental neurobiology of transplanted hNT neurons warrants further investigation. Although, at this time, there is no unequivocal evidence for morphologically and functionally bona fide synaptic interaction between rat neurons and this line of human neuronallike cells in the spinal cord or elsewhere in the CNS, Hartley et al. *(117)* did describe punctate synaptic protein immunoreactivity near HO-14 immunoreactive processes in host gray matter. Further, confocal double-label immunolabeling studies with hSYN and HO-14 revealed robust punctate hSYN immunoreactivity within the grafts and in the surrounding host. It was cautioned, however, that the hSYN immunoreactivity in the host in regions adjacent to the graft may represent connections between graft processes.

This issue of hNT neuron connectivity with the host CNS is of considerable interest regarding the fundamental neurobiology of the hNT neuron. Although some evidence indicates that human primary fetal CNS xenografts are capable of forming putative synaptic interactions with neurons in the rodent CNS *(153,154)*, this aspect of neurotransplantation has received limited attention. This topic also has considerable bearing on the interpretation of recent findings suggesting functional effects of hNT cell grafts after spinal contusion injuries *(155)*. In that study, rats were subjected to mid-thoracic contusion that resulted in a complete loss of motor-evoked potentials (MEPs). These animals then received grafts of hNT neurons either at the time of injury or 2 wk later following an evaluation for the loss of MEPs. At 8 wk, the graft recipients demonstrated a return of MEPs. Three interpretations were offered for these findings, two of which focused on some form of restored neural circuitry or the re-establishment of functional cellular relays. However, until more definitive evidence showing host–graft connectivity is demonstrated in an hNT neuron transplantation paradigm, these considerations have to be viewed with caution. The authors also speculate on a trophic effect; however, the site or sites of action of such an effect were not discussed, and many possible explanations exist ranging from activation of endogenous oligodendroglial precursors *(156)* to some unknown enhancement of the surrounding tissue milieu resulting in improved action potential propagation and synaptic efficacy. Although these findings raise interesting neurobiological questions, further analysis and attention to possible technical caveats related to several facets of that study are necessary before the implications of those findings relative to clinical translation in SCI can be appropriately considered.

10. CLOSING REMARKS

In this chapter, we have presented a comprehensive survey of an extensive body of research conducted over the last 15 yr looking at the characterization and basic biology of a multipotent human cell line. Those collective efforts have led to Phase I and II clinical trials using LBS-Neurons (i.e., hNT neurons prepared for human use). In that regard, our intent was to provide a balanced review of this experience in translational neurobiology without trying to argue either against or defend the timeliness of this advance to the clinical arena. Rather, it is more useful and informative to extract basic science and clinical insights from the overall impressions of that effort that could be vital to other bench-to-bedside applications of stem cell biology for neurological and even non-neurological diseases.

Unquestionably, a basic prerequisite for stem cell translation will be the need for well-grounded, objective, and standardized in vitro and in vivo characterizations of various cell populations before clinical applications are entertained. In the case of the hNT neuron line, research involving tissue culture and animal models showed those cells having considerable utility as an attractive paradigm for investigating the neurobiology of transplanted human neurons and the discovery of new transplant therapies for several neurological disorders. Whereas from a commercial perspective, hNT neurons possess several distinct advantages *(126)*, the more relevant issue related to the present discussion is the ability of grafted hNT neurons to recapitulate some key features of mature human neurons in vivo.

As in the case of fetal CNS grafts and immortalized cell lines, hNT neurons can be instrumental in delineating properties of the diseased or injured CNS. They also can be used to identify mechanisms of repair that might prove useful in the development of alternative cell sources and strategies, irrespective of their overall therapeutic potential. A dramatic example of this relates to the long-distance growth potential of these cells in the brain and spinal cord. This is reminiscent of the type of axonal elongation associated with grafts of mouse or human fetal CNS tissue in the adult rat *(99–101,157)*. Such growth in CNS white matter is considered to be under potent inhibitory influences of myelin-associated inhibitors *(158)*, which may entail a common receptor *(159,160)*. Whether these or other human and mouse neurons are lacking in the associated receptor(s) is unknown. It also appears that hNT neurons, and perhaps other human neurons as well, may be able to circumvent the effects of other inhibitory molecules, such as certain extracellular matrix proteins *(161–164)*.

As far as clinical applications are concerned, the hNT neurons also present many intriguing possibilities; however, several fundamental questions remain. Foremost among the issues to consider, as mentioned throughout this chapter, is whether the hNT neurons actually have the potential for cell replacement via synaptic integration in the rodent CNS. In terms of preclinical studies, the same considerations also might apply to other human-derived cell populations having the potential for CNS repair. Although there is some precedent for human fetal CNS cells forming morphologically distinct synapses with rat neurons, these examples are rare and rigorous electrophysiological confirmation is not yet available. This review of the hNT neuron transplantation literature thus underscores the importance of a line of very fundamental investigation that can be overshadowed by looking for functional outcomes before any tangible *a priori* basis for such results is established.

Furthermore, in contrast to the very detailed in vitro characterizations that have been conducted on the hNT cell line, the level and range of phenotypic differentiation that these cells achieve after transplantation has not been as fully evaluated. Unlike grafts of some neural precursors *(13,165)*, transplanted hNT neurons do not appear to have the developmental plasticity required for certain types of site-specific differentiation. Thus, even in the rodent focal ischemia model leading to the clinical trials in stroke, there is no evidence to date that hNT neurons in vivo acquire properties of striatal neurons. Therefore, a poorly understood trophic effect is a common interpretational theme put forth to explain functional improvements purportedly mediated by hNT neuron grafts in this and other experimental models of neurodegenerative diseases. It is certainly plausible that the functional capacity of some neurons surviving in the region of the stroke cavity might be augmented by sustained interactions with the grafted NT2N (hNT) neurons. The same principle could apply to recent observations of an apparent functional effect of hNT neuron transplants in the injured spinal cord, as noted earlier. Although other approaches (e.g., gene delivery) could possibly produce the same effect, hNT neurons could be biologically more ideally programmed to achieve the same end point. Therefore, identification of the nature of possible molecule(s) involved and the mechanism(s) by which the observed functional improvements evolve has profound significance to the evolution of preclinical investigations and the interpretation of outcomes. It also has commercial ramifications that will apply to the plethora of other precursor cells being tested or under development.

At this stage, the issue is not a question of whether or not a clinical trial using hNT neurons should or should not have been conducted with FDA approval based on data available 5 yr ago. Retrospective analysis of this judgment is almost impossible given the benefit of knowledge acquired since that time. The more cogent issue is what has been learned from the extrapolation to human use. In that regard, we need to be reminded that clinical trials usually do not translate into immediate cures; rather, they represent part of a learning curve. Indeed, although offering the possibility of patient benefits years before the FDA approval of a new therapy for the general population, clinical trials also can provide critical information to clarify future directions for the development of other new therapies that might benefit larger numbers of patients in the future. Such information also could help shape further meaningful laboratory preclinical studies by uncovering issues about the biology of the human condition and associated clinical circumstances that otherwise might not have been as readily apparent or as fully appreciated. For example, a recently completed trial involving intraspinal transplantation of human fetal spinal cord tissue into a small, selected cohort of spinal injured subjects *(140,141)* not only provided a measure of procedural safety and feasibility but also drew attention to the significant neuroplastic potential of the human spinal cord that had been previously ignored *(4)*. Similarly, the recent follow-up PET imaging study on the patients receiving LBS-Neuron grafts *(125)* pointed out that part of the problem with interpreting the data was the fact that virtually everything known about altered brain metabolic activity in humans after a stroke is limited to the acute and subacute stages. Yet there is an extensive literature indicating significant neuroplasticity or the potential for such after many neurological insults, including stroke for which complementary PET and other data would be invaluable.

In terms of procedural safety, the evidence obtained from the stroke trials to date is in register with the animal data. Although some would argue that neoplastic growth may take longer to evolve, the early analysis of tissue from one of the hNT neuron recipients suggests that at least by 2 yr postgrafting, no signs of incipient tumorigenesis were apparent *(126)*. That evaluation provides an important reminder of the challenges that will be involved in assessing graft survival in human subjects, even if postmortem tissue becomes available. In that study, identification of human donor cells was aided by the polyploidal nature of the LBS-Neuron (aka hNT cell). In other cases, however, the identification of autologous cells without either a biological marker or other methods to eliminate possible contributions of inflammation and other host-related cellular responses will be even more difficult. More challenging will be the unequivocal demonstration of subtle regions of host–graft integration. Even with fetal CNS grafts to the spinal cord, it became apparent that clinical-grade MRI could not readily define host–graft borders *(141)*, and the early posttransplantation MR images from the Phase I stroke trial are unremarkable in that regard. How effectively any diagnostic approach can be employed in human subjects will undoubtedly require an in-depth understanding of how cells behave in vivo via well-defined, and perhaps multiple, animal models of any given neurological disorder. Finally, these and other clinical trials involving CNS repair bring additional attention to the complex issue of placebo controls, as well as the need to re-evaluate various functional scoring methods that were not optimally designed to take into account behavioral changes resulting from neuroplasticity or regeneration.

In SCI, these human cell grafts have many attributes that more closely resemble primary fetal CNS neuron grafts than any other cell line described in the literature thus far. This is especially true in terms of their capacity to replace large expanses of tissue destruction with three-dimensional integrated transplants and their high neuronal yields. Other cell lines in this rapidly evolving field are now being characterized which also have a similar fetal donor tissue characteristic *(166)*. These initial observations with hNT neurons establish an essential framework for multidisciplinary investigations of the in vivo neurobiology and functional benefits of a human cell line that does not originate from embryos or fetuses but which exhibits primary fetal-like and neural-stem-cell-like features in a clinically relevant model of SCI.

The results also present an interesting point of discussion relative to the timeliness of a clinical trial. Safety–feasibility is clearly a primary issue relative to any clinical advance. The findings obtained with hNT neurons in rodent contusion models, coupled with completed or ongoing clinical trials with other sources of donor tissue, argues strongly for preclinical establishment of safety and the feasibility in SCI applications of LBS-neurons or other donor cell lines. However, a counterargument to this view could be offered based on the previous discussion, which is that no "method of use" involving hNT neurons has yet been clearly defined for SCI. By extension, it could be argued further that without this, the possible risk of even a safety–feasibility study would be significant relative to any patient benefit. This raises an interesting dilemma, however, because a well-conceived and limited patient trial could actually provide a more immediate definition of benefit that could then be used to better model potential applications at the experimental level. Certainly, should demonstrations of some form of functional circuit repair not be obtained even at the level of subhuman primate testing, this issue assumes even greater philosophical significance.

Although the intent of this discussion is to raise several practical and theoretical issues that will be repeatedly encountered as more advances are made with stem cell therapies, it is obvious that many points of debate will not be easily, if at all, completely resolved. At the same time, if ever the delicate balance is to be reached relative to clinical urgency, political and ethical debates, judicious scientific progress, and entrepreneurial interests, then the need for an active dialogue between basic scientists and clinicians is acutely apparent. An important closing thought is that although the concept of cellular replacement therapies has rightfully sparked an enormous amount of enthusiasm, the fact remains that science has still not reached the level of knowing how to effectively rebuild a CNS, much less a microcosm of the brain or spinal cord in terms of specific neural circuitries. In that sense, the situation is much like having bricks and mortar for new construction but no blueprint. One can be optimistic, however, that greater insights into this complex issue will evolve with further development and experimental applications of neural progenitors, in parallel with continuing experimental use of more traditional cell sources (e.g., primary fetal CNS tissue) as an important frame of reference. Finally, it should not be overlooked that all aspects of life are fraught with risks, and for patients with debilitating CNS diseases, time is certainly of the essence. Accordingly, this places "sins of omission" and "sins of commission" on a nearly equal footing with respect to the deleterious consequences both can have for neurological diseases that are relentlessly progressive or without any other realistic alternative therapeutic intervention. Thus, inaction or delayed action provides no "safe haven" for researchers or patients when it comes to the development of new therapies for CNS diseases. At the same time, every measure must be taken to avoid placing patients at undue risk in efforts to move research from the bench to the bedside.

DISCLOSURE

Drs. Trojanowski and Lee are founding scientists of Layton Bioscience, Inc. Dr. Reier and Dr. Velardo are collaborators of Layton Bioscience, Inc. through the NIH SBIR funding mechanism. Neither Dr. Reier nor Dr.

Velardo has any other direct or indirect financial ties to or other vested interests in Layton Bioscience, Inc. As noted in the text, the spirit of this chapter was to provide a comprehensive and unbiased overview of translational research related to a human neuronal precursorlike cell line. The experience with these cells can serve as an instructional template for related issues that are likely to apply for to other available or evolving candidate sources of donor tissue. The authorship of this chapter was thus designed to ensure an accurate presentation of the science, while objectively addressing various conceptual and interpretational issues bearing on bench-to-bedside translation. This review is not intended to establish any position with implied or otherwise construed commercial interests.

ACKNOWLEDGMENTS

The authors wish to express their appreciation to Dr. Ronald Mandel for his thoughtful comments during preparation of this manuscript. The studies summarized here from the Center for Neurodegenerative Disease Research (CNDR) were supported by grants from the National Institutes of Health as well as the Alzheimer's Association, and members of CNDR are thanked for their contributions to work on these studies. For more information on this work, see the CNDR website (www.uphs.upenn.edu/cndr). Other original studies (PJR and MJV) noted in this review were supported by NIH SBIR 1 R443NS38828, the State of Florida Brain and Spinal Cord Injury Research Trust Fund, and the Mark F. Overstreet Chair for Spinal Cord Injury and Regeneration Research.

REFERENCES

1. Fricker-Gates, R. A., Lundberg, C., and Dunnett, S. B. (2001) Neural transplantation: restoring complex circuitry in the striatum, *Restor. Neurol. Neurosci.* **19,** 119–138.
2. Dunnett, S. B., Björklund, A., and Lindvall, O. (2001) Cell therapy in Parkinson's disease—stop or go?, *Nat. Rev. Neurosci.* **2,** 365–369.
3. Dunnett, S. B. and Björklund, A. (1994) *Functional Neural Transplantation* Raven, New York.
4. Reier, P. J., Thompson, F. J., Fessler, R. G., Anderson, D. K., and Wirth, E. D., III (2001) Spinal cord injury and fetal CNS tissue transplantation: an initial "bench-to-bedside" translational research experience, in *Regeneration in the Central Nervous System* (Ingoglia, N. A. and Murray, M., eds.), Marcel Dekker, New York, pp. 603–647.
5. Björklund, A. and Lindvall, O. (2000) Cell replacement therapies for central nervous system disorders, *Nature Neurosci.* **3,** 537–544.
6. Björklund, L. M., Sanchez-Pernaute, R., Chung, S., et al. (2002) Embryonic stem cells develop into functional dopaminergic neurons after transplantation in a Parkinson rat model, *Proc. Natl. Acad. Sci. USA* **99,** 2344–2349.

7. Cao, Q., Benton, R. L., and Whittemore, S. R. (2002) Stem cell repair of central nervous system injury, *J. Neurosci. Res.* **68**, 501–510.

8. Cao, Q. L., Zhang, Y. P., Howard, R. M., Walters, W. M., Tsoulfas, P., and Whittemore, S. R. (2001) Pluripotent stem cells engrafted into the normal or lesioned adult rat spinal cord are restricted to a glial lineage, *Exp. Neurol.* **167**, 48–58.

9. Castellanos, D. A., Tsoulfas, P., Frydel, B. R., Gajavelli, S., Bes, J. C., and Sagen, J. (2002) TrkC overexpression enhances survival and migration of neural stem cell transplants in the rat spinal cord, *Cell Transplant.* **11**, 297–307.

10. Chow, S. Y., Moul, J., Tobias, C. A., et al. (2000) Characterization and intraspinal grafting of EGF/bFGF-dependent neurospheres derived from embryonic rat spinal cord, *Brain Res.* **874**, 87–106.

11. Liu, Y., Himes, B. T., Solowska, J., et al. (1999) Intraspinal delivery of neurotrophin-3 using neural stem cells genetically modified by recombinant retrovirus, *Exp. Neurol.* **158**, 9–26.

12. Magnuson, D. S., Zhang, Y. P., Cao, Q. L., Han, Y., Burke, D. A., and Whittemore, S. R. (2001) Embryonic brain precursors transplanted into kainate lesioned rat spinal cord, *Neuroreport* **12**, 1015–1019.

13. Flax, J. D., Aurora, S., Yang, C., et al. (1998) Engraftable human neural stem cells respond to developmental cues, replace neurons, and express foreign genes, *Nature Biotechnol.* **16**, 1033–1039.

14. Fricker, R. A., Carpenter, M. K., Winkler, C., Greco, C., Gates, M. A., and Björklund, A. (1999) Site-specific migration and neuronal differentiation of human neural progenitor cells after transplantation in the adult rat brain, *J Neurosci.* **19**, 5990–6005.

15. Shihabuddin, L. S., Horner, P. J., Ray, J., and Gage, F. H. (2000) Adult spinal cord stem cells generate neurons after transplantation in the adult dentate gyrus, *J. Neurosci.* **20**, 8727–8735.

16. McDonald, J. W., Liu, X. Z., Qu, Y., et al. (1999) Transplanted embryonic stem cells survive, differentiate and promote recovery in injured rat spinal cord, *Nature Med.* **5**, 1410–1412.

17. Gage, F. H., Coates, P. W., Palmer, T. D., et al. (1995) Survival and differentiation of adult neuronal progenitor cells transplanted to the adult brain, *Proc. Natl. Acad. Sci. USA* **92**, 11,879–11,883.

18. Cao, Q. L., Howard, R. M., Dennison, J. B., and Whittemore, S. R. (2002) Differentiation of engrafted neuronal-restricted precursor cells is inhibited in the traumatically injured spinal cord, *Exp. Neurol.* **177**, 349–359.

19. Han, S. S. W., Kang, D. Y., Mujtaba, T., Rao, M. S., and Fischer, I. (2002) Grafted lineage-restricted precursors differentiate exclusively into neurons in the adult spinal cord, *Exp. Neurol.* **177**, 360–375.

20. Ogawa, Y., Sawamoto, K., Miyata, T., et al. (2002) Transplantation of in vitro-expanded fetal neural progenitor cells results in neurogenesis and functional recovery after spinal cord contusion injury in adult rats, *J. Neurosci. Res.* **69**, 925–933.

21. Redmond, D. E. and Freeman, T. (2001) The American society for neural transplantation and repair considerations and guidelines for studies of human

subjects. the practice committee of the society approved by council, *Cell Transplant.* **10,** 661–664.

22. Madrazo, I., Drucker-Colin, R., Diaz, V., Martinez-Marta, J., Torres, C., and Becerril, J. J. (1987) Open microsurgical autograft of adrenal medulla to the right caudate nucleus in Parkinson's disease: a report of two cases, *N. Engl. J. Med.* **316,** 831–834.

23. Forno, L. S. and Langston, J. W. (1991) Unfavorable outcome of adrenal medullary transplant for Parkinson's disease, *Acta Neuropathol.* **81,** 691–694.

24. Freed, W. J., Poltorak, M., and Becker, J. B. (1990) Intracerebral adrenal medulla grafts: a review, *Exp. Neurol.* **110,** 139–166.

25. Hurtig, H., Joyce, J., Sladek, J. R., Jr., and Trojanowski, J. Q. (1989) Postmortem analysis of adrenal-medulla-to-caudate autograft in a patient with Parkinson's disease, *Ann. Neurol.* **25,** 607–614.

26. Penn, R. D., Goetz, C. G., Tanner, C. M., et al. (1988) The adrenal medullary transplant operation for Parkinson's disease: clinical observations in five patients, *Neurosurgery* **22,** 999–1004.

27. Quinn, N. P. (1990) The clinical application of cell grafting techniques in patients with Parkinson's disease, *Prog. Brain Res.* **82,** 619–625.

28. Freed, C. R., Greene, P. E., Breeze, R. E., et al. (2001) Transplantation of embryonic dopamine neurons for severe Parkinson's disease, *N. Engl. J. Med.* **344,** 710–719.

29. Freeman, T. B., Willing, A., Zigova, T., Sanberg, P. R., and Hauser, R. A. (2001) Neural transplantation in Parkinson's disease, *Adv. Neurol.* **86,** 435–445.

30. Dawson, T. M., Mandir, A. S., and Lee, M. K. (2002) Animal models of PD: pieces of the same puzzle?, *Neuron* **35,** 219–222.

31. Andrews, P. W. (1988) Human teratocarcinomas, *Biochim. Biophys. Acta* **948,** 17–36.

32. Andrews, P. W. (1998) Teratocarcinomas and human embryology: pluripotent human EC cell lines, *APMIS* **106,** 158–167.

33. Andrews, P. W., Bronson, D. L., Benham, F., Strickland, S., and Knowles, B. B. (1980) A comparative study of eight cell lines derived from human testicular teratocarcinoma, *Int. J. Cancer* **26,** 269–280.

34. Andrews, P. W., Damjanov, I., Simon, D., et al. (1984) Pluripotent embryonal carcinoma clones derived from the human teratocarcinoma cell line Tera-2: differentiation in vivo and in vitro, *Lab. Invest.* **50,** 147–162.

35. Cheung, W. M., Fu, W. Y., Hui, W. S., and Ip, N. Y. (1999) Production of human CNS neurons from embryonal carcinoma cells using a cell aggregation method, *Biotechniques* **26,** 946–954.

36. Cook, D. G., Lee, V. M., and Doms, R. W. (1994) Expression of foreign proteins in a human neuronal system, *Methods Cell Biol.* **43(Pt. A),** 289–303.

37. Guillemain, I., Alonso, G., Patey, G., Privat, A., and Chaudieu, I. (2000) Human NT2 neurons express a large variety of neurotransmission phenotypes in vitro, *J. Comp. Neurol.* **422,** 380–395.

38. Bani-Yaghoub, M., Bechberger, J. F., Underhill, T. M., and Naus, C. C. (1999) The effects of gap junction blockage on neuronal differentiation of human NTera2/clone D1 cells, *Exp. Neurol.* **156,** 16–32.

39. Kruk, P. A., Balajee, A. S., Rao, K. S., and Bohr, V. A. (1996) Telomere reduction and telomerase inactivation during neuronal cell differentiation, *Biochem. Biophys. Res. Commun.* **224,** 487–492.

40. Ostenfeld, T., Caldwell, M. A., Prowse, K. R., Linskens, M. H., Jauniaux, E., and Svendsen, C. N. (2000) Human neural precursor cells express low levels of telomerase in vitro and show diminishing cell proliferation with extensive axonal outgrowth following transplantation, *Exp. Neurol.* **164,** 215–226.

41. Lee, V. M. and Andrews, P. W. (1986) Differentiation of NTERA-2 clonal human embryonal carcinoma cells into neurons involves the induction of all three neurofilament proteins, *J. Neurosci.* **6,** 514–521.

42. Pleasure, S. J. and Lee, V. M. (1993) NTera 2 cells: a human cell line which displays characteristics expected of a human committed neuronal progenitor cell, *J. Neurosci. Res.* **35,** 585–602.

43. Pleasure, S. J., Page, C., and Lee, V. M. (1992) Pure, postmitotic, polarized human neurons derived from NTera 2 cells provide a system for expressing exogenous proteins in terminally differentiated neurons, *J. Neurosci.* **12,** 1802–1815.

44. Svendsen, C. N., Bhattacharyya, A., and Tai, Y. T. (2001) Neurons from stem cells: preventing an identity crisis, *Nat. Rev. Neurosci.* **2,** 831–834.

45. Almaas, R., Saugstad, O. D., Pleasure, D., and Rootwelt, T. (2002) Neuronal formation of free radicals plays a minor role in hypoxic cell death in human NT2-N neurons, *Pediatr. Res.* **51,** 136–143.

46. Mantione, J. R., Kleppner, S. R., Miyazono, M., Wertkin, A. M., Lee, V. M., and Trojanowski, J. Q. (1995) Human neurons that constitutively secrete A beta do not induce Alzheimer's disease pathology following transplantation and long-term survival in the rodent brain, *Brain Res.* **671,** 333–337.

47. Satoh, J. I. and Kuroda, Y. (2001) Alpha-synuclein expression is up-regulated in NTera2 cells during neuronal differentiation but unaffected by exposure to cytokines and neurotrophic factors, *Parkinsonism. Relat. Disord.* **8,** 7–17.

48. Wertkin, A. M., Turner, R. S., Pleasure, S. J., et al. (1993) Human neurons derived from a teratocarcinoma cell line express solely the 695-amino acid amyloid precursor protein and produce intracellular beta-amyloid or A4 peptides, *Proc. Natl. Acad. Sci. USA* **90,** 9513–9517.

49. Trojanowski, J. Q., Kleppner, S. R., Hartley, R. S., et al. (1997) Transfectable and transplantable postmitotic human neurons: a potential "platform" for gene therapy of nervous system diseases, *Exp. Neurol.* **144,** 92–97.

50. Hartley, R. S., Margulis, M., Fishman, P. S., Lee, V. M., and Tang, C. M. (1999) Functional synapses are formed between human NTera2 (NT2N, hNT) neurons grown on astrocytes, *J. Comp. Neurol.* **407,** 1–10.

51. Houldsworth, J., Heath, S. C., Bosl, G. J., Studer, L., and Chaganti, R. S. (2002) Expression profiling of lineage differentiation in pluripotential human embryonal carcinoma cells, *Cell Growth Differ.* **13,** 257–264.

52. Jorgensen, M., Bevort, M., Kledal, T. S., Hansen, B. V., Dalgaard, M., and Leffers, H. (1999) Differential display competitive polymerase chain reaction: an optimal tool for assaying gene expression, *Electrophoresis* **20,** 230–240.

53. Leypoldt, F., Lewerenz, J., and Methner, A. (2001) Identification of genes up-regulated by retinoic-acid-induced differentiation of the human neuronal precursor cell line NTERA-2 cl.D1, *J. Neurochem.* **76,** 806–814.
54. Satoh, J. and Kuroda, Y. (2000) Differential gene expression between human neurons and neuronal progenitor cells in culture: an analysis of arrayed cDNA clones in NTera2 human embryonal carcinoma cell line as a model system, *J. Neurosci. Methods* **94,** 155–164.
55. Mavilio, F., Simeone, A., Boncinelli, E., and Andrews, P. W. (1988) Activation of four homeobox gene clusters in human embryonal carcinoma cells induced to differentiate by retinoic acid, *Differentiation* **37,** 73–79.
56. Smith, S. C., Pzyborski, S., Aston, C., et al. (1999) Gene expression profiling in a model of human neuronal differentiation using oligonucleotide arrays, *Nature Genet.* **23,** 74.
57. Cheung, W. M., Chu, A. H., Leung, M. F., and Ip, N. Y. (1996) Induction of trk receptors by retinoic acid in a human embryonal carcinoma cell line, *Neuroreport.* **7,** 1204–1208.
58. Gao, Z. Y., Xu, G., Stwora, W. M., Matschinsky, F. M., Lee, V. M., and Wolf, B. A. (1998) Retinoic acid induction of calcium channel expression in human NT2N neurons, *Biochem. Biophys. Res. Commun.* **247,** 407–413.
59. Gower, W. R., Jr., Sanberg, P. R., Brown, P. G., et al. (2002) hNT neurons express an immunosuppressive protein that blocks T-lymphocyte proliferation and interleukin-2 production, *J. Neuroimmunol.* **125,** 103–113.
60. Matsuoka, T., Kondoh, T., Tamaki, N., and Nishizaki, T. (1997) The GABAA receptor is expressed in human neurons derived from a teratocarcinoma cell line, *Biochem. Biophys. Res. Commun.* **237,** 719–723.
61. Sanderson, K. L., Butler, L., and Ingram, V. M. (1997) Aggregates of a beta-amyloid peptide are required to induce calcium currents in neuron-like human teratocarcinoma cells: relation to Alzheimer's disease, *Brain Res.* **744,** 7–14.
62. Satoh, J. I. and Kuroda, Y. (2002) Cytokines and neurotrophic factors fail to affect Nogo-A mRNA expression in differentiated human neurones: implications for inflammation-related axonal regeneration in the central nervous system, *Neuropathol. Appl. Neurobiol.* **28,** 95–106.
63. Segars, J. H., Nagata, T., Bours, V., et al. (1993) Retinoic acid induction of major histocompatibility complex class I genes in NTera-2 embryonal carcinoma cells involves induction of NF-kappa B (p50–p65) and retinoic acid receptor beta-retinoid X receptor beta heterodimers, *Mol. Cell Biol.* **13,** 6157–6169.
64. Iacovitti, L. and Stull, N. D. (1997) Expression of tyrosine hydroxylase in newly differentiated neurons from a human cell line (hNT), *Neuroreport* **8,** 1471–1474.
65. Saporta, S., Willing, A. E., Colina, L. O., et al. (2000) In vitro and in vivo characterization of hNT neuron neurotransmitter phenotypes, *Brain Res. Bull.* **53,** 263–268.
66. Younkin, D. P., Tang, C. M., Hardy, M., et al. (1993) Inducible expression of neuronal glutamate receptor channels in the NT2 human cell line, *Proc. Natl. Acad. Sci. USA* **90,** 2174–2178.

67. Zeller, M. and Strauss, W. L. (1995) Retinoic acid induces cholinergic differentiation of NTera 2 human embryonal carcinoma cells, *Int. J. Dev. Neurosci.* **13,** 437–445.

68. Zigova, T., Barroso, L. F., Willing, A. E., et al. (2000) Dopaminergic phenotype of hNT cells in vitro, *Brain Res. Dev. Brain Res.* **122,** 87–90.

69. Zigova, T., Willing, A. E., Tedesco, E. M., et al. (1999) Lithium chloride induces the expression of tyrosine hydroxylase in hNT neurons, *Exp. Neurol.* **157,** 251–258.

70. Hurlbert, M. S., Gianani, R. I., Hutt, C., Freed, C. R., and Kaddis, F. G. (1999) Neural transplantation of hNT neurons for Huntington's disease, *Cell Transplant.* **8,** 143–151.

71. Yoshioka, A., Yudkoff, M., and Pleasure, D. (1997) Expression of glutamic acid decarboxylase during human neuronal differentiation: studies using the NTera-2 culture system, *Brain Res.* **767,** 333–339.

72. Hardy, M., Younkin, D., Tang, C. M., et al. (1994) Expression of non-NMDA glutamate receptor channel genes by clonal human neurons, *J. Neurochem.* **63,** 482–489.

73. Beczkowska, I. W., Gracy, K. N., Pickel, V. M., and Inturrisi, C. E. (1997) Inducible expression of *N*-methyl-D-aspartate receptor, and delta and mu opioid receptor messenger RNAs and protein in the NT2-N human cell line, *Neuroscience* **79,** 855–862.

74. Rendt, J., Erulkar, S., and Andrews, P. W. (1989) Presumptive neurons derived by differentiation of a human embryonal carcinoma cell line exhibit tetrodotoxin-sensitive sodium currents and the capacity for regenerative responses, *Exp. Cell. Res.* **180,** 580–584.

75. Dunlop, J., Beal, M. H., Lou, Z., and Franco, R. (1998) The pharmacological profile of L-glutamate transport in human NT2 neurones is consistent with excitatory amino acid transporter 2, *Eur. J. Pharmacol.* **360,** 249–256.

76. Rootwelt, T., Dunn, M., Yudkoff, M., Itoh, T., Almaas, R., and Pleasure, D. (1998) Hypoxic cell death in human NT2-N neurons: involvement of NMDA and non-NMDA glutamate receptors, *J. Neurochem.* **71,** 1544–1553.

77. Leypoldt, F., Flajolet, M., and Methner, A. (2002) Neuronal differentiation of cultured human NTERA-2cl.D1 cells leads to increased expression of synapsins, *Neurosci. Lett.* **324,** 37–40.

78. Neelands, T. R., Zhang, J., and Macdonald, R. L. (1999) GABA(A) receptors expressed in undifferentiated human teratocarcinoma NT2 cells differ from those expressed by differentiated NT2-N cells, *J. Neurosci.* **19,** 7057–7065.

79. Konobu, T., Sessler, F., Luo, L. Y., and Lehmann, J. (1998) The hNT human neuronal cell line survives and migrates into rat retina, *Cell Transplant.* **7,** 549–558.

80. Kleppner, S. R., Robinson, K. A., Trojanowski, J. Q., and Lee, V. M. (1995) Transplanted human neurons derived from a teratocarcinoma cell line (NTera-2) mature, integrate, and survive for over 1 year in the nude mouse brain, *J. Comp. Neurol.* **357,** 618–632.

81. Trojanowski, J. Q., Mantione, J. R., Lee, J. H., et al. (1993) Neurons derived from a human teratocarcinoma cell line establish molecular and structural

polarity following transplantation into the rodent brain, *Exp. Neurol.* **122,** 283–294.

82. Trojanowski, J. Q., Walkenstein, N., and Lee, V. M. (1986) Expression of neurofilament subunits in neurons of the central and peripheral nervous system: an immunohistochemical study with monoclonal antibodies, *J. Neurosci.* **6,** 650–660.

83. Shaw, G., Winialski, D., and Reier, P. J. (1988) The effect of axotomy and deafferentation on phosphorylation dependent antigenicity of neurofilaments in superior cervical ganglion neurons, *Brain Res.* **460,** 227–234.

84. Hedreen, J. C. and Koliatsos, V. E. (1994) Phosphorylated neurofilaments in neuronal perikarya and dendrites in human brain following axonal damage, *J. Neuropathol. Exp. Neurol.* **53,** 663–671.

85. Koliatsos, V. E., Applegate, M. D., Kitt, C. A., Walker, L. C., DeLong, M. R., and Price, D. L. (1989) Aberrant phosphorylation of neurofilaments accompanies transmitter-related changes in rat septal neurons following transection of the fimbria-fornix, *Brain Res.* **482,** 205–218.

86. Armstrong, R. J., Watts, C., Svendsen, C. N., Dunnett, S. B., and Rosser, A. E. (2000) Survival, neuronal differentiation, and fiber outgrowth of propagated human neural precursor grafts in an animal model of Huntington's disease, *Cell Transplant.* **9,** 55–64.

87. Belkadi, A. M., Geny, C., Naimi, S., Jeny, R., Peschanski, M., and Riche, D. (1997) Maturation of fetal human neural xenografts in the adult rat brain, *Exp. Neurol.* **144,** 369–380.

88. Brundin, P., Strecker, R. E., Widner, H., et al. (1988) Human fetal dopamine neurons grafted in a rat model of Parkinson's disease: immunological aspects, spontaneous and drug-induced behavior, and dopamine release, *Exp. Brain Res.* **70,** 192–208.

89. Naimi, S., Jeny, R., Hantraye, P., Peschanski, M., and Riche, D. (1996) Ontogeny of human striatal DARPP-32 neurons in fetuses and following xenografting to the adult rat brain, *Exp. Neurol.* **137,** 15–25.

90. Geny, C., Naimi-Sadaoui, S., Jeny, R., El Majid Belkadi, A., Juliano, S. L., and Peschanski, M. (1994) Long-term delayed vascularization of human neural transplants to the rat brain, *J. Neurosci.* **14,** 7553–7562.

91. Daadi, M. M., Saporta, S., Willing, A. E., Zigova, T., McGrogan, M. P., and Sanberg, P. R. (2001) In vitro induction and in vivo expression of bcl-2 in the hNT neurons, *Brain Res. Bull.* **56,** 147–152.

92. Miyazono, M., Lee, V. M., and Trojanowski, J. Q. (1995) Proliferation, cell death, and neuronal differentiation in transplanted human embryonal carcinoma (NTera2) cells depend on the graft site in nude and severe combined immunodeficient mice, *Lab. Invest.* **73,** 273–283.

93. Miyazono, M., Nowell, P. C., Finan, J. L., Lee, V. M., and Trojanowski, J. Q. (1996) Long-term integration and neuronal differentiation of human embryonal carcinoma cells (NTera-2) transplanted into the caudoputamen of nude mice, *J. Comp. Neurol.* **376,** 603–613.

94. Kesari, S., Randazzo, B. P., Valyi-Nagy, T., et al. (1995) Therapy of experimental human brain tumors using a neuroattenuated herpes simplex virus mutant, *Lab. Invest.* **73,** 636–648.

95. Zigova, T., Pencea, V., Sanberg, P. R., and Luskin, M. B. (2000) The properties of hNT cells following transplantation into the subventricular zone of the neonatal forebrain, *Exp. Neurol.* **163,** 31–38.
96. Englund, U., Fricker-Gates, R. A., Lundberg, C., Bjorklund, A., and Wictorin, K. (2002) Transplantation of human neural progenitor cells into the neonatal rat brain: extensive migration and differentiation with long-distance axonal projections, *Exp. Neurol.* **173,** 1–21.
97. Flax, J. D., Aurora, S., Yang, C., et al. (1998) Engraftable human neural stem cells respond to developmental cues, replace neurons, and express foreign genes, *Nature Biotechnol.* **16,** 1033–1039.
98. Turner, R. S., Suzuki, N., Chyung, A. S., Younkin, S. G., and Lee, V. M. (1996) Amyloids β_{40} and β_{42} are generated intracellularly in cultured human neurons and their secretion increases with maturation, *J. Biol. Chem.* **271,** 8966–8970.
99. Wictorin, K. and Björklund, A. (1992) Axon outgrowth from grafts of human embryonic spinal cord in the lesioned adult rat spinal cord, *Neuroreport* **3,** 1045–1048.
100. Wictorin, K., Brundin, P., Gustavii, B., Lindvall, O., and Björklund, A. (1990) Reformation of long axon pathways in adult rat central nervous system by human forebrain neuroblasts, *Nature* **347,** 556–558.
101. Wictorin, K., Brundin, P., Sauer, H., Lindvall, O., and Björklund, A. (1992) Long distance directed axonal growth from human dopaminergic mesencephalic neuroblasts implanted along the nigrostriatal pathway in 6-hydroxydopamine lesioned adult rats, *J. Comp. Neurol.* **323,** 475–494.
102. Borlongan, C. V., Saporta, S., Poulos, S. G., Othberg, A., and Sanberg, P. R. (1998) Viability and survival of hNT neurons determine degree of functional recovery in grafted ischemic rats, *Neuroreport.* **9,** 2837–2842.
103. Borlongan, C. V., Tajima, Y., Trojanowski, J. Q., Lee, V. M., and Sanberg, P. R. (1998) Cerebral ischemia and CNS transplantation: differential effects of grafted fetal rat striatal cells and human neurons derived from a clonal cell line, *Neuroreport* **9,** 3703–3709.
104. Borlongan, C. V., Tajima, Y., Trojanowski, J. Q., Lee, V. M., and Sanberg, P. R. (1998) Transplantation of cryopreserved human embryonal carcinoma-derived neurons (NT2N cells) promotes functional recovery in ischemic rats, *Exp. Neurol.* **149,** 310–321.
105. Nishino, H., Koide, K., Aihara, N., Kumazaki, M., Sakurai, T., and Nagai, H. (1993) Striatal grafts in the ischemic striatum improve pallidal GABA release and passive avoidance, *Brain Res. Bull.* **32,** 517–520.
106. Aihara, N., Mizukawa, K., Koide, K., Mabe, H., and Nishino, H. (1994) Striatal grafts in infarct striatopallidum increase GABA release, reorganize GABAA receptor and improve water-maze learning in the rat, *Brain Res. Bull.* **33,** 483–488.
107. Borlongan, C. V., Koutouzis, T. K., Jorden, J. R., et al. (1997) Neural transplantation as an experimental treatment modality for cerebral ischemia, *Neurosci. Biobehav. Rev.* **21,** 79–90.
108. Hodges, H., Nelson, A., Virley, D., Kershaw, T. R., and Sinden, J. D. (1997) Cognitive deficits induced by global cerebral ischaemia: prospects for transplant therapy, *Pharmacol. Biochem. Behav.* **56,** 763–780.

109. Netto, C. A., Hodges, H., Sinden, J. D., et al. (1993) Effects of fetal hippocampal field grafts on ischaemic-induced deficits in spatial navigation in the water maze, *Neuroscience* **54,** 69–92.

110. Netto, C. A., Hodges, H., Sinden, J. D., et al. (1993) Foetal grafts from hippocampal regio superior alleviate ischaemic-induced behavioral deficits, *Behav. Brain Res.* **58,** 107–112.

111. Sinden, J. D., Rashid-Doubell, F., Kershaw, T. R., et al. (1997) Recovery of spatial learning by grafts of a conditionally immortalized hippocampal neuroepithelial cell line into the ischaemia-lesioned hippocampus, *Neuroscience* **81,** 599–608.

112. Sinden, J. D., Stroemer, P., Grigoryan, G., Patel, S., French, S. J., and Hodges, H. (2000) Functional repair with neural stem cells, *Novartis Found. Symp.* **231,** 270–283.

113. Saporta, S., Borlongan, C. V., and Sanberg, P. R. (1999) Neural transplantation of human neuroteratocarcinoma (hNT) neurons into ischemic rats. A quantitative dose-response analysis of cell survival and behavioral recovery, *Neuroscience* **91,** 519–525.

114. Dunnett, S. B. (1999) Striatal reconstruction by striatal grafts, *J. Neural Transm.* **55(Suppl.),** 115–129.

115. Dunnett, S. B. (2000) Functional analysis of fronto-striatal reconstruction by striatal grafts, *Novartis Found. Symp.* **231,** 21–41.

116. Savitz, S. I., Rosenbaum, D. M., Dinsmore, J. H., Wechsler, L. R., and Caplan, L. R. (2002) Cell transplantation for stroke, *Ann. Neurol.* **52,** 266–275.

117. Hartley, R. S., Trojanowski, J. Q., and Lee, V. M. (1999) Differential effects of spinal cord gray and white matter on process outgrowth from grafted human NTERA2 neurons (NT2N, hNT), *J. Comp. Neurol.* **415,** 404–418.

118. Kondziolka, D., Wechsler, L., Goldstein, S., et al. (2000) Transplantation of cultured human neuronal cells for patients with stroke, *Neurology* **55,** 565–569.

119. Buchan, A. M., Warren, D., Burnstein, R., and Kondziolka, D. (2001) Transplantation of cultured human neuronal cells for patients with stroke, *Neurology* **56,** 821–823.

120. Zivin, J. A. (2000) Cell transplant therapy for stroke: hope or hype, *Neurology* **55,** 467.

121. Freed, C. R., Breeze, R. E., and Rosenberg, N. L. (1991) Brain tissue transplantation without immunosuppression—reply, *Arch. Neurol.* **48,** 260–262.

122. Freed, C. R., Breeze, R. E., Rosenberg, N. L., et al. (1992) Survival of implanted fetal dopamine cells and neurologic improvement 12 to 46 months after transplantation for Parkinson's disease, *N. Engl. J. Med.* **327,** 1549–1555.

123. Freeman, T. B., Olanow, C. W., Hauser, R. A., et al. (1995) Bilateral fetal nigral transplantation into the postcommissural putamen in Parkinson's disease, *Ann. Neurol.* **38,** 379–388.

124. Kordower, J. H., Freeman, T. B., Chen, E. Y., et al. (1998) Fetal nigral grafts survive and mediate clinical benefit in a patient with Parkinson's disease, *Mov. Disord.* **13,** 383–393.

125. Meltzer, C. C., Kondziolka, D., Villemagne, V. L., et al. (2001) Serial [^{18}F] fluorodeoxyglucose positron emission tomography after human neuronal implantation for stroke, *Neurosurgery* **49**, 586–591.

126. Nelson, P. T., Kondziolka, D., Wechsler, L., et al. (2002) Clonal human (hNT) neuron grafts for stroke therapy: neuropathology in a patient 27 months after implantation, *Am. J. Pathol.* **160**, 1201–1206.

127. Baker, K. A., Hong, M., Sadi, D., and Mendez, I. (2000) Intrastriatal and intranigral grafting of hNT neurons in the 6-OHDA rat model of Parkinson's disease, *Exp. Neurol.* **162**, 350–360.

128. Garbuzova-Davis, S., Willing, A. E., Milliken, M., et al. (2002) Positive effect of transplantation of hNT neurons (NTera 2/D1 cell-line) in a model of familial amyotrophic lateral sclerosis, *Exp. Neurol.* **174**, 169–180.

129. Willing, A. E., Garbuzova-Davis, S., Saporta, S., Milliken, M., Cahill, D. W., and Sanberg, P. R. (2001) hNT neurons delay onset of motor deficits in a model of amyotrophic lateral sclerosis, *Brain Res. Bull.* **56**, 525–530.

130. Garbuzova-Davis, S., Willing, A. E., Milliken, M., et al. (2001) Intraspinal implantation of hNT neurons into SOD1 mice with apparent motor deficit, *Amyotroph. Lateral. Sclerosis Other Motor Neuron Disord.* **2**, 175–180.

131. Lin, S. Z., Hayashi, T., Su, T.-P., et al. (2000) Detection of GDNF in cultured and grafted hNT neurons, *Exp. Neurol.* **164**, 447.

132. Weima, S. M., van Rooijen, M. A., Mummery, C. L., et al. (1988) Differentially regulated production of platelet-derived growth factor and of transforming growth factor beta by a human teratocarcinoma cell line, *Differentiation* **38**, 203–210.

133. Muir, J. K., Raghupathi, R., Saatman, K. E., et al. (1999) Terminally differentiated human neurons survive and integrate following transplantation into the traumatically injured rat brain, *J. Neurotrauma* **16**, 403–414.

134. Philips, M. F., Muir, J. K., Saatman, K. E., et al. (1999) Survival and integration of transplanted postmitotic human neurons following experimental brain injury in immunocompetent rats, *J. Neurosurg.* **90**, 116–124.

135. Zhang, C., Royo, N., Schouten, J., Saatman, K. E., and McIntosh, T. K. (2003) Long-term survival of human neurons (HNT) engrafted in the chronic postinjury period following experimental traumatic brain injury in rats, *Exp. Neurol.* **181**, 111 (abstract).

136. Velardo, M. J., McGrogan, M. P., Silver, X. S., Mareci, T. H., and Reier, P. J. (2002) Long-term engraftment of a human neuronal precursor cell line in acute and chronic cervical contusion lesions of the rat spinal cord, *Exp. Neurol.* **175**, 416–417.

137. Velardo, M. J., O'Steen, B. E., McGrogan, M. P., and Reier, P. J. (1999) Transplantation of hNT cells into rat cervical spinal cord contusions, *J. Neurotrauma* **16**, 1013.

138. Velardo, M. J., O'Steen, B. E., McGrogan, M. P., and Reier, P. J. (2000) hNT cells and transplantation repair of cervical contusions of the rat spinal cord, *Exp. Neurol.* **164**, 454.

139. Velardo, M. J., O'Steen, B. E., McGrogan, M. P., and Reier, P. J. (2000) Survival, maturation and innervation of hNT cell transplants in the contused spinal cord of the adult rat, *Soc. Neurosci. Abstr.* **26**, 1104.

140. Thompson, F. J., Reier, P. J., Uthman, B., et al. (2001) Neurophysiological assessment of the feasibility and safety of neural tissue transplantation in patients with syringomyelia, *J. Neurotrauma* **18,** 931–945.

141. Wirth, E. D., III, Reier, P. J., Fessler, R. G., et al. (2001) Feasibility and safety of neural tissue transplantation in patients with syringomyelia, *J. Neurotrauma* **18,** 911–929.

142. Lee, V. M., Hartley, R. S., and Trojanowski, J. Q. (2000) Neurobiology of human neurons (NT2N) grafted into mouse spinal cord: implications for improving therapy of spinal cord injury, *Prog. Brain Res.* **128,** 299–307.

143. Tohyama, T., Lee, V. M., Rorke, L. B., and Trojanowski, J. Q. (1991) Molecular milestones that signal axonal maturation and the commitment of human spinal cord precursor cells to the neuronal or glial phenotype in development, *J. Comp. Neurol.* **310,** 285–299.

144. Reier, P. J., Golder, F. J., Bolser, D. C., et al. (2002) Gray matter repair in the cervical spinal cord, in *Spinal Cord Trauma: Regeneration, Neural Repair and Functional Recovery* (McKerracher, L., Doucet, G., and Rossignol, S., eds.), Elsevier, New York, pp. 49–70.

145. Reier, P. J., Anderson, D. K., and Stokes, B. T. (1992) Neural tissue transplantation and CNS trauma: anatomical and functional repair of the injured spinal cord, in *Central Nervous System Status Report* (Jane, J. A., Anderson, D. K., Torner, J. C., and Young, W., eds.), *J. Neurotrauma* **9(Supp. 1),** S223–S248.

146. Anderson, D. K., Reier, P. J., Wirth, E. D., III, Theele, D. P., and Brown, S. A. (1991) Transplants of fetal CNS grafts in chronic compression lesions of the adult cat spinal cord, *Restor. Neurol. Neurosci.* **2,** 309–325.

147. Reier, P. J., Anderson, D. K., Schrimsher, G. W., et al. (1994) Neural cell grafting: anatomical and functional repair of the spinal cord., in *The Neurobiology of Central Nervous System Trauma* (Salzman, S. K. and Faden, A. I., eds.), Oxford University Press, New York, pp. 288–311.

148. Reier, P. J., Stokes, B. T., Thompson, F. J., and Anderson, D. K. (1992) Fetal cell grafts into resection and contusion/compression injuries of the rat and cat spinal cord, *Exp. Neurol.* **115,** 177–188.

149. Stokes, B. T. and Reier, P. J. (1992) Fetal grafts alter chronic behavioral outcome after contusion damage to the adult rat spinal cord, *Exp. Neurol.* **116,** 1–12.

150. Houlé, J. D. and Reier, P. J. (1988) Transplantation of fetal spinal cord into the chronically injured adult rat spinal cord, *J. Comp. Neurol.* **269,** 535–547.

151. Akesson, E., Kjaeldgaard, A., and Seiger, A. (1998) Human embryonic spinal cord grafts in adult rat spinal cord cavities: survival, growth, and interactions with the host, *Exp. Neurol.* **149,** 262–276.

152. Giovanini, M. A., Reier, P. J., Eskin, T. A., and Anderson, D. K. (1997) MAP2 expression in the developing human fetal spinal cord and following xenotransplantation, *Cell Transplant.* **6,** 339–346.

153. Mahalik, T. J., Stromberg, I., Gerhardt, G. A., et al. (1989) Human ventral mesencephalic xenografts to the catecholamine-depleted striata of athymic rats: ultrastructure and immunocytochemistry, *Synapse* **4,** 19–29.

154. Strömberg, I., Almqvist, P., Bygdeman, M., et al. (1989) Human fetal mesencephalic tissue grafted to dopamine-denervated striatum of athymic rats: light- and electron-microscopical histochemistry and in vivo chronoamperometric studies, *J. Neurosci.* **9,** 614–624.

155. Saporta, S., Makoui, A. S., Willing, A. E., Daadi, M., Cahill, D. W., and Sanberg, P. R. (2002) Functional recovery after complete contusion injury to the spinal cord and transplantation of human neuroteratocarcinoma neurons in rats, *J. Neurosurg.* **97,** 63–68.

156. McTigue, D. M., Horner, P. J., Stokes, B. T., and Gage, F. H. (1998) Neurotrophin-3 and brain-derived neurotrophic factor induce oligodendrocyte proliferation and myelination of regenerating axons in the contused adult rat spinal cord, *J. Neurosci.* **18,** 5354–5365.

157. Li, Y. and Raisman, G. (1993) Long axon growth from embryonic neurons transplanted into myelinated tracts of the adult rat spinal cord, *Brain Res.* **629,** 115–127.

158. Woolf, C. J. and Bloechlinger, S. (2002) Neuroscience. It takes more than two to Nogo, *Science* **297,** 1132–1134.

159. Domeniconi, M., Cao, Z., Spencer, T., et al. (2002) Myelin-associated glycoprotein interacts with the Nogo66 receptor to inhibit neurite outgrowth, *Neuron* **35,** 283–290.

160. Watkins, T. and Barres, B. (2002) Nerve regeneration: regrowth stumped by shared receptor, *Curr. Biol.* **12,** R654.

161. Bradbury, E. J., Moon, L. D., Popat, R. J., et al. (2002) Chondroitinase ABC promotes functional recovery after spinal cord injury, *Nature* **416,** 636–640.

162. Davies, S. J., Fitch, M. T., Memberg, S. P., Hall, A. K., Raisman, G., and Silver, J. (1997) Regeneration of adult axons in white matter tracts of the central nervous system, *Nature* **390,** 680–683.

163. Lemons, M. L., Howland, D. R., and Anderson, D. K. (1999) Chondroitin sulfate proteoglycan immunoreactivity increases following spinal cord injury and transplantation [in process citation], *Exp. Neurol.* **160,** 51–65.

164. Lemons, M. L., Sandy, J. D., Anderson, D. K., and Howland, D. R. (2001) Intact aggrecan and fragments generated by both aggrecanse and metalloproteinase-like activities are present in the developing and adult rat spinal cord and their relative abundance is altered by injury, *J. Neurosci.* **21,** 4772–4781.

165. Shihabuddin, L. S., Hertz, J. A., Holets, V. R., and Whittemore, S. R. (1995) The adult CNS retains the potential to direct region-specific differentiation of a transplanted neuronal precursor cell line, *J. Neurosci.* **15,** 6666–6678.

166. Velardo, M. J., Williams, P. R., Hazel, T. G., Johe, K. K., and Reier, P. J. (2003) Human stem cell xenografts exhibit high neuronal yields and long distance neuritic outgrowth in the contused spinal cord, *Exp. Neurol.* **181,** 108 (abstract).

Appendix I

CELL LINES AND COMPANIES INVOLVED WITH HUMAN EMBRYONIC STEM CELL RESEARCH

Appendix I
Cell Lines and Companies Involved with Human Embryonic Stem Cell Research

As of May 2002, a total of 78 cell lines that fulfilled the criteria for federal funding were identified from 14 sources (*see* below). At the time, these cells were at different stages of development; many were not characterized and/or ready for distribution to investigators.

Company	Number of eligible lines
BresaGen, Inc., Athens, GA	4
CyThera, Inc., San Diego, CA	9
ES Cell International, Australia	6
Geron Corp., Menlo Park, CA	7
Goteborg University, Sweden	19
Karolinska Institute, Sweden	6
Maria Biotech Co Ltd., Korea	3
MizMedi Hospital–Seoul University, Korea	1
Tata Institute, India	3
Pochon CHA University, Korea	2
Reliance Life Sciences, India	7
Technion University, Israel	4
University of California, San Francisco, CA	2
Wisconsin Alumni Research Foundation, WI	5

To be eligible for federally funded research in the United States, the human ES cells must meet the following criteria:

• Derived from an embryo that was created for reproductive purposes.
• The embryo was no longer needed for these purposes.
• Informed consent must have been obtained for the donation of the embryo.
• No financial inducements were provided for the donation of the embryo.
• Derived before August 9, 2002.

Cell lines and companies involved with human ES cell research can be found on the NIH Human Embryonic Stem Cell Registry (www.escr.nih.gov), which was set up to serve two functions. It provides a unique NIH code for each eligible cell line that must be used in grant applications to the federal

From: *Human Embryonic Stem Cells*
Edited by: A. Chiu and M. S. Rao © Humana Press Inc., Totowa, NJ

government. It also provides contact information for investigators to help them acquire the cells for their investigations, and it links to the providers' websites. The NIH Registry was updated in November 2002 to list eligible lines that are currently available for shipping and also provides available information on cell characterization about each of the available lines.

Company	Cell line
BresaGen, Inc (www.bresagen.com)	NIH Code: BG01 Provider's Code: hESBGN-01
ES Cell International (www.escellinterenatinal.com/stem/celltable)	NIH Code: ES01 Provider's Code: HES-1
	NIH Code: ES02 Provider's Code: HES-2
	NIH Code: ES03 Provider's Code: HES-3
	NIH Code: ES04 Provider's Code: HES-4
	NIH Code: ES06 Provider's Code: HES-6
University of California at San Francisco (www.escells.ucsf.edu)	NIH Code: UC06 Provider's Code: HSF-6
Wisconsin Alumni Research Foundation (WiCell Research Institute) (www.wicell.org)	NIH Code: WA01 Provider's Code: H1
	NIH Code: WA09 Provider's Code: H9

Appendix II
USEFUL WEBSITES

Appendix II
Useful Websites

NIH POLICY AND IMPLEMENTATION

NIH Stem Cell Information is provided at (www.nih.gov/news/stemcell/index). This site provides links to information on issues of conducting research on human ES cells, Stem Cell Material Transfer Agreements, and awards.

In late 2002, the Director of NIH established the NIH Stem Cell Task Force to enable and accelerate the pace of stem cell research and to seek the advice of scientific leaders in stem cell research (www.stemcelltaskforce.nih.gov).

An NIH web page provides more information for grant applicants and research institutions interested in human embryonic stem cell research. Updated NIH Guide announcements, the NIH Human Embryonic Stem Cell Registry, archives of Frequently Asked Questions (FAQ), and pertinent grants policy statements can all be found at (www.grants.nih.gov/grants/stem_cells).

NIH Human Embryonic Stem Cell Registry currently lists cell lines that meet the eligibility criteria for federally funded research and are available for shipping (www.escr.nih.gov).

The criteria for federal funding of stem cell research and the establishment of the NIH Human Embryonic Stem Cell Registry can be found at (www.grants.nih.gov/grants/guide/notice-files/NOT-OD-02-005).

NIH Strategies for Implementing Human Embryonic Stem Cell Research describes the action NIH has taken to enable hES cell research to flourish (www.nih.gov/news/stemcell/022802implement).

National Institutes of Health guidelines for research using human pluripotent stem cells (Effective August 25, 2000, 65 FR 51976) (Corrected November 21, 2000, 65 FR 69951; and partially withdrawn November 7, 2001, NOT-OD-02-007). Can be found at (www.nih.gov/news/stemcell/stemcellguidelines and www.grants.nih.gov/grants/guide/notice-files/NOT-OD-02-007).

Implementation issues for human embryonic stem cell research—Frequently Asked Questions (NOT-OD-02-014) can be found at (www.grants.nih.gov/grants/guide/notice-files/NOT-OD-02-014 and www.grants.nih.gov/grants/stem_cell_faqs).

From: *Human Embryonic Stem Cells*
Edited by: A. Chiu and M. S. Rao © Humana Press Inc., Totowa, NJ

Information on the Federal Government Clearances for Receipt of International Shipment of Human Embryonic Stem Cells can be found at (www.grants.nih.gov/grants/guide/notice-files/NOT-OD-02-013).

OHRP Guidance for Investigators and Institutional Review Boards Regarding Research Involving Human Embryonic Stem Cells and Human Embryonic Germ Cells is found at (www.ohrp.osophs.dhhs.gov/humansubjects/guidance/stemcell).

The General Stem Cell Information Site of NIH is (www.nih.gov/news/stemcell/index)

The NINDS Stem Cell Information website is (www.ninds.nih.gov/stemcells).

For information on calculating and parsing Facilities and Administrative (F&A) costs for eligible research from ineligible research, check the Frequently Asked Questions website (www.grants.nih.gov/grants/stem_cell_faqs).

RESEARCH AGREEMENTS AND PATENT OFFICES

WiCell Research Institute Research Agreement with the NIH is published at (www.nih.gov/news/stemcell/WicellMOU).

BreSagen, Inc.Research Agreement with the NIH is published at (www.nih.gov/news/stemcell/BresaGenMOU).

ES Cell International Research Agreement with the NIH is published at (www.nih.gov/news/stemcell/ESIMOU).

University of California, San Francisco Research Agreement with the NIH is published at (www.nih.gov/news/stemcell/UCSF).

The United States Patent and Trademark Office (www.uspto.gov).

For U.S. patent searches (www.uspto.gov/patft/index).

The European Patent Office (www.european-patent-office.org).

For European patent searches (www.european-patent-office.org/espacenet/info/access.htm).

FOOD AND DRUG ADMINISTRATION (FDA) INFORMATION AND REGULATORY GUIDANCES

Within the FDA, the Center for Biologics Evaluation and Research (CBER) branch has the responsibility of regulating the safety of biological products that includes cellular therapies such as cells for transplantation (www.fda.gov/cber).

CBER Guidances/Guidelines/Points to Consider: A comprehensive listing of documents that provides information detailing CBER's regulatory approach for a various categories of biological products can be found at (www.fda.gov/cber/guidelines).

FDA/CBER Tissue Action Plan details CBER's strategy for the regulation of cellular and tissue-based products (www.fda.gov/cber/tissue/tissue).

Cellular therapies developed from human embryonic stem cell lines initially maintained by culturing on murine feeder cell layers are considered xeno-transplantation products. The FDA/CBER Xenotransplantation Action Plan details a comprehensive regulatory approach for xenotransplantation regulation (www.fda.gov/cber/xap/xap).

Information on submitting an Investigational New Drug (IND) Application to FDA/CBER can be found at (www.fda.gov/cber/ind/ind).

Appendix III

Research Agreements and Material Transfer Agreements Between Investigator and Stem Cell Provider

Appendix III
Research Agreements and Material Transfer Agreements Between Investigator and Stem Cell Provider

To enable intramural researchers at the National Institutes of Health (NIH) to conduct research on eligible hES cell lines, the NIH signed Memoranda of Agreement with the providers for research use of their cells. The first Memorandum of Understanding (MOU) was signed on September 5th 2001 between the NIH and the WiCell Research Institute, Inc. (http://www.nih.gov/news/pr/sep2001/od-05.htm.)

Similar research agreements between the NIH and three other providers, ES Cell International, BreSagen, Inc. and the Regents of the University of California were signed in April 2002. These documents can be found on the NIH Stem Cell Information Website (http://www.nih.gov/news/stemcell/index.htm). Each can also be read directly.

> WiCell Research Institute Research Agreement: (http://www.nih.gov/news/stemcell/WicellMOU.pdf)
>
> BreSagen, Inc. Research Agreement: (http://www.nih.gov/news/stemcell/BresaGenMOU.pdf)
>
> ES Cell International Research Agreement (http://www.nih.gov/news/stemcell/ESIMOU.pdf)
>
> University of California San Francisco Research Agreement (http://www.nih.gov/news/stemcell/UCSF.pdf)

We have reproduced this agreement below as an example of such an MOU.

Memorandum of Understanding

between

The Regents of the University of California

and

Public Health Service

U.S. Department of Health and Human Services

From: *Human Embryonic Stem Cells*
Edited by: A. Chiu and M. S. Rao © Humana Press Inc., Totowa, NJ

This Memorandum of Understanding (hereinafter "Agreement"), effective April 26, 2002 by and between the Public Health Service of the U.S. Department of Health and Human Services as represented by the Office of Technology Transfer, having an address at National Institutes of Health, 6011 Executive Boulevard, Suite 325, Rockville, Maryland 20852, United States ("PHS") and The Regents of the University of California on behalf of the University of California San Francisco, acting through its Office of Technology Management having an address at 1294 Ninth Avenue, Suite 1, Box 1209, San Francisco, CA 94143-1209 ("UC"). PHS and UC are referred to herein as the "Parties."

WHEREAS specific human embryonic stem cell line materials, their unmodified and undifferentiated progeny and unmodified and undifferentiated derivatives ("Material") have been derived consistent with the Presidential Statement of August 9, 2001, from the research efforts of Dr. Meri Firpo, Dr. Roger Pedersen, and Juanito Meneses of the University of California San Francisco with funds provided by the State of California under the University of California Discovery Grant/BioSTAR Project and Geron Corporation ("Sponsor") under a Sponsored Research Agreement; and

WHEREAS the Material was made using funds provided by the State of California and Sponsor, and as such, their ownership is not subject to any rights or obligations previously granted to the NIH; and

WHEREAS property rights in the Material are owned by UC, and Sponsor has an exclusive license to UC's property rights and intellectual property rights embodied in or directly related to the Material for commercial use; and

WHEREAS the Material is or may be subject to patent rights owned by the WiCell Research Institute ("WiCell") concerning primate embryonic stem cells and their cultivation as claimed in U.S. Patent 5,843,780, U.S. Patent 6,200,806, U.S. Patent Application 09/522,030 and corresponding U.S. or foreign patent rights and any patents granted on any divisional and continuation applications, reissues and reexaminations ("Wisconsin Patent Rights"); and

WHEREAS UC has entered into a Memorandum of Understanding with WiCell effective April 24, 2002, under which UC has the requisite permission from Wisconsin to make the Material available to investigators at non-profit institutions under this Agreement; and

WHEREAS PHS has a basic mission on behalf of the U.S. Government for the conduct and support of health research performed at its own facilities or through funding agreements to other institutions ("Recipient Institutions"); and

WHEREAS PHS funded primate research studies at the University of Wisconsin–Madison that led to certain discoveries claimed in Wisconsin Patent

Rights, and therefore the Government has certain use and other rights to the intellectual property comprising the Wisconsin Patent Rights granted by law and regulation that may be applicable for the use of the Material; and

WHEREAS PHS has received permission to use Wisconsin Patent Rights under the terms and conditions of its September 5, 2001 Memorandum of Understanding with WiCell; and

WHEREAS UC desires to serve the public interest by making the Material widely available to PHS and researchers at other non-profit institutions.

NOW, THEREFORE, the Parties hereby agree to the following terms and conditions regarding use of UC Material:

1. The Parties agree that the Material is to be made available by UC for use in specified PHS biomedical research programs, either by PHS or on behalf of PHS by its specified contractors under the terms and conditions specified in the attached Institution Material Transfer Agreement.
2. UC agrees that it shall make the Material available for use by non-profit Recipient Institutions and offer such Institutions the attached Institution Material Transfer Agreement or terms no more restrictive to govern the exchange of the Material.
3. Notwithstanding any terms of this Agreement, nothing herein shall be construed to diminish or supercede any rights or authorities available to PHS as a U.S. government agency. The provisions of this Agreement and the obligations hereunder with respect to the Material shall continue as long as the Material continues to be used by PHS or its Contractors.
4. Nothing contained herein shall be considered to be the grant of a commercial license or right under the Material. Furthermore, nothing contained herein shall be construed to be a waiver of UC property rights in the Material.

IN WITNESS WHEREOF, the Parties agree to the foregoing and have caused this Agreement to be executed by their duly authorized representatives.

The Regents of the University of California
By:
Name:
Title:

Public Health Service
By:
Name:
Title:

INSTITUTION MATERIAL TRANSFER AGREEMENT

The Regents of the University of California, on behalf of the University of California San Francisco, ("UCSF") agrees to provide the following institution: [   ] ("Recipient") with human embryonic stem cell line ___, requested by Recipient for use by its scientist, [   ] ("Recipient Scientist"), subject to the terms and conditions set forth in this Institution Material Transfer Agreement (the "Agreement").

Background

1. Human embryonic stem cell line [   ] (together with its unmodified and undifferentiated progeny and unmodified and undifferentiated derivatives, the "Material") has been derived consistent with the Presidential Statement of August 9, 2001, from the research efforts of Dr. Meri Firpo, Dr. Roger Pedersen, and Juanito Meneses (collectively, "Investigator") of UCSF with funds provided by the State of California under the University of California Discovery Grant/BioSTAR Project and Geron Corporation ("Sponsor") under a Sponsored Research Agreement (the "Research Agreement").

2. The Material is or may be subject to patent rights owned by the University of Wisconsin or its affiliates ("Wisconsin") concerning primate embryonic stem cells and their cultivation as claimed in U.S. Patent 5,843,780, U.S. Patent 6,200,806, U.S. Patent Application 09/522,030 and corresponding U.S. or foreign patent rights and any patents granted on any divisional and continuation applications, reissues and reexaminations ("Wisconsin Patent Rights"). Sponsor has an exclusive license to Wisconsin Patent Rights for commercial use in certain fields of use.

3. Property rights in the Material are owned by UCSF, and Sponsor has an exclusive license to UCSF's property rights and intellectual property rights embodied in or directly related to the Material for commercial use. Sponsor is an intended and express third party beneficiary of this Agreement.

4. The Regents of the University of California has entered into a Memorandum of Understanding with WiCell Research Institute effective April 25, 2002, under which UCSF has the requisite permission from Wisconsin to make the Material available to academic investigators under this Agreement.

5. UCSF, Wisconsin and Sponsor would like to make the Material broadly available to academic investigators for research purposes.

Terms and Conditions

1. Recipient agrees to use the Material solely for use in Recipient Scientist's research as set forth in the descriptive title attached as **Exhibit A** (the "Research"), which will be shared with Sponsor by UCSF.

2. The Material is the property of UCSF subject to the commercial license rights of Sponsor, and is made available as a service to the research community by

UCSF. Legal title to the Material shall be unaffected by this Agreement or the transfer made hereunder, and nothing in this Agreement grants Recipient any rights under any patents or other intellectual property rights or tangible property rights of UCSF, Wisconsin, or Sponsor, including, but not limited to, Wisconsin Patent Rights. Wisconsin is and shall be an intended and express third party beneficiary of this Agreement, and nothing contained in this Agreement shall constitute a grant of a commercial license or right under the Wisconsin Patent Rights. Furthermore, nothing contained herein shall be construed to be a waiver by Wisconsin of the Wisconsin Patent Rights. This Agreement does not restrict UCSF's or Sponsor's rights to distribute the Material to other researchers at academic or non-profit institutions.

3. Recipient and Recipient Scientist shall maintain the confidentiality of any proprietary information that is provided by UCSF under this Agreement. Such information will be labeled as "confidential" or "proprietary" at the time it is provided to Recipient or Recipient Scientist. The confidentiality obligations set forth in this Agreement shall not apply to information that Recipient or Recipient Scientist can document: (a) was known to Recipient or Recipient Scientist prior to the disclosure of such information under this Agreement; (b) is or becomes generally available to the public through no fault of Recipient or Recipient Scientist; (c) is subsequently made available to Recipient or Recipient Scientist from any third party that is not under any obligation of confidentiality to UCSF; (d) is developed independently by Recipient or Recipient Scientist, without use of any proprietary information of UCSF; and (e) Recipient or Recipient Scientist is required to disclose by a court of law, or under federal regulations, or applicable laws, in which case, Recipient or Recipient Scientist agrees to provide prompt written notice to UCSF and to take reasonable steps to enable UCSF to seek a protective order or otherwise prevent disclosure of such information.

4. The transfer of the Material constitutes a non-exclusive license to use the Material solely for conduct of the Research under the direction of Recipient Scientist. Recipient will not use the Material (including any Material wholly or partially contained or incorporated in any Derivative Material (as defined below)) in experiments other than the Research. *Commercial use of the Material or any Derivative Material which wholly or partially contains or incorporates the Material is strictly prohibited*. Use of the Material and any Derivative Material which wholly or partially contains or incorporates the Material for commercial purposes may only be undertaken pursuant to licenses from a) Wisconsin and b) Sponsor, which neither of these parties shall have an obligation to grant.

5. In the course of the Research, Recipient may create Derivative Material. "Derivative Material" shall be defined as any materials or products a) that are physically derived from the Materials, including, without limitation, differentiated cell lines or populations derived from the Material, and any unmodified products of such cell lines or populations (i.e., RNA, DNA, proteins, cellular

components or fragments thereof), or b) that wholly or partially contain or incorporate the Material. Inventorship of any Derivative Material, whether or not patentable, will be determined according to the principles of U.S. patent law. Subject to any obligation of the inventors to assign ownership of inventions or Derivative Materials to Recipient in accordance with applicable regulations or Recipient's policies, ownership will follow inventorship.

6. Recipient and Recipient Scientist agree that the Material and any Derivative Material which wholly or partially contains or incorporates the Material will not be used in research that is or becomes subject to consulting, licensing, or similar obligations to any commercial entity, except pursuant to a license obtained in accordance with the last sentence of Paragraph 4. In no event will the Material or any Derivative Material which wholly or partially contains or incorporates the Material be used in research a) sponsored by a for-profit entity, either through supply of materials or funding, except as such research sponsor i) is permitted to use Wisconsin Patent Rights under a separate written agreement with Wisconsin; or ii) receives no rights, whether actual or contingent, in or to the results of the sponsored research; or b) sponsored by a private non-profit entity, either through supply of materials or funding, except as such sponsor is i) permitted to use Wisconsin Patent Rights under a separate written agreement with Wisconsin; or ii) receives no commercial rights, whether actual or contingent, in or to the results of the sponsored research. Notwithstanding the foregoing, no such research sponsor shall receive any rights to the Material. Further, the provider of third party proprietary research material may restrict the use of Derivative Material, including the use of Derivative Material which contains or incorporates the Material, by Recipient, UCSF and Sponsor, solely to the extent that the proprietary research material of the provider is contained or incorporated in such Derivative Material.

7. Recipient and Recipient Scientist agree to use the Material and any Derivative Material which wholly or partially contains or incorporates the Material under suitable handling, containment, and storage conditions and in compliance with all applicable statutes, regulations and guidelines relating to their handling, use or disposal. Material and any Derivative Material which wholly or partially contains or incorporates the Material shall not be used for diagnostic or therapeutic purposes. Recipient and Recipient Scientist agree not to (i) mix the Materials with an intact human embryo; (ii) implant the Materials in a human uterus; or (iii) attempt to make whole human embryos with Materials or any Derivative Material which wholly or partially contains or incorporates the Material by any method. An Annual Certification Statement in the form attached hereto as **Exhibit B**, confirming compliance with the restrictions on the use of Material and any Derivative Material which wholly or partially contains or incorporates the Material, shall be supplied to UCSF by Recipient and Recipient Scientist.

8. Neither Recipient Scientist nor Recipient nor any other person authorized to use the Material under the Agreement shall make available any portion of the

Material or any Derivative Material which wholly or partially contains or incorporates the Material to any person or entity other than laboratory personnel under the Recipient Scientist's immediate and direct control. The Material and any Derivative Material which wholly or partially contains or incorporates the Material will not be distributed to others without UCSF's written consent after a Material Transfer Agreement in this form has been executed.

9. The Recipient shall refer any request for the Material to UCSF. To the extent supplies are available, UCSF agrees to make the Material available to other scientists for teaching or non-commercial research purposes only, under a separate Institution Material Transfer Agreement in this form and in any event under terms no more restrictive than those set forth herein. For Derivative Material in which UCSF has an ownership interest pursuant to the last sentence of Paragraph 5, UCSF and Recipient will jointly prepare a mutually acceptable material transfer agreement for transfer of such Derivative Materials to other scientists. UCSF will inform Sponsor of, and Sponsor shall be entitled to receive a copy of, any material transfer agreement used for the transfer of any Derivative Material which wholly or partially contains or incorporates the Material.

10. UCSF or Recipient may terminate this Agreement at any time, in which case Recipient and Recipient Scientist will discontinue, within thirty (30) days after notice of such termination, its use of the Material. On or before the expiration of such thirty (30) day period, Recipient and Recipient Scientist agree, upon direction of UCSF, to return or destroy the Material. Any further research use by Recipient or Recipient Scientist of Derivative Materials which wholly or partially contain or incorporate the Materials shall be subject to all terms and conditions of this Agreement. If this Agreement is terminated by UCSF for material breach by Recipient of this Agreement, Recipient agrees that any Material and/or Derivative Material which wholly or partially contains or incorporates the Material will immediately be returned to UCSF or destroyed. The provisions of Paragraphs 2, 3, 6, 9–14, 16 and 17 of the Terms and Conditions shall survive any termination of this Agreement.

11. The Material is experimental in nature and shall be used with prudence and appropriate caution, since not all of its characteristics are known. THE MATERIAL IS PROVIDED WITHOUT WARRANTY OF MERCHANTABILITY OR FITNESS FOR A PARTICULAR PURPOSE OR ANY OTHER WARRANTY, EXPRESS OR IMPLIED. UCSF MAKES NO REPRESENTATION OR WARRANTY THAT THE USE OF THE MATERIAL WILL NOT INFRINGE ANY PATENT OR OTHER PROPRIETARY RIGHT.

12. In the event that a journal publication or scientific article is published based on use of the Material or Derivative Material, Recipient Scientist will make reasonable efforts to provide Sponsor with a copy of such publication, or the publication cite, promptly after it becomes available to Recipient Scientist, by sending it to Sponsor at the following address:

Geron Corporation, 230
Constitution Drive, Menlo Park, CA 94025
Attn: Corporate Development,
Fax number (650) 566-7181.

13. Recipient Scientist and Recipient shall acknowledge in any publication of results of the Research (a) UCSF as the source of the Material and (b) that the Material resulted from research conducted by Investigator under the Research Agreement with Sponsor.

14. Upon UCSF's or Sponsor's written request, Recipient agrees to provide, without cost other than reasonable handling and shipping costs, reasonable quantities of any Derivative Materials that wholly or partially contain or incorporate the Materials to UCSF or Sponsor for research purposes only at UCSF or Sponsor under the terms and conditions of a material transfer agreement to be negotiated between Recipient and UCSF, or Recipient and Sponsor, respectively. Such obligation shall apply only after Recipient has publicly disclosed or reasonably characterized such Derivative Materials, and is subject to the last sentence of Paragraph 6, above, with respect to Derivative Materials that contain or incorporate proprietary research material owned and provided by a third party. It is understood that neither the provision by Recipient to UCSF or Sponsor of any Derivative Material that wholly or partially contains or incorporates the Material nor anything else in this Agreement constitutes a grant by Recipient to UCSF or Sponsor of any license or other right which may be necessary to commercialize any Derivative Material created by Recipient.

15. Recipient Scientist and Recipient shall, at the request of UCSF, return or destroy all unused Material.

16. Except to the extent prohibited by applicable law, the Recipient assumes all liability for damages which may arise from the use, storage, handling or disposal of the Material or Derivative Material by Recipient or Recipient Scientist. Further, unless Recipient is an agency or department of the U.S. Government, and except to the extent prohibited by applicable law, Recipient agrees to indemnify and hold harmless UCSF and Sponsor, and their respective trustees, regents, officers, agents, and employees, and Sponsor, from any liability, loss, or damage they may suffer as a result of claims, demands, costs, or judgments against them arising out of the use, storage or disposal of the Material or Derivative Material by the Recipient or Recipient Scientist. Neither UCSF nor Sponsor will be liable to the Recipient for any loss, claim or demand made by the Recipient or the Recipient Scientist, or made against the Recipient or the Recipient Scientist by any other party, due to or arising from the use of the Material or Derivative Material by the Recipient or the Recipient Scientist.

17. Neither UCSF nor Sponsor will release or provide to Recipient or Recipient Scientist any identifiable private information about the Material or the donors of the embryos used to derive the Material, under any circumstances.

18. This Agreement is not assignable or transferable by Recipient.

19. The Material is provided with a transmittal fee solely to reimburse UCSF for its preparation and distribution costs. The amount of the fee for this transfer of Material will be $5,000/cell line. Payment should be made by check or Purchase Order to "The Regents of the University of California" and should be sent to:

> UCSF Office of Technology Management
> 1294 Ninth Avenue, Suite 1, Box 1209
> San Francisco, CA 94143-1209
> Attention: ES Cell MTA Coordinator

FOR THE REGENTS OF THE UNIVERSITY OF CALIFORNIA

Signature:
Name:
Title:
Date:

FOR RECIPIENT

Note: must be signed by authorized institution official

Signature:
Name:
Title:
Date:

RECIPIENT SCIENTIST

Signature:
Name:
Date:

Address to which the cells should be shipped:

Contact information for the individual responsible for receipt of the cells:

Name:
Phone number:
E-mail address:

Exhibit A
Descriptive Title

Exhibit B
ANNUAL CERTIFICATION STATEMENT

Annual Certification of Recipient Scientist: I have read and understood the conditions outlined in this Agreement and I understand my obligation as an employee of Recipient to abide by them in the receipt and use of the Material and any Derivative Material which wholly or partially contains or incorporates the Material. I further certify that I have not engaged and will not engage in commercial research using the Material or any Derivative Material which wholly or partially contains or incorporates the Material, where such research is a) sponsored by a for-profit entity, either through supply of materials or funding, except as such research sponsor has i) received a license from Wisconsin under the Wisconsin Patent Rights; or ii) receives no rights, whether actual or contingent, to the results of the sponsored research; or b) sponsored by a private nonprofit entity, either through supply of materials or funding, except as such research sponsor has i) received a license from Wisconsin under the Wisconsin Patent Rights; or ii) receives no commercial rights, whether actual or contingent, to the results of the sponsored research.

Name of Institution:
Signature of Recipient Scientist:
Print Name of Recipient Scientist:
Date:

Upon completion of the Research, please complete the following certification:

I have discontinued or completed the Research outlined in Exhibit A and have destroyed all remaining Material and any Derivative Material which wholly or partially contains or incorporates the Material.

Name of Institution:
Signature of Recipient Scientist:
Print Name of Recipient Scientist:
Date:

Please sign this form and send it to the following address on or before each anniversary of this Agreement:

> University of California, San Francisco
> Office of Technology Management
> 1294 Ninth Avenue, Suite 1, Box 1209
> San Francisco, CA 94143-1209
> Attention: ES Cell MTA Coordinator

Appendix IV
STEM CELL PATENTS

Appendix IV
Stem Cell Patents

Two issued patents for embryonic stem cells are presented here and reproduced in their entirety in the following pages: one filed by Brigid Hogan for nonmurine pluripotential cells and a second filed by James Thomson for primate embryonic stem cells. More than fifty others have been filed and are available through the USPTO (United States Patent Office). The two patents reproduced below represent the primary patents describing the generation and use of human ES cells

Pluripotential Embryonic Cells and Methods of Making Same

Abstract

The claimed invention is directed toward nonmurine pluripotential cells that have the ability to be passaged in vitro for at least 20 passages and which differentiate in culture into a variety of tissues. The scope of the claimed cells includes any nonmurine *ES* cells and particular claims are drawn to *human* pluripotential cells.

Inventors:	**Hogan; Brigid L. M.** (Brentwood, TN)
Assignee:	**Vanderbilt University** (Nashville, TN)
Appl. No.:	**217921**
Filed:	**March 25, 1994**

United States Patent	5,690,926
Hogan	November 25, 1997

Primate Embryonic Stem Cells

Abstract

A purified preparation of primate embryonic stem cells is disclosed. This preparation is characterized by the following cell surface markers: SSEA-1 (–); SSEA-4 (+); TRA-1-60 (+); TRA-1-81 (+); and alkaline phosphatase (+). In a particularly advantageous embodiment, the cells of the preparation

From: *Human Embryonic Stem Cells*
Edited by: A. Chiu and M. S. Rao © Humana Press Inc., Totowa, NJ

are human embryonic stem cells, have normal karyotypes, and continue to proliferate in an undifferentiated state after continuous culture for 11 months. The embryonic stem cell lines also retain the ability, throughout the culture, to form trophoblast and to differentiate into all tissues derived from all three embryonic germ layers (endoderm, mesoderm and ectoderm). A method for isolating a primate embryonic stem cell line is also disclosed.

Inventors:	**Thomson; James A.** (Madison, WI)
Assignee:	**Wisconsin Alumni Research Foundation** (Madison, WI)
Appl. No.:	**106390**
Filed:	**June 26, 1998**

United States Patent	6,200,806
Thomson	March 13, 2001

United States Patent [19]

Hogan

[11] Patent Number: 5,690,926

[45] Date of Patent: Nov. 25, 1997

[54] **PLURIPOTENTIAL EMBRYONIC CELLS AND METHODS OF MAKING SAME**

[75] Inventor: **Brigid L. M. Hogan**, Brentwood, Tenn.

[73] Assignee: **Vanderbilt University**, Nashville, Tenn.

[21] Appl. No.: **217,921**

[22] Filed: **Mar. 25, 1994**

Related U.S. Application Data

[63] Continuation-in-part of Ser. No. 958,562, Oct. 8, 1992, Pat. No. 5,453,357.

[51] Int. Cl.^{6} **C12N 5/06**; C12N 5/16; C12N 15/09; A01N 63/00

[52] **U.S. Cl.** **424/93.1**; 424/9.1; 424/93.21; 435/172.3; 435/325; 435/352; 435/353; 435/366

[58] **Field of Search** 800/2; 424/93.21, 424/9.1, 93.1; 435/172.3, 69.1, 240.21, 240.2; 935/40, 71

[56] **References Cited**

U.S. PATENT DOCUMENTS

5,166,065 11/1992 Williams et al. 435/240.1

FOREIGN PATENT DOCUMENTS

WO 90/03432 4/1990 WIPO .

OTHER PUBLICATIONS

Bradley et al., 1992. "Modifying the mouse: Design and Desire". Biotechnology 10:534–539, Mar. 1992.

McMahon et al., 1990, "The Wnt–1 (int–1) proto–oncogen is required for development of a large region of the mouse brain", Cell 62:1073–1085, Sep. 21, 1990.

H.P.M. Pratt, 1987, "Isolation, culture, and manipulations of preimplantation mouse embryos", in Mammalian Development: A practical Approach, M. Monk. ed., IRL Press. Washington, DC., pp. 13–42.

Lapidot et al., Science 255:1137–1141 (1992).

Anne McLaren, Nature 359:482–483 (1992).

Reynolds & Weiss, Science, 1707–1710(1992).

Snouwaert t al., Science, 257:1083–1088 (1992).

Erwin F. Wagner, The EMBO Journal, 9:3025–3032 (1990).

Stemple and Anderson, Cell 71:973–985 (1992).

Jones et al., Nature 347:188–189 (1990).

Resnick et al., Nature 359:550–551 (1992).

Evans and Kaufman, nNture 292:154–156 (1981).

Martin, PNAS 78:7634–7638 (1981).

Bradley et al., Nature 309:255–256 (1984).

Rathjen, P.D. Genes Dev. 4:2308–2318 (1990).

Rathjen, Cell 62:1105–1114 (1990).

De Felici and McLaren. Exp. Cell Res. 144:417–427 (1983).

Wabik–Sliz and McLaren, Exp. Cell. Res. 154:530–536 (1984).

Donovan et al., Cell 44:831–838 (1986).

Dolci et al., Nature 352:809–811 (1991).

Matsui et al., Nature 353:750–752 (1991).

Stewart and Mintz, Proc. Nat'l. Acad. Sci. U.S.A. 78:6314–6318 (1981).

Mintz et al., Proc. Nat'l. Acad. Sci. U.S.A. 75:2834–2838 (1978).

Noguchi and Stevens, J. Nat'l Cancer Inst. 69:907–913 (1982).

Ginsburg et al., Development 110:521–528 (1990).

Monk et al., Development 99:371–382 (1987).

Godin et al., Nature 352:807–808 (1991).

Matsui et al., Cell 70:841–847 (1992).

Notarianni et al., Journals of Reproduction & Fertlity Supplement 41:51–56 (1990).

(List continued on next page.)

Primary Examiner—Brian R. Stanton
Attorney, Agent, or Firm—Needle & Rosenberg, P.C.

[57] **ABSTRACT**

The claimed invention is directed towards non-murine pluripotential cells that have the ability to be passaged in vitro for at least 20 passages and which differentiate in culture into a variety of tissues. The scope of the claimed cells includes any non-murine ES cells and particular claims are drawn to human pluripotential cells.

7 Claims, 6 Drawing Sheets

5,690,926

Page 2

OTHER PUBLICATIONS

Smith et al., *Nature* 336:688–690 (Dec. 15, 1988).

Mummery et al., *Cell Differentiation and Development* 30:195–206 (Jun. 1, 1990).

Piedrahita et al., *Theriogenology* 29:286 (Jan. 1988).

Handyside et al., *Roux's Arch. Dev. Biol.* 196:185–190 (1987).

Flake et al., *Science* 233:776–778 (Aug. 15, 1986).

Ware et al., Biology of Reproduction Supp. 38:129 (1988).

Robertson et al. *Nature* 323:445–448 (1986).

Labosky et al., *1994 Germline development*, Wiley, Chichester (Ciba Foundation Symposium 182) pp. 1557–178.

Patent application abstract, U.S. Serial No. 07/958,009, Apr. 1, 1993.

Mar. 25, 1994 letter from Judith Plesset, Ph.D., Technology Licensing Specialist, Office of Technology Transfer, Department of Health and Human Services, National Institutes of Health, reporting abandonment of U.S. Serial No. 07/958,009.

418

1

PLURIPOTENTIAL EMBRYONIC CELLS AND METHODS OF MAKING SAME

This application is a continuation-in-part of U.S. Ser. No. 07/958,562, filed Oct. 8, 1992, now U.S. Pat. No. 5,453,357.

This invention was made with government support under grant number HD25580-04 from the National Institute of Health Child Health and Development. The United States government has certain rights in the invention.

BACKGROUND OF THE INVENTION

1. Field of the Invention

This invention relates to pluripotential embryonic stem cells and methods and compositions for making pluripotential embryonic stem cells.

2. Background Art

Primordial germ cells (PGCs) in the mouse are thought to be derived from a small population of embryonic ectoderm (epiblast) cells set aside at the egg cylinder stage prior to gastrulation (Lawson and Pederson, 1992), or even earlier (Soriano and Jaenisch, 1986). By 7 days post coitum (p.c.) about 100 alkaline phosphatase (AP) positive PGCs can be detected in the extra embryonic mesoderm just posterior to the definitive primitive streak (Ginsberg et al., 1990). These cells continue to proliferate and their number increases rapidly to around 25,000 at 13.5 days p.c. (Mintz and Russell, 1957; Tam and Snow, 1981). At the same time the PGCs migrate from the base of the allantois along the hind gut and reach the genital ridges by 11.5 days p.c. In the genital ridge, PGCs stop dividing at around 13.5 days p.c., and enter either mitotic arrest in the developing testis or meiosis in the ovary. In a few strains of mice, e.g. 129, this normal program can be disrupted if the male genital ridge from an 11.5 to 12.5 days p.c. embryo is grafted to an ectopic site such as the testis or kidney capsule. Under these conditions some PGCs give rise to teratomas and transplantable teratocarcinomas containing pluripotential embryonal carcinoma (EC) stem cells (Stevens and Makerisen, 1961; Stevens, 1983; Noguchi and Stevens, 1982).

Previous studies have shown that steel factor (SF) and leukemia inhibitory factor (LIF) synergistically promote the survival and in some cases the proliferation of mouse PGCs in culture (Godin et al., 1991; Dolci et al., 1991; Matsui et al., 1991). However, under these conditions, PGCs have a finite proliferative capacity that correlates with their cessation of division in vivo. A similar finite proliferative capacity has been reported for oligodendrocyte-type 2 astrocyte (O-2A) progenitor cells in the rat optic nerve. In this case, PDGF is involved in the self renewal growth of O-2A cells (Noble et al., 1988; Raft et al., 1988). After a determined number of cell divisions, O-2A cells may lose their responsiveness to PDGF and start differentiating into oligodendrocytes. If both PDGF and basic fibroblast growth factor (bFGF) are added in culture, O-2A progenitor cells keep growing without differentiation (Bogler et al., 1990).

Since pluripotential embryonic stem cells (ES) can give rise to virtually any mature cell type they are of great value for uses such as creating genetically manipulated animals. However, according to the published scientific literature, it has previously been possible only to obtain ES cells from mice. These murine ES cells were obtained from cultures of early blastocysts. Attempts at isolating ES cells from other animals apparently have failed. One patent publication, Evans et al., published Apr. 5, 1990 under PCT Publication WO 90/03432, claims that pluripotential ES cells can be obtained from ungulate blastocysts in vitro. The application

2

claims that these cells are expected to be epithelial and to have a very different morphology to mouse ES cells because ungulate embryos normally form an "embryonic disc". This appears to be the basis of the allegation that the cells which they grow out of pig and cow blastocysts and which have a more epithelial morphology than mouse ES cells are, in fact, ES cells. However, mouse embryos also develop an epithelial layer of pluripotential embryonic ectoderm or epiblast cells. This layer is called an "egg cylinder" rather than an "embryonic disc". Therefore, there is apparently no strong embryological reason why the ungulate ES cells should have a different morphology to mouse ES cells. In addition, the evidence presented in the Evans application for the differentiation of the cow and pig putative ES cell lines into differentiated cell types in monolayer culture, in embryoid bodies and in tumors, is not convincing. Therefore, there is a great need to produce and maintain ES cells from a variety of different animals.

The present invention satisfies this need by demonstrating that, in the presence of bFGF, SF and LIF, PGCs continue to proliferate in culture and give rise to colonies of ES cells. These stem cells can give rise to a wide variety of mature, differentiated cell types both in vitro and when injected into nude mice and when combined with embryos to form a chimera.

SUMMARY OF THE INVENTION

The present invention provides a non-mouse, including human, pluripotential embryonic stem cell which can:

(a) be maintained on feeder layers for at least 20 passages; and

(b) give rise to embryoid bodies and multiple differentiated cell phenotypes in monolayer culture.

In addition, in non-humans, the cells can form chimeras when combined with host embryos and give rise to mature sperm.

The invention further provides a method of making a pluripotential embryonic stem cell comprising culturing primordial germ cells, embryonic ectoderm cells and/or germ cell progenitors in a composition comprising a growth enhancing amount of basic fibroblast growth factor, leukemia inhibitory factor, membrane associated steel factor, and soluble steel factor to primordial germ cells under cell growth conditions, thereby making a pluripotential embryonic stem cell.

Also provided are compositions useful to produce the pluripotent embryonic stem cells and methods of screening associated with the method of making the embryonic stem cell.

BRIEF DESCRIPTION OF THE DRAWINGS

FIGS. 1A–C show the effect of growth factors on murine PGCs in culture.

(A) PGCs from 8.5 day p.pc. embryos were seeded into wells containing SISl[4] feeder cells either alone (open circles) or with soluble rSF (closed circles), soluble rSF and LIF (closed squares), or soluble rSF, LIF and bFGF (closed triangles). Cultures were fixed and the number of AP positive cells counted.

(B) As in (A) except that cells were cultured without added factors (open circles), with soluble rSF (closed circles), with bFGF (closed triangles) or with soluble rSF and bFGF (open triangles).

(C) As in (A) except that cells were cultured on Sl[4]-m220 cells either alone (open circles) or with soluble rSF (closed

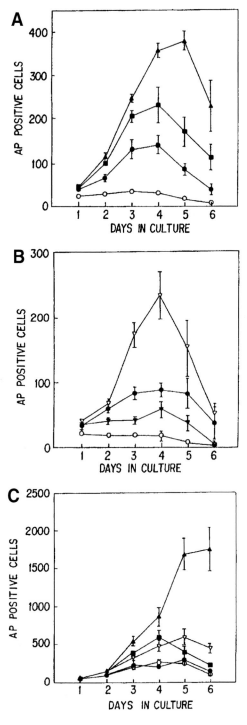

Fig. 1.

5,690,926

3

circles), soluble rSF and LIF (closed squares), soluble rSF and bFGF (open triangles) and soluble rSF, LIF and bFGF (closed triangles).

Each experiment was carried out with duplicate wells and numbers are the means+s.e.m. of three separate experiments.

FIGS. 2A–I show the morphology of primary and secondary cultures of PGCs and their descendants. PGCs from 8.5 d p.c. embryos (A–E, G,H) or 12.5 d p.c. male genital ridges (F) were cultured on Sl⁴-m220 cells as described and stained for AP activity.

(A) Primary culture after 4 days in the presence of LIF. Note that the AP positive cells are scattered among the feeder cells.

(B) Primary culture after 4 days in the presence of soluble rSF, LIF and bFGF. Note that the AP positive cells now form tight clumps.

(C) As for B, but after 6 days in culture.

(D) Secondary culture after 6 days in the presence of soluble rSF, LIF, and bFGF In this colony all the cells are AP positive.

(E) As for D except that cells at the edges of the colony are AP negative.

(F) PGCs from 12.5 day p.c. male genital ridge were cultured for 6 days in the presence of soluble rSF, LIF and bFGF. Colonies of tightly packed AP positive cells are present.

(G) Colony of ES-like cells in a secondary culture with soluble SF, LIF and bFGF stained with SSEA-1 monoclonal antibody and for AP activity. Phase contrast microscopy.

(H) The same colony as in G viewed by fluorescence microscopy. AP positive cells also express SSEA-1.

(I) Colony grown under same conditions as (G) but stained without primary antibody Scale bars=200/μm.

FIG. 3 shows the effect of growth factors on male and female PGCs in culture. Cells were dissociated from either male (squares) or female (circles) genital ridges from 12.5 day p.c. mouse embryos and cultured on Sl⁴-m220 feeder cells either alone (empty symbols) or with soluble rSF, LIF and bFGF (filled symbols). Cells were fixed and the number of AP positive cells counted. The experiment was carried out three times, with duplicate wells.

FIGS. 4A–E shows the morphology of undifferentiated PGC derived ES cells and their differentiated derivatives.

(A) Colony of densely packed ES-like cells obtained from PGCs of an 8.5 day p.c. embryo grown on Sl⁴-m220 cells in the presence of soluble rSF, LIF and bFGF for 6 days. Scale bar=100 μm.

(B) Simple embryoid bodies with an outer layer of endoderm (arrows) obtained after culturing PGC-derived ES cells for 4 days in suspension.

(C) Section of a teratoma obtained by injecting ES-like cells derived from PGCs of an 8.5 day p.c. embryo into a nude mouse. The region shown here contains neural tissue and pigmented epithelium. Scale bar=200 μm.

(D) Region of the same tumor as in (C) showing a dermoid cyst and secretory epithelium.

(E) Region of the same tumor as in C and D, showing bone and cartilage. The differentiated tissues shown in C-E were seen in addition to other tissue types in multiple tumors from all three lines tested.

FIGS. 5A-C shows a photomicrograph of a colony of alkaline phosphatase positive cells derived in culture from an approximately 10.5 week old human embryonic testis. Following dissociation, testis cells were seeded into wells of

4

a 24-well plate containing irradiated feeder cells (Sl⁴ h220) and growth factors 10 ng/ml bFGF, 60 ng/ml soluble SCF and 10 ng/ml LIF). After 5 days the cells were subcultured at a dilution of 1:4 into wells containing a feeder layer of irradiated mouse embryo fibroblasts with the same cocktail of growth factors. After 10 days the cultures were fixed and stained for alkaline phosphatase activity. The colony shown here (one of many) closely resembles colonies of alkaline phosphatase positive cells derived from primordial germ cells of the mouse embryo (see FIG. 2C). In particular, the human cells associate into tightly packed clusters (see arrow in FIG. 5C). FIGS. 5A, 5B and 5C are different magnifications of the same colony. The shadow in FIG. 5A is the edge of the well in which the cells were growing.

DETAILED DESCRIPTION OF THE INVENTION

The term "embryonic ectoderm" is used herein. "Embryonic ectoderm" and "epiblast" can be used interchangeably to refer to the same cell type.

A "pluripotential embryonic stem cell" as used herein means a cell which can give rise to many differentiated cell types in an embryo or adult, including the germ cells (sperm and eggs). Pluripotent embryonic stem cells are also capable of self-renewal. Thus, these cells not only populate the germ line and give rise to a plurality of terminally differentiated cells which comprise the adult specialized organs, but also are able to regenerate themselves. This cell type is also referred to as an "ES cell" herein.

A "fibroblast growth factor" (FGF) as used herein means any suitable FGF. There are presently seven known FGFs (Yamaguchi et al. (1992)). These FGFs include FGF-1 (acidic fibroblast growth factor), FGF-2 (basic fibroblast growth factor), FGF-3 (int-2), FGF-4 (hst/K-FGF), FGF-5, FGF-6, FGF-7 and FGF-8. Each of the suitable factors can be utilized directly in the methods taught herein to produce or maintain ES cells. Each FGF can be screened in the methods described herein to determine if the FGF is suitable to enhance the growth of or allow continued proliferation of ES cells or their progenitors. Various examples of FGF and methods of producing an FGF are well known; see, for example, U.S. Pat. Nos. 4,994,559; 4,956,455; 4,785,079; 4,444,760; 5,026,839; 5,136,025; 5,126,323; and 5,155,214.

"Steel factor" (SF) is used herein. SF is also called stem cell factor, mast cell growth factor and c-kit ligand in the art. SF is a transmembrane protein with a cytoplasmic domain and an extracellular domain. Soluble SF refers to a fragment cleaved from the extracellular domain at a specific proteolytic cleavage site. Membrane associated SF refers to both normal SF before it has been cleaved or the SF which has been altered so that proteolytic cleavage cannot take place. SF is well known in the art; see European Patent Publication No. 0 423 980 A1, corresponding to European Application No. 90310889.1.

"Leukemia Inhibitory Factor" (LIF) is also used herein. LIF is also known as DIA or differentiation inhibiting activity. LIF and uses of LIF are also well known; see for example U.S. Pat. Nos. 5,187,077 and 5,166,065.

It should be recognized that FGF, SF and LIF are all proteins and as such certain modifications can be made to the proteins which are silent and do not remove the activity of the proteins as described herein. Such modifications include additions, substitutions and deletions. Methods modifying proteins are well established in the art (Sambrook et al., *Molecular Cloning: A Laboratory Manual*, 2nd Ed., Cold Spring Harbor Laboratory, Cold Spring Harbor, N.Y., 1989).

421

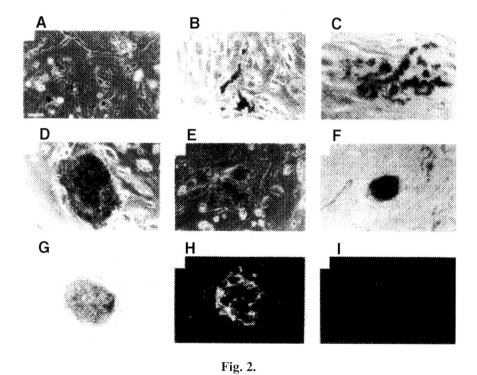

Fig. 2.

This invention provides a non-mouse pluripotential ES cell which can be maintained on feeder layers for at least 20 passages, and give rise to embryoid bodies and multiple differentiated cell phenotypes in monolayer culture. Only those non-mouse animals which can be induced to form ES cells by the described methods are within the scope of the invention. Given the methods described herein, an ES cell can be made for any animal. However, mammals are preferred since many beneficial uses of mammalian ES cells exist. Mammalian ES cells such as those from rats, rabbits, guinea pigs, goats, pigs, cows, and humans can be obtained. Alternatively, embryos from these animals can be screened for the ability to produce ES cells.

The ES cells of this invention can be maintained for at least 20 passages. However, the ES cells can be capable of indefinite maintenance. Typically, after about 10 passages the cells are frozen so that the starting population is not altered by minor chromosomal alterations.

Once the non-mouse ES cells are established, they can be genetically manipulated to produce a desired characteristic. For example, the ES cells can be mutated to render a gene non-functional, e.g. the gene associated with cystic fibrosis or an oncogene. Alternatively, functional genes can be inserted to allow for the production of that gene product in an animal, e.g. growth hormones or valuable proteins. Such methods are very well established in the art (Sedivy and Joyner (1992)).

The invention also provides a composition comprising:

(a) pluripotential ES cells; and

(b) an FGF, LIF, membrane associated SF, and soluble SF in amounts to enhance the growth of and allow the continued proliferation of the cell. Thus, this composition represents the composition after primordial germ cells, embryonic ectoderm or germ cells have become pluripotential ES cells. The pluripotential ES cells can continue to be maintained in this composition or alternatively they can be maintained on a feeder layer. Optimally, HF can be added to the feeder layer.

Also provided is a composition comprising an FGF, LIF, membrane associated SF, and soluble SF in amounts to enhance the growth of, and allow the continued proliferation of primordial germ cells and the formation of pluripotent ES cells from the primordial germ cell. This composition need not include primordial germ cells but comprises the various growth factors in amounts that promote the growth, proliferation and formation of pluripotent ES cells. Thus, this composition can be sufficient for the establishment of pluripotent ES cells from embryonic ectoderm cells or germ cells.

Also provided is a composition comprising: (a) mammalian primordial germ cells and/or germ cells and/or embryonic ectoderm cells; and (b) a fibroblast growth factor, leukemia inhibitory factor, membrane associated steel factor and soluble steel factor in amounts to enhance the growth and allow the continued proliferation of the cells and the formation of pluripotent embryonic stem cells.

Typically, the compositions of the invention include a feeder layer. Feeder layers can either be cells or cell lines cultured for the purpose of culturing pluripotent ES cells. Alternatively, feeder layers can be derived from or provided by the organ or tissue in which the primordial germ cells, embryonic ectoderm cells or germ cells are located, e.g. the gonad. Thus, if the somatic cells of the tissue or organ in which the desired cells are located are sufficient to provide the appropriate culture environment, a separate feeder layer is not required. Alternatively, the feeder cells could be

substituted with extracellular matrix plus bound growth factors. Feeder layers which are representative of those which can be utilized are set forth in the Examples. Naturally, the membrane associated SF can be contained on the cells of such a feeder layer.

The compositions arise from the fact that FGF, LIF and SF are used either to enhance the growth and proliferation of primordial germ cells and/or embryonic ectoderm cells to become ES cells. Growth and proliferation enhancing amounts can vary depending on the species or strain of the cells, and type or purity of the factors. Generally, 0.5 to 500 ng/ml of each factor within the culture solution is adequate. In a more narrow range, the amount is between 10 to 20 ng/ml for bFGF and LIF and between 10 to 100 ng/ml for SF. Regardless of whether the actual amounts are known, the optimal concentration of each factor can be routinely determined by one skilled in the art. Such determination is performed by titrating the factors individually and in combination until optimal growth is obtained. Additionally, other factors can also be tested to determine their ability to enhance the effect of FGF, LIF and SF on ES cell proliferation. As described below, such other factors, or combinations of factors when used to enhance ES cell proliferation can be included within the above compositions. Also, compounds and fragments of FGF, LIF and SF which mimic the function of these factors can be used to enhance the growth and proliferation of the cells to become ES cells and are included within the scope of the invention.

The factors are essential to the formation of pluripotent ES cells. Thus, the amount of the factors utilized is determined by the end result of the pluripotent ES cells. However, the factors also serve to enhance the growth and allow the continued proliferation of the cells. Relatedly, the factors also appear to help the cells survive.

Alternatively, FGF, LIF, and SF can be used to maintain ES cells. The amounts of FGF, LIF and SF necessary to maintain ES cells can be much less than that required to enhance growth or proliferation to become ES cells. However, the cells may be maintained on a feeder layer without the addition of growth factors. Optimally, LIF can be added to enhance maintenance.

In general, FGF or LIF from a species different from the source of the ES, primordial germ cell, germ cell or embryonic ectoderm cell can be utilized. However, all the factors utilized and especially the SF utilized are preferably from the same species as the utilized cell type. However, FGF, LIF or SF from various species can be routinely screened and selected for efficacy with a cell from a different species. Recombinant fragments of FGF, LIF or SF can also be screened for efficacy as well as organic compounds derived from, for example, chemical libraries.

The invention also provides a method of making a pluripotential ES cell comprising administering a growth enhancing amount of FGF, LIF, membrane associated SF, and soluble SF to primordial germ cells and/or embryonic ectoderm cells under cell growth conditions, thereby making a pluripotential ES cell. Thus, primordial germ cells and embryonic ectoderm cells can be cultured as a composition in the presence of these factors to produce pluripotent ES cells. As noted above, typically the composition includes a feeder layer.

The invention also provides a method of making a mammalian pluripotential embryonic stem cell comprising culturing a germ cell or a composition from postnatal mammalian testis in a composition comprising a growth enhancing amount of basic fibroblast growth factor, leukemia inhibitory factor, membrane associated steel factor, and soluble steel factor, thereby making a pluripotential embryonic stem cell from a germ cell. "Germ cells" as used herein means the cells which exist in neonatal or postnatal testis and are the progenitors of gametes. In the testis, these germ cells represent a small population of stem cells capable of both self-renewal and differentiation into mature spermatogonia. Thus, "germ cells" are the postnatal equivalent to the prenatal primordial germ cells and can include primitive or immature spermatogonia such as type A spermatogonia or any undifferentiated early stage cell that can form a pluripotent embryonic stem cell.

These methods can be practiced utilizing any animal cell, especially mammal cells including mice, rats, rabbits, guinea pigs, goats, cows, pigs, humans, etc. The ES cell produced by this method is also contemplated.

Also provided is a method of screening cells which can be promoted to become an ES cell comprising contacting the cells with FGF, LIF, membrane associated SF, and soluble SF in amounts to enhance the growth of and allow proliferation of the cells and determining which cells become ES cells. Utilizing this method, cells other than primordial germ cells, germ cells, and embryonic ectoderm cells can be selected as a source of ES cells.

Since the invention provides ES cells generated from virtually any animal, the invention provides a method of using the ES cells to contribute to chimeras in vivo in non-humans comprising injecting the cell into a blastocyst and growing the blastocyst in a foster mother. Alternatively, aggregating the cell with a non-human morula stage embryo and growing the embryo in a foster mother can be used to produce a chimera. Chimeric animals can subsequently be bred to obtain germ line transmission of ES cell traits. As discussed above, the ES cells can be manipulated to produce a desired effect in the chimeric animal. The methods of producing such chimeric animals are well established (Robertson (1987)).

Alternatively, the ES cells can be used to derive cells for therapy to treat an abnormal condition. For example, derivatives of human ES cells could be placed in the brain to treat a neurodegenerative disease. Relatedly, ES cells can be used to screen factors to determine which factors produce derivative (more differentiated) cells. Many standard means to determine the presence of a more differentiated cell are well known in the art.

FGF, SF and LIF have been shown herein to be critical for making ES cells. However, as noted above for FGF, other members of the respective growth factor family could also be used to make ES cells. Thus, later discovered members of each family can merely be substituted to determine if the new factor enhances the growth and allows the continued proliferation of PGCs or embryonic ectoderm cells to form ES cells. For example, if a new member of the LIF family is discovered, the new LIF is merely combined with SF and FGF to determine if the new family member enhances the growth and allows the continued proliferation of PGCs or embryonic ectoderm cells. Thus, this invention provides the use of family members and a method of screening family members for activity.

Likewise, additional growth factors may be found useful in enhancing the growth and proliferation of PGCs or embryonic ectoderm cells from various animals. This invention provides combining FGF, SF and LIF with other growth factors to obtain or enhance the production of ES cells. Thus, a method of screening other growth factors for the ability to promote PGCs and embryonic ectoderm cells to form ES

cells is also provided. In this regard, IL-11, IL-6, CNTF, NGF, IGFII, flt3/flk-2 ligand and members of the Bone Morphogenetic Protein family are good screening candidates and can be used to promote ES cell formation.

EXAMPLES

All the cell types and other materials listed below can be obtained through available sources and/or through routine methods.

MATERIALS AND METHODS

Feeder cells

The Sl/Sl[4] cell line, derived from a homozygous null Sl/Sl mouse embryo, and its derivative, Sl[4]-m220, which stably expresses only membrane bound murine SF lacking exon 6 encoding the proteolytic cleavage site, were obtained from Dr. David Williams (Howard Hughes Medical Institute, Indiana University Medical School). Other cell lines which produce adequate SF can be substituted for Sl/Sl[4], for example mouse or human embryo fibroblasts or cell lines or somatic cell lines from gonads or genital ridges. Combinations of feeder cells can also be utilized. They were maintained in DMEM with 10% calf serum and 50 ug/ml gentamicin. For making feeder layers they were irradiated (500 rads) and plated at a density of 2×105 per well of 24-well plates (Falcon) in the same medium, 24 hrs before use. Wells were pre-treated with 1% gelatin. STO cells stably transfected with human LIF and the bacterial neor gene (SLN) were obtained from Dr. Allan Bradley.

Primary cultures of PGCs

Embryos were from ICR females mated with (C57BLxDBA)F1 males. Noon of the day of plug is 0.5 day post coitum (p.c.). The caudal region of 8.5 day p.c. embryos (between the last somite and the base of the allantois) was dissociated into single cells by incubation at 37° C. with 0.05% trypsin, 0.02% EDTA in Ca^{++}/Mg^{++} free Dulbecco's phosphate-buffered saline (PBS) for about 10 mins with gentle pipetting. At this stage there are between about 149 and 379 PGCs in each embryo (Mintz and Russell, 1957). Cells from the equivalent of 0.5 embryo were seeded into a well containing feeder cells as above and 1 ml of DMEM, 2 mM glutamine, 1 mM sodium pyruvale, 100 i.u./ml penicillin and 100 ug/ml streptomycin and 15% fetal bovine serum (PGC culture medium). Finely minced fragments of genital ridges from 11.5 and 12.5 day p.c. embryos were trypsinized as above and plated at a concentration of 0.1 embryo per well. Growth factors were added at the time of seeding, usually at the following concentrations, which were shown to be optimal for PGC proliferation; recombinant human LIF and bFGF (10–20 ng/ml) and soluble rat SF (60 ng/ml). The medium was changed every day.

Secondary culture of PGC

Primary cultures were trypsinized and reseeded into wells containing Sl[4]-m220 feeder layers in PGC culture medium. For further subculture, rounded colonies of densely packed ES-like cells were carefully picked up in a finely drawn pipette and trypsinized in a microdrop under mineral oil before seeding into wells containing feeder cells as above. After several subcultures in this way, cultures were passaged without picking individual colonies.

Alkaline phosphatase (AP) staining

This was carried out as described (Matsui et al. 1991). After staining, AP positive cells were counted using an inverted microscope.

SSEA-1 staining

PGC cultures on Sl[4]-m220 feeder cells on a chamber slide (Nunc) were washed twice with PBS containing 2% calf serum, 0.1% sodium azide and then incubated with mouse monoclonal antibody SSEA-1 (1:100 dilution) on ice for 30 min. After washing with PBS, cells were incubated for 30 mins with FITC-conjugated Fab' fragment of goat anti mouse IgG (H+L) (Cappell, 1:5 dilution). After washing in PBS, cells were fixed in 4% paraformaldehyde before staining for AP.

Tumors in nude mice

Approximately 2×10^6 cells from three independent lines were injected subcutaneously into nude mice (three mice per line). After three weeks tumors were fixed in Bouin's fixative, processed for histology and sections stained with haematoxylin and eosin.

Chimera formation

Ten to fifteen cells from two independent lines derived from 8.5 day p.c. embryos were injected into the blastocoel of 3.5 day p.c. blastocysts of either ICR or C57BL/6 mice. These were returned to the uteri of 2.5 day p.c. pseudopregnant foster mothers.

Culture of murine PGCs in the presence of growth factors

Initial experiments used Sl/Sl[4] cells derived from a homozygous null Sl/Sl mutant mouse as a feeder layer for the culture of cells dissociated from the posterior of 8.5 days p.c. embryos, and AP staining as a marker for PGCs (FIG. 1A). As shown previously (Matsui et al., 1991), soluble SF and LIF act synergistically on PGCs. Addition of bFGF further enhances growth, and the cells continue to increase in number until day 5 in culture, i.e. one day longer than usual. The effect of bFGF alone is small, and both SF and LIF are needed in addition to bFGF for maximal effect on PGC growth (FIG. 1A, B). A variety of other growth factors, including human activin, Bone Morphogenetic Protein-4, βNGF, and PDGF at 10 and 50 ng/ml had no effect in the presence of SF and LIF.

Membrane associated SF seems to play an important role in PGC proliferation since Sl[d] mouse mutants which make only soluble SF have a reduced number of PGCs in vivo, and membrane associated SF is more effective than soluble SF in supporting PGC growth and survival in culture (Dolci et al., 1991; Matsui et al., 1991). To test the effect of added factors in the presence of membrane associated SF, 8.5 day p.c. PGCs were cultured on Sl[4]-m220 feeder cells, which express only membrane associated SF (Matsui et al., 1991; Toksoz et al., 1991). Both LIF and bFGF separately enhance PGC growth on Sl[4]-m220 feeder cells with added soluble rSF. However, when LIF and bFGF are added together, PGC growth is dramatically stimulated and the cells continue to proliferate through to day 6 in culture (FIG. 1C). The cells survive until day 8, at which time the feeder layer deteriorates, but they can be trypsinized and subcultured (see below).

Pregonadal PGCs are motile in vivo, and when cultured with LIF on a Sl[4]-m220 feeder layer they form burst colonies of cells with a flattened and polarized morphology, characteristic of motile cells (FIG. 2A). In contrast, PGCs cultured on a Sl[4]-m220 feeder layer with soluble SF and bFGF or with bFGF and LIF (FIG. 2B, C), form discrete colonies of tightly packed cells. These colonies increase in size over day 6 in culture only when both bFGF and LIF are present (FIG. 2C).

To determine whether PGCs and their descendants continue to proliferate in culture, primary colonies of PGCs were trypsinized after 6 days in culture and replated on a fresh Sl[4]-m220 feeder layer with added growth factors. By day 6 in secondary culture, large colonies of densely packed AP positive cells resembling embryonic stem (ES) cells are present (FIG. 2D,E; FIG. 4, A), with an overall plating

efficiency of about 5%. These colonies are also positive for the expression of the antigen SSEA-1, a characteristic of PGCs (Donovan et al., 1986) and undifferentiated embryonal carcinoma and ES cells (Solter and Knowles, 1978) (FIG. 2 G, H). Although the growth of primary cultures is strictly dependent on the presence of LIF and bFGF, secondary colonies can form in the absence of these factors (Table 1), indicating a reduced exogenous growth factor requirement for the descendants of PGCs after subculture. Most of the colonies show strong, uniform AP staining. However, some colonies contain only a small number of strongly stained cells, surrounded by cells which are weakly stained or negative (FIG. 2E). In many cases these negative cells are larger and have a more flattened morphology than the AP positive cells. For further subculture, individual colonies of cells with a distinctive, tightly packed, ES cell-like morphology are picked up in a micropipet, trypsinized and replated on a fresh feeder layer with added factors. Such colonies can be subcultured at least ten times and continue to give rise to colonies of similar morphology. In later passages, these cultures were transferred to feeder layers of STO cells in medium without added factors normally used for blastocyst-derived ES cell culture (Robertson, 1987). Under these conditions they continue to proliferate in an undifferentiated state, for a total of at least 20 passages.

Two independent lines at passage 14 (1/14, 2/14) and one at passage 20 (3/20) were karyotyped. Most cells had a normal or near normal XY karyotype, but in two lines (2/14 and 3/20) there was a significant proportion of trisomic cells. Long term culture of PGC-derived cells from genital ridges

Since transplantable teratocarcinomas can be induced experimentally by grafting genital ridges from 11.5 or 12.5; days p.c. male embryos of the 129 strain to an ectopic site, we tested the possibility that ES-like cells can be obtained from genital ridges in culture. Genital ridges were trypsinized and the cells plated on an Sl[4]-m220 feeder layer with soluble SF, LIF and bFGF. The number of PGCs initially declines but increases after 3 days, and by 6 days colonies of densely packed, AP positive cells can be seen (FIG. 2F). If cells from male and female 12.5 days p.c. genital ridges are cultured separately, male PGCs increase in number and form colonies. In contrast, only a few female PGCs form colonies (FIG. 3). The differentiation capacity of genital ridge-derived colonies has not so far been tested.

Differentiation of PGC-derived ES cells in vitro and in nude mice

Four independent lines of undifferentiated cells derived from 8.5 day embryos and cultured onto STO feeder layers were trypsinized and pipetted gently to generate small clumps of cells which were then placed in bacteriological plastic dishes. After five to seven days most of the clumps differentiated into typical simple or cystic embryoid bodies (EBs), with a clear outer layer of extraembryonic endoderm cells (FIG. 4, B). When these EBs were returned to tissue culture plastic dishes they rapidly attached and over two weeks gave rise to a variety of cell types, including extraembryonic endoderm, spontaneously contracting muscle, nerve and endothelial and fibroblast-like cells.

Three of these four lines, at passages 9 and 15 on STO cells, were injected subcutaneously into nude mice. Each line gave rise to multiple, well-differentiated teratocarcinomas, containing a wide variety of tissues, including keratinized, secretory and ciliated epithelium, neuroepithelium and pigmented epithelium, cartilage, bone, and muscle, as well as nests of undifferentiated embryonic cells (FIG. 4, C–E).

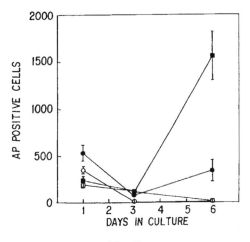

Fig. 3.

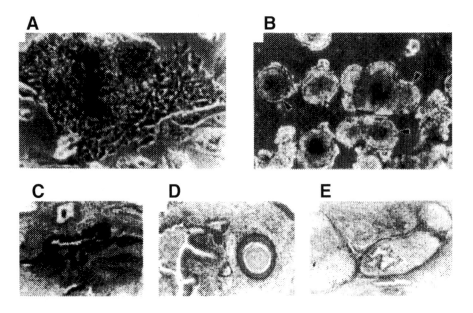

Fig. 4.

5,690,926

PGC-derived ES cells can contribute to chimeras in vivo

To test whether the descendants of PGCs in culture are able to contribute to chimeras in vivo, 10–15 cells with an ES-like morphology from two independent early passage cultures derived from 8.5 day embryos and cultured on either Sl⁴m220 cells or STO cells were injected into host ICR or C57BL/6 blastocysts. From a total of 21 pups born, four were chimeric, as judged by coat color, but only two were extensive, with approximately 50 and 90% chimerism. The 50% coat color chimera, generated by injecting cells from the 4th passage on STO cells into an ICR blastocyst, died at 11 days after birth and showed stunted growth and skeletal abnormalities. The 90% coat color chimera, obtained by injecting cells from the 6th passage on STO cells into a C57BL/6 blastocyst, had no obvious abnormalities.

Germ line transmission

Materials and methods

PGC culture

Cultures are initiated as described above by dissecting C57BL/6 8.5 days p.c. embryos free of extraembryonic tissues. Fragments comprising the posterior third of the embryo (from the base of the allantois to the first somite) are then pooled, rinsed with Dulbecco's Ca⁺⁺.Mg⁺⁺ free phosphate buffered saline (PBS) and dissociated with 0.25% trypsin, 1 mM EDTA (GIBCO) and gentle pipetting. This single cell suspension is then plated in 0.1% gelatin coated 24 well dishes (Corning) with irradiated Sl/Sl⁴ m220 cells as feeder layers at a concentration of approximately 0.5 embryo equivalents per well. The cultures were grown in Dulbecco's modified Eagle's medium (DMEM) (Specialty Media, Lavallette, N.J.) supplemented with 0.01 mM non-essential amino acids (GIBCO), 2 mM glutamine (GIBCO), 50 µg/ml gentamycin (Sigma), 15% fetal bovine serum (selected batches, Hyclone) and 0.1 mM 2-mercaptoethanol (Sigma). For these primary cultures, the medium is additionally supplemented with soluble recombinant rat SF at 60 ng/ml, bFGF at 20 ng/ml (GIBCO), and LIF at 20 ng/ml. After 6 days some of the cultures are stained for alkaline phosphatase (AP) as described above in order to assess the survival and proliferation of PGCs. After 10 days, parallel cultures are dissociated into single cells and plated onto mouse embryo fibroblast (reef) feeder layers with LIF (ESGRO, GIBCO 1000 U/ml). These cultures are monitored for the appearance of colonies of EG cells. Individual EG colonies are isolated with a micropipette and lines established. EG cultures are then maintained in the same manner as ES cell lines with irradiated mefs as feeder cells and LIF (Smith et al., 1988 and Williams et al., 1988).

Blastocyst injection

Ten to twenty EG cells at passage numbers 6 to 10 were injected into 3.5 days p.c. blastocysts from BALFl/c mice. Foster mothers were (C57BL/6×DBA)_F1 females mated to vasectomized Swiss Webster males. Injected blastocysts were transferred to the uterus of 2.5 days p.c. foster mothers (Hogan et al., 1986) and chimeric pups were identified by their coat color. Chimeras were bred to either BALB/c or ICR mice and germ line transmission was judged on the day of birth by the presence of eye pigment.

Results

Two ES lines derived from 8.5 day C57BL/6 embryos produced chimeras which transmitted the C57BL/6 genome through mature sperm (germ line transmission). This was judged by mating male chimeras with albino females and observing the production of pigmented pups. With the ES cell line known as TGC⁸⁵10, nine chimeric males were obtained and four produced pigmented pups. With the ES

cell line known as TGC⁸⁵19 six chimeric males were obtained and one of these produced pigmented pups.

Generation of ES cells from other mammals

ES cells from other mammals can be produced using the methods described above for murine. The mammalian cell of choice is simply substituted for murine and the murine methods are duplicated. The appropriate species specific growth factors (e.g. SF) can be substituted for murine growth factors as necessary. Any additional growth factors which can promote the formation of ES cells can be determined by adding the growth factors to FGF, LIF, and SF as described above and monitored for an affect on ES formation.

Method for the isolation of pluripotential stem cells from human primordial germ cells and human embryonic (fetal) gonads

The above methods for isolation of ES cells from murine embryos were repeated for isolation of ES cells from human embryos. Specifically, testes were dissected from a 10.5 week human embryo. Younger or older embryos represent alternative sources. The preferred age range is between 8.5 weeks and 22 weeks. Tissue was rinsed in buffered saline, and incubated in trypsin solution (0.25% trypsin, 1 mM EDTA in Ca³⁰ ⁺/Mg⁺⁺ free HEPES buffered saline) for 10 minutes at 37° C. The tissue was dissociated by pipetting and the cells plated into wells of a 24 well tray containing irradiated feeder cells. In this experiment the feeder cells were Sl/Sl mouse fibroblasts transfected with human membrane associated Stem Cell Factor (Sl⁴h220 cells from Dr. David Williams, HHMI, Indiana State University School of Medicine). An alternative feeder layer would consist of a mixture of mouse or human embryo fibroblasts and Sl⁴ʰ²²⁰ cells, to provide a more coherent layer for long term cell attachment. The culture medium consists of Dulbecco's modified Eagle's medium (DMEM) with 10% fetal bovine serum supplemented with 10 ng/ml human bFGF, 60 ng/ml human Stem Cell Factor and 10 ng/ml human LIF. Alternatively, the amounts of bFGF can be increased (e.g. 20 ng/ml). Other additional or additional supplements can be added at this time, for example IF-6, IL-11, CNTF, NGF, IGFII, flt3/flk2 ligand, and/or members of the Bone Morphogenetic Protein family. The cultures were maintained for 5 days, with daily addition of fresh growth factors. Longer culture could also be utilized, e.g. 5 to 20 days.

After 5 days, cultures were dissociated with trypsin solution as before and seeded into wells containing a feeder layer of irradiated mouse embryo fibroblasts. The medium was supplemented with growth factors daily as above. The addition of growth factors to the culture medium at this stage can be utilized, and a feeder layer of a mixture of mouse or human fibroblast and Sl⁴h220 cells can be substituted.

After 10 days the cultures were fixed and stained for alkaline phosphatase activity. Colonies of cells expressing high levels of alkaline phosphatase and closely resembling primordial germ cells of the mouse embryo were detected in many wells (see FIG. 5). Closely packed clusters of cells were present in some colonies (arrow in FIG. 5). In cultures of mouse embryo germ cells these colonies give rise to lines of pluripotential embryonic stem cells. Therefore, the identified human cells can give rise to cell lines.

Method for the isolation of embryonic stem cell lines form postnatal mammalian testis

Testes are dissected and the tunica removed. The testes are then incubated at 32° C. with mild shaking in buffered saline containing bovine serum albumin and collagenase (final concentration approximately 0.5 mg/ml). When the tissue has dissociated the tubules are allowed to settle out and then

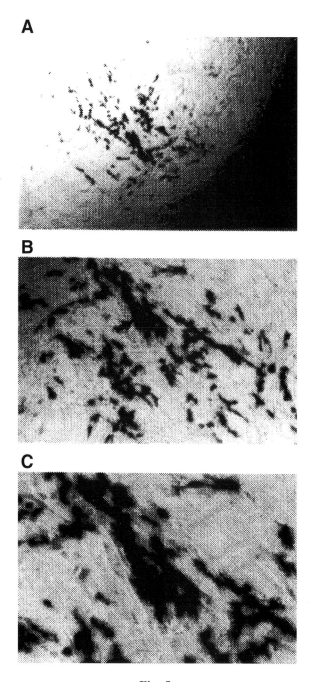

Fig. 5.

washed in saline several times. The collagenase treatment is repeated to remove all the cells surrounding the tubules (Leydig cells and connective tissue). The tubules are then washed and treated with hyaluronidase in buffered saline (final concentration approximately 0.5 mg/ml) at 32° C. until the tubules are free of adherent material. The tubules are washed and placed onto tissue culture dishes coated with Poly-L-lysine. The Sertoli cells attach strongly to the dish and spread out, while the germ cells remain in suspension. The germ cells are collected and plated onto a layer of irradiated feeder cells comprising membrane bound and soluble stem cell factor, LIF and basic FGF as described above.

Generation of chimeras using non-murine ES cells

Chimeras utilizing non-murine non-human ES cells can likewise be produced utilizing the methods for murine described above and simply substituting the appropriate non-murine blastocyst for the species of ES utilized.

Throughout this application various publications are referenced. The disclosures of these publications in their entireties are hereby incorporated by reference into this application in order to more fully describe the state of the art to which this invention pertains.

The preceding examples are intended to illustrate but not limit the invention. While they are typical of those that might be used, other procedures known to those skilled in the art may be alternatively employed.

TABLE 1

Growth Factor Requirements for Secondary Cultures of PCG-Derived Cells

Days in Culture	SF → SF + LIF + bFGF	SF + LIF + bFGF → S F + LIF + bFGF	SF + LIF + bFGF → S F
1	112 ± 16 cells	116 ± 20 cells	142 ± 18 cells
3	0.9 ± 0.6 colonies	4.6 ± 1.1 colonies	5.6 ± 0.8 colonies
5	0.5 ± 0.4 colonies	6.9 ± 1.2 colonies	6.6 ± 1.3 colonies

PGCs from 8.5 dpc embryos were cultured for 6 days on SI^4-m220 cells in the presence of either soluble rat SF alone or with soluble rat SF, LIF, an bFGF. Cultures were trypsinized and seeded into wells containing SI^4-m220 feeder cells with either soluble rat SF alone or soluble rat SF, LIF, and bFGF. Cultures were fixed and AP-positive cells (day 1) or colonies (days 2 and 5) counted. Numbers are mean ± SEM from four experiments. Secondary cultures show a reduced growth factor requirement compared with primary cultures.

REFERENCES

Bogler, O., Wren, D., Barnett, S. C., Land, H. and Noble, M. (1990). Cooperation between two growth factors promotes extended self-renewal and inhibits differentiation of oligodendrocyte-type-2 astrocyte (O-2A) progenitor cells. Proc. Natl. Acad. Sci. USA. 87, 6368–6372.

Dolci, S., Williams, D. E., Ernst, M. K., Resnick, J. L., Brannan, C. I., Lock, L. F., Lyman, S. D., Boswell, H. S. and Donovan, P. J. (1991). Requirement for mast cell growth factor for primordial germ cell survival in culture. Nature 352, 809–811.

Donovan, P. J., Stott, D., Cairns, L. A., Heasman, J. and Wylie, C. C. (1986). Migratory and postmigratory mouse primordial germ cells behave differenetly in culture. Cell 44, 831–838.

Ffrench-Constant, C., Hollingsworth, A., Heasman, J. and Wylie, C. C. (1991). Response to fibronectin of mouse primordial germ cells before, during and after migration. Development 113, 1365–1373.

Godin, I. and Wylie, C. C. (1991). TGFβ-1 inhibits proliferation and has a chemotropic effect on mouse primordial germ cells in culture. Development 113, 1451–1457.

Godin, I., Deed, R., Cooke, J., Zsebo, K., Dexter, M. and Wylie, C. C. (1991). Effects of the steel gene product on mouse primordial germ cells in culture. Nature 352, 807–809.

Ginsberg, M., Snow, M. H. L. and McLaren, A. (1990). Primordial germ cells in the mouse embryo during gastrulation. Development 110, 521–528.

Hogan, B. L. M., Costantini, F., and Lacy, E. (1986). Manipulating the Mouse Embryo: A Laboratory Manual. Cold Spring Harbor Publications, Cold Spring Harbor, N.Y.

Lawson, K. A. and Pederson, R. A. (1992). Clonal analysis of cell fate during gastrulation and early neurulation in the mouse in CIBA Foundation Symposium 165 Post Implantation Development in the Mouse John Wiley and Sons.

Mann, J. R., Gadi, I., Harbison, M. L., Abbondanzo, S. J. and Stewart, C. L. (1990). Androgenetic mouse embryonic stem cells are pluripotent and cause skeletal defects in chimeras: implications for genetic imprinting. Cell 62, 251–260.

Manova, K. and Bachvarova, R. F. (1991). Expression of c-kit encoded at the W locus of mice in developing embryonic germ cells and presumptive melanoblasts. Dev. Biol. 146, 312–324.

Matsui, Y., Toksok, D., Nishikawa, S., Nishikawa, S-I, Williams, D., Zsebo, K. and Hogan, B. L. M. (1991). Effect of steel factor and leukemia inhibitory factor on murine primordial germ cells in culture. Nature 353, 750–752.

Mintz, B. and Russell, E. S. (1957). Gene-induced embryological proliferation of primordial germ cells in the mouse. J. Exp. Zoology 134, 207–230.

Noble, M., Murray, K., Stroobant, P., Waterfield, M. D. and Riddle, P. (1988). Platelet-derived growth factor promotes division and motilty and inhibits premature differentiation of the oligodendrocyte/type-2 astrocyte progenitor cell. Nature 333, 560–562.

Noguchi, T. and Stevens, L. C. (1982). Primordial germ cell proliferation in fetal testes in mouse strains with high and low incidences of congenital testicular teratomas J. Natl. Cancer Inst. 69, 907–913.

Raff, M. C., Lillien, L. E., Richardson, W. D., Burne, J. F. and Noble, M. D. (1988). Platelet-derived growth factor from astrocytes drives the clock that times oligodendrocyte development in culture. Nature 333, 562–565.

Robertson, E. J. (1987). Embryo-derived stem cell lines in teratocarcinomas and embryonic stem cells: a practical approach. Ed. E. J. Robertson, IRL Press, Oxford pp. 71–112.

Robertson, E. J. (1987). Teratocarcinomas and embryonic stem cells: a practical approach, Ed. E. J. Robertson, IRL Press, Oxford.

Sedivy, J. M., and A. Joyner (1992). Gene Targeting, W. H. Freeman and Co.

Solter, D., Adams, N., Damjanov, I. and Koprowski, H. (1975). Control of teratocarcinogenesis in teratomas and differentiation, Eds. M. I. Sherman and D. Solter, Academic Press.

Solter, D. and Knowles, B. B. (1978). Monoclonal antibody defining a stage-specific mouse embryonic antigen (SSEA-1) Proc. Natl. Acad. Sci. USA 75, 5565–5569.

Soriano, P. and Jaenisch, R. (1986). Retroviruses as probes for mammalian development: allocation of cells to the somatic and germ cell lineages. Cell 46, 19–29.

15

Stevens, L. C. (1983). The origin and development of testicular, ovarian, and embryo-derived teratomas. Cold Spring Harbor Conferences on Cell Proliferation Vol. 10 Teratocarcinoma Stem Cells. Eds. Silver, L. M., Martin, G. R. and S. Strickland. 10, 23–36.

Stevens, L. C. and Makensen, J. A. (1961). Genetic and environmental influences on teratogenesis in mice. J. Natl. Cancer Inst. 27, 443–453.

Surani, M. A., Kothary, R., Allen, N. D., Singh, P. B., Fundele, R., Ferguson-Smith, A. C. and Barton, S. C. (1990). Genome imprinting and development in the mouse development. Suppl., 89–98.

Tam, P. P. L. and Snow, M. H. L. (1981). Proliferation and migration of primordial germ cells during compensatory growth in mouse embryos. J. Embryol. Exp. Morph. 64, 133–147.

Toksoz, D., Zsebo, K. M., Smith, K. A., Hu, S., Brankow, D., Suggs, S. V., Martin, F. H. and Williams, D. A. (1992). Support of human hematopoiesis in long-term bone marrow cultures by murine stromal cells selectively expressing the membrane-bound and secreted forms of the human homolog of the steel product, stem cell factor. Proc. Natl. Acad. Sci. USA, in press.

Yamaguchi, T. P., Conlon, R. A., and J. Rossant (1992). Expression of the fibroblast growth factor receptor FGFR-1/flg during gastrulation and segmentation in the mouse embryo. Development 152:75–88.

What is claimed is:

1. An isolated non-murine mammalian pluripotential cell wherein said cell exhibits the following characteristics:

(a) can be maintained on feeder layers for at least 20 passages; and

(b) gives rise to embryoid bodies and differentiated cells of multiple phenotypes in monolayer culture; and wherein said cell is derived from a primordial germ cell by the process of:

(1) culturing a non-murine mammalian primordial germ cell in a composition comprising basic fibroblast growth factor, leukemia inhibitory factor, membrane associated steel factor, and soluble steel factor;

(2) selecting cells that have characteristics (a) and (b), above, and

(3) isolating said non-murine pluripotential cell.

2. The pluripotential cell of claim 1, having a mutation which renders a gone non-functional.

3. The pluripotential cell of claim 1, having an insertion of a functional gene.

4. A method of using a non-murine and non-human pluripotential cell of claim 1 to contribute to chimeras in vivo comprising:

injecting the cell into a blastocyst of the same species of said cell;

16

implanting the resultant chimeric blastocyst into a foster mother of the same species; and

allowing the chimeric blastocyst to grow within the foster mother.

5. A method of using a non-murine and non-human pluripotential cell of claim 1 to contribute to chimeras in vivo comprising:

aggregating the cell with a morula stage embryo of the same species of said cell;

implanting the resultant chimeric embryo into a foster mother of the same species; and

allowing the chimeric embryo to grow within the foster mother.

6. An isolated human pluripotential cell wherein said cell exhibits the following characteristics:

(a) can be maintained on feeder layers for at least 20 passages; and

(b) gives rise to embryoid bodies and differentiated cells of multiple phenotypes in monolayer culture; and wherein said cell is derived from a human primordial germ cell by the process of:

(1) culturing a human primordial germ cell in a composition comprising basic fibroblast growth factor, leukemia inhibitory factor, membrane associated steel factor, and soluble steel factor;

(2) selecting cells that have characteristics (a) and (b), above, and

(3) isolating said human pluripotential cell.

7. A composition comprising:

(A) a human pluripotential cell derived from a primordial germ cell wherein said cell exhibits the following characteristics:

(1) can be maintained on feeder layers for at least 20 passages; and

(2) gives rise to embryoid bodies and differentiated cells of multiple phenotypes in monolayer culture; and wherein said cell is derived from a primordial germ cell by the process of:

(a) culturing a human primordial germ cell in a composition comprising basic fibroblast growth factor, leukemia inhibitory factor, membrane associated steel factor, and soluble steel factor;

(b) selecting cells that have characteristics (a) and (b), above, and

(c) isolating said human pluripotential cell; and

(B) a fibroblast growth factor, leukemia inhibitory factor, membrane associated steel factor, and soluble steel factor, each in amounts sufficient to permit continued proliferation of said cell.

* * * * *

(12) **United States Patent** (10) **Patent No.:** **US 6,200,806 B1**
Thomson (45) **Date of Patent:** **Mar. 13, 2001**

(54) **PRIMATE EMBRYONIC STEM CELLS**

(75) Inventor: **James A. Thomson**, Madison, WI (US)

(73) Assignee: **Wisconsin Alumni Research Foundation**, Madison, WI (US)

(*) Notice: Subject to any disclaimer, the term of this patent is extended or adjusted under 35 U.S.C. 154(b) by 0 days.

This patent is subject to a terminal disclaimer.

(21) Appl. No.: **09/106,390**

(22) Filed: **Jun. 26, 1998**

Related U.S. Application Data

(60) Division of application No. 08/591,246, filed on Jan. 18, 1996, now Pat. No. 5,843,780, and a continuation-in-part of application No. 08/376,327, filed on Jan. 20, 1995, now abandoned.

(51) **Int. Cl.**[7] **C12N 5/08**; C12N 5/06

(52) **U.S. Cl.** **435/366**; 435/325

(58) **Field of Search** 800/8; 435/325, 435/366

(56) **References Cited**

U.S. PATENT DOCUMENTS

5,061,620	10/1991	Tsukamoto et al.	435/7.21
5,166,065	11/1992	Williams et al.	435/325
5,340,740	8/1994	Petitte et al.	435/325
5,449,620	9/1995	Khillan	435/325
5,453,357	9/1995	Hogan	435/7.21
5,523,226	* 6/1996	Wheeler	435/325
5,589,376	12/1996	Anderson et al.	435/325
5,591,625	1/1997	Gerson et al.	435/325
5,843,780	* 12/1998	Thomson	435/363

FOREIGN PATENT DOCUMENTS

WO 94/03585   2/1994   (WO) .

OTHER PUBLICATIONS

Andrews, P., et al., "Cell lines from human germ cell tumours," *Teratocarcinoms and Embryonic Stem Cells; A Practical Approach*, Oxford: IRL Press, Ch. 8:207–248 (1987).

Andrews, P., et al., "Pluripotent Embryonal Carcinoma Clones Derived from the Human Teratocarcinoma Cell Line Tera–2," *Lab. Invest.*, 50(2):147–162 (1984).

Bongso, A., et al., "Isolation and culture of inner cell mass cells from human blastocysts," *Human Reprod.*, 9(1):2110–2117 (1994).

Bongso, A., et al., "The Growth of Inner Cell Mass Cells from Human Blastocysts," *Theriogenology*, 41:167 (1994).

Brown, D.G., et al., "Criteria that Optimize the Potential of Murine Embryonic Stem Cells for In Vitro and In Vivo Developmental Studies," *In Vitro Cell. Dev. Bio.*, 28(A)773–778 (Nov., Dec. 1992).

Damjanov, Ivan., et al., Retinoic Acid–Induced Differentiation of the Developmentally Pluripotent Human Germ Cell Tumor–Derived Cell Line, NCCIT, *Laboratory Investigation*, 68(2):220–232 (1993).

Doetschman, T., et al., Establishment Of Hamster Blastocyst–Derived Embryonic Stem (ES) Cells, *Developmental Biology*, 127:224–227 (1988).

Doetschman, T., et al., "The *in vitro* development of blastocyst–derived embryonic stem cell lines: formation of visceral yolk sac, blood islands and myocardium," *J. Embryol. exp. Morph.*, 87:27–45 (1985).

Evans, M., et al., "Establishment in culture of pluripotential cells from mouse embryos," *Nature*, 292:154–156 (1981).

Evans, M., et al., "Derivation and Preliminary Characterization of Pluripotent Cell Lines from Porcine and Bovine Blastocysts," *Theriogenology*, 33(1):125–128 (1990).

Giles, J., et al., "Pluripotency of Cultured Rabbit Inner Cell Mass Cells Detected by Isozyme Analysis and Eye Pigmentation of Fetuses Following Injection into Blastocysts or Morulae," *Mol. Reprod. Dev.*, 36:130–138 (1993).

Golos, T., et al., "Cloning of Four Growth Hormone/Chorionic Somatomammotropin–related Complementary Deoxyribonucleic Acids Differentially Expressed during Pregnancy in the Rhesus Monkey Placenta," *Endocrinology*, 133(4):1744–1752 (1993).

Graves, K., et al., "Derivation and Characterization of Putative Pluripotential Embryonic Stem Cells from Preimplantation Rabbit Embryos," *Mol. Reprod. Dev.*, 36:424–433 (1993).

Lapidot, T., et al., "Modeling Human Hematopoiesis in Immunodeficient Mice," *Lab. Animal Sci.*, 43(2):147–149 (1993).

Marshall, E., "Rules on Embryo Research Due Out," *Science*, 265:1024–1026 (1994).

Notarianni, E., et al., "Maintenance and differentiation in culture of pluripotential embryonic cell lines from pig blastocysts," *J. Reprod. Fert. Suppl.*, 41:51–56 (1990).

Notarianni, E., et al., "Derivation Of Pluripotent, Embryonic Cell Lines From The Pig And Sheep," *J. Rep. & Fert.* 43 255–260 (1991).

(List continued on next page.)

Primary Examiner—Deborah J. R. Clark
(74) *Attorney, Agent, or Firm*—Quarles & Brady LLP

(57) **ABSTRACT**

A purified preparation of primate embryonic stem cells is disclosed. This preparation is characterized by the following cell surface markers: SSEA-1 (–); SSEA-4 (+); TRA-1-60 (+); TRA-1-81 (+); and alkaline phosphatase (+). In a particularly advantageous embodiment, the cells of the preparation are human embryonic stem cells, have normal karyotypes, and continue to proliferate in an undifferentiated state after continuous culture for eleven months. The embryonic stem cell lines also retain the ability, throughout the culture, to form trophoblast and to differentiate into all tissues derived from all three embryonic germ layers (endoderm, mesoderm and ectoderm). A method for isolating a primate embryonic stem cell line is also disclosed.

11 Claims, 8 Drawing Sheets

OTHER PUBLICATIONS

Piedrahita, et al., "On The Isolation Of Embryonic Stem Cells: Comparative Behavior Of Murine, Porcine And Ovine Embryos," *Theriogenology*, 34(5):879–901 (1990).

Piedrahita, J. Dissertation On "Studies On The Isolation Of Embryonic Stem (ES) Cells: Comparative Behavior Of Murine, Porcine, And Ovine Species," University of California Davis, 1989.

Rossant, J., et al., "The relationship between embryonic, embryonal carcinoma and embryo–derived stem cells," *Cell Diff.*, 15:155–161 (1984).

Seshagiri, P., et al., "Non–Surgical Uterine Flushing for the Recovery of Preimplantation Embryos in Rhesus Monkeys: Lack of Seasonal Infertility," *Am. J. Primatol.*, 29:81–91 (1993).

Strojek, R. et al., "A Method For Cultivating Morphologically Undifferentiated Embryonic Stem Cells From Porcine Blastocysts," Theriogenology 33 901–913 (1990 .

Sukoyan, M., et al., "Isolation and Cultivation of Blastocyst–derived Stem Cell Lines from American Mink (*Mustela vision*)," *Mol. Reprod. Dev.*, 33:418–431 (1992).

Sukoyan, M., et al., "Embryonic Stem Cells Derived from Morulae, Inner Cell Mass, and Blastocysts of Mink: Comparisons of Their Pluripotencies," *Mol. Reprod. Deve.*, 36:148–158 (1993).

Talbot, et al. "Culturing The EpiBlast Cells Of The Pig Blastocyst", , *In Vitro Cell. Dev. Bio.*, 29(Λ):543–554 (1993).

Thomson, J., et al., "Nonsurgical uterine stage preimplantation embryo collection from the common marmoset," *J. Med. Primatol.*, 23:333–336 (1994).

Thomson, James A., et al., "Pluripotent Cell Lines Derived from Common Marmoset (*Callithrix jacchus*) Blastocysts," *Biology of Reproduction*, 55:254–259 (1996).

Ware, et al., "Development Of Embryonic Stem Cell Lines From Farm Animals", *Biol. Reprod.*, 38(Suppl. 1):129 (1988).

Wenk, J., et al., "Glycolipods of Germ Cell Tumors: Extended Globo–series Glycolipods are a Hallmark of Human Embryonal Carcinoma Cells," *Int. J. Can*, 58:108–115 (1994).

Williams, R., et al., "Myeloid leukaemia inhibitory factor maintains the developmental potential of embryonic stem cells," *Nature*, 336:684–692 (1988).

Associated Press Milwaukee Journal Article dated Nov. 4, 1994 "Embryonic Monkey Cells Isolated" (Nov. 4, 1994).

Thomson et al. Embryonic Stem Cell Lines Derived from Human Blastocysts. Science, vol. 282, pp. 1145–1147, Nov. 6, 1998.*

Cruz et al. Origin of Embryonic and Extraembryonic Cell Lineages in Mammalian Embryos. Current Communications, vol. 4, pp. 147–204, 1991.*

Nichols et al. Establishment of germ–line–competent embryonic stem (ES) cells using differentiation inhibiting activity. Development, vol. 110, pp. 1341–1348, 1990.*

Clark et al. Germ line manipulation: applications in agriculture and biotechnology. Transgenic Animals. Grosveld et al. eds. p. 250, 1993.*

* cited by examiner

PRIMATE EMBRYONIC STEM CELLS

CROSS REFERENCES TO RELATED APPLICATIONS

This application is a divisional of U.S. Ser. No. 08/591, 246 which was filed on Jan. 18, 1996, issued as U.S. Pat. No. 5,843,780, Dec. 1, 1998 and is a continuation-in-part of U.S. Ser. No. 08/376,327 which was filed on Jan. 20, 1995, abandoned.

STATEMENT REGARDING FEDERALLY SPONSORED RESEARCH

Not Applicable.

BACKGROUND OF THE INVENTION

In general, the field of the present invention is stem cell cultures. Specifically, the field of the present invention is primate embryonic stem cell cultures.

In general, stem cells are undifferentiated cells which can give rise to a succession of mature functional cells. For example, a hematopoietic stem cell may give rise to any of the different types of terminally differentiated blood cells. Embryonic stem (ES) cells are derived from the embryo and are pluripotent, thus possessing the capability of developing into any organ or tissue type or, at least potentially, into a complete embryo.

One of the seminal achievements of mammalian embryology of the last decade is the routine insertion of specific genes into the mouse genome through the use of mouse ES cells. This alteration has created a bridge between the in vitro manipulations of molecular biology and an understanding of gene function in the intact animal. Mouse ES cells are undifferentiated, pluripotent cells derived in vitro from preimplantation embryos (Evans, et al. *Nature* 292:154–159, 1981; Martin, *Proc. Natl. Acad. Sci. USA* 78:7634–7638, 1981) or from fetal germ cells (Matsui, et al., *Cell* 70:841–847, 1992). Mouse ES cells maintain an undifferentiated state through serial passages when cultured in the presence of fibroblast feeder layers in the presence of Leukemia Inhibitory Factor (LIF) (Williams, et al., *Nature* 336:684–687, 1988). If LIF is removed, mouse ES cells differentiate.

Mouse ES cells cultured in non-attaching conditions aggregate and differentiate into simple embryoid bodies, with an outer layer of endoderm and an inner core of primitive ectoderm. If these embryoid bodies are then allowed to attach onto a tissue culture surface, disorganized differentiation occurs of various cell types, including nerves, blood cells, muscle, and cartilage (Martin, 1981, supra; Doetschman, et al., *J. Embryol. Exp. Morph.* 87:27–45, 1985). Mouse ES cells injected into syngeneic mice form teratocarcinomas that exhibit disorganized differentiation, often with representatives of all three embryonic germ layers. Mouse ES cells combined into chimeras with normal preimplantation embryos and returned to the uterus participate in normal development (Richard, et al., *Cytogenet. Cell Genet.* 65:169–171, 1994).

The ability of mouse ES cells to contribute to functional germ cells in chimeras provides a method for introducing site-specific mutations into mouse lines. With appropriate transfection and selection strategies, homologous recombination can be used to derive ES cell lines with planned alterations of specific genes. These genetically altered cells can be used to form chimeras with normal embryos and chimeric animals are recovered. If the ES cells contribute to

the germ line in the chimeric animal, then in the next generation a mouse line for the planned mutation is established.

Because mouse ES cells have the potential to differentiate into any cell type in the body, mouse ES cells allow the in vitro study of the mechanisms controlling the differentiation of specific cells or tissues. Although the study of mouse ES cells provides clues to understanding the differentiation of general mammalian tissues, dramatic differences in primate and mouse development of specific lineages limits the usefulness of mouse ES cells as a model of human development. Mouse and primate embryos differ meaningfully in the timing of expression of the embryonic genome, in the formation of an egg cylinder versus an embryonic disc (Kaufman, *The Atlas of Mouse Development*, London: Academic Press, 1992), in the proposed derivation of some early lineages (O'Rahilly & Muller, *Developmental Stages in Human Embryos*, Washington: Carnegie Institution of Washington, 1987), and in the structure and function in the extraembryonic membranes and placenta (Mossman, *Vertebrate Fetal Membranes*, New Brunswick: Rutgers, 1987). Other tissues differ in growth factor requirements for development (e.g. the hematopoietic system (Lapidot et al., *Lab An Sci* 43:147–149, 1994)), and in adult structure and function (e.g. the central nervous system). Because humans are primates, and development is remarkably similar among primates, primate ES cells lines will provide a faithful model for understanding the differentiation of primate tissues in general and human tissues in particular.

The placenta provides just one example of how primate ES cells will provide an accurate model of human development that cannot be provided by ES cells from other species. The placenta and extraembryonic membranes differ dramatically between mice and humans. Structurally, the mouse placenta is classified as labyrinthine, whereas the human and the rhesus monkey placenta are classified as villous. Chorionic gonadotropin, expressed by the trophoblast, is an essential molecule involved in maternal recognition of pregnancy in all primates, including humans (Hearn, *J Reprod Fertil* 76:809–819, 1986; Hearn et al., *J Reprod Fert* 92:497–509, 1991). Trophoblast secretion of chorionic gonadotropin in primates maintains the corpus luteum of pregnancy and, thus, progesterone secretion. Without progesterone, pregnancy fails. Yet mouse trophoblast produces no chorionic gonadotropin, and mice use entirely different mechanisms for pregnancy maintenance (Hearn et al., "Normal and abnormal embryo-fetal development in mammals," In: Lamming E, ed. *Marshall's Physiology of Reproduction*. 4th ed. Edinburgh, N.Y.: Churchill Livingstone, 535–676, 1994). An immortal, euploid, primate ES cell line with the developmental potential to form trophoblast in vitro, will allow the study of the ontogeny and function of genes such as chorionic gonadotropin which are critically important in human pregnancy. Indeed, the differentiation of any tissue for which there are significant differences between mice and primates will be more accurately reflected in vitro by primate ES cells than by mouse ES cells.

The major in vitro models for studying trophoblast function include human choriocarcinoma cells, which are malignant cells that may not faithfully reflect normal trophectoderm; short-term primary cultures of human and non-human primate cytotrophoblast, which in present culture conditions quickly form non-dividing syncytial trophoblast; and in vitro culture of preimplantation non-human primate embryos (Hearn, et al., *J. Endocrinol.* 119:249–255, 1988; Coutifaris, et al., Ann. NY Acad. Sci. 191–201, 1994). An immortal, euploid, non-human primate embryonic stem (ES) cell line with the developmental potential to form trophec-toderm offers significant advantages over present in vitro models of human trophectoderm development and function, as trophoblast-specific genes such as chorionic gonadotropin could be stably altered in the ES cells and then studied during differentiation to trophectoderm.

The cell lines currently available that resembles primate ES cells most closely are human embryonic carcinoma (EC) cells, which are pluripotent, immortal cells derived from teratocarcinomas (Andrews, et al., *Lab. Invest.* 50(2) :147–162, 1984; Andrews, et al., in: Robertson E., ed. *Teratocarcinomas and Embryonic Stem Cells: A Practical Approach.* Oxford: IRL press, pp. 207–246, 1987). EC cells can be induced to differentiate in culture, and the differentiation is characterized by the loss of specific cell surface markers (SSEA-3, SSEA-4, TRA-1-60, and TRA-1-81) and the appearance of new markers (Andrews, et al., 1987, supra). Human EC cells will form teratocarcinomas with derivatives of multiple embryonic lineages in tumors in nude mice. However, the range of differentiation of these human EC cells is limited compared to the range of differentiation obtained with mouse ES cells, and all EC cell lines derived to date are aneuploid (Andrews, et al., 1987, supra). Similar mouse EC cell lines have been derived from teratocarcinomas, and, in general their developmental potential is much more limited than mouse ES cells (Rossant, et al., *Cell Differ.* 15:155–161, 1984). Teratocarcinomas are tumors derived from germ cells, and although germ cells (like ES cells) are theoretically totipotent (i.e. capable of forming all cell types in the body), the more limited developmental potential and the abnormal karyotypes of EC cells are thought to result from selective pressures in the terato-carcinoma tumor environment (Rossant & Papaioannou, *Cell Differ* 15:155–161, 1984). ES cells, on the other hand, are thought to retain greater developmental potential because they are derived from normal embryonic cells in vitro, without the selective pressures of the teratocarcinoma environment. Nonetheless, mouse EC cells and mouse ES cells share the same unique combination of cell surface markers (SSEA-1 (+), SSEA-3 (−), SSEA-4 (−), and alkaline phosphatase (+)).

Pluripotent cell lines have also been derived from preimplantation embryos of several domestic and laboratory animals species (Evans, et al., *Theriogenology* 33(1):125–128, 1990; Evans, et al., *Theriogenology* 33(1):125–128, 1990; Notarianni, et al., *J. Reprod. Fertil.* 41(Suppl.):51–56, 1990; Giles, et al., *Mol. Reprod. Dev.* 36:130–138, 1993; Graves, et al., *Mol. Reprod. Dev.* 36:424–433, 1993; Sukoyan, et al., *Mol. Reprod. Dev.* 33:418–431, 1992; Sukoyan, et al., *Mol. Reprod. Dev.* 36:148–158, 1993; Iannaccone, et al., *Dev. Biol.* 163:288–292, 1994).

Whether or not these cell lines are true ES cells lines is a subject about which there may be some difference of opinion. True ES cells should: (i) be capable of indefinite proliferation in vitro in an undifferentiated state; (ii) maintain a normal karyotype through prolonged culture; and (iii) maintain the potential to differentiate to derivatives of all three embryonic germ layers (endoderm, mesoderm, and ectoderm) even after prolonged culture. Strong evidence of these required properties have been published only for rodents ES cells including mouse (Evans & Kaufman, *Nature* 292:154–156, 1981; Martin, *Proc Natl Acad Sci USA* 78:7634–7638, 1981) hamster (Doetschmanet al. *Dev Biol* 127:224–227, 1988), and rat (Iannaccone et al. *Dev Biol* 163:288–292, 1994), and less conclusively for rabbit ES cells (Giles et al. *Mol Reprod Dev* 36:130–138, 1993; Graves & Moreadith, *Mol Reprod Dev* 36:424–433, 1993).

However, only established ES cell lines from the rat (Iannaccone, et al., 1994, supra) and the mouse (Bradley, et al., *Nature* 309:255–256, 1984) have been reported to participate in normal development in chimeras. There are no reports of the derivation of any primate ES cell line.

BRIEF SUMMARY OF THE INVENTION

The present invention is a purified preparation of primate embryonic stem cells. The primate ES cell lines are true ES cell lines in that they: (i) are capable of indefinite proliferation in vitro in an undifferentiated state; (ii) are capable of differentiation to derivatives, of all three embryonic germ layers (endoderm, mesoderm, and ectoderm) even after prolonged culture; and (iii) maintain a normal karyotype throughout prolonged culture. The true primate ES cells lines are therefore pluripotent.

The present invention is also summarized in that primate ES cell lines are preferably negative for the SSEA-1 marker, preferably positive for the SSEA-3 marker, and positive for the SSEA-4 marker. The primate ES cell lines are also positive for the TRA-1-60, and TRA-1-81 markers, as well as positive for the alkaline phosphatase marker.

It is an advantageous feature of the present invention that the primate ES cell lines continue to proliferate in an undifferentiated state after continuous culture for at least one year. In a particularly advantageous embodiment, the cells remain euploid after proliferation in an undifferentiated state.

It is a feature of the primate ES cell lines in accordance with the present invention that the cells can differentiate to trophoblast in vitro and express chorionic gonadotropin.

The present invention is also a purified preparation of primate embryonic stem cells that has the ability to differentiate into cells derived from mesoderm, endoderm, and ectoderm germ layers after the cells have been injected into an immunocompromised mouse, such as a SCID mouse.

The present invention is also a method of isolating a primate embryonic stem cell line. The method comprises the steps of isolating a primate blastocyst, isolating cells from the inner cellular mass (ICM) of the blastocyst, plating the ICM cells on a fibroblast layer (wherein ICM-derived cell masses are formed) removing an ICM-derived cell mass and dissociating the mass into dissociated cells, replating the dissociated cells on embryonic feeder cells and selecting colonies with compact morphology containing cells with a high nucleus/cytoplasm ratio, and prominent nucleoli. The cells of the selected colonies are then cultured.

It is an object of the present invention to provide a primate embryonic stem cell line.

It is an object of the present invention to provide a primate embryonic stem cell line characterized by the following markers: alkaline phosphatase(+); SSEA-1(−); preferably SSEA-3(+); SSEA-4(+); TRA-1-60(+); and TRA-1-81(+).

It is an object of the present invention to provide a primate embryonic stem cell line capable of proliferation in an undifferentiated state after continuous culture for at least one year. Preferably, these cells remain euploid.

It is another object of the present invention to provide a primate embryonic stem cell line wherein the cells differentiate into cells derived from mesoderm, endoderm, and ectoderm germ layers when the cells are injected into an immunocompromised mouse.

Other objects, features, and advantages of the present invention will become obvious after study of the specification, drawings, and claims.

BRIEF DESCRIPTION OF THE DRAWINGS

FIG. 1 is a photomicrograph illustrating normal XY karyotype of rhesus ES cell line R278.5 after 11 months of continuous culture.

FIGS. 2A–2D are a set of phase-contrast photomicrographs demonstrating the morphology of undifferentiated rhesus ES (R278.5) cells and of cells differentiated from R278.5 in vitro (bar=100μ). Photograph 2A demonstrates the distinct cell borders, high nucleus to cytoplasm ratio, and prominent nucleoli, of undifferentiated rhesus ES cells. Photographs 2B–2D shows differentiated cells eight days after plating R278.5 cells on gel treated tissue culture plastic (with 10^3 units/ml added human LIF). Cells of these three distinct morphologies are consistently present when R278.5 cells are allowed to differentiate at low density without plating R278.5 cells on gel treated tissue culture plastic fibroblasts either in the presence or absence of soluble human LIF.

FIGS. 3A–F are photomicrographs demonstrating the expression of cell surface markers on undifferentiated rhesus ES (R278.5) cells (bar=100μ). Photograph 3A shows Alkaline Phosphatase (+); Photograph 3B shows SSEA-1 (–); Photograph 3C shows SSEA-3 (+); Photograph 3D shows SSEA-4 (+); Photograph 3E shows TRA-1-60 (+); and Photograph 3F shows TRA-1-81 (+).

FIGS. 4A–4B are photographs illustrating expression of α-fetoprotein mRNA and α- and β-chorionic gonadotrophin mRNA expression in rhesus ES cells (R278.5) allowed to differentiate in culture.

FIGS. 5A–5F include six photomicrographs of sections of tumors formed by injection of 0.5×10^6 rhesus ES (R278.5) cells into the hindleg muscles of SCID mice and analyzed 15 weeks later. Photograph 5A shows a low power field demonstrating disorganized differentiation of multiple cell types. A gut-like structure is encircled by smooth muscle(s), and elsewhere foci of cartilage (c) are present (bar=400μ); Photograph 5B shows striated muscle (bar=40μ); Photograph 5C shows stratified squamous epithelium with several hair follicles. The labeled hair follicle (f) has a visible hair shaft (bar=200μ); Photograph 5D shows stratified layers of neural cells in the pattern of a developing neural tube. An upper "ventricular" layer, containing numerous mitotic figures (arrows), overlies a lower "mantle" layer. (bar=100μ); Photograph 5E shows ciliated columnar epithelium (bar=40μ); Photograph 5F shows villi covered with columnar epithelium with interspersed mucus-secreting goblet cells (bar=200μ).

FIGS. 6A–6B include photographs of an embryoid Body. This embryoid body was formed from a marmoset ES cell line (Cj62) that had been continuously passaged in vitro for over 6 months. Photograph 6A (above) shows a section of the anterior ⅓ of the embryonic disc. Note the primitive ectoderm (E) forms a distinct cell layer from the underlying primitive endoderm (e), with no mixing of the cell layers. Note also that amnion (a) is composed of two distinct layers; the inner layer is continuous with the primitive ectoderm at the margins. Photograph 6B (below) shows a section in the caudal ⅓ of embryonic disc. Note central groove (arrow) and mixing of primitive ectoderm and endoderm representing early primitive streak formation, indicating the beginning of gastrulation. 400×, toluidine blue stain.

DETAILED DESCRIPTION OF THE PREFERRED EMBODIMENTS

(1) In General

(a) Uses of Primate ES Cells

The present invention is a pluripotent, immortal euploid primate ES cell line, as exemplified by the isolation of ES

cell lines from two primate species, the common marmoset (*Callithrix jacchus*) and the rhesus monkey (*Macaca mulatta*). Primate embryonic stem cells are useful for:

(i) Generating transgenic non-human primates for models of specific human genetic diseases. Primate embryonic stem cells will allow the generation of primate tissue or animal models for any human genetic disease for which the responsible gene has been cloned. The human genome project will identify an increasing number of genes related to human disease, but will not always provide insights into gene function. Transgenic nonhuman primates will be essential for elucidating mechanisms of disease and for testing new therapies.

(ii) Tissue transplantation. By manipulating culture conditions, primate ES cells, human and non-human, can be induced to differentiate to specific cell types, such as blood cells, neuron cells, or muscle cells. Alternatively, primate ES cells can be allowed to differentiate in tumors in SCID mice, the tumors can be disassociated, and the specific differentiated cell types of interest can be selected by the usage of lineage specific markers through the use of fluorescent activated cell sorting (FACS) or other sorting method or by direct microdissection of tissues of interest. These differentiated cells could then be transplanted back to the adult animal to treat specific diseases, such as hematopoietic disorders, endocrine deficiencies, degenerative neurological disorders or hair loss.

(b) Selection of Model Species

Macaques and marmosets were used as exemplary species for isolation of a primate ES cell line. Macaques, such as the rhesus monkey, are Old World species that are the major primates used in biomedical research. They are relatively large (about 7–10 kg). Males take 4–5 years to mature, and females have single young. Because of the extremely close anatomical and physiological similarities between humans and rhesus monkeys, rhesus monkey true ES cell lines provide a very accurate in vitro model for human differentiation. Rhesus monkey ES cell lines and rhesus monkeys will be particularly useful in the testing of the safety and efficacy of the transplantation of differentiated cell types into whole animals for the treatment of specific diseases or conditions. In addition, the techniques developed for the rhesus ES cell lines model the generation, characterization and manipulation of human ES cell lines.

The common marmoset (*Callithrix jacchus*) is a New World primate species with reproductive characteristics that make it an excellent choice for ES cell derivation. Marmosets are small (about 350–400 g), have a short gestation period (144 days), reach sexual maturity in about 18 months, and routinely have twins or triplets. Unlike in macaques, it is possible to routinely synchronize ovarian cycles in the marmoset with prostaglandin analogs, making collection of age-matched embryos from multiple females possible, and allowing efficient embryo transfer to synchronized recipients with 70%–80% of embryos transferred resulting in pregnancies. Because of these reproductive characteristics that allow for the routine efficient transfer of multiple embryos, marmosets provide an excellent primate species in which to generate transgenic models for human diseases.

There are approximately 200 primate species in the world. The most fundamental division that divides higher primates is between Old World and New world species. The evolutionary distance between the rhesus monkey and the common marmoset is far greater than the evolutionary distance between humans and rhesus monkeys. Because it is here demonstrated that it is possible to isolate ES cell lines from

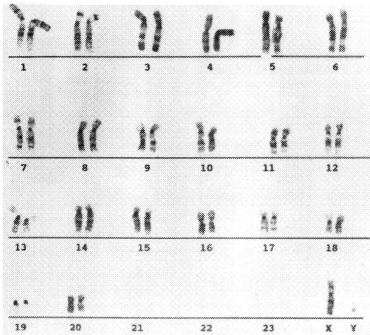

Fig. 1.

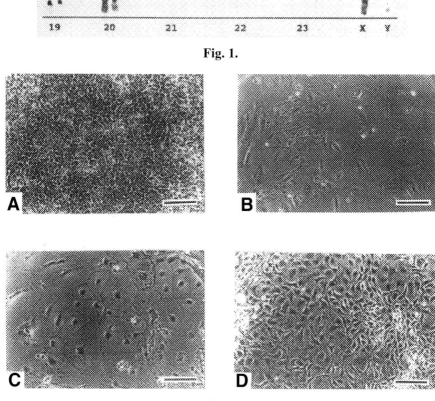

Fig. 2.

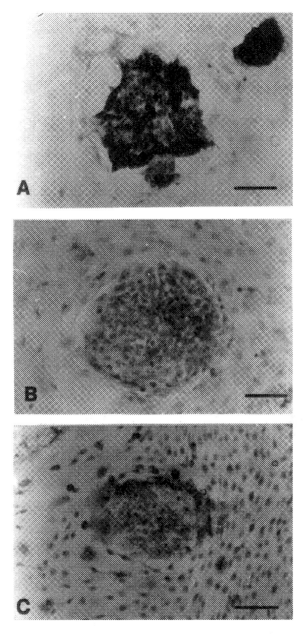

Fig. 3.

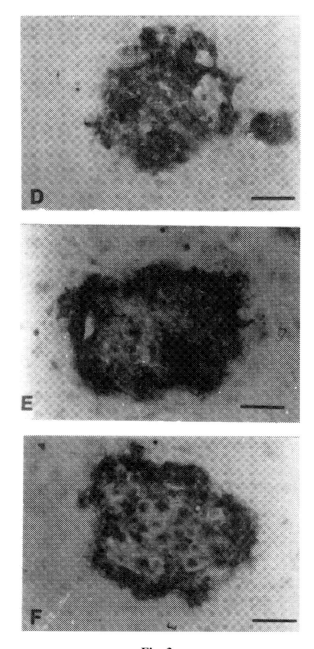

Fig. 3.

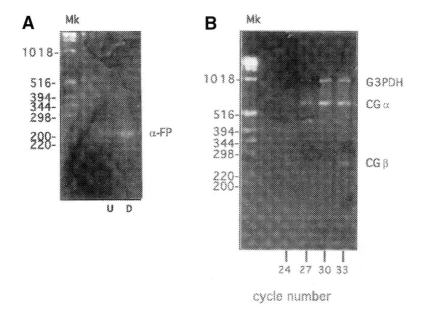

Fig. 4.

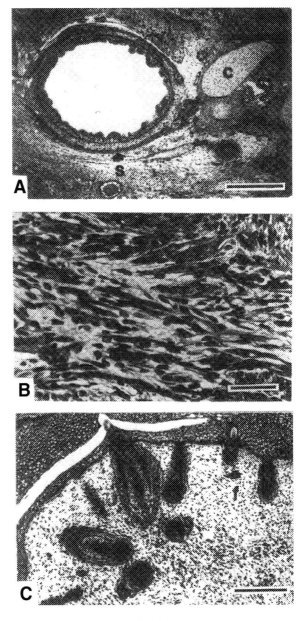

Fig. 5.

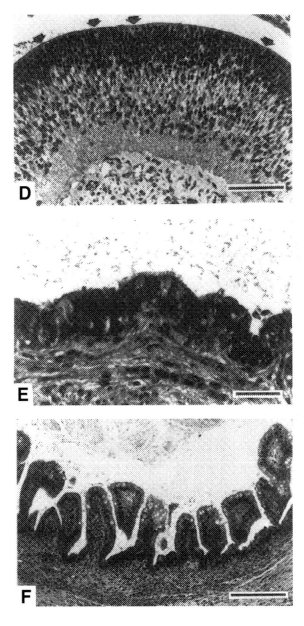

Fig. 5.

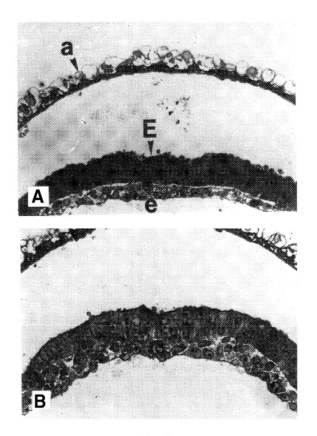

Fig. 6.

a representative species of both the Old World and New World group using similar conditions, the techniques described below may be used successfully in deriving ES cell lines in other higher primates as well. Given the close evolutionary distance between rhesus macaques and humans, and the fact that feeder-dependent human EC cell lines can be grown in conditions similar to those that support primate ES cell lines, the same growth conditions will allow the isolation and growth of human ES cells. In addition, human ES cell lines will be permanent cell lines that will also be distinguished from all other permanent human cell lines by their normal karyotype and the expression of the same combination of cell surface markers (alkaline phosphotase, preferably SSEA-3, SSEA-4, TRA-1-60 and TRA-1-81) that characterize other primate ES cell lines. A normal karyotype and the expression of this combination of cell surface markers will be defining properties of true human ES cell lines, regardless of the method used for their isolation and regardless of their tissue of origin.

No other primate (human or non-human) ES cell line is known to exist. The only published permanent, euploid, embryo-derived cell lines that have been convincingly demonstrated to differentiate into derivatives of all three germ layers have been derived from rodents (the mouse, rat, and hamster), and possibly from rabbit. The published reports of embryo-derived cell lines from domestic species have failed to convincingly demonstrate differentiation of derivatives of all three embryonic germ layers or have not been permanent cell lines. Research groups in Britain and Singapore are informally reported, later than the work described here, to have attempted to derive human ES cell lines from surplus in vitro fertilization-produced human embryos, although they have not yet reported success in demonstrating pluripotency of their cells and have failed to isolate permanent cell lines. In the only published report on attempts to isolate human ES cells, conditions were used (LIF in the absence of fibroblast feeder layers) that the results below will indicate will not result in primate ES cells which can remain in an undifferentiated state. It is not surprising, then that the cells grown out of human ICMs failed to continue to proliferate after 1 or 2 subcultures, Bongso et al. *Hum. Reprod.* 9:2100–2117 (1994).

(2) Embryonic Stem Cell Isolation

A preferable medium for isolation of embryonic stem cells is "ES medium." ES medium consists of 80% Dulbecco's modified Eagle's medium (DMEM; no pyruvate, high glucose formulation, Gibco BRL), with 20% fetal bovine serum (FBS; Hyclone), 0.1 mM β-mercaptoethanol (Sigma), 1% non-essential amino acid stock (Gibco BRL). Preferably, fetal bovine serum batches are compared by testing clonal plating efficiency of a low passage mouse ES cell line (ES_{f3}), a cell line developed just for the purpose of this test. FBS batches must be compared because it has been found that batches vary dramatically in their ability to support embryonic cell growth, but any other method of assaying the competence of FBS batches for support of embryonic cells will work as an alternative.

Primate ES cells are isolated on a confluent layer of murine embryonic fibroblast in the presence of ES cell medium. Embryonic fibroblasts are preferably obtained from 12 day old fetuses from outbred CF1 mice (SASCO), but other strains may be used as an alternative. Tissue culture dishes are preferably treated with 0.1% gelatin (type I; Sigma).

For rhesus monkey embryos, adult female rhesus monkeys (greater than four years old) demonstrating normal ovarian cycles are observed daily for evidence of menstrual

bleeding (day 1 of cycle=the day of onset of menses). Blood samples are drawn daily during the follicular phase starting from day 8 of the menstrual cycle, and serum concentrations of luteinizing hormone are determined by radioimmunoassay. The female is paired with a male rhesus monkey of proven fertility from day 9 of the menstrual cycle until 48 hours after the luteinizing hormone surge; ovulation is taken as the day following the luteinizing hormone surge. Expanded blastocysts are collected by non-surgical uterine flushing at six days after ovulation. This procedure routinely results in the recovery of an average 0.4 to 0.6 viable embryos per rhesus monkey per month, Seshagiri et al. *Am J Primatol* 29:81–91, 1993.

For marmoset embryos, adult female marmosets (greater than two years of age) demonstrating regular ovarian cycles are maintained in family groups, with a fertile male and up to five progeny. Ovarian cycles are controlled by intramuscular injection of 0.75 g of the prostaglandin PGF2a analog cloprostenol (Estrumate, Mobay Corp, Shawnee, KS) during the middle to late luteal phase. Blood samples are drawn on day 0 (immediately before cloprostenol injection), and on days 3, 7, 9, 11, and 13. Plasma progesterone concentrations are determined by ELISA. The day of ovulation is taken as the day preceding a plasma progesterone concentration of 10 ng/ml or more. At eight days after ovulation, expanded blastocysts are recovered by a non-surgical uterine flush procedure, Thomson et al. "Non-surgical uterine stage preimplantation embryo collection from the common marmoset," *J Med Primatol,* 23:333–336 (1994). This procedure results in the average production of 1.0 viable embryos per marmoset per month.

The zona pellucida is removed from blastocysts by brief exposure to pronase (Sigma). For immunosurgery, blastocysts are exposed to a 1:50 dilution of rabbit anti-marmoset spleen cell antiserum (for marmoset blastocysts) or a 1:50 dilution of rabbit anti-rhesus monkey (for rhesus monkey blastocysts) in DMEM for 30 minutes, then washed for 5 minutes three times in DMEM, then exposed to a 1:5 dilution of Guinea pig complement (Gibco) for 3 minutes.

After two further washes in DMEM, lysed trophectoderm cells are removed from the intact inner cell mass (ICM) by gentle pipetting, and the ICM plated on mouse inactivated (3000 rads gamma irradiation) embryonic fibroblasts.

After 7–21 days, ICM-derived masses are removed from endoderm outgrowths with a micropipette with direct observation under a stereo microscope, exposed to 0.05% Trypsin-EDTA (Gibco) supplemented with 1% chicken serum for 3–5 minutes and gently dissociated by gentle pipetting through a flame polished micropipette.

Dissociated cells are replated on embryonic feeder layers in fresh ES medium, and observed for colony formation. Colonies demonstrating ES-like morphology are individually selected, and split again as described above. The ES-like morphology is defined as compact colonies having a high nucleus to cytoplasm ratio and prominent nucleoli. Resulting ES cells are then routinely split by brief trypsinization or exposure to Dulbecco's Phosphate Buffered Saline (without calcium or magnesium and with 2 mM EDTA) every 1–2 weeks as the cultures become dense. Early passage cells are also frozen and stored in liquid nitrogen.

Cell lines may be karyotyped with a standard G-banding technique (such as by the Cytogenetics Laboratory of the University of Wisconsin State Hygiene Laboratory, which provides routine karyotyping services) and compared to published karyotypes for the primate species.

Isolation of ES cell lines from other primate species would follow a similar procedure, except that the rate of development to blastocyst can vary by a few days between species, and the rate of development of the cultured ICMs will vary between species. For example, six days after ovulation, rhesus monkey embryos are at the expanded blastocyst stage, whereas marmoset embryos don't reach the same stage until 7–8 days after ovulation. The Rhesus ES cell lines were obtained by splitting the ICM-derived cells for the first time at 7–16 days after immunosurgery; whereas the marmoset ES cells were derived with the initial split at 7–10 days after immunosurgery. Because other primates also vary in their developmental rate, the timing of embryo collection, and the timing of the initial ICM split will vary between primate species, but the same techniques and culture conditions will allow ES cell isolation.

Because ethical considerations in the U.S. do not allow the recovery of human in vivo fertilized preimplantation embryos from the uterus, human ES cells that are derived from preimplantation embryos will be derived from in vitro fertilized (IVF) embryos. Experiments on unused (spare) human IVF-produced embryos are allowed in many countries, such as Singapore and the United Kingdom, if the embryos are less than 14 days old. Only high quality embryos are suitable for ES isolation. Present defined culture conditions for culturing the one cell human embryo to the expanded blastocyst are suboptimal but practicable, Bongso et al., *Hum Reprod* 4:706–713, 1989. Co-culturing of human embryos with human oviductal cells results in the production of high blastocyst quality. IVF-derived expanded human blastocysts grown in cellular co-culture, or in improved defined medium, will allow the isolation of human ES cells with the same procedures described above for nonhuman primates.

(3) Defining Characteristics of Primate ES Cells

Primate embryonic stem cells share features with the primate ICM and with pluripotent human embryonal carcinoma cells. Putative primate ES cells may therefore be characterized by morphology and by the expression of cell surface markers characteristic of human EC cells. Additionally, putative primate ES cells may be characterized by developmental potential, karyotype and immortality.

(a) Morphology

The colony morphology of primate embryonic stem cell lines is similar to, but distinct from, mouse embryonic stem cells. Both mouse and primate ES cells have the characteristic features of undifferentiated stem cells, with high nuclear/cytoplasmic ratios, prominent nucleoli, and compact colony formation. The colonies of primate ES cells are flatter than mouse ES cell colonies and individual primate ES cells can be easily distinguished. In FIG. 2, reference character A indicates a phase contrast photomicrograph of cell line R278.5 demonstrating the characteristic primate ES cell morphology.

(b) Cell Surface Markers

A primate ES cell line of the present invention is distinct from mouse ES cell lines by the presence or absence of the cell surface markers described below.

One set of glycolipid cell surface markers is known as the Stage-specific embryonic antigens 1 through 4. These antigens can be identified using antibodies for SSEA 1, preferably SSEA-3 and SSEA-4 which are available from the Developmental Studies Hybridoma Bank of the National Institute of Child Health and Human Development. The cell surface markers referred to as TRA-1-60 and TRA-1-81 designate antibodies from hybridomas developed by Peter Andrews of the University of Sheffield and are described in Andrews et al., "Cell lines from human germ cell tumors," In: Robertson E, ed. *Teratocarcinomas and Embryonic Stem*

10

Cells: A Practical Approach. Oxford: IRL Press, 207–246, 1987. The antibodies were localized with a biotinylated secondary antibody and then an avidin/biotinylated horseradish peroxidase complex (Vectastain ABC System, Vector Laboratories). Alternatively, it should also be understood that other antibodies for these same cell surface markers can be generated. NTERA-2 cl. D1, a pluripotent human EC cell line (gift of Peter Andrews), may be used as a negative control for SSEA-1, and as a positive control for SSEA-3, SSEA-4, TRA-1-60, and TRA-1-81. This cell line was chosen for positive control only because it has been extensively studied and reported in the literature, but other human EC cell lines may be used as well.

Mouse ES cells (ES_{D3}) are used as a positive control for SSEA-1, and for a negative control for SSEA-3, SSEA-4, TRA-1-60, and TRA-1-81. Other routine negative controls include omission of the primary or secondary antibody and substitution of a primary antibody with an unrelated specificity.

Alkaline phosphatase may be detected following fixation of cells with 4% para-formaldehyde using "Vector Red" (Vector Laboratories) as a substrate, as described by the manufacturer (Vector Laboratories). The precipitate formed by this substrate is red when viewed with a rhodamine filter system, providing substantial amplification over light microscopy.

Table 1 diagrams a comparison of mouse ES cells, primate ES cells, and human EC cells. The only cells reported to express the combination of markers SSEA-3; SSEA-4, TRA-1-60, and TRA-1-81 other than primate ES cells are human EC cells. The globo-series glycolipids SSEA-3 and SSEA-4 are consistently present on human EC cells, and are of diagnostic value in distinguishing human EC cell tumors from human yolk sac carcinomas, choriocarcinomas, and other lineages which lack these markers, Wenk et al., *Int J Cancer* 58:108–115, 1994. A recent survey found SSEA-3 and SSEA-4 to be present on all of over 40 human EC cell lines examined, Wenk et al. TRA-1-60 and TRA-1-81 antigens have been studied extensively on a particular pluripotent human EC cell line, NTERA-2 CL. D1, Andrews et al, supra. Differentiation of NTERA-2 CL. D1 cells in vitro results in the loss of SSEA-3, SSEA-4, TRA-1-60, and TRA-1-81 expression and the increased expression of the lacto-series glycolipid SSEA-1, Andrews et al, supra. This contrasts with undifferentiated mouse ES cells, which express SSEA-1, and neither SSEA-3 nor SSEA-4. Although the function of these antigens are unknown, their shared expression by R278.5 cells and human EC cells suggests a close embryological similarity. Alkaline phosphatase will also be present on all primate ES cells. A successful primate ES cell culture of the present invention will correlate with the cell surface markers found in the rhesus macaque and marmoset cell lines described in Table 1.

As disclosed below in Table 1, the rhesus macaque and marmoset cell lines are identical to human EC cell lines for the 5 described markers. Therefore, a successful primate ES cell culture will also mimic human EC cells. However, there are other ways to discriminate ES cells from EC cells. For example, the primate ES cell line has a normal karyotype and the human EC cell line is aneuploid.

In FIG. 3, the photographs labelled A through F demonstrate the characteristic staining of these markers on a rhesus monkey ES cell line designated R278.5.

TABLE 1

	Mouse ES	C. jacchus ES	M. mulatta ES	Human EC (NTERA-2 cl.D1)
SSEA-1	+	–	–	–
SSEA-3	–	+	+	+
SSEA-4	–	+	+	+
Tra-1-60	–	+	+	+
Tra-1-81	–	+	+	+

(c) Developmental Potential

Primate ES cells of the present invention are pluripotent. By "pluripotent" we mean that the cell has the ability to develop into any cell derived from the three main germ cell layers or an embryo itself. When injected into SCID mice, a successful primate ES cell line will differentiate into cells derived from all three embryonic germ layers including: bone, cartilage, smooth muscle, striated muscle, and hematopoietic cells (mesoderm); liver, primitive gut and respiratory epithelium (endoderm); neurons, glial cells, hair follicles, and tooth buds (ectoderm).

This experiment can be accomplished by injecting approximately $0.5–1.0 \times 10^6$ primate ES cells into the rear leg muscles of 8–12 week old male SCID mice. The resulting tumors can be fixed in 4% paraformaldehyde and examined histologically after paraffin embedding at 8–16 weeks of development. In FIG. 4, photomicrographs designated A–F are of sections of tumors formed by injection of rhesus ES cells into the hind leg muscles of SCID mice and analyzed 15 weeks later demonstrating cartilage, smooth muscle, and striated muscle (mesoderm); stratified squamous epithelium with mantle layers (ectoderm); neural tube with ventricular, intermediate, and mantle layers (ectoderm); ciliated columnar epithelium and villi lined by absorptive enterocytes and mucus-secreting goblet cells (endoderm).

A successful nonhuman primate ES cell line will have the ability to participate in normal development when combined in chimeras with normal preimplantation embryos. Chimeras between preimplantation nonhuman primate embryos and nonhuman primate ES cells can be formed by routine methods in several ways. (i) injection chimeras: 10–15 nonhuman primate ES cells can be microinjected into the cavity of an expanded nonhuman primate blastocyst; (ii) aggregation chimeras: nonhuman primate morulae can be co-cultured on a lawn of nonhuman primate ES cells and allowed to aggregate; and (iii) tetraploid chimeras: 10–15 nonhuman primate ES cells can be aggregated with tetraploid nonhuman primate morulae obtained by electrofusion of 2-cell embryos, or incubation of morulae in the cytoskeletal inhibitor cholchicine. The chimeras can be returned to the uterus of a female nonhuman primate and allowed to develop to term, and the ES cells will contribute to normal differentiated tissues derived from all three embryonic germ layers and to germ cells. Because nonhuman primate ES can be genetically manipulated prior to chimera formation by standard techniques, chimera formation followed by embryo transfer can lead to the production of transgenic nonhuman primates.

(d) Karyotype

Successful primate ES cell lines have normal karyotypes. Both XX and XY cells lines will be derived. The normal karyotypes in primate ES cell lines will be in contrast to the abnormal karyotype found in human embryonal carcinoma (EC), which are derived from spontaneously arising human germ cell tumors (teratocarcinomas). Human embryonal carcinoma cells have a limited ability to differentiate into multiple cell types and represent the closest existing cell

lines to primate ES cells. Although tumor-derived human embryonal carcinoma cell lines have some properties in common with embryonic stem cell lines, all human embryonal carcinoma cell lines derived to date are aneuploid. Thus, primate ES cell lines and human EC cell lines can be distinguished by the normal karyotypes found in primate ES cell lines and the abnormal karyotypes found in human EC lines. By "normal karyotype" it is meant that all chromosomes normally characteristic of the species are present and have not been noticeably altered.

Because of the abnormal karyotypes of human embryonal carcinoma cells, it is not clear how accurately their differentiation reflects normal differentiation. The range of embryonic and extra-embryonic differentiation observed with primate ES cells will typically exceed that observed in any human embryonal carcinoma cell line, and the normal karyotypes of the primate ES cells suggests that this differentiation accurately recapitulates normal differentiation.

(e) Immortality

Immortal cells are capable of continuous indefinite replication in vitro. Continued proliferation for longer than one year of culture is a sufficient evidence for immortality, as primary cell cultures without this property fail to continuously divide for this length of time (Freshney, Culture of animal cells. New York: Wiley-Liss, 1994). Primate ES cells will continue to proliferate in vitro with the culture conditions described above for longer than one year, and will maintain the developmental potential to contribute all three embryonic germ layers. This developmental potential can be demonstrated by the injection of ES cells that have been cultured for a prolonged period (over a year) into SCID mice and then histologically examining the resulting tumors. Although karyotypic changes can occur randomly with prolonged culture, some primate ES cells will maintain a normal karyotype for longer than a year of continuous culture.

(f) Culture Conditions

Growth factor requirements to prevent differentiation are different for the primate ES cell line of the present invention than the requirements for mouse ES cell lines. In the absence of fibroblast feeder layers, Leukemia inhibitory factor (LIF) is necessary and sufficient to prevent differentiation of mouse ES cells and to allow their continuous passage. Large concentrations of cloned LIF fail to prevent differentiation of primate ES cell lines in the absence of fibroblast feeder layers. In this regard, primate ES stem cells are again more similar to human EC cells than to mouse ES cells, as the growth of feeder-dependent human EC cells lines is not supported by LIF in the absence of fibroblasts.

(g) Differentiation to Extra Embryonic Tissues

When grown on embryonic fibroblasts and allowed to grow for two weeks after achieving confluence (i.e., continuously covering the culture surface), primate ES cells of the present invention spontaneously differentiate and will produce chorionic gonadotropin, indicating trophoblast differentiation (a component of the placenta) and produce α-fetoprotein, indicating endoderm differentiation. Chorionic gonadotropin activity can be assayed in the medium conditioned by differentiated cells by Leydig cell bioassay, Seshagiri & Hearn, Hum Reprod 8:279–287, 1992. For mRNA analysis, RNA can be prepared by guanidine isothiocyanate-phenol/chloroform extraction (1) from approximately 0.2×10^6 differentiated cells and from 0.2×10^6 undifferentiated cells. The relative levels of the mRNA for α-fetoprotein and the α- and β-subunit of chorionic gonadotropin relative to glyceraldehyde-3-phosphate dehydrogenase can be determined by semi-quantitative Reverse

13

Transcriptase-Polymerase Chain Reaction (RT-PCR). The PCR primers for glyceraldehyde 3-phosphate dehydrogenase (G3PDH), obtained from Clontech (Palo Alto, Calif.), are based on the human cDNA sequence, and do not amplify mouse G3PDH mRNA under our conditions. Primers for the α-fetoprotein mRNA are based on the human sequence and flank the 7th intron (5′ primer=(5′) GCTGGATTGTCTG-CAGGATGGGGAA (SEQ ID NO: 1); 3′ primer=(5′) TCCCCTGAAGAAAATTGGTTAAAAT (SEQ ID NO: 2)). They amplify a cDNA of 216 nucleotides. Primers for the β-subunit of chorionic gonadotropin flank the second intron (5′ primer=(5′) ggatc CACCGTCAACACCACCATCT-GTGC (SEQ ID NO: 3); 3′ primer=(5′) ggatc CACAGGT-CAAAGGGTGGTCCTTGGG (SEQ ID NO: 4)) (nucleotides added to the hCGb sequence to facilitate subcloning are shown in lower case italics). They amplify a cDNA of 262 base pairs. The primers for the CGα subunit can be based on sequences of the first and fourth exon of the rhesus gene (5′ primer=(5′) gggaattc GCAGTTACT-GAGAACTCACAAG (SEQ ID NO: 5); 3′ primer=(5′) gggaattc GAAGCATGTCAAAGTGGTATGG (SEQ ID NO: 6) and amplify a cDNA of 556 base pairs. The identity of the α-fetoprotein, CGα and CGβ cDNAs can be verified by subcloning and sequencing.

For Reverse Transcriptase-Polymerase Chain Reaction (RT-PCR), 1 to 5 μl of total R278.5 RNA can be reverse transcribed as described Golos et al. *Endocrinology* 133(4) :1744–1752, 1993, and one to 20 μl of reverse transcription reaction was then subjected to the polymerase chain reaction in a mixture containing 1–12.5 pmol of each G3PDH primer, 10–25 pmol of each mRNA specific primer, 0.25 mM dNTPs (Pharmacia, Piscataway, N.J.), 1× AmpliTaq buffer (final reaction concentrations=10 mM Tris, pH 8.3, 50 mM KCl, 1.5 mM MgCl2, 0.001% (w/v) gelatin) 2.5 μCi of deoxycytidine 5′a[32P]triphosphate (DuPont, Boston, Mass.), 10% glycerol and 1.25 U of AmpliTaq (Perkin-Elmer, Oak Brook, Ill.) in a total volume of 50 μl. The number of amplification rounds which produced linear increases in target cDNAs and the relation between input RNA and amount of PCR product is empirically determined as by Golos et al. Samples were fractionated in 3% Nusieve (FMC, Rockland, Me.) agarose gels (1× TBE running buffer) and DNA bands of interest were cut out, melted at 65° C. in 0.5 ml TE, and radioactivity determined by liquid scintillation counting. The ratio of counts per minute in a specific PCR product relative to cpm of G3PDH PCR product is used to estimate the relative levels of a mRNAs among differentiated and undifferentiated cells.

The ability to differentiate into trophectoderm in vitro and the ability of these differentiated cells to produce chorionic gonadotropin distinguishes the primate ES cell line of the present invention from all other published ES cell lines.

EXAMPLES

(1) Animals and Embryos

As described above, we have developed a technique for non-surgical, uterine-stage embryo recovery from the rhesus macaque and the common marmoset.

To supply rhesus embryos to interested investigators, The Wisconsin Regional Primate Research Center (WRPRC) provides a preimplantation embryo recovery service for the rhesus monkey, using the non-surgical flush procedure described above. During 1994, 151 uterine flushes were attempted from rhesus monkeys, yielding 80 viable embryos (0.53 embryos per flush attempt).

By synchronizing the reproductive cycles of several marmosets, significant numbers of in vivo produced, age-matched, preimplantation primate embryos were studied in

controlled experiments for the first time. Using marmosets from the self-sustaining colony (250 animals) of the Wisconsin Regional Primate Research Center (WRPRC), we recovered 54 viable morulae or blastocysts, 7 unfertilized oocytes or degenerate embryos, and 5 empty zonae pellucidae in a total of 54 flush attempts (1.0 viable embryo-flush attempt). Marmosets have a 28 day ovarian cycle, and because this is a non-surgical procedure, females can be flushed on consecutive months, dramatically increasing the embryo yield compared to surgical techniques which require months of rest between collections.

(2) Rhesus Macaque Embryonic Stem Cells

Using the techniques described above, we have derived three independent embryonic stem cell lines from two rhesus monkey blastocysts (R278.5, R366, and R367). One of these, R278.5, remains undifferentiated and continues to proliferate after continuous culture for over one year. R278.5 cells have also been frozen and successfully thawed with the recovery of viable cells.

The morphology and cell surface markers of R278.5 cells are indistinguishable from human EC cells, and differ significantly from mouse ES cells. R278.5 cells have a high nucleus/cytoplasm ratio and prominent nucleoli, but rather than forming compact, piled-up colonies with indistinct cell borders similar to mouse ES cells, R278.5 cells form flatter colonies with individual, distinct cells (FIG. 2A). R278.5 cells express the SSEA-3, SSEA-4, TRA-1-60, and TRA-81 antigens (FIG. 3 and Table 1), none of which are expressed by mouse ES cells. The only cells known to express the combination of markers SSEA-3, SSEA-4, TRA-1-60, and TRA-1-81 other than primate ES cells are human EC cells. The globo-series glycolipids SSEA-3 and SSEA-4 are consistently present on human EC cells, and are of diagnostic value in distinguishing human EC cell tumors from yolk sac carcinomas, choriocarcinomas and other stem cells derived from human germ cell tumors which lack these markers, Wenk et al, *Int J Cancer* 58:108–115, 1994. A recent survey found SSEA-3 and SSEA-4 to be present on all of over 40 human EC cell lines examined (Wenk et al.).

TRA-1-60 and TRA-1-81 antigens have been studied extensively on a particular pluripotent human EC cell line, NTERA-2 CL. D1 (Andrews et al.). Differentiation of NTERA-2 CL. D1 cells in vitro results in the loss of SSEA-3, SSEA-4, TRA-1-60, and TRA-1-81 expression and the increased expression of the lacto-series glycolipid SSEA-1. Undifferentiated mouse ES cells, on the other hand, express SSEA-1, and not SSEA-3, SSEA-4, TRA-1-60 or TRA-1-81 (Wenk et al.). Although the function of these antigens is unknown, their expression by R278.5 cells suggests a close embryological similarity between primate ES cells and human EC cells, and fundamental differences between primate ES cells and mouse ES cells.

R278.5 cells also express alkaline phosphatase. The expression of alkaline phosphatase is shared by both primate and mouse ES cells, and relatively few embryonic cells express this enzyme. Positive cells include the ICM and primitive ectoderm (which are the most similar embryonic cells in the intact embryo to ES cells), germ cells, (which are totipotent), and a very limited number of neural precursors, Kaufman M H. *The atlas of mouse development*. London: Academic Press, 1992. Cells not expressing this enzyme will not be primate ES cells.

Although cloned human LIF was present in the medium at cell line derivation and for initial passages, R278.5 cells grown on mouse embryonic fibroblasts without exogenous LIF remain undifferentiated and continued to proliferate. R278.5 cells plated on gelatin-treated tissue culture plates without fibroblasts differentiated to multiple cell types or failed to attach and died, regardless of the presence or absence of exogenously added human LIF (FIG. 2). Up to 10^4 units/ml human LIF fails to prevent differentiation. In addition, added LIF fails to increase the cloning efficiency or proliferation rate of R278.5 cells on fibroblasts. Since the derivation of the R278.5 cell line, we have derived two additional rhesus ES cell lines (R366 and R367) on embryonic fibroblasts without any exogenously added LIF at initial derivation. R366 and R367 cells, like R278.5 cells, continue to proliferate on embryonic fibroblasts without exogenously added LIF and differentiate in the absence of fibroblasts, regardless of the presence of added LIF. RT-PCR performed on mRNA from spontaneously differentiated R278.5 cells revealed α-fetoprotein mRNA (FIG. 4). α-fetoprotein is a specific marker for endoderm, and is expressed by both extra-embryonic (yolk sac) and embryonic (fetal liver and intestines) endoderm-derived tissues. Epithelial cells resembling extraembryonic endoderm are present in cells differentiated in vitro from R278.5 cells (FIG. 2). Bioactive CG (3.89 ml units/ml) was present in culture medium collected from differentiated cells, but not in medium collected from undifferentiated cells (less than 0.03 ml units/ml), indicating the differentiation of trophoblast, a trophectoderm derivative. The relative level of the CGα mRNA increased 23.9-fold after differentiation (FIG. 4).

All SCID mice injected with R278.5 cells in either intra-muscular or intra-testicular sites formed tumors, and tumors in both sites demonstrated a similar range of differentiation. The oldest tumors examined (15 weeks) had the most advanced differentiation, and all had abundant, unambiguous derivatives of all three embryonic germ layers, including gut and respiratory epithelium (endoderm); bone, cartilage, smooth muscle, striated muscle (mesoderm); ganglia, glia, neural precursors, and stratified squamous epithelium (ectoderm), and other unidentified cell types (FIG. 5). In addition to individual cell types, there was organized development of some structures which require complex interactions between different cell types. Such structures included gut lined by villi with both absorptive enterocytes and mucus-secreting goblet cells, and sometimes encircled by layers of smooth muscle in the same orientation as muscularis mucosae (circular) and muscularis (outer longitudinal layer aid inner circular layer); neural tubes with ventricular, intermediate, and mantle layers; and hair follicles with hair shafts (FIG. 5).

The essential characteristics that define R278.5 cells as ES cells include: indefinite (greater than one year) undifferentiated proliferation in vitro, normal karyotype, and potential to differentiate to derivatives of trophectoderm and all three embryonic germ layers. In the mouse embryo, the last cells capable of contributing to derivatives of both trophectoderm and ICM are early ICM cells. The timing of commitment to ICM or trophectoderm has not been established for any primate species, but the potential of rhesus ES cells to contribute to derivatives of both suggests that they most closely resemble early totipotent embryonic cells. The ability of rhesus ES cells to form trophoblast in vitro distinguishes primate ES cell lines from mouse ES cells. Mouse ES cell have not been demonstrated to form trophoblast in vitro, and mouse trophoblast does not produce gonadotropin. Rhesus ES cells and mouse ES cells do demonstrate the similar wide range of differentiation in tumors that distinguishes ES cells from EC cells. The development of structures composed of multiple cell types such as hair follicles, which require inductive interactions between the embryonic epidermis and underlying mesenchyme, demonstrates the

ability of rhesus ES cells to participate in complex developmental processes.

The rhesus ES lines R366 and R367 have also been further cultured and analyzed. Both lines have a normal XY karyotype and were proliferated in an undifferentiated state for about three months prior to freezing for later analysis. Samples of each of the cell lines R366 and R367 were injected into SCID mice which then formed teratomas identical to those formed by R278.5 cells. An additional rhesus cell line R394 having a normal XX karyotype was also recovered. All three of these cell lines, R366, R367 and R394 are identical in morphology, growth characteristics, culture requirements and in vitro differentiation characteristics, i.e. the trait of differentiation to multiple cell types in the absence of fibroblasts, to cell line 278.5.

It has been determined that LIF is not required either to derive or proliferate these ES cultures. Each of the cell lines R366, R367 and R394 were derived and cultured without exogenous LIF.

It has also been demonstrated that the particular source of fibroblasts for co-culture is not critical. Several fibroblast cell lines have been tested both with rhesus line R278.5 and with the marmoset cell lines described below. The fibroblasts tested include mouse STO cells (ATCC 56-X), mouse 3T3 cells (ATCC 48-X), primary rhesus monkey embryonic fibroblasts derived from 36 day rhesus fetuses, and mouse Sl/Sl⁴ cells, which are deficient in the steel factor. All these fibroblast cell lines were capable of maintaining the stem cell lines in an undifferentiated state. Most rapid proliferation of the stem cells was observed using primary mouse embryonic fibroblasts.

Unlike mouse ES cells, neither rhesus ES cells nor feeder-dependent human EC cells remain undifferentiated and proliferate in the presence of soluble human LIF without fibroblasts. The factors that fibroblasts produce that prevent the differentiation of rhesus ES cells or feeder-dependent human EC cells are unknown, but the lack of a dependence on LIF is another characteristic that distinguishes primate ES cells from mouse ES cells. The growth of rhesus monkey ES cells in culture conditions similar to those required by feeder-dependent human EC cells, and the identical morphology and cell surface markers of rhesus ES cells and human EC cells, suggests that similar culture conditions will support human ES cells.

Rhesus ES cells will be important for elucidating the mechanisms that control the differentiation of specific primate cell types. Given the close evolutionary distance and the developmental and physiological similarities between humans and rhesus monkeys, the mechanisms controlling the differentiation of rhesus cells will be very similar to the mechanisms controlling the differentiation of human cells. The importance of elucidating these mechanisms is that once they are understood, it will be possible to direct primate ES cells to differentiate to specific cell types in vitro, and these specific cell types can be used for transplantation to treat specific diseases.

Because ES cells have the developmental potential to give rise to any differentiated cell type, any disease that results in part or in whole from the failure (either genetic or acquired) of specific cell types will be potentially treatable through the transplantation of cells derived from ES cells. Rhesus ES cells and rhesus monkeys will be invaluable for testing the efficacy and safety of the transplantation of specific cell types derived from ES cells. A few examples of human diseases potentially treatable by this approach with human ES cells include degenerative neurological disorders such as Parkinson's disease (dopaergic neurons), juvenile onset

17

diabetes (pancreatic β-islet cells) or Acquired Immunodeficiency Disease (lymphocytes). Because undifferentiated ES cells can proliferate indefinitely in vitro, they can be genetically manipulated with standard techniques either to prevent immune rejection after transplantation, or to give them new genetic properties to combat specific diseases. For specific cell types where immune rejection can be prevented, cells derived from rhesus monkey ES cells or other non-human primate ES cells could be used for transplantation to humans to treat specific diseases.

(3) Marmoset Embryonic Stem Cells

Our method for creating an embryonic stem cell line is described above. Using isolated ICM's derived by immunosurgery from marmoset blastocysts, we have isolated 7 putative ES cell lines, each of which have been cultured for over 6 months.

One of these, Cj11, was cultured continuously for over 14 months, and then frozen for later analysis. The Cj11 cell line and other marmoset ES cell lines have been successfully frozen and then thawed with the recovery of viable cells. These cells have a high nuclear/cytoplasmic ratio, prominent nucleoli, and a compact colony morphology similar to the pluripotent human embryonal carcinoma (EC) cell line NT2/D2.

Four of the cell lines we have isolated have normal XX karyotypes, and one has a normal XY karyotype (Karyotypes were performed by Dr. Charles Harris, University of Wisconsin). These cells were positive for a series of cell surface markers (alkaline phosphatase, SSEA-3, SSEA-4, TRA-1-60, and TRA-1-81) that in combination are definitive markers for undifferentiated human embryonal carcinoma cells (EC) and primate ES cells. In particular, these markers distinguish EC cells from the earliest lineages to differentiate in the human preimplantation embryo, trophectoderm (represented by BeWO choriocarcinoma cells) and extraembryonic endoderm (represented by 1411H yolk sac carcinoma cells).

When the putative marmoset ES cells were removed from fibroblast feeders, they differentiated into cells of several distinct morphologies. Among the differentiated cells, trophectoderm is indicated by the secretion of chorionic gonadotropin and the presence of the chorionic gonadotropin β-subunit mRNA. 12.7 mIU/ml luteinizing hormone (LH) activity was measured in the WRPRC core assay lab using a mouse Leydig cell bioassay in medium conditioned 24 hours by putative ES cells allowed to differentiate for one week. Note that chorionic gonadotropin has both LH and FSH activity, and is routinely measured by LH assays. Control medium from undifferentiated ES cells had less than 1 mIU/ml LH activity.

Chorionic gonadotropin β-subunit mRNA was detected by reverse transcriptase-polymerase chain reaction (RT-PCR). DNA sequencing confirmed the identity of the chorionic gonadotropin β-subunit.

Endoderm differentiation (probably extraembryonic endoderm) was indicated by the presence of α-fetoprotein mRNA, detected by RT-PCR.

18

When the marmoset ES cells were grown in high densities, over a period of weeks epithelial cells differentiated and covered the culture dish. The remaining groups of undifferentiated cells rounded up into compact balls and then formed embryoid bodies (as shown in FIG. 6) that recapitulated early development with remarkable fidelity. Over 3–4 weeks, some of the embryoid bodies formed a bilaterally symmetric pyriform embryonic disc, an amnion, a yolk sac, and a mesoblast outgrowth attaching the caudal pole of the amnion to the culture dish.

Histological and ultrastructural examination of one of these embryoid bodies (formed from a cell line that had been passaged continuously for 6 months) revealed a remarkable resemblance to a stage 6–7 post-implantation embryo. The embryonic disc was composed of a polarized, columnar epithelial epiblast (primitive ectoderm) layer separated from a visceral endoderm (primitive endoderm) layer. Electron microscopy of the epiblast revealed apical junctional complexes, apical microvilli, subapical intermediate filaments, and a basement membrane separating the epiblast from underlying visceral endoderm. All of these elements are features of the normal embryonic disc. In the caudal third of the embryonic disc, there was a midline groove, disruption of the basement membrane, and mixing of epiblast cells with underlying endodermal cells (early primitive streak). The amnion was composed of an inner squamous (ectoderm) layer continuous with the epiblast and an outer mesoderm layer. The bilayered yolk sac had occasional endothelial-lined spaces containing possible hematopoietic precursors.

The morphology, immortality, karyotype, and cell surface markers of these marmoset cells identify these marmoset cells as primate ES cells similar to the rhesus ES cells. Since the last cells in the mammalian embryo capable of contributing to both trophectoderm derivatives and endoderm derivatives are the totipotent cells of the early ICM, the ability of marmoset ES cells to contribute to both trophoblast and endoderm demonstrates their similarities to early totipotent embryonic cells of the intact embryo. The formation of embryoid bodies by marmoset ES cells, with remarkable structural similarities to the early post-implantation primate embryo, demonstrates the potential of marmoset ES cells to participate in complex developmental processes requiring the interaction of multiple cell types.

Given the reproductive characteristics of the common marmoset described above (efficient embryo transfer, multiple young, short generation time), marmoset ES cells will be particularly useful for the generation of transgenic primates. Although mice have provided invaluable insights into gene function and regulation, the anatomical and physiological differences between humans and mice limit the usefulness of transgenic mouse models of human diseases. Transgenic primates, in addition to providing insights into the pathogenesis of specific diseases, will provide accurate animal models to test the efficacy and safety of specific treatments.

SEQUENCE LISTING

<160> NUMBER OF SEQ ID NOS: 6

<210> SEQ ID NO 1
<211> LENGTH: 25
<212> TYPE: DNA
<213> ORGANISM: Artificial Sequence

-continued

```
<220> FEATURE:
<223> OTHER INFORMATION: Description of Artificial Sequence:
      Oligonucleotide Primer

<400> SEQUENCE: 1

gctggattgt ctgcaggatg gggaa                                         25


<210> SEQ ID NO 2
<211> LENGTH: 25
<212> TYPE: DNA
<213> ORGANISM: Artificial Sequence
<220> FEATURE:
<223> OTHER INFORMATION: Description of Artificial Sequence:
      Oligonucleotide Primer

<400> SEQUENCE: 2

tcccctgaag aaaattggtt aaaat                                         25


<210> SEQ ID NO 3
<211> LENGTH: 29
<212> TYPE: DNA
<213> ORGANISM: Artificial Sequence
<220> FEATURE:
<223> OTHER INFORMATION: Description of Artificial Sequence:
      Oligonucleotide Primer

<400> SEQUENCE: 3

ggatccaccg tcaacaccac catctgtgc                                     29


<210> SEQ ID NO 4
<211> LENGTH: 30
<212> TYPE: DNA
<213> ORGANISM: Artificial Sequence
<220> FEATURE:
<223> OTHER INFORMATION: Description of Artificial Sequence:
      Oligonucleotide Primer

<400> SEQUENCE: 4

ggatccacag gtcaaagggt ggtccttggg                                    30


<210> SEQ ID NO 5
<211> LENGTH: 30
<212> TYPE: DNA
<213> ORGANISM: Artificial Sequence
<220> FEATURE:
<223> OTHER INFORMATION: Description of Artificial Sequence:
      Oligonucleotide Primer

<400> SEQUENCE: 5

gggaattcgc agttactgag aactcacaag                                    30


<210> SEQ ID NO 6
<211> LENGTH: 30
<212> TYPE: DNA
<213> ORGANISM: Artificial Sequence
<220> FEATURE:
<223> OTHER INFORMATION: Description of Artificial Sequence:
      Oligonucleotide Primer

<400> SEQUENCE: 6

gggaattcga agcatgtcaa agtggtatgg                                    30
```

21

I claim:

1. A purified preparation of pluripotent human embryonic stem cells which (i) will proliferate in an in vitro culture for over one year, (ii) maintains a karyotype in which the chromosomes are euploid and not altered through prolonged culture, (iii) maintains the potential to differentiate to derivatives of endoderm, mesoderm, and ectoderm tissues throughout the culture, and (iv) is inhibited from differentiation when cultured on a fibroblast feeder layer.

2. The preparation of claim 1, wherein the stem cells will spontaneously differentiate to trophoblast and produce chorionic gonadotropin when cultured to high density.

3. A purified preparation of pluripotent human embryonic stem cells wherein the cells are negative for the SSEA-1 marker, positive for the SSEA-4 marker, express alkaline phosphatase activity, are pluripotent, and have euploid karyotypes and in which none of the chromosomes are altered.

4. The preparation of claim 3, wherein the cells are positive for the TRA-1-60, and TRA-1-81 markers.

5. The preparation of claim 3, wherein the cells continue to proliferate in an undifferentiated state after continuous culture for at least one year.

6. The preparation of claim 3, wherein the cells will differentiate to trophoblast when cultured beyond confluence and will produce chorionic gonadotropin.

7. The preparation of claim 3, wherein the cells remain euploid for more than one year of continuous culture.

22

8. The preparation of claim 3, wherein the cells differentiate into cells derived from mesoderm, endoderm and ectoderm germ layers when the cells are injected into a SCID mouse.

9. A method of isolating a pluripotent human embryonic stem cell line, comprising the steps of:

(a) isolating a human blastocyst;

(b) isolating cells from the inner cell mass of the blastocyte of (a);

(c) plating the inner cell mass cells on embryonic fibroblasts, wherein inner cell mass-derived cell masses are formed;

(d) dissociating the mass into dissociated cells;

(e) replating the dissociated cells on embryonic feeder cells;

(f) selecting colonies with compact morphologies and cells with high nucleus to cytoplasm ratios and prominent nucleoli; and

(g) culturing the cells of the selected colonies to thereby obtain an isolated pluripotent human embryonic stem cell line.

10. A method as claimed in claim 9, further comprising maintaining the isolated cells on a fibroblast feeder layer to prevent differentiation.

11. A cell line developed by the method of claim 9.

*   *   *   *   *

Index